T0216759

Lecture Notes in Computer Science 557

Edited by G. Goos and J. Hartmanis

Advisory Board: W. Brauer   D. Gries   J. Stoer

W. L. Hsu   R. C. T. Lee (Eds.)

ISA '91 Algorithms

2nd International Symposium on Algorithms
Taipei, Republic of China, December 16-18, 1991
Proceedings

Springer-Verlag

Berlin Heidelberg New York
London Paris Tokyo
Hong Kong Barcelona
Budapest

Series Editors

Gerhard Goos
Universität Karlsruhe
Postfach 69 80
Vincenz-Priessnitz-Straße 1
W-7500 Karlsruhe, FRG

Juris Hartmanis
Department of Computer Science
Cornell University
5148 Upson Hall
Ithaca, NY 14853, USA

Volume Editors

Wen-Lian Hsu
Institute of Information Science, Academia Sinica
Nankang, Taipei, Taiwan 115, Republic of China

R. C. T. Lee
Institute of Computer Science, Tsing Hua University
Hsin Chu, Republic of China

CR Subject Classification (1991): F.1-2, G.2-3, I.1.2, I.3.5

ISBN 3-540-54945-5 Springer-Verlag Berlin Heidelberg New York
ISBN 0-387-54945-5 Springer-Verlag New York Berlin Heidelberg

This work is subject to copyright. All rights are reserved, whether the whole or part of
the material is concerned, specifically the rights of translation, reprinting, re-use of
illustrations, recitation, broadcasting, reproduction on microfilms or in any other way,
and storage in data banks. Duplication of this publication or parts thereof is permitted
only under the provisions of the German Copyright Law of September 9, 1965, in its
current version, and permission for use must always be obtained from Springer-Verlag.
Violations are liable for prosecution under the German Copyright Law.

© Springer-Verlag Berlin Heidelberg 1991
Printed in Germany

Typesetting: Camera ready by author
Printing and binding: Druckhaus Beltz, Hemsbach/Bergstr.
45/3140-543210 - Printed on acid-free paper

Preface

The papers in this volume were presented at the Second Annual International Symposium on Algorithms. The conference took place December 16 – 18, 1991, at Academia Sinica in Taipei, Taiwan, Republic of China. It was organized by the Institute of Information Science, Academia Sinica and the National Tsing Hua University, in cooperation with the Special Interest Group on Algorithms (SIGAL) of the Information Processing Society of Japan (IPSJ) and the Technical Group on Theoretical Foundation of Computing of the Institute of Electronics, Information and Communication Engineers (IEICE).

The goal of the annual International Symposium on Algorithms (ISA) is to provide a forum for researchers in the Pacific rim as well as other parts of the world to exchange ideas on computing theory. The first such symposium was held 1990 in Tokyo, Japan as "SIGAL International Symposium on Algorithms".

In response to the program committee's call for papers, 90 papers were submitted. From these submissions, the committee selected 36 for presentation at the symposium. In addition to these contributed papers, the symposium included 5 invited presentations.

December 1991
Wen-Lian Hsu
R. C. T. Lee

Symposium Chairs:
R. C. T. Lee (Tsing Hua Univ.)
Wen-Lian Hsu (Academia Sinica)

Program Committee:
Takao Asano (Sophia Univ.)
Tetsuo Asano (Osaka Electro-Comm. Univ.)
Francis Y. L. Chin (Univ. of Hong Kong)
Wen-Lian Hsu (Academia Sinica)
Toshihide Ibaraki (Kyoto Univ.)
Hiroshi Imai (Univ. of Tokyo)
D. T. Lee (Northwestern Univ.)
Sing-Ling Lee (Chung Cheng Univ.)
C. L. Liu (Univ. of Illinois)
Kurt Mehlhorn (Univ. Saarland)
Takao Nishizeki (Tohoku Univ.)
Jia-Shung Wang (Tsing Hua Univ.)

Finance:
Ting-Yi Sung (Academia Sinica)

Local Arrangement:
R. C. Chang (Chiao Tung Univ.)
Ming-Tat Ko (Academia Sinica)

Publication, Publicity:
Jan-Ming Ho (Academia Sinica)

Supported by
National Science Council of ROC
Institute for Information Industry
Telecommunication Laboratories

Contents

Contents

DECISION–MAKING WITH INCOMPLETE INFORMATION

Christos H. Papadimitriou
University of California at San Diego

ABSTRACT

We must often make decisions in situations such as routing, scheduling, compiling, and resource allocation, without crucial information about (respectively) the terrain topography, the execution times, the run–time environment, and the future requests. Information may be unavailable because of its temporal or distributed nature. A reasonable way to assess the effectiveness of a decision rule in such situations is to compare its outcome to the ideal optimum solution, attainable only if we had complete information. An algorithm is considered effective (or "competitive") if its performance is a multiplicative constant away from the ideal solution. We review some recent and on–going work on this active area.

Maximum Independet Set of a Permutation Graph in k Tracks [1]

D. T. Lee   and   Majid Sarrafzadeh
Department of Electrical Engineering
and Computer Science
Northwestern University
Evanston, IL 60208

Abstract

A maximum independent set of a permutation graph is a maximum subset of noncrossing chords in a matching diagram Φ. (Φ consists of a set of chords with end-points on two horizontal lines.) The problem of finding, among all noncrossing subsets of Φ with density at most k, one with maximum size is considered, where the density of a subset is the maximum number of chords crossing a vertical line and k is a given parameter. A $\Theta(n \log n)$ time and $\Theta(n)$ space algorithm, for solving the problem with n chords, is proposed. As an application, we solve the problem of finding, among all proper subsets with density at most k of an interval graph, one with maximum number of intervals.

1   Introduction

Consider two rows of distinct integer points in the xy-plane. The *lower row* is at $y = b_1$ and the *upper row* at $y = b_2$, for some $b_1 < b_2$. A *chord* $N_i = (\ell_i, u_i)$ is a line segment passing through points (ℓ_i, b_1) and (u_i, b_2) of the plane where ℓ_i is the *lower terminal* and u_i is the *upper terminal* of N_i. A *matching diagram* consists of a collection $\Phi = \{N_1, \ldots, N_n\}$ of chords. A subset of Φ is called an *independent set* if the corresponding chords are pairwise noncrossing. An independent set of maximum cardinality, among all independent sets, is called a *maximum independent set* (MIS). The density d_i^a of a subset Φ_a of Φ at a half-integer column $x = i + \frac{1}{2}$ (where i is an integer) is defined as follows.

$$d_i^a = \left| \{ N = (\ell, u) \in \Phi_a \mid \ell < i + \tfrac{1}{2} < u \ \ or \ \ u < i + \tfrac{1}{2} < \ell \} \right|$$

The *density* d_a of a subset Φ_a of Φ is $\max_i(d_i^a)$, for all i.

Throughout this paper, we assume the input is a collection of (unsorted) pairs (ℓ_i, u_i). A geometric algorithm for obtaining a maximum independent set of Φ in $\Theta(n \log n)$ time has been obtained, where $n = |\Phi|$ [AK]. However the density of the resulting MIS can be as large as n. MIS-k problem is to find, among all independent subsets of Φ with density at most k, one with maximum cardinality (Figure 1). Note that MIS-n is the traditional MIS problem and MIS-1 problem is that of finding a maximum independent set of a collection of intervals

[1]Supported in part by the National Science Foundation under grants CCR-8901815 and MIP-8921540.

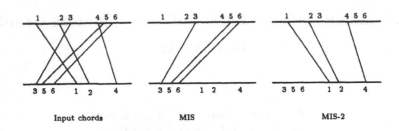

| Input chords | MIS | MIS-2 |

Figure 1: Density of maximum independent sets

– both problems can be solved in $\Theta(n \log n)$ time [AK,GLL]. The converse problem – to find among all maximum independent sets in Φ one with minimum density – can be easily solved (by selecting at each "maxima level" [LSL], in a greedy fashion, a chord that minimizes density).

In Section 2, we study fundamental properties of MIS-k and discuss the main steps of the proposed algorithm. We present details of our implementation in Section 3. Finally, an application of the proposed algorithms in interval graphs is presented in Section 4.

2 Basic Concepts

In the traditional MIS problem, where the goal is to find a maximum independent set irrespective of its density, only relative positions of the terminals are important. However in MIS-k, absolute positions of the terminals are important and cannot be arbitrarily changed. For example, in Figure 1, by changing absolute positions of some terminals, while maintaining their relative positions, we can increase the size of MIS for a fixed k. (Indeed, many fundamental properties employed in MIS-n problem do not hold for MIS-k, as we will describe.)

Consider an instance $\Phi = \{N_1, \ldots, N_n\}$ of MIS-k problem. To each net $N_i = (\ell_i, u_i)$ we associate a point $P_i = (x_i, y_i)$ in the xy-plane, where $x_i = \ell_i$ and $y_i = u_i$. In the point set $\Psi = \{P_1, \ldots, P_n\}$ we assume $x_i > 0$, $y_i > 0$ (otherwise we can shift the origin) and $x_i > x_j$ for $i > j$. Clearly, x- and y-coordinates of the points are distinct. Following [PS] we say P_i dominates P_j if $x_i > x_j$ and $y_i > y_j$. Points P_i and P_j are crossing of $x_i > x_j$ and $y_j > y_i$ or vice versa; otherwise, they are noncrossing. A subset of Ψ is called a chain if its points are pairwise noncrossing. A maximum chain (or simply, maxchain) is a chain with the maximum number of points among all chains. A legal chain is a chain with density (of the corresponding chords) at most k. A legal maxchain is a legal chain of maximum cardinality. Note that a maxchain of Ψ corresponds to a maximum independent set of Φ. Thus, to solve MIS-k, we need to obtain a chain of Ψ attaining certain properties – to account for its density. The transformation just described has been previously employed to solve various problems in a permutation graph (being the intersection graph of chords of Φ), for example, see [AK, LLS, LS, LSL, SL1, SL2].

With each point $P_i = (x_i, y_i)$ we associate an interval $I_i = (x'_i, y'_i)$, where $x'_i = \min(x_i, y_i)$, and $y'_i = \max(x_i, y_i)$. I_i is called the spanning interval of P_i. Consider a legal chain Ψ_{a_m} of Ψ, where $\Psi_{a_m} = \{P_{a_1}, \ldots, P_{a_m}\}$, with $x_{a_1} < \ldots < x_{a_m}$. We have the following properties.

Non-containment: No spanning interval of a point in Ψ_{a_m} contains that of the other, *i.e.*, $x'_{a_1} < x'_{a_2} < \ldots < x'_{a_m}$ and $y'_{a_1} < y'_{a_2} < \ldots < y'_{a_m}$.

Density: For any value x, $x_{a_1} \leq x \leq x_{a_m}$ the number of spanning intervals containing x in its interior is at most k.

The following lemma is readily established.

Lemma 1 *A chain is legal if and only if it satisfies both the non-containment and the density property.*

We partition the set Ψ of points into two sets. P_i is called a *positive point* if $x_i \leq y_i$ and a *negative point* if $x_i > y_i$. For each chain $\Psi_{a_m} = \{P_{a_1}, \ldots, P_{a_m}\}, m \geq 1$, where $y'_{a_1} < y'_{a_2} < \cdots < y'_{a_m}$, we define a *header k-subchain* to consist of the last k points, $1 \leq k \leq m$, denoted $H(\Psi_{a_m})$. The point P_{a_m} is referred to as the *head* and the point $P_{a_{m-k+1}}$ the *tail* of the header subchain. If $k \geq m$, the header k-subchain is the chain Ψ_{a_m} itself, and P_{a_1} is the tail. Let $\mathcal{H}(\Psi_{a_m}) = (y'_{a_{m-k+1}}, y'_{a_{m-k+2}}, \ldots, y'_{a_m})$, *i.e.*, $\mathcal{H}(\Psi_{a_m})$ consists of the list of right endpoints of the spanning intervals associated with the header k-subchain. As each chain has a unique header k-subchain which has a unique *head* and *tail*, we shall refer to the head and tail of the header k-subchain simply as the head and tail of the chain. We will also omit the parameter k if it is clearly understood, and the word *legal*, as we are only interested in legal chains.

Associated with the head P_i of the header subchain $H(\Psi_i)$ we define a *critical value*, denoted, $K(P_i)$, as follows. Let d_i denote the density of the spanning intervals of the points in $H(\Psi_i)$. If $d_i < k$, $K(P_i) = x'_i$, the left endpoint of spanning interval I_i; otherwise $(d_i = k)$ $K(P_i) = y'_j$, the right endpoint of spanning interval I_j of the tail P_j of Ψ_i. In the former case the chain is said to be *non-critical*, and in the latter it is said to be *critical*.

As each chain has a unique head, the critical value associated with the head is also referred to as the critical value associated with the chain. We say that the chain Ψ_i is *extendible* by point P_j, or P_j *matches* Ψ_i, if P_j dominates P_i, and $x'_j \geq K(P_i)$, *i.e.*, P_j can be appended to chain Ψ_i to obtain a new chain $\Psi_j = \Psi_i \cup \{P_j\}$. P_j becomes the new head of Ψ_j and the new tail will be the next point in $\tilde{H}(\Psi_i)$ following the tail in the header subchain of Ψ_i, if $|H(\Psi_i)| = k$, and is the same as the old one, otherwise.

Let $\mathcal{R}$ denote a set of chains $\mathcal{R} = \{\Psi_{i_1}, \Psi_{i_2}, \ldots, \Psi_{i_t}\}$ extendible by point P_j. The chain Ψ_{i_q} and P_j are said to be a *canonical match*, if Ψ_{i_q} is such that $\mathcal{H}(\Psi_{i_q})$ is the smallest lexicographically. We say that the chain $\Psi_j = \Psi_{i_q} \cup \{P_j\}$ is obtained by *canonical* extension of Ψ_{i_q} to P_j. A chain Ψ is said to be a *canonical chain*, if it satisfies one of the following:

- Ψ contains only one point.

- If Ψ contains more than one point, *i.e.*, $\Psi = \{P_{i_1}, P_{i_2}, \ldots, P_{i_t}\}$, $i_1 < i_2 < \ldots < i_t$, and $t > 1$, then each chain Ψ_{i_j} is obtained by canonical extension of $\Psi_{i_{j-1}}$ to P_{i_j}, where chain Ψ_{i_j} contains points $P_{i_1}, P_{i_2}, \ldots, P_{i_j}$ in that order, $j = 1, 2, \ldots, t$, *i.e.*, P_{i_j} and chain $\Psi_{i_{j-1}}$ are a canonical match.

Among all chains with P_i as their head let Ψ_i^* be the one with maximum cardinality. We say P_i is at level $|\Psi_i^*|$, or, *level number* of P_i is $|\Psi_i^*|$. Thus, the set Ψ of points can be (uniquely) partitioned into a collection of points at levels 1 through s, where s is the size of a legal maxchain

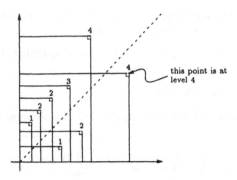

Figure 2: Level numbers when $k = 2$

(see Figure 2).[2] We shall consider an iterative method that extends a chain one point at a time. We claim that when considering chains to be extended by a new point P_i, the canonical extension will produce a solution as good as any other strategy. This is known as the *greedy* method.

We summarize it in the following lemma whose proof is omitted.

Lemma 2 *Let S_i denote the set of points to the left of P_i and including P_i. Let Ψ_i denote a chain whose head $P_i \in S_i$ is at level ℓ. There always exists a canonical chain Ψ of size ℓ containing P_i as head.*

Note that in a non-critical chain, the density of the header subchain is always less than k, so the critical value of the head P_i is $x_i' = min(x_i, y_i)$. We thus have the following lemma.

Lemma 3 *Any point that dominates the head of a non-critical chain, matches it.*

3 Details of Implementation

In this section we present a detailed algorithm for solving MIS-k problem. We employ the plane-sweep technique [PS] by processing the points $\Psi = \{P_1, \ldots, P_n\}$ from left to right. When P_i is encountered, we assume that the level numbers of previously scanned points have been computed. We proceed to find, among all chains with head P_i, one with maximum cardinality, which gives the level number of P_i. As shown in Lemma 2, it suffices to find a canonical chain Ψ_j for some $j < i$ of maximum cardinality such that Ψ_j and P_i are a canonical match.

We assume further that for each previously scanned point P_j, we have also calculated its critical value $K(P_j)$ (of the canonical chain Ψ_j) and obtained a *linked list* representation of the header subchain $H(\Psi_j)$. Note that each header subchain $H(\Psi_j)$, consists of points arranged in *ascending* order of their y' values, *i.e.*, the right endpoints of their spanning intervals. For

---

[2]In MIS-n and related problems [AK,SL1], the points belonging to the same level are pairwise crossing. This fact was exploited in designing algorithms for problems related to MIS-n. However, in MIS-k two points belonging to the same level may not cross each other.

ease of reference, the list is represented as $\hat{\mathcal{H}}(\Psi_j) = (y'_{j_1}, y'_{j_2}, \ldots, y'_{j_k})$, where y'_{j_1} corresponds to tail($\mathcal{H}(P_j)$) and y'_{j_k} corresponds to head($\mathcal{H}(P_j)$).

Now, consider point P_i and the set Γ_i of points dominated by P_i. In Γ_i we find a chain headed by point P_j such that $K(P_j) \leq x_i$, and its level number λ_j is maximum. If more than one chain satisfies the condition, select the one whose associated header list $\hat{\mathcal{H}}(P_j)$ is *lexicographically minimum* – canonical extension. The level number of P_i is $\lambda_i = \lambda_j + 1$ and we have a pointer from P_i to P_j. We obtain the canonical chain $\Psi_i = \Psi_j \cup \{P_i\}$, and $\hat{\mathcal{H}}(P_i)$ becomes $\hat{\mathcal{H}}(P_j) -$ tail($\mathcal{H}(P_j)$) $\cup \{y'_i\}$, i.e., the tail of $\mathcal{H}(P_j)$ is deleted and y'_i is appended. Set $K(P_i)$ to be $max(x'_i, Y)$, where $x'_i = min(x_i, y_i)$, and $Y =$ tail($\mathcal{H}(P_i)$). We distinguish two cases.

Case 1) P_i **is a positive point:**

We store $\hat{\mathcal{H}}(P_i)$, and λ_i, in a structure T at location $y'_i = max(x_i, y_i) = y_i$. Suppose $\Psi_i = \Psi_j \cup \{P_i\}$. If $x_i < Y$, where $Y =$ tail($\mathcal{H}(P_i)$), then $K(P_i) = Y$. Note that in this case, the density of Ψ_i must be k and Ψ_i is critical. Thus the leftmost point that matches Ψ_i must have x-coordinate not smaller than Y. See Figure 3a. To ensure the above matching condition, we need to do the following. we temporarily set λ_i in T to zero and when the sweep-line reaches $x = Y$ we will restore the actual value of λ_i. This is equivalent to saying in the x-interval $[x_i, Y)$ no point matches Ψ_i. In this case the tail of $\mathcal{H}(P_i) = P_q$ will be entered in another structure I at location $y_q (= Y)$ so that when we scan passing the value, y_q, i.e., $x_l \geq K(P_i) = y_q$ for some l, we need to restore the value λ_i in T. Otherwise, $(x_i \geq Y)$ Ψ_i is non-critical, the actual value λ_i is stored directly in T at y'_i. Therefore T should support *insertions, updates* and *Max* operations, i.e., given a value y_i, the point P_j with the maximum λ_j value and minimum $\hat{\mathcal{H}}(P_j)$ can be located efficiently.

Case 2) P_i **is a negative point:**

This case is slightly different from the previous one in the treatment of λ values. We store $\hat{\mathcal{H}}(P_i)$ and λ_i at location $K(P_i)$ in the structure T. See Figure 3b. To see this, let P_q be the tail of $\mathcal{H}(P_i)$. If Ψ_i is critical, i.e., $y_i < x_q$, and $K(P_i) = x_q$, the bottommost point that matches Ψ_i must have y-coordinate not smaller than x_q. As the next point to be examined lies to the right of x_i and $x_i > x_q$, we can store λ_i in T at $K(P_i)$ directly, since $y_i < x_i$, and $y_q < x_q$.

The structure T is a height-balanced tree with n leaves and is organized with respect to y-coordinates of the points. To be more precise, T is a *priority* queue with values $(\lambda_i, -\hat{\mathcal{H}}(\Psi_i))$ stored at the leaf corresponding to a point P_i at location Y, where $Y = y_i$ if P_i is a positive point, and $Y = K(P_i)$ otherwise. Values $(\lambda_v, -\hat{\mathcal{H}}_v)$ are stored at internal node v such that $(\lambda_v, -\hat{\mathcal{H}}_v)$ is lexicographically maximum among all such values in the leaves of the subtree rooted at v. In T we need to spend $O(k)$ time at each node to compare two header sub-chains. The next lemma shows that $\hat{\mathcal{H}}(\Psi_i)$ does not have to be stored explicitly.

Consider processing a point P_i. Let $\Psi_\pi = \{P_{\pi_1}, \ldots P_{\pi_l}\}$ and $\Psi_\sigma = \{P_{\sigma_1}, \ldots P_{\sigma_l}\}$ be two canonical maxchains that are extensible by P_i. First we observe that if $P_{\pi_l} = P_{\sigma_l}$ then $\hat{\mathcal{H}}(P_{\pi_l}) = \hat{\mathcal{H}}(P_{\sigma_l})$, for each canonical chain that matches P_{π_l} – by definition – has the lexicographically smallest header subchain.

Lemma 4 *If $y'_{\pi_l} < y'_{\sigma_l}$ then $\hat{\mathcal{H}}(P_{\pi_l})$ is lexicographically smaller than $\hat{\mathcal{H}}(P_{\sigma_l})$.*

Proof: (Omitted.)

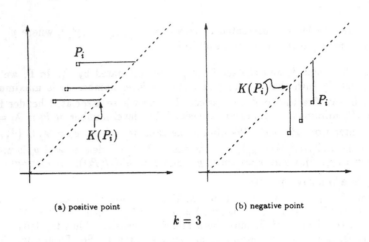

(a) positive point   (b) negative point

$$k = 3$$

Figure 3: Cases 1 and 2

Based on Lemma 4, T is a *priority* queue with values $(\lambda_i, -y_i')$ stored at the leaf corresponding to a point P_i at location Y, where $Y = y_i$ if P_i is a positive point, and $Y = K(P_i)$ otherwise. Values $(\lambda_v, -y_v')$ stored at internal node v are such that the pair, $(\lambda_v, -y_v')$, is lexicographically maximum among all such pairs of values stored in the leaves of the subtree rooted at v. By virtue of Lemma 4, a lexicographically minimum $(\lambda_v, -y_v')$ implies a lexicographically minimum $(\lambda_v, -\hat{\mathcal{H}}(\Psi_v))$. Each leaf node points to the previous point in its header subchain, if its header subchain is non-empty.

When we process a positive point P_i, the value λ_i in $(\lambda_i, -y_i')$ stored in the corresponding leaf node at y_i in T may be set to contain a temporary value Temp-i (being zero), which is made permanent and contains the actual value λ_i at a later time. To efficiently obtain the canonical chain with P_i as the head we also store a pointer to the head of the chain that P_i matches (*i.e.*, P_j in the above discussion). I is a linear array that keeps track of those positive points whose λ values need to be reset to permanant. It is organized with respect to increasing order of y-coordinate; each entry y_j is a list of points P_i whose $K(P_i) = y_j$ and whose λ values need to be reset.

When we process a negative point P_i we store the pair $(\lambda_i, -y_i')$ at location $K(P_i)$, and a pointer to the head of the chain that P_i matches (*i.e.*, P_j in the above discussion).

We perform the following operations on the structures just described.

MAX(y_i) :   Find the maximum value $(\lambda_j, -y_j')$ in $[1, y_i]$ of T and return the result (*max*, *index*), where *max* contains λ_j, and *index* contains j.

INS(y_i, λ_i, y_i'): Insert (in Temp location) at y_i of T the value $(\lambda_i, -y_i')$.

SET(y_i) :   Set the temporary value of y_i in T into its permanent value.

IN.I(y_j, P_i) :   Insert P_i at location y_j of I.

$OT.I(x_i)$:   Return all points in the list located at position $y_i < x_i$ in I.

$PTR(P_i, P_j)$:   Add a pointer from P_i to P_j.

tail(P_i) :   Find $tail(\hat{\mathcal{H}}(P_i))$.

Next, we will describe in details how to update the structures. All values are initially set to zero. We assume points $P_1, \ldots, P_{i-1}$ have been processed and the structures have been accordingly updated.

Algorithm LEVEL (Ψ):
for each point P_i do

$$\{P_{a_1}, \ldots, P_{a_m}\} := OT.I(x_i)$$
for $1 \le j \le m$ do $SET(y_{a_j})$
$y_i' := max(x_i, y_i)$
$(max, index) := MAX(y_i)$
if P_i is a positive point
then begin
    $INS(y_i, max + 1, y_i')$
    $j := index$
    $PTR(P_i, P_j)$
    $Y := tail(P_i)$:
    if $x_i < Y$ then IN.I(Y, P_i) else $SET(y_i)$
end
else begin $\{P_i$ is a negative point$\}$
    $j := index$
    $PTR(P_i, P_j)$
    $Y := tail(P_i)$:
    $INS(Y, max + 1, y_i')$
    $SET(Y)$
end

Lemma 5 *Algorithm LEVEL correctly computes the level number of all points.*

Proof: Assume level number of points $P_1, \ldots, P_{i-1}$ have been correctly computed and each point P_t is associated with a lexicographically minimum $\hat{\mathcal{H}}(P_t)$, $t = 1, 2, \ldots, i-1$. Consider the set $\Gamma_i = \{P_{a_1}, \ldots, P_{a_m}\}$ of points dominated by P_i. Let λ_{max} denote the maximum level of points in Γ_i, that is, $\lambda_{max} = \max_{j=1}^{m}(\lambda_{a_j})$.

Let P_i match a chain Ψ_t of size λ_{long} with head P_t and P_i is a canonical match of Ψ_t. We have $\lambda_i = \lambda_{long} + 1$ where $\lambda_{long} \in \{\lambda_{max} - 1, \lambda_{max}\}$ for there exists at least one chain of size λ_{max} whose head is in Γ_i and we can replace the head with P_i.

CASE 1) $\lambda_i = \lambda_{max} + 1$
    This is the maximum possible level number for P_i since we have assumed all points in Γ_i have the correct level number. Since P_i matches Ψ_t, i.e., $min(x_i, y_i) > K(P_t)$, $\lambda_t = \lambda_{max}$

must have been set permanant before we process P_i. Since MAX(y_i) returns the point P_t with minimum $\hat{\mathcal{H}}(\Psi_t)$, $\hat{\mathcal{H}}(\Psi_i)$ and λ_i will be computed and correctly entered into T.

CASE 2) $\lambda_i = \lambda_{max}$

In this case for all points P_j in Γ_i with level number λ_{max}, $K(P_j) > x_i'$, i.e., their λ values either remain temporary or are not involved when MAX(y_i) is called. Since for all points P_t with level number $\lambda_{max} - 1$, $K(P_t) < x_i'$, they must all be permanant and hence available when MAX(y_i) is called. $\hat{\mathcal{H}}(\Psi_i)$ and λ_i are computed and correctly entered into T.

□

Operations described in Algorithm LEVEL are straightforward except for the operation $tail(P_i)$. It requires a special treatment of the tree T. One approach is to follow the pointers, starting at the leaf corresponding to P_i, to obtain Y. This approach requires $O(\log n + k)$ time to process each point P_i. That is, the algorithm runs in $O(n(\log n + k))$ time and requires $O(n)$ space. Next we show how to improve the time complexity of this direct approach.

Consider a path $\mathcal{P}$ consisting of m vertices. The vertex with out-degree zero is at level one. A vertex with distance l from v_1 is at level $l + 1$, $1 \leq l \leq m - 1$. The vertex at level l is denoted by v_l. Edges of $\mathcal{P}$ are called *solid edges*. We add a collection of *dashed edges* to $\mathcal{P}$ as follows. The set of vertices between levels 1 and m is called a *block* and is denoted by $B(1, m)$. Each block $B(a, b)$, $b - a > 3$, is partitioned into two subblocks $B(a + 1, \lfloor \frac{a+b}{2} \rfloor)$ and $B(\lfloor \frac{a+b}{2} \rfloor + 1, b - 1)$, and there is a dashed edge of length $b - a$ between vertex v_a to vertex v_b of block $B(a, b)$. See Figure 4. The new graph, being the union of solid and dashed edges, is denoted by $\mathcal{P}^*$. Note that there is a path from any vertex v_i to any other vertex v_j, $i \geq j$, of length $O(\log m)$. In $\mathcal{P}^*$ each vertex has degree at most two: one contributed by a dashed edge and one by a solid edge. Next, we will use the just described technique for finding the tail of a point in Algorithm LEVEL.

Consider a *forest* with vertices corresponding to points $\{P_1, \ldots, P_n\}$ in which the total number of levels is m. Its edges correspond to pointers added in Algorithm LEVEL. That is, when $PTR(P_i, P_j)$ is executed, an edge from P_i to P_j is introduced. By introducing a point at $(-\infty, -\infty)$, the given forest becomes a tree (certainly, that point will not be considered when calculating the levels). We denote this tree by T_{ptr}. Edges of the tree are the *solid edges*. When we process P_i, we add a *dashed edge* from P_i at level l_1 to the point in the chain containing P_i, at level l_2, $l_1 > l_2$, if there is a dashed edge between v_{l_1} to v_{l_2} in $\mathcal{P}^*$. Thus, dashed edges will be added dynamically. To add this dashed edge, we need to follow exactly two other dashed edges and three solid edges, if $l_1 - l_2 > 3$. In case $l_1 - l_2 = 3$, we need to follow exactly three solid edges (see Figure 4). Thus, it takes $O(1)$ to add a dashed edge. The new tree, containing solid and dashed edges is denoted by T_{ptr}^*. We can find the tail of a point P_i in T_{ptr}^* in $O(\log n)$ time. Note that for our purpose, dashes edges of length greater than k need not be added to T_{ptr}^*. Thus, it takes $O(\log k)$ time to find the tail of a point. Processing P_i requires $O(\log n + \log k)$ time. Thus we have the following.

Theorem 1 *Algorithm LEVEL correctly computes the level number of all points in Ψ in $\Theta(n \log n)$ time and $\Theta(n)$ space, where $n = |\Psi|$.*

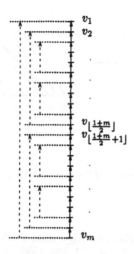

Figure 4: Dashed and solid edges in $\mathcal{P}^*$

4 Discussion and Conclusion

Here, we consider an application of MIS-k. Consider a set of intervals $\mathcal{I} = \{I_1, \ldots I_n\}$, where $I_i = (\ell_i, r_i)$. A *proper* subset of $\mathcal{I}$ is a subset where no interval contains any other intervals. That is, there are no two intervals I_i and I_j with $\ell_i < \ell_j < r_j < r_i$. MPS-$k$ problem is to find, among all proper subsets of $\mathcal{I}$ with density at most k, one with maximum cardinality. We can transform MPS-k problem into MIS-k problem. To each interval $I_i = (\ell_i, r_i)$ we associate a chord $N_i = (\ell_i, r_i)$ (see Figure 5). Let $\mathcal{N}$ denote $\{N_1, \ldots N_n\}$. Certainly a subset of chords in $\mathcal{N}$ are independent if and only if the corresponding subset of intervals in $\mathcal{I}$ are proper. We conclude:

Theorem 2 *An arbitrary instance $\mathcal{I}$ of MPS-k can be solved in $\Theta(n \log n)$ time and $\Theta(n)$ space, where $n = |\mathcal{I}|$.*

A related problem of finding a maximum subset with density at most k of a collection of intervals was recently solved in $O(nlogk)$ time [YG]. Intervals in the subset are arbitrary, that is, they are not pairwise proper. It is assumed that intervals are pre-sorted; otherwise, an additional $O(nlogn)$ time is required for pre-processing.

The complexity of the weighted version of MIS-k remains to be seen. We are also investigating the problem of finding among all two-chains (see [LSL] for a definition) with density k, one with maximum cardinatlity.

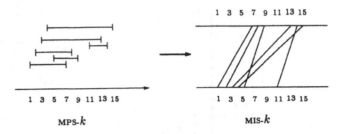

MPS-k MIS-k

Figure 5: Transforming MPS-k into MIS-k

References

[AK]   Atallah, M.J., Kosaraju, S.R., "An Efficient Algorithm for Maxdominance, with Applications," *Algorithmica*, Vol. 4, No. 2, pp. 221-236, 1989.

[GLL]   Gupta, U., Lee, D.T., Leung, J., "An Optimal Solution for the Channel Assignment Problem," *IEEE Transactions on Computers*, November 1979, pp. 807-810.

[LLS]   Liao, K.F., Lee, D.T., and Sarrafzadeh, M., "Planar Subset of Multi-terminal Nets," *INTEGRATION: The VLSI Journal*, No. 1, Vol. 10, Sep. 1990, pp. 19-37.

[LS]   Lou, R.D. and Sarrafzadeh, M., " Circular Permutation Graph Family," *International Workshop on Discrete Algorithms and Complexity*, Fukuoka, Japan, November 1989, pp. 107-114; also to appear in *Discrete Applied Mathematics*.

[LSL]   Lou, R.D., Sarrafzadeh, M., and Lee, D.T., "An Optimal Algorithm for the Maximum Two-chain Problem," *Proceedings of the First SIAM-ACM Conference on Discrete Algorithms*, San Francisco, CA, January 1990, pp. 149-15; also to appear in *SIAM Journal on Discrete Mathematics*.

[SL1]   Sarrafzadeh, M. and Lee, D.T., "A New Approach to Topological Via Minimization" *IEEE Transactions on Computer-Aided Design*, Vol. 8, No. 8, August 1989, pp. 890-900.

[SL2]   Sarrafzadeh, M. and Lee, D.T., "Topological Via Minimization Revisited," manuscript, 1989; to appear in *IEEE Transactions on Computers*.

[YG]   M. Yannakakis anf F. Gavril, "The Maximum K-colorable Subgraph Problem for Chordal Graphs" *Information Processing Letters*, Vol. 24, 1987, pp. 133-137.

Algorithms for Square Roots of Graphs
(Extended Abstract)

Yaw-Ling Lin
Steven S. Skiena*

Department of Computer Science
State University of New York
Stony Brook, NY 11794-4400
e-mail: {yawlin,skiena}@sbcs.sunysb.edu

1   Introduction

Given a graph G, the *distance* between vertices u and v, denoted by $d_G(u,v)$, is the length of the shortest path from u to v in G. The *kth power* ($k \geq 1$) of G, written G^k, is defined to be the graph having $V(G)$ as its vertex set with two vertices u, v adjacent in G^k if and only if there exists a path of length at most k between them, i.e., $d_G(u,v) \leq k$. Similarly, graph H has an *kth root* G if $G^k = H$. For the case of $k = 2$, we say that G^2 is the *square* of G and G is the *square root* of G^2. The square of a graph can be computed by squaring its adjacency matrix.

Powers of graphs have many interesting properties. For example, Fleischner [4] proved that the square of a biconnected graph is always hamiltonian. Although the hamiltonian cycle problem is NP-complete for general graphs [5], the biconnectivity of a graph can be tested in linear time. Thus an efficient algorithm for finding the square root of a graph could be useful for finding hamiltonian cycles in square graphs.

Mukhopadhyay [15] showed that a connected undirected graph G with vertices $v_1, ..., v_n$ has a square root if and only if G contains a collection of n complete subgraphs $G_1, ..., G_n$ such that for all $1 \leq i, j \leq n$: (1) $\bigcup_{1 \leq i \leq n} G_i = G$, (2) $v_i \in G_i$, and (3) $v_i \in G_j$ if and only if $v_j \in G_i$. Characterizations of squares of digraphs were given by Geller [7] and of nth power of graphs and digraphs was given by Escalante, Montejano, and Rojano [3]. Unfortunately, these characterizations do not lead to efficient algorithms.

This paper studies two distinct classes of problems concerning powers of graphs. In Section 2, we concentrate on the problem of finding the square roots of graphs, presenting efficient algorithms for finding the tree square roots of G^2, finding square roots when

---

*The work of the second author was partially supported by NSF Research Initiation Award CCR-9109289

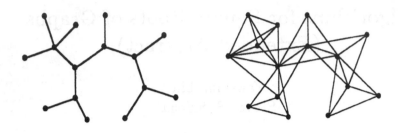

Figure 1: A tree and its square.

G^2 is a planar graph, and finding any square roots which are subdivision graphs. This last problem provides an efficient algorithm for inverting total graphs. In Section 3, we consider the complexity of finding maximum cliques and hamiltonian cycles in G^k given G. We conclude with a list of open problems.

2   Finding Square Roots of Graphs

In this section, we concentrate on the problem of finding a square root of a given graph. Not all graphs are squares, the smallest example being a simple path on three vertices. Further, a graph may have exponentially many distinct square roots. For example, any graph containing a vertex of degree $n - 1$ is a square root of the complete graph K_n.

We give an $O(m)$ time algorithm for finding the tree square roots of a graph in Section 2.1, and an $O(n)$ algorithm for finding the square roots of a planar graph in Section 2.2. In Section 2.3, we provide an $O(m^2)$ time algorithm for the inversion of total graphs.

2.1   Tree Square Roots

In this section, we present an $O(|V| + |E|)$ algorithm for finding the tree square root T of a given graph $G = T^2 = (V, E)$. Tree square roots were first considered by Ross and Harary [16], who showed that they are unique up to isomorphism. Figure 1 presents a tree and its square. Later, in Section 3, we will consider algorithmic problems on tree squares and prove the graph theoretic result that all powers of trees are chordal.

Our algorithm for inversion proceeds by identifying the leaves of the tree square root, and then trimming all leaves from G to obtain a graph which is the square of a smaller tree.

Now consider a graph $T^2 = (V, E)$ which is known to be the square of some tree $T = (V, E')$. The *degree* of a vertex v in G, written $deg_G(v)$, is the size of its neighborhood in G. A vertex v is a *leaf* (or *endpoint*) of a graph G if $deg_G(v) = 1$.

The *diameter* of a graph G is the longest distance defined between a pair of vertices. A tree with diameter less than or equal to 2 is called a *star*. Note that the square of a star is a complete graph. The *neighborhood* of a vertex v in G, denoted by $N_G(v)$, is the set containing all vertices that are adjacent to v. Since $v \notin N_G(v)$, we denote $\{v\} \cup N_G(v)$ by $N'_G(v)$. A vertex v is *simplicial* if the induced subgraph of its neighborhood forms a clique.

Lemma 2.1 *v is a simplicial point of T^2 if and only if v is a leaf of T or T is a star.*

Proof. Assume that v is a leaf of T. Since $N'_{T^2}(v) = \{u : d_T(u,v) \leq 2\}$, the induced subgraph $\langle N'_{T^2}(v) \rangle_T$ can be obtained by starting from a leaf V and walking at most two steps in T. The resulting graph has a diameter less than or equal to two, therefore $\langle N'_{T^2}(v) \rangle_T$ is a star. The square of a star is a complete graph, so $\langle N'_{T^2}(v) \rangle_{T^2}$ is complete, meaning v is simplicial. Now assume v is not a leaf and T is not a star. Then there must be a path (x,v,y,z) in T. Since $d_T(x,z) = 3$ implies $xz \notin E(T^2)$ and $x, z \in N_{T^2}(v)$, v is not simplicial. ∎

For each leaf v of T, we can partition $N_{T^2}(v)$ into two sets L_v and M_v, where L_v denotes all leaves of T in $N_{T^2}(v)$ and M_v the set $N_{T^2}(v) - L_v$. According to Lemma 2.1, a leaf node has the lowest degree of any vertex in its neighborhood of T^2. The following lemma characterizes when the degree of a leaf will be strictly less than the degree of its neighboring internal nodes.

Lemma 2.2 *Let T be a non-star tree and v a leaf of T, then*

1. *For all $u \in L_v, deg_{T^2}(u) = deg_{T^2}(v)$.*

2. *For all $w \in M_v, deg_{T^2}(w) > deg_{T^2}(v)$.*

Proof. For part 1, let $u \in L_v$. Since $\langle N'_{T^2}(v) \rangle_T$ is a star, we know that $N_{T^2}(v) = N_{T^2}(u)$. So $deg_{T^2}(u) = deg_{T^2}(v)$. For part 2, let $w \in M_v$. Since T is not a star, without loss of generality, there is either a path (v,w,x,y) or a path (v,x,w,y) in T. For either case, $y \in N_{T^2}(w)$ but $y \notin N_{T^2}(v)$. We conclude that $N'_{T^2}(w) \not\subset N'_{T^2}(v)$. Since, by Lemma 2.1, $N'_{T^2}(v) \subset N'_{T^2}(w), deg_{T^2}(w) > deg_{T^2}(v)$. ∎

The *center* of star S is the central vertex v such that for all other vertices u in S, $d_S(u,v) = 1$. If $|S| \neq 2$, there is only one center v in S, which we denote by $center(S)$. Let K_n denotes the complete graph of size n. The cardinality of M_v, $|M_v|$, provides valuable information about the structure of the graph.

Lemma 2.3 *Let v be a leaf of T, then*

1. *if $|M_v| \leq 1$, T is a star;*

2. *if $|M_v| = 2$, say $M_v = \{x,y\}$, then $xy \in E(T)$;*

3. *if $|M_v| \geq 3$, $(v, center(M_v)) \in E(T)$.*

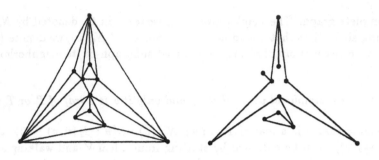

Figure 2: A planar graph and its square root.

Proof. For part 1, either $|M_v| = 0$, meaning $T = K_2$, or $|M_v| = 1$, where again, T is a star. For part 2, let $M_v = \{x, y\}$. Since v is a leaf, Lemma 2.1 implies that $\langle N'_{T^2}(v) \rangle_T$ is a star. Since $|N_{T^2}(v)| \geq |M_v| = 2$ meaning $|N'_{T^2}(v)| \geq 3$, $center(\langle N'_{T^2}(v) \rangle_T)$ is well-defined. Further, $center(\langle N'_{T^2}(v) \rangle_T) \notin L_v$. So, without loss of generality, we can let x be the center of $\langle N'_{T^2}(v) \rangle_T$. Again, since $\langle N'_{T^2}(v) \rangle_T$ is a star, y must be connected to x. So $xy \in E(T)$. For part 3, $|M_v| \geq 3$. Since $\langle N_{T^2}(v) \rangle_T$ is a star and v is the center, we know that $center(\langle N_{T^2}(v) \rangle_T) = center(\langle M_v \rangle_T)$. Clearly $(v, center(\langle N_{T^2}(v) \rangle_T)) \in E(T)$. ∎

Given a tree T, we can delete all the leaves of T resulting in a smaller tree T'. This trimming operation defines a function $trim(T) = T'$.

Theorem 2.4 *The tree square root of a graph G can be found in $O(m)$ time, where m denotes the number of edges of the given tree square graph.*

Proof. The proof appears in the complete paper. In summary, it takes $O(m)$ time to identify all leaves of T, and another $O(m)$ time to trim the tree and determine the structure of T. With a final $O(m)$ confirmation step, we conclude that the tree square root of a square graph can be identified in linear time. ∎

2.2   Square Roots of Planar Graphs

In this section, we present an $O(n)$ algorithm for finding the square root of a given planar graph, based on the characterization of planar squares found by Harary, Karp, and Tutte [10] and the linear time triconnected components algorithm given by Hopcroft and Tarjan [13]. Planar roots are not necessarily unique up to isomorphism. Figure 2 presents a planar graph and its square root.

Theorem 2.5 (Harary, Karp, and Tutte [10]) *A graph G has a planar square if and only if*

1. *every point of G has degree less than or equal to three,*

2. *every block of G with more than four points is a cycle of even length, and*

3. G does not have three mutually adjacent articulation vertices.

An *endline* of G is an edge uv of G such that either u or v is a leaf. A *burr* of G is a maximal connected (induced) subgraph of G in which every bridge is an end line. By removing all leaves of a burr B in G, the remaining subgraph B' is called the *central block* of B. Since B does not contain inner bridge, B' must be a biconnected component, a single vertex, or the null graph. As shown by Theorem 2.5, if G^2 is a planar graph, then for each burr B of G, we have three cases:

Case 1. The central block B' is an even length cycle with more than four vertices. Let C_n denote a chordless cycle of length n. That is, $B' = C_{2n}$, for some n greater than 2. Further, each vertex of B' is adjacent to at most one endline.

Case 2. The central block B' is a block of size at most four. Each vertex of B' can be linked with at most one leaf, providing it does not violate the conditions of Theorem 2.5. Such kind of burr has at most eight vertices. That is, a burr consists of a central block C_4 with each vertex adjacent to exactly one leaf.

Case 3. The central block B' consists of a single vertex v, which can be adjacent to at most three other leaves. That is, B is a star of size at most four.

Lemma 2.6 *The square of a burr is triconnected or complete; Further, each triconnected component of a planar square G^2 is the square of a burr in G.*

Proof. The proof appears in the complete paper. ∎

Hopcroft and Tarjan [13] shows that $O(m + n)$ time suffices to find all triconnected components of a given graph. For planar graphs, the number of edges $m \leq 3n - 6$, so we can find all triconnected components in $O(n)$ time.

For G^2 to be planar, each large burr in G must be an even length cycle, perhaps including some number of endlines. Otherwise, the central block will just be a finite graph with size less than five. The following lemma characterizes the leaves of a large burr.

Lemma 2.7 *Given a planar square graph G^2 and a burr B of G such that $|B| > 5$, then v is a leaf of B if and only if $deg_{B^2}(v) = 3$.*

Proof. Since $|B| > 5$ and B^2 is planar, the central block B' must have at least four vertices. Otherwise the condition of Theorem 2.5 will be violated. First we show that for each non-leaf v of B, $deg_{B^2}(v) > 3$. If B' is an even length cycle C_{2n} such that $n \geq 3$, then $deg_{B^2}(v) \geq 4$. Otherwise, the central block is a block of size four. There are at least two leaves outside this size-four block. Again, $deg_{B^2}(v) \geq 4$.

Now we show the each leaf $v \in B$ has degree three in B^2. Let $uv \in B$ meaning $u \in B'$. Since B^2 is planar, $deg_B(u) \leq 3$. Since B' is a block, $deg_{B'}(u) \geq 2$, meaning $deg_B(u) \geq 3$. Since $deg_B(u) = 3$, $deg_{B^2}(v) = 3$. ∎

Now we show that we can efficiently find the original burr given its square.

Lemma 2.8 *Given a planar graph B^2, where B is a burr, the structure of B can be computed in $O(|B|)$ time.*

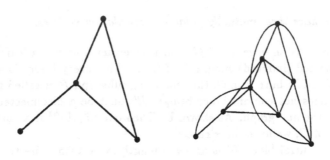

Figure 3: A graph and its total graph

Proof. The proof appears in the complete paper. ∎

Theorem 2.9 *The square root of a planar graph G^2 can be found in $O(n)$ time.*

Proof. By Lemma 2.6, the square burrs of G^2 can be found in $O(n)$ time. By Lemma 2.8, the structure of each burr B_i can be found in $O(|B_i|)$ time. The rest of the proof, which shows that to indentify the inner bridges between burrs can be done in $O(n)$ time also, appears in the complete paper. It then follows that finding the square root of a planar graph can be done in linear time. ∎

2.3 Inversion of Total Graphs

The *total graph* $T(G)$ of a graph $G = (V, E)$ has vertex set $V \cup E$ with two vertices of $T(G)$ adjacent whenever they are neighbors in G. If uv is an edge of G, and w not a vertex of G, then uv is *subdivided* when it is replaced by two edges uw and wv. If every edge of G is subdivided, the resulting graph is the *subdivision graph* $S(G)$. [9] Behzad [2] showed that, given a graph G, the total graph $T(G)$ is isomorphic to the square of its subdivision graph $S(G)$. In this section, we present an efficient algorithm for inverting total graphs by reducing the problem to finding square roots of the squares of subdivision graphs. Figure 3 presents a graph with its total graph.

Recall that a *maximal clique* of a graph is an induced complete subgraph such that no other vertex can be added to form a larger clique. We can identify whether an edge uv is in $S(G)$ by examining these maximal cliques in the square containing both vertices u and v.

Lemma 2.10 *Given a graph $G = (V, E)$ and an edge uv of G, there are at most two maximal cliques of size 3 containing both u and v in G^2.*

Proof. Let $K = \{u, v, w\}$ be a maximal clique in G^2. Note that it is not possible that $d_G(u, w) = d_G(v, w) = 2$ in G since that will imply that K is not maximal. So, without loss of generality, we can assume that uw is an edge of G. Note that v and w are the only vertices allowed to be adjacent to u since, otherwise, K will not be maximal. It is possible that v can be adjacent to a vertex $w' \neq w$ such that $N_G(v) = \{u, w'\}$, implying $\{u, v, w'\}$ is also a maximal clique, but that covers all possible maximal cliques of size 3 containing both u and v. ∎

Given a graph $G = (V, E)$ and its subdivision graph $S(G)$, we call the set of newly added vertices W, such that $V(S(G)) = V \cup W$. By Behzad's result, we know that $T(G) = [S(G)]^2$. Given $v \in V$, we call $Inner(v) = N_{S(G)}(v)$, the *inner neighborhoods* of vertex v, and $Outer(v) = N_G(v)$, the *outer neighborhoods*. Note that $N_{T(G)}(v) = N_{[S(G)]^2}(v) = Inner(v) \cup Outer(v)$. All inner nodes are elements of W, and all outer nodes are elements of V. We observe that

Lemma 2.11 *Let uv be an edge of G and w the subdivided vertex of uv, implying $uw, wv \in S(G)$. Then*

(i) $Inner(v) = N_{T(G)}(v) \cap N_{T(G)}(w) - \{u\}$

(ii) *For each pair* $(w', u') \in Inner(v) \times Outer(v)$, $w'u' \in S(G)$ *if and only if* $N_{T(G)}(w') \cap N_{T(G)}(v) - Inner(v) = \{u'\}$

Proof. (i) Since, for each $x \in Outer(v) - \{u\}$, $d_{S(G)}(w, x) = 3$, it follows that $Inner(v) = N_{T(G)}(v) \cap N_{T(G)}(w) - \{u\}$.

(ii) The only if part is trivial by (i). Now assume that $\{u'\} = N_{T(G)}(w') \cap N_{T(G)}(v) - Inner(v)$. Note that $d_{S(G)}(u', w') \leq 2$ and $d_{S(G)}(u', v) = 2$. It follows that $u'v' \in S(G)$. ∎

With these results, now we are ready to present a polynomial time algorithm for the inversion of total graphs.

Theorem 2.12 *The inversion of a total graph H can be performed in $O(m^2)$ time where m denotes the number of edges of H.*

Proof. The proof appears in the complete paper. Denote the lowest degree of all vertices in H by δ and the number of degree δ vertices in $T(G)$ by α. In summary, it will spend at most $O(m)$ time for each possible triple $\{u, v, w\}$, while the number of candidate triples is bounded by $O(\alpha\delta)$. Since we need the time to initialize the adjacency matrix of H, the total time bound will be $O(n^2 + \alpha\delta m)$. In the worst case, this will take $O(m^2)$ time, although typically $\alpha\delta$ will be much smaller than m, which collapses the time complexity to $O(n^2)$. ∎

3 Algorithms on Powers of Graphs

In this section, we present several results concerning the complexity of two optimization problems for squares and higher powered graphs.

Hamiltonian cycles in powers of graphs have received considerable attention in the literature. In particular, Fleischner [4] proved that the square of a biconnected graph is always hamiltonian. Harary and Schwenk [11] proved that the square of a tree T is hamiltonian if and only if T does not contain $S(K_{1,3})$ as its induced subgraph. Here $S(K_{1,3})$ denotes the subdivision graph of the complete bipartite graph $K_{1,3}$. Harary and Schwenk's result leads to an optimal algorithm.

Theorem 3.1 *For any tree T, the problem of determining whether T^2 is hamiltonian can be answered in $O(n)$ time.*

Proof. The problem reduces to testing whether $S(K_{1,3})$ is a subgraph of T. Starting from a leaf v of T, we perform a depth-first search, for each node x computing the distance from v as well as the diameter of the subtree rooted at x. T^2 is hamiltonian unless there exists a vertex x which has three decendants of diameter at least two, or is a distance of at least two from v and has two decendants of diameter two. ∎

A graph G is *hamiltonian connected* if every two distinct vertices are connected by a hamiltonian path. Sekanina [17] proved G^3 is hamiltonian connected if G is connected, which implies that if the size of G is greater or equal to 3, then G^3 is hamiltonian. Here we present a linear time algorithm (in terms of the number of edges of input G) for finding a hamiltonian cycle in G^k, where $k \geq 3$.

Theorem 3.2 *Given a connected graph $G = (V, E)$ with size at least 3 and an integer $k \geq 3$, we can find a hamiltonian cycle in G^k within $O(|V| + |E|)$ time.*

Proof. The proof appears in the complete paper. ∎

Recall that a clique in a graph G is *maximum* if it is the largest induced complete subgraph of G. Here we prove that finding maximum cliques in powered graphs remains NP-complete by a reduction from the general problem of finding the maximum cliques in arbitrary graphs. In the subsequent section, we provide a linear algorithm for finding the maximum clique of a tree square.

Theorem 3.3 *Let $G = (V, E)$ be a graph. Then, for any fixed integer $k \geq 1$, finding the maximum clique of G^k is NP-complete.*

Proof. The proof appears in the complete paper. ∎

Graphs whose every simple cycle of length strictly greater 3 possesses a chord are called *chordal graphs*. In the literature, chordal graphs have also been called *triangulated*, *rigid-circuit*, *monotone transitive*, and *perfect elimination* graphs. Chordal graphs, which can be reconginized in linear time, are perfect graphs, and the maximum clique in chordal graphs can be determined in linear time [8].

Gavril [6] proved that a graph G is chordal if and only if G is the intersection graph of a family of subtrees of a tree. Figure 4 shows that the square of a chordal graph is not necessarily chordal. However, we will show that arbitrary powers of trees are always chordal.

Theorem 3.4 *All powers of trees are chordal.*

Proof. Let T be a tree. We want to prove that T^k is chordal for any positive integer k. Let $S(G)$ be the subdivision graph of a graph G as defined in Section 2.3, and $N_G^k(v)$ denotes the set $\{u \in G : d(u, v) \leq k\}$ where $d(u, v)$ is the distance between vertices u and v in G. Note that $uv \in G^k$ if and only if $N_{S(G)}^k(u) \cap N_{S(G)}^k(v) \neq \emptyset$, implying that G^k is the intersection graph of the family $S = \{N_{S(G)}^k(v) : \text{for all } v \in G\}$. It follows that

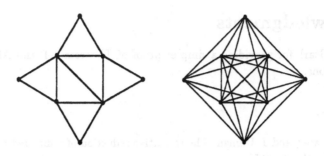

Figure 4: A chordal graph and its nonchordal square.

T^k is the intersection graph of the family $\{N^k_{S(T)}(v) : \text{ for all } v \in T\}$. Since $S(T)$ is a tree and $\langle N^k_{S(T)}(v)\rangle_{S(T)}$ is a subtree of $S(T)$ with center vertex v and radius k, T^k is the intersection graph of a family of subtrees of the tree $S(T)$. Thus T^k is chordal. ∎

Corollary 3.5 *The maximum clique in an arbitrary power of a tree can be determined in linear time.*

4 Conclusions

We have presented efficient algorithms for finding the square roots of graphs in three interesting special cases: tree squares, planar graphs and subdivided graphs. Further, we have studied the complexity of finding hamiltonian cycles and maximum cliques in powers of graphs. Several interesting open problems remain:

- What is the complexity of recognizing square graphs? We conjecture that the general problem is NP-complete.

- Let A be the adjacency matrix of a graph, made reflexive by adding a self-loop to each vertex. Given A^2, can we determine one of its $(0,1)$-matrix square roots in polynomial time? This problem is potentially easier than the finding the square root of a graph, since we are also given the number of paths of length at most two between each pair of vertices.

- What is the complexity of determining whether G^2 is hamiltonian, given G? Relevant results appear in [12, 14].

- What is the complexity of recognizing the squares of directed-acyclic graphs? Clearly, the square of DAG is a DAG. All square roots of a DAG G^2 contain the transitive reduction of G^2 as a subgraph. The time complexity of finding both the transitive closure and reduction of a digraph is equivalent to boolean matrix multiplication [1].

- Give an $o(m^2)$ algorithm for inverting total graphs.

5 Acknowledgments

We thank Gene Stark for providing a simpler proof of Theorem 3.4, and Alan Tucker for valuable discussions.

References

[1] A. Aho, M. Garey, and J. Ullman. The transitive reduction of a directed graph. *SIAM J. Computing*, 1:131–137, 1972.

[2] M. Behzad. A criterion for the planarity of a total graph. *Proc. Cambridge Philos. Soc.*, 63:679–681, 1967.

[3] F. Escalante, L. Montejano, and T. Rojano. Characterization of n-path graphs and of graphs having nth root. *J. Combin. Theory B*, 16:282–289, 1974.

[4] H. Fleischner. The square of every two-connected graph is Hamiltonian. *J. Combin. Theory B*, 16:29–34, 1974.

[5] M. R. Garey and D. S. Johnson. *Computers and Intractability – A guide to the Theory of NP-Completeness*. Freeman, New York, 1979.

[6] F. Gavril. The intersection graphs of subtrees in trees are exactly the chordal graphs. *J. Combin. Theory B*, 16:47–56, 1974.

[7] D. P. Geller. The square root of a digraph. *J. Combin. Theory*, 5:320–321, 1968.

[8] M. C. Golumbic. *Algorithmic Graph Theory and Perfect Graphs*. Academic Press, New York, 1980.

[9] F. Harary. *Graph Theory*. Addison-Wesley, Massachusetts, 1972.

[10] F. Harary, R. M. Karp, and W. T. Tutte. A criterion for planarity of the square of a graph. *J. Combin. Theory*, 2:395–405, 1967.

[11] F. Harary and A. Schwenk. Trees with hamiltonian square. *Mathematika*, 18:138–140, 1971.

[12] G. Hendry and W. Vogler. The square of a connected $S(K_{1,3})$-free graph is vertex pancyclic. *J. Graph Theory*, 9:535–537, 1985.

[13] J. E. Hopcroft and R. E. Tarjan. Dividing a graph into triconnected components. *SIAM J. Computing*, 2:135–158, 1973.

[14] M. Matthews and D. Summer. Hamiltonina results in $S(K_{1,3})$-free graphs. *J. Graph Theory*, 8:139–146, 1984.

[15] A. Mukhopadhyay. The square root of a graph. *J. Combin. Theory*, 2:290–295, 1967.

[16] I. C. Ross and F. Harary. The square of a tree. *Bell System Tech. J.*, 39:641–647, 1960.

[17] M. Sekanina. *On an ordering of the set of vertices of a connected graph.* Technical Report No. 412, Publ. Fac. Sci. Univ. Brno, 1960.

Distributed k-Mutual Exclusion Problem and k-Coteries

Satoshi FUJITA, Masafumi YAMASHITA, and Tadashi AE
Department of Electrical Engineering, Faculty of Engineering,
Hiroshima University, Higashi-Horoshima, 724 Japan
E-mail : {fujita,mak,ae}@csl.hiroshima-u.ac.jp

Abstract

The distributed k-mutual exclusion problem is the problem of guaranteeing that at most k processes are in a critical section simultaneously. This problem can be solved using the k-coterie: We first prepare a set (k-coterie) C of sets (quorums) Q of processes such that each k-set $\{Q_1, \cdots, Q_k\}$ of quorums in C contains a pair Q_i and Q_j ($i \neq j$) intersecting each other. A process wishing to enter a critical section is required to collect a permission from each member of a quorum in C. Then at most k processes can be in a critical section because of the intersection property of k-coterie, and the (average) number of messages necessary for entering a critical section is in proportion to the (average) quorum size of C.

This paper proposes a new scheme for constructing a k-coterie with small quorums; the size of each quorum is $O((1/\epsilon)n^{(1+3\epsilon)/2})$ when $k = n^\epsilon$ ($0 < \epsilon < 1/3$), and $O(\sqrt{n} \log n)$ when $k = O(1)$.

1 Introduction

The *distributed k-mutual exclusion problem* (in short, *k-mutex problem*) is the problem of guaranteeing that at most k processes are in a critical section simultaneously in a distributed system. The distributed mutual exclusion extensively investigated (e.g., [1][2][3][4]) is a special case of the k-mutex problem for $k = 1$.

1.1 k-Mutex Problem and k-Coterie

The k-mutex problem can be solved in a centralized manner: A process wishing to enter a critical section sends a request to the process (coordinator) who is in charge of the access to the critical section, and the coordinator sends back a permission if at most $k - 1$ processes are in the critical section. However, such a centralized approach is not favorable, since a failure of the coordinator may become fatal and the graceful degradation is violated.

A distributed approach for solving the k-mutex problem uses the k-coterie.

Definition 1.1 *Let S and k be an n-set[1] and a natural number $\leq n$, respectively. Then a set C of sets $Q(\subseteq S)$ which satisfies all of the following three conditions is called a k-coterie under S:*

Intersection Property: *For each $k + 1$-set $X = \{Q_1, \cdots, Q_{k+1}\} \subseteq C$, there exist two elements Q_i and Q_j ($i \neq j$) such that $Q_i \cap Q_j \neq \phi$.*

---

[1] An n-set is a set whose size is n.

Non-intersection Property: *For any $h(\leq k-1)$, if an h-set $X = \{Q_1, \cdots, Q_h\} \subseteq C$ satisfies $Q_i \cap Q_j = \phi$ for all $i \neq j$, $1 \leq i, j \leq h$, then there exists an element $Q \in C - X$ such that $Q \cap Q_i = \phi$ for all $1 \leq i \leq h$.*

Minimality Property: *For any two distinct elements $Q_i, Q_j \in C$, $Q_i \not\subseteq Q_j$.*

An element Q of C is called a quorum. □

By definition, the 1-coterie is the coterie introduced by Garcia-Molina and Barbara [7]. Figure 1 shows an example of 3-coterie under $S = \{a, b, \cdots, i\}$. We would like to explain a flavor how to use the k-coterie for solving the k-mutex problem. Let C be a k-coterie under the set $S = \{p_1, p_2, \cdots, p_n\}$ of all processes.

1. A process p_i wishing to enter a critical section sends a request to each member of a quorum Q in C.

2. Upon receipt of a request, process p_j checks if it has sent a permission to some other process and has not yet received a release message from it. If not, p_j sends a permission to p_i.

3. If p_i has received a permission from each member in Q, then p_i can enter the critical section.

4. Upon leaving from the critical section, p_i sends a release message to all members in Q.

By definition of k-coterie, since any $k+1$-set of quorums contains two quorums intersecting each other, at most k processes can enter a critical section simultaneously, and therefore, k-mutex problem is solved. Note that this explanation does not include how to guarantee the starvation and deadlock freeness.[2]

Let us observe two important facts: First, the (average) number of messages necessary to enter a critical section seems to be in proportion to the (average) size of quorums. Next, the larger the size of k-coterie is, the faster a process could enter a critical section, since usually the process should compete with other processes wishing to enter the same critical section for "available" quorums. Therefore, in this paper, we measure the goodness of a k-coterie using two measures; the size of k-coterie and the size of quorum.

1.2 Naive Schemes for Constructing k-Coteries

Let S be an n-set. The simplest k-coterie under S is a k-set of singlton sets. Obviously, it is best among all k-coteries in terms of the size of quorum. However, we cannot consider that it is the best k-coterie, because the size of this k-coterie is k, which is worst among all k-coteries.

The set of all $\lfloor n/(k+1) \rfloor + 1$-sets $Q \subset S$ also forms a k-coterie under S, which is a natural extension of the majority coterie. We call this scheme for constructing a k-coterie *Maj*. The size of each quorum of k-coterie produced by *Maj* is $O(n/k)$, which is worst when k is a constant. In terms of the number of quorums, however, it is best.

Although constructing a k-coterie seems to be difficult, constructing a (1-)coterie should be easier. A scheme for constructing a k-coterie from coteries is to divide the set of processes into k clusters, and then to associate a coterie with each cluster. Now, the union of all associated

---

[2]Recently, Kakugawa et al. proposed a k-mutex algorithm which avoids both deadlocks and starvations, based on this idea [8].

coteries forms a k-coterie. We call this scheme *Div*. Figure 2 (b) illustrates this scheme. S is divided into three clusters and the size of each quorum is two, which is smaller than that of *Maj*. In general, *Div* produces a better k-coterie than *Maj* in the sense of quorum size, although a fatal drawback of *Div* is that since clusters are fixed, the possibility of making the number quorums increase by adding quorums which contain processes from different clusters must be abandoned.

This paper will propose a new scheme *Rec* for constructing a k-coterie such that the size of each quorum is $O((1/\epsilon)n^{(1+3\epsilon)/2})$ when $k = n^{\epsilon}$ $(0 < \epsilon < 1/3)$, and $O(\sqrt{n}\log n)$ when $k = O(1)$. We will also discuss the number of quorums.

This paper is organized as follows: Section 2 explains our scheme *Rec* for constructing k-coteries, for limited k. In Section 3, we extend *Rec* for arbitrary k, and both the sizes of k-coterie and quorum are estimated Finally, Section 4 concludes this paper by we summarizing the properties of *Rec* and showing some future problems.

2 Basic Algorithm

We first explain our scheme called *Rec* for constructing a k-coterie, and then present an algorithm Algorithm 1 for producing a quorum of the k-coterie based on *Rec*.[3] The k-coterie can be produced by invoking Algorithm 1 repeatedly, as will be explained later. In what follows, without loss of generality, we only discuss k-coteries under set $S = \{1, 2, \cdots, n\}$. In this section, we assume that k is an integer in form $n^{(1/3)^i}$ for simplicity. This restriction will be removed in the next section.

We start from the following simple observations. Let $K = \{(x, y, z) \mid x, y, z \in \{1, \cdots, n^{1/3}\}\}$ be a cube of volume $(n^{1/3})^3 (= n)$ and fix a one-to-one mapping ψ from S to K, i.e., we identify elements in S with elements in K. For each element $\alpha = (x_1, y_1, z_1)$ in K, we associate a set of elements $Q(\alpha) = \{(x, y, z) \in K \mid ((x = x_1) \wedge (y = y_1)) \vee ((y = y_1) \wedge (z = z_1)) \vee ((z = z_1) \wedge (x = x_1))\}$. We say that $Q(\alpha)$ is *generated* by α, or α is the *generator* of $Q(\alpha)$. Let $C = \{Q(\alpha) \mid \alpha \in K\}$.

Observation 2.1 *Let* $\alpha = (x_1, y_1, z_1)$ *and* $\beta = (x_2, y_2, z_2)$ *be in* K. *Then* $Q(\alpha)$ *and* $Q(\beta)$ *intersect each other iff either* $x_1 = x_2$, $y_1 = y_2$, *or* $z_1 = z_2$ *holds. Or equivalently,* $Q(\alpha)$ *and* $Q(\beta)$ *do not intersect iff* $(x_1 \neq x_2) \wedge (y_1 \neq y_2) \wedge (z_1 \neq z_2)$. □

Observation 2.2 C *is an* $n^{1/3}$*-coterie, and the size of each quorum* $Q(\alpha)$ *is* $3n^{1/3} - 2$. □

Now we obtain a procedure for constructing an $n^{1/3}$-coterie. An $n^{(1/3)^i}$-coterie can be constructed by recursive applications of this procedure, whose correctness is suggested by the following observations. For $i = 1, 2, 3$, let C_i be any k-coterie under set $\{1, \cdots, m\}$, where $m = n^{1/3}$. Define a coterie $C(C_1, C_2, C_3)$ under set K (therefore, under set S) as follows:

$$C(C_1, C_2, C_3) = \{Q(q_1, q_2, q_3) : q_i \in C_i, i = 1, 2, 3\},$$
$$\text{where } Q(q_1, q_2, q_3) = \bigcup_{x \in q_1, y \in q_2, z \in q_3} Q(x, y, z).$$

Observation 2.3 $Q(q_1, q_2, q_3) \cap Q(r_1, r_2, r_3) = \phi$ *iff* $q_i \cap r_i = \phi$ *for all* $i = 1, 2, 3$. □

Observation 2.4 $C(C_1, C_2, C_3)$ *is a* k*-coterie under* K. □

---

[3] Algorithm 1 produces a quorum, since for a process, knowing a quorum (not the k-coterie) is important.

The problem of constructing a k-coterie for set S is reduced to the same problem for set $\{1, \cdots, n^{1/3}\}$, and it shows that the recursive procedure works well since $n^{1/3}$-coterie under set S is constructible.

A formal description of algorithm Algorithm 1 for producing a quorum based on Rec described above is given in Figure 3. k-coterie $\mathcal{C}$ can be produced by invoking Algorithm 1 for each seed $s \in S$ at step 1. Note that since several ceiling functions appearing in Algorithm 1 are for the case k is not in form $n^{(1/3)^i}$, which will be discussed in the next section, we may ignore them in this section.

Let us show the quorum size of $\mathcal{C}$. If we can ignore lower order terms, the size of each quorum becomes $3^i n^{(1-(1/3)^i)/2}$ when $k = n^{(1/3)^i}$. Here the formula of geometrical series is used. Then the size of each quorum is $9n^{4/9}$ when $i = 2$, and $27n^{13/27}$ when $i = 3$. The exponent of n approaches to 0.5 as i increases.

3 Algorithms and Analysis

In this section, we first extend Rec , in order to handle a general k, which may not have form $n^{(1/3)^i}$, and then propose two algorithms Algorithm 2 and 3, based on the extended Rec, for producing a quorum of a k-coterie.[4] We can produce the k-coterie by repeated invokations of Algorithm 2 (or 3) like the case of Algorithm 1. We say that Algorithm 2 (or 3) produces a k-coterie, although it produces its quorum. Algorithm 2 produces a k-coterie, the quorum size of which is $O(\sqrt{n} \log n)$, for any constant k, and Algorithm 3 produces a k-coterie, the quorum size of which is $O((1/\epsilon)n^{(1+3\epsilon)/2})$, for any k in form n^ϵ, where $0 < \epsilon < 1/3$ is a fixed constant.

The k-coteries produced by Algorithm 2 (and 3) will be evaluated by two measures: (1) the quorum size, and (2) the number of quorums which do not intersect with each of fixed $h(< k)$ quorums. For explaining the meaning of the second measure, consider a situation that when a process p wishes to enter a critical section, h other processes p_i have already been in the critical section. Suppose that process p_i collected permissions from all members in quorum Q_i for entering the critical section. Then p must select an "available" quorum which does not intersect with each of Q_i's as Q. Although it is guaranteed that there are at least $k - h$ available quorums by the non-intersection property of k-coterie, the larger the number of available quorums is, the easier p could enter the critical section, since p would compete with other processes wishing to enter the same critical section for available quorums. From a fault tolerant point of view, the larger the second measure is, the more robust the system would be, since if the number of available quorums is large, even if some process q is faulty and available quorums including q cannot be used as Q, p could find another available quorum as Q. Note that the second measure simply indicates the size of coterie when $k = 1$.

3.1 Algorithms for Arbitrary k

In order for extending scheme Rec to arbitrary k, we combine it with other schemes for constructing k-coteries such as Div or Maj. We select Maj as the partner in this paper.[5] By recursive applications of Rec, we reduce the size t of the set T for which a k-coterie is required until t becomes sufficiently small, and, at the deepest (largest) recursion level, we use Maj for constructing a k-coterie under T.

An issue we have to discuss is that since k is not in form $n^{(1/3)^j}$, $n^{(1/3)^j}$ may not be an integer. This problem is simply solved by introducing ceiling functions (e.g., step 1 of function

---

[4]See the previous footnote, for the reason Algorithm 2 (or 3) produces a quorum (not the k-coterie).

[5]Another scheme such as Div can be selected as the partner, as well.

getset in Algorithm 1): Suppose that T_j is the set for which a k-coterie is required at recursion level j, and let t_j be the size of T_j. If $(t_j)^{1/3}$ is not an integer, we consider a cube K_j of edge length $\lceil t^{1/3} \rceil$, instead of that with volume exactly t_j. Suppose that k-coterie $C*$ under K_j is constructed using the k-coteries returned (by recursively called procedures). Define C_j by

$$C_j = \{Q \cap T_j : Q \in C*\}.$$

Then function dec guarantees that C_j is a k-coterie under T_j.

Algorithm for constant k

First, consider the case in which k is a fixed integer, and give an algorithm Algorithm 2 for this case. Algorithm 2 reduces by Rec the size t of set T for which a k-coterie is required until $t < k$, and uses Maj for constructing a k-coterie under T. Algorithm 2 is the same as Algorithm 1 except for the following two points. Recall that Algorithm 1 proceeds until t becomes smaller than a fixed constant $const$. First, k is taken as $const$ in Algorithm 2. Second, step 6 of function getset in Algorithm 1 should be modified to implement Maj as follows:

$$\text{else}\{ \quad T := \{1, 2, \cdots, (isize)\};$$
$$S_x := \{x \mid i \in S_x,\ |S_x| = \lfloor (isize)/(k+1) \rfloor + 1,\ and\ S_x \subset T\};$$
$$S_y := \{y \mid j \in S_y,\ |S_y| = \lfloor (isize)/(k+1) \rfloor + 1,\ and\ S_y \subset T\};$$
$$S_y := \{z \mid k \in S_z,\ |S_z| = \lfloor (isize)/(k+1) \rfloor + 1,\ and\ S_z \subset T\}\}$$

It is easy to see by the argument above that Algorithm 2 produces a k-coterie for arbitrary fixed k. Next, let us estimate the size of quorum produced by Algorithm 2.

Let t_i be the size of the set T_i for which a k-coterie is constructed at recursion level i in Algorithm 2. By definition, $t_0 = n$. If $k = n^{(1/3)^j}$ for some j, then $t_i = n^{(1/3)^i}$ for any $i (\geq j)$ as in Algorithm 1. When k is not in form $n^{(1/3)^j}$, $t_i (= \lceil (t_{i-1})^{1/3} \rceil)$ is in general greater than $n^{(1/3)^i}$ by definition of Algorithm 2. The following lemma, however, ensures that the difference is at most $3/2$, if $n^{(1/3)^i} \geq 1$.

Lemma 3.1 If $n^{(1/3)^i} \geq 1$, then $t_i < n^{(1/3)^i} + 3/2$.

Proof. First, we prove that for any i such that $n^{(1/3)^i} \geq 1$, $t_i < n^{(1/3)^i} + (\sum_{j=1}^{i} 3^j)/3^i$ $\cdots$ (1),

holds, by induction on i.

Base Case: Since $t_1 = \lceil n^{1/3} \rceil < n^{1/3} + 1$, it is trivial.

Induction Step: Suppose that inequality (1) holds for i. Since $t_{i+1} < (t_i)^{1/3} + 1$, the following inequality holds :

$$t_{i+1} < \{n^{(1/3)^i} + (\sum_{j=1}^{i} 3^j)/3^i\}^{1/3} + 1$$

$$< n^{(1/3)^{i+1}} + (1/3)(\sum_{j=1}^{i} 3^j/3^i) + 1$$

$$= n^{(1/3)^{i+1}} + (\sum_{j=1}^{i+1} 3^j/3^{i+1}).$$

Here we use the following property : if $x^{1/3} \geq 1$ then $(x + \alpha)^{1/3} < x^{1/3} + \alpha/3$, since $(x^{1/3} + \alpha/3)^3 > x + \alpha$. Therefore, inequality (1) holds for $i + 1$.

Thus, for any i such that $n^{(1/3)^i} \geq 1$, $t_i < n^{(1/3)^i} + (\sum_{j=1}^{i} 3^j)/3^i$. Now because $\sum_{j=1}^{i} 3^j = (3/2)(3^i - 1)$, we have $t_i < n^{(1/3)^i} + (3/2)\{1 - (1/3)^i\} < n^{(1/3)^i} + 3/2$. $\square$

Lemma 3.2 *The size of each quorum of t_i-coterie produced by Algorithm 1 is $3^i n^{(1-(1/3)^i)/2}$, when lower order terms are ignored.*

Proof. The size of each quorum is at most $3t_1 \times \cdots \times 3t_i = 3^i \prod_{j=1}^{i} t_j$, when lower order terms are ignored. Since $t_i < n^{(1/3)^i} + 3/2$ by Lemma 3.1, the right hand side is $3^i \prod_{j=1}^{i} n^{(1/3)^j}$ when lower order terms are ignored. Hence the theorem holds. $\square$

Theorem 3.3 *For any constant k (i.e., $k = O(1)$), Algorithm 2 produces a k-coterie such that the size of each quorum is $O(\sqrt{n} \log n)$.*

Proof. Let t_i be the size of set T for which k-coterie is constructed by Algorithm 2 at recursion level i. Let h be the deepest (largest) level of recursion that Algorithm 2 goes into (and uses *Maj* there). By Lemma 3.1, the smallest j satisfying $t_{j+1} < k$ is smaller than the smallest j satisfying $n^{(1/3)^{j+1}} + 3/2 < k$. Thus, $\log n / \log(k - 3/2) < 3^{h+1}$, that is, $h + 1 > \{\log \log n - \log \log(k - 3/2)\}/\log 3$. It implies that $h = \log \log n / \log 3 + const$. By Lemma 3.2, since $3^h = 3^{\log \log n / \log 3 + const} = O(\log n)$, the size of each quorum is given by $3^h n^{(1-(1/3)^h)/2} = O(\sqrt{n} \log n)$, when lower order terms are ignored. $\square$

Algorithm for $k = n^\epsilon$

Next, consider the case in which $k = n^\epsilon$ for an arbitrarily fixed ϵ ($0 < \epsilon < 1/3$). An algorithm Algorithm 3, which is another modification of Algorithm 1, for producing such a k-coterie is given. The way of the modification is similar to that for Algorithm 2. In fact, the only difference between Algorithm 2 and 3 is the difference of the recursion level at which *Maj* is applied. In Algorithm 3, unlike Algorithm 2, we apply *Maj* at the recursion level h satisfying $t_{h+1} < n^\epsilon \leq t_h$. Then since h satisfies $n^{(1/3)^{h+1}} \leq t_{h+1} < n^\epsilon \leq t_h < n^{(1/3)^h} + 3/2$, we have $\log(n^\epsilon - 3/2) < (1/3)^h \log n$, which implies that $(1/3)^h > \epsilon$ for sufficiently large n.

Theorem 3.4 *For any k in form n^ϵ for some constant $0 < \epsilon < 1/3$, Algorithm 3 produces a k-coterie such that the size of each quorum is $O((1/\epsilon)n^{(1+3\epsilon)/2})$.*

Proof. Let $h + 1$ be the recursion level at which *Maj* is applied. For sufficiently large n, $t_h = 3^i n^{(1-(1/3)^h)/2} < (1/\epsilon)n^{(1-\epsilon)/2}$. Since we must select $\lceil t_h/n^\epsilon \rceil (< n^{2\epsilon})$ additional elements at $h+1$-st level, the total number of elements is at most $O((1/\epsilon)n^{(1-\epsilon)/2} \times n^{2\epsilon}) = O((1/\epsilon)n^{(1+3\epsilon)/2})$. $\square$

3.2 Comparison by the Number of Available Quorums

In this subsection, we compare three schemes *Maj*, *Div* and *Rec*, in terms of the second measure, the number of available quorums. Let $\mathcal{C}$ be a k-coterie. A subset $\mathcal{H}$ of $\mathcal{C}$ is said to be *element disjoint* if all elements in $\mathcal{H}$ are mutually disjoint. A quorum Q in $\mathcal{C}$ is said to be *available* with respect to an element disjoint h-set $\mathcal{H} \subseteq \mathcal{C}$ if $\mathcal{H} \cup \{Q\}$ is element disjoint. In general, the number of available quorums with respect to an element disjoint h-set decreases, as h increases. Let $r(h)$ be the minimum number which is [6] achievable as the number of available quorums with respect

---

[6] More strictly, we estimate a lower bound on the number.

to an element disjoint h-set $\mathcal{H}$. We represent the type of k-coterie by subscript like $r_{Maj}(h)$ and $r_{Rec}(i)$.

In case of Maj, since the size of each quorum is $\lfloor n/(k+1) \rfloor + 1$, there are $m = n - h \times \lfloor n/(k+1) \rfloor + 1$ available quorums with respect to any element disjoint h-set. Therefore, $r_{Maj}(h)$ is $_m C_{\lfloor n/(k+1) \rfloor + 1}$. The table below shows the values $r_{Maj}(h)$ for $0 \le h < k$, when $n = 125$ and $k = 5$:

$$r_{Maj}(0) = 3.578 \times 10^{23}$$
$$r_{Maj}(1) = 5.109 \times 10^{21}$$
$$r_{Maj}(2) = 2.454 \times 10^{19}$$
$$r_{Maj}(3) = 1.841 \times 10^{16}$$
$$r_{Maj}(4) = 2.691 \times 10^{11}.$$

Next, consider the case of Div. In this paper, we examine the case in which Maekawa's $\sqrt{n}$ coterie [5] is used as the coterie for each cluster. Note that both the size of each quorum and the number of quorums of Maekawa's coterie are $\sqrt{n}$. In Div, since the original set S is divided into k subsets of size n/k, $r_{Div}(h)$ is given by $r_{Div}(h) = (k - h) \times n/k = n - in/k$. By easy calculation, when $n = 125$ and $k = 5$, $r_{Div}(h) = 125 - 25h$.

Finally, consider the case of Rec. In what follows, we examine the k-coteries C produced by Algorithm 1, assuming that k is in form $n^{(1/3)^j}$ for some j, in order to make the effect of the basic idea clear, and to exclude the effect of other schemes applied at the deepest recursion level in Algorithm 2 or 3.

Algorithm 1 produces 3^{j-1} k-coteries C' under sets T with size k^3 at (the deepest) recursion level $j - 1$. Consider $\mathcal{H}/T = \{Q: Q = Q' \cap T$ for some $Q' \in \mathcal{H} \}$. Then by definition, $\mathcal{H}/T \subseteq C'$, and $\mathcal{H}$ is an element disjoint h-set iff so does $\mathcal{H}/T$. Since C' is a k-coterie, there are at least $k - h$ quorums available with respect to $\mathcal{H}/T$. Recall Rec. It is easy to see that for each selection of an available quorum from each k-coterie C', a quorum Q in C is constructed, and the mapping from the set of selections to C is a bijection. Moreover, all quorums corresponding to selections are available with respect to $\mathcal{H}$. Consequently, $r_{Rec}(h)$ is given by $r_{Rec}(h) = (k - h)^{3^j} = (k - h)^{\log n/\log k}$. By easy calculation, when $n = 125$ and $k = 5$, $r_{Rec}(h) = (5 - h)^3$.

4 Concluding Remarks

In this paper, we have proposed two algorithms Algorithm 2 and 3 for producing k-coteries, based on new scheme Rec. Algorithm 2 produces a k-coterie such that the size of each quorum is $O(\sqrt{n} \log n)$ for an arbitrary fixed constant k, and Algorithm 3 produces a k-coterie such that the size of each quorum is $O((1/\epsilon)n^{(1+3\epsilon)/2})$ for any $k = n^\epsilon$, where $0 < \epsilon < 1/3$ is a fixed constant.

It is also worthy of note that both the size of each quorum and the degree of contribution of each process to quorums are identical in each k-coterie produced by Algorithm 2 or 3 like Maekawa's coterie [5], which are considered as desirable properties of $(k-)$coteries (see [5], for details).

For each of Maj, Div and Rec, we have estimated the number of available quorums with respect to an element disjoint h-set, which indicates the degree of ease for entering a critical section, provided that h processes have already been in the same critical section. As a result, it has turned out that Rec is worse than Maj, as far as this measure is concerned (although it is better than Maj, in terms of the quorum size). The problem of modifying the proposed algorithms in such a way that produced k-coteries contain more available quorums is a major problem remained.

No actual distributed systems consist of identical processors and a complete connection as was assumed here. Therefore, the results obtained in this paper may not be applied immediately to real systems. Although considering good k-coteries in more natural systems is an obvious open problem, mainly from a theoretical viewpoint, however, further investigation of the k-mutex problem under the uniformity assumption using the concept of k-coteries is required.

Some other problems left open as future works are:

1. To find an optimal k-coterie in terms of the quorum size. We have been unable to find out a tight lower bound on it.

2. To estimate the robustness against system failures using a measure similar to the availability for coteries.

3. To increase the number of quorums by modifying the proposed algorithms without increasing the quorum size.

References

[1] K.Raymond, "A tree-based algorithm for distributed mutual exclusion," *ACM Trans. Computer Systems*, 7, 1 (Feb. 1989) pp.61-77.

[2] G.Ricart and A.K.Agrawala, "An optimal algorithm for mutual exclusion in computer network," *Comm. of ACM*, 24, 1 (Jan. 1981) pp.9-17.

[3] I.Suzuki and T.Kasami, "A distributed mutual exclusion algorithm," *ACM Trans. Computer Systems*, 3, 4 (Nov. 1985) pp.344-349.

[4] L.Lamport, "Time, clocks, and the ordering of events in a distributed system," *Comm. of ACM*, 21, 7 (July 1978) pp.558-565.

[5] M. Maekawa, "A $\sqrt{n}$ Algorithm for Mutual Exclusion in Decentralized Systems," *ACM Trans. on Computer Systems*, 3, 2 (May 1985) pp.145-159.

[6] D.Agrawal and A.El Abbadi, "An Efficient Solution to the Distributed Mutual Exclusion Problem (Preliminary Report)," *Proc. of 8th PODC*, (Aug. 1989) pp.193-200.

[7] H.Garcia-Molina and D.Barbara, "How to assign votes in a distributed system," *Journal of the ACM*, 32, 4 (Oct. 1985) pp.841-860.

[8] H.Kakugawa, S.Fujita, M.Yamashita, and T.Ae, "A protocol for distributed k-mutual exclusion," *IEICE Technical Report* (to appear), in Japanese.

$$S = \{ a, b, c, d, e, f, g, h, i \}$$

$$C = \{ \{a,b,c\}, \{d,e,f\}, \{g,h,i\}, \{a,d\}, \{b,e\} \}$$

Figure 1   An example of 3-coterie ($n = 9$).

$$S = \{ a, b, c, d, e, f, g, h, i \}$$

$$
\begin{aligned}
C &= \{ Q \mid |Q| = 3 \text{ and } Q \subset S \} \\
&= \{ \{a,b,c\}, \{a,b,d\}, \{a,b,e\}, ..., \{f,g,h\}, \{f,g,i\}, \{g,h,i\} \}
\end{aligned}
$$

(a) 3-coterie based on majority (*Maj*).

$$
\begin{aligned}
S &= \{ a, b, c, d, e, f, g, h, i \} \\
&\Rightarrow S_1 = \{ a, b, c \}, \; S_2 = \{ d, e, f \}, \; S_3 = \{ g, h, i \}
\end{aligned}
$$

$$
\begin{aligned}
C_1 &= \{ \{a,b\}, \{b,c\}, \{c,a\} \} \\
C_2 &= \{ \{d,e\}, \{e,f\}, \{f,d\} \} \\
C_3 &= \{ \{g,h\}, \{h,i\}, \{i,g\} \}
\end{aligned}
$$

$$
\begin{aligned}
C &= C_1 \cup C_2 \cup C_3 \\
&= \{ \{a,b\}, \{b,c\}, \{c,a\}, \{d,e\}, \{e,f\}, \{f,d\}, \{g,h\}, \{h,i\}, \{i,g\} \}
\end{aligned}
$$

(b) 3-coterie based on partitioning(*Div*).

Figure 2   *k*-coeries *Maj* and *Div*.

Algorithm 1
   begin
1:        select a seed s randomly from $\{1, 2, \cdots, n\}$;
2:        $size := \lceil n^{1/3} \rceil$;
3:        $(i, j, k) := \mathbf{dec}(s, size)$;
4:        $S := \mathbf{getset}(i, j, k, size)$;
5:        output S as a subset to be sucured
   end.

function getset($i,j,k,isize$) {
1:      $nsize := \lceil (isize)^{1/3} \rceil$;   // size of the next level //
2:      **if** $(nsize > constant)$ {
3:        $(i_1, i_2, i_3) := \mathbf{dec}(i, nsize)$; $S_x := \mathbf{getset}(i_1, i_2, i_3, nsize)$;
4:        $(j_1, j_2, j_3) := \mathbf{dec}(j, nsize)$; $S_y := \mathbf{getset}(j_1, j_2, j_3, nsize)$;
5:        $(k_1, k_2, k_3) := \mathbf{dec}(k, nsize)$; $S_z := \mathbf{getset}(k_1, k_2, k_3, nsize)$}
6:      **else** $\{S_x := \{i\}$; $S_y := \{j\}$; $S_z := \{k\}\}$;
    // reconstruct and expand //
7:      $S := \phi$;
8:      **while** (S_x is not empty) {
9:      pick up α, β, γ from S_x, S_y, S_z, respectively;
10:     $S := S$
           $\cup \{(isize)^2 x + (isize)\beta + \gamma \mid 1 \le x \le isize\}$
           $\cup \{(isize)^2 \alpha + (isize)y + \gamma \mid 1 \le y \le isize\}$
           $\cup \{(isize)^2 \alpha + (isize)\beta + z \mid 1 \le z \le isize\}\}$
11:    **return** S }

function dec(i,K) {
1:      $z := i(mod K)$;
2:      $x := \lfloor i/K^2 \rfloor$;
3:      $y := \lfloor i/K \rfloor - xK$;
4:      **return** (x, y, z) }   // $i = xK^2 + yK + z$ //

Figure 3 Algorithm 1 to output a quorum of k-coterie.

input : the set of identifier of processors $\{1, 2, \cdots, n\}$
output: the identifiers of processors in a quorum

Is the Shuffle-Exchange Better than the Butterfly?

Arvind Raghunathan *   Huzur Saran †

ABSTRACT

This paper deals with the relative powers of two popular interconnection networks, the butterfly and the shuffle-exchange. Under a certain model of comparison, we show that the shuffle-exchange is no worse than the butterfly. While we do not have concrete evidence that the shuffle-exchange is strictly superior to the butterfly under this model, we isolate a "shuffle-exchange-like" graph that is strictly superior to the butterfly.

1. Introduction

Efficient construction and functioning of a multiprocessor system depend crucially on the type of interconnection network chosen. The interconnection network affects hardware cost, software cost, network reliability and congestion, to name a few of the parameters of system performance. Currently, there exist numerous choices of interconnection networks in the literature [3], [10], each with its own merits and demerits. How does one choose a network that is best suited for one's purposes?

To answer this question, researchers have started studying the relative powers of some of the most popular networks: the boolean hypercube, the butterfly, the shuffle-exchange, the two-dimensional mesh, and the tree machine [1], [2], [4], [6], [8]. We will define each of these networks and certain others of interest in turn. All graphs in this paper are simple and undirected.

The m-dimensional *boolean hypercube* $Q(m)$ is the graph whose vertex set is the set $\{0,1\}^m$ of length-m binary strings and whose edges connect just those pairs of vertices/strings that differ in precisely one bit-position. The hypercube so defined emerges as a particularly versatile network for parallel computation. However, it is impractical to build large-sized hypercubes because it is not a bounded-degree network.

The following two networks we describe are both bounded-degree networks, and are frequently used in practice as interconnection networks of large systems.

The m-level *butterfly graph* $B(m)$ has vertex set

$$V_m = \{0, 1, \ldots, m-1\} \times \{0,1\}^m$$

*Division of Computer Science, University of California, Davis, CA 95616. The work of this author was supported in part by the National Science Foundation under grant CCR-9107847

†Indian Institute of Technology, Delhi

The edges of $B(m)$ form butterflies between consecutive levels of vertices, with wraparound in the sense that level 0 and level m are identical. Each butterfly connects vertices

$$< l, \beta_0 \beta_1 \ldots \beta_{l-1} 0 \beta_{l+1} \ldots \beta_{m-1} >$$

and

$$< l, \beta_0 \beta_1 \ldots \beta_{l-1} 1 \beta_{l+1} \ldots \beta_{m-1} >$$

$(0 \le l < m;$ each $\beta_i \in \{0, 1\})$ with vertices

$$< l+1, \beta_0 \beta_1 \ldots \beta_{l-1} 0 \beta_{l+1} \ldots \beta_{m-1} >$$

and

$$< l+1, \beta_0 \beta_1 \ldots \beta_{l-1} 1 \beta_{l+1} \ldots \beta_{m-1} >$$

It is easy to see that the m-level butterfly has $m \cdot 2^m$ vertices, where each vertex has degree 4.

The m-vertex *shuffle-exchange* $S(m)$ has vertex set $V_m = \{0, 1, \ldots, m-1\}$, where $m = 2^i$ for some positive integer i. There are two types of edges in the shuffle-exchange.

1. *The exchange edges:* Each vertex i, with i even, is connected to vertex $i+1$. (Here and throughout this paper, 0 is considered even.)

2. *The perfect shuffle edges:* Each vertex i, with $1 \le i \le m-2$, is connected with vertex $2 \cdot i \bmod m - 1$. Vertices 0 and $m-1$ are connected to themselves.

Each vertex in a shuffle-exchange has degree three, and hence it is a bounded degree graph.

We now describe two more bounded degree networks of interest in this paper.

The height-h *complete binary tree* $T(h)$ is the graph whose vertices comprise all binary strings of length at most h, except binary string 0, and whose edges connect each vertex x with binary strings of length less than h with vertices $x0$ and $x1$. The lth level of $T(h)$ $(1 \le l \le h)$ consists of all vertices/strings of length l.

The height-h *X-tree* $X(h)$ [3] is obtained from the complete binary tree $T(h)$ by adding edges that connect the vertices at each level of $T(h)$ in a path, with the vertices in the standard lexicographic order.

In order to compare the various interconnection networks, the following model has been proposed and used [1], [2], [4], [6], [8]. Let G and H be simple, undirected graphs. G is called the guest graph and H is called the host graph. An *embedding* of G in H is a one-to-one association of the vertices of G with the vertices of H, together with a specification of paths in H connecting the images of the endpoints of each edge in G. Note that these paths need not be vertex disjoint. The *dilation* of the embedding is the maximum length of any of these G-edge-routing paths. The *expansion* of the embedding is the ratio, $|H|/|G|$, of the number of vertices in H to the number of vertices in G. We need to consider both dilation and expansion, since one can sometimes decrease dilation by embedding G in a larger relative of H. The *congestion* of the embedding is the maximum number of edges of G that are routed through a single edge of H. In this paper, we will not be concerned with congestion. We note, however, that every embedding provided in this paper has $O(1)$ congestion.

It has been known for many years [7] that a large class of algorithms run as fast on the butterfly as on the hypercube. It is this observation that motivates much of the interest in butterfly-like architectures. However, since the butterfly is a bounded-degree network, and the hypercube is not, one would believe that the hypercube is a strictly more powerful network. Recently, Bhatt, Chung, Hong, Leighton and Rosenberg [2] found a proof that this is indeed the case: they showed that any embedding of the X-tree $X(h)$ in a butterfly must have dilation $\Omega(\log h) = \Omega(\log \log |X(h)|)$. Since the X-tree is embeddable in the hypercube with simultaneous $O(1)$ dilation and expansion, they located the first class of graphs that can be efficiently embedded in the hypercube but not the butterfly.

The work of Bhatt, Chung, Hong, Leighton and Rosenberg raise two natural questions. First, can one show similar results separating the hypercube and the shuffle-exchange? Perhaps more important from the practical viewpoint is the second question: can one show a separation theorem for the butterfly and the shuffle-exchange? A large body of researchers have privately held views about the superiority of óne over the other without proper mathematical justification.

- In this paper, we show that every n-vertex graph that is embeddable in the butterfly with dilation $O(1)$ can also be embedded in the shuffle-exchange with $O(1)$ dilation. We do this by providing an embedding of the n-vertex butterfly $B(m)$ (here, $n = m \cdot 2^m$) on a shuffle-exchange with $O(1)$ dilation and $O(n)$ expansion. A similar result was obtained independently by Koch, Leighton, Maggs, Rao and Rosenberg [5], as they indicate in their paper (see page 239 of their paper).

Motivated by this, we asked ourselves: are there n-vertex graphs that are embeddable in "shuffle-exchange-like" networks with $O(1)$ dilation, but require $\Omega(\log \log n)$ dilation in any embedding on the butterfly? To this, we answer in the affirmative. Let us now introduce an interconnection network called the *generalized shuffle-exchange*. The m-vertex generalized shuffle-exchange, $gS(m)$, is defined exactly as $S(m)$, except that we relax the condition that m be a power of 2. Thus, $gS(m)$ has vertex set $V_m = \{0, 1, \ldots, m-1\}$. There are two types of edges in the generalized shuffle-exchange.

1. *The exchange edges:* Each vertex i, with i even, is connected to vertex $i + 1$.

2. *The perfect shuffle edges:* Each vertex i, with $1 \leq i \leq m - 2$ is connected with vertex $2 \cdot i \bmod m - 1$. Vertices 0 and $m - 1$ are connected to themselves.

Each vertex in a generalized shuffle-exchange has degree three, and hence it is a bounded degree graph. In this paper, we will be concerned with $gS(m)$ with m even and with $m/2$ odd.

We can show the following surprising results.

- $S(m)$ is embeddable in $gS(m+2)$ with $O(1)$ dilation (and, of course, $O(1)$ expansion and congestion).

- The X-tree, $X(h)$, is embeddable in the generalized shuffle-exchange with simultaneous $O(1)$ dilation and expansion. In particular, $X(h)$ is embeddable in $gS(2^{h+1} + 2)$ with $O(1)$ dilation.

The first result shows that the generalized shuffle exchange, $gS(m+2)$ is at least as powerful as the shuffle exchange $S(m)$. Hence, every parallel algorithm that is implemented on the shuffle-exchange can likewise be implemented on the generalized shuffle-exchange, with only a penalty of two processors in hardware cost and with no more than a constant factor delay in time cost. The second result exhibits the first class of graphs that can be efficiently embedded on "shuffle-exchange-like" networks, but not in the butterfly.

In conclusion, we show that the butterfly can be embedded with constant dilation in the shuffle-exchange, but every embedding of the generalized shuffle-exchange $gS(m)$ on the butterfly requires $\Omega(\log \log m)$ dilation.

2. A Family of Graphs Equivalent to the Butterfly

In this section, we provide an alternate view of the butterfly network that we use later in the paper. We first need the following definition.

Definition 1. *Two graph families* $G_1, G_2, \ldots$ *and* $H_1, H_2, \ldots$ *are said to be embedding equivalents, or simply, equivalents, of each other, if any graph from one family can be embedded in some graph of the other family with simultaneous $O(1)$ dilation and expansion.*

The m-level **FFT graph** $F(m)$ is defined identically to $B(m)$, with the exception that there are no wraparound edges. Thus, vertices at levels 0 and $m - 1$ are of degree 2 each. The following lemma shows that the butterfly and FFT graph families are equivalents.

Lemma 1. *The butterfly and FFT graph families are embedding equivalents.*

Proof Sketch: $F(m)$ is clearly a subgraph of $B(m)$ and hence embeds in it with $O(1)$ dilation and expansion.

We now show that $B(m)$ embeds in $F(m + 1)$ with $O(1)$ dilation. ($O(1)$ expansion follows.) For the purpose of this proof sketch, let m be odd. The other case is similar.

We define a injective function f mapping vertices of $B(m)$ to vertices of $F(m + 1)$ below. The basic idea is to embed consecutive levels of the butterfly in alternate levels of the FFT until level $m - 1$ of the FFT is reached, and to then embed the remaining levels of the butterfly by moving in the opposite direction.

$$f(< 0, \beta_0, \ldots, \beta_{m-1} >) = < 0, \beta_{m-1}, \beta_0, \beta_{m-2}, \beta_1, \ldots, \beta_{\frac{m-1}{2}-2}, \beta_{\frac{m-1}{2}-1}, \beta_{\frac{m-1}{2}} >$$

When $2 \cdot i \leq m - 1$,

$$f(< i, \beta_0, \ldots, \beta_{m-1} >) = < 2 \cdot i, \beta_{m-1}, \beta_0, \beta_{m-2}, \beta_1, \ldots, \beta_{\frac{m-1}{2}-2}, \beta_{\frac{m-1}{2}-1}, \beta_{\frac{m-1}{2}} >$$

When $2 \cdot i > m - 1$,

$$f(< i, \beta_0, \ldots, \beta_{m-1} >) = < 2 \cdot m - 2 \cdot i - 1, \beta_{m-1}, \beta_0, \beta_{m-2}, \beta_1, \ldots, \beta_{\frac{m-1}{2}-2}, \beta_{\frac{m-1}{2}-1}, \beta_{\frac{m-1}{2}} >$$

Such an f guarantees that every edge in $B(m)$ is routed by a path of length at most three in $F(m+1)$. We leave out the details of checking this statement. ∎

The height-h *recursive tree* $R(h)$ is defined as follows. $R(1)$ is a single vertex labeled 1. This vertex is also called the leaf of $R(1)$. $R(h)$ is constructed from two copies of $R(h-1)$ in the following manner. Let x be the binary string label of a leaf of $R(h-1)$. We create two new vertices labeled $x0$ and $x1$. $x0$ and $x1$ are adjacent in $R(h)$ to the two leaves labeled x, one in each copy of $R(h-1)$. We repeat this construction for every leaf of $R(h-1)$ to obtain $R(h)$. The leaves of $R(h)$ are exactly the new vertices we created in this construction. The lth level of $R(h)$, $1 \le l \le h$, consists of all vertices with labels of length l.

Lemma 2. *The level-m FFT graph is isomorphic to $R(m)$.*

Proof Sketch: Inductively assume that $F(m-1)$ is isomorphic to $R(m-1)$, with the vertices at level $m-2$ as the leaves. $F(1)$ is clearly a single vertex, thus establishing the base case. From $F(m)$, delete the vertices at level $m-1$. We get two connected components, each isomorphic to $F(m-1)$. In one connected component, the label of every vertex has $\beta_{m-1} = 0$ and in the other connected component, the label of every vertex has $\beta_{m-1} = 1$. Let us discard β_{m-1} from the labels of all vertices in level $m-2$. Now, the vertices at level $m-1$ have the required labels. Further, each vertex at level $m-2$ with new label $< (m-2)x >$ is adjacent to vertices $< (m-1)x0 >$ and $< (m-1)x1 >$ at level $m-1$. ∎

3. Embedding the Butterfly in the Shuffle-Exchange with $O(1)$ Dilation

This section describes our embedding of the n-vertex butterfly in the shuffle-exchange with $O(1)$ dilation and $O(n)$ expansion. A similar result was independently obtained by Koch et al. [5]. We do this by first defining a graph family that is embedding equivalent to the shuffle-exchange, and then embedding the recursive tree $R(h)$ in a graph of this family. Since the recursive tree is embedding equivalent to the butterfly, the result would follow.

The m-vertex *paired tree* $P(m)$ is defined on the vertex set $V_m = \{0, 1, \ldots, m-1\}$, where $m = 2^i$ for some nonnegative integer i. The edges of $P(m)$ are partitioned into three groups.

1. *Even tree edges.* Let i be even, with $2 \le i \le m/2 - 2$. Then i is adjacent to $2 \cdot i$ and $2 \cdot i + 2$.

2. *Odd tree edges.* Let i be odd, with $m - 3 \le i \le m/2 + 1$. Then, i is adjacent to $2 \cdot i - m + 1$ and $2 \cdot i - m - 1$.

3. *Exchange edges.* Let i be even. Then i is connected to $i+1$.

Lemma 3. *The paired tree and the shuffle-exchange graph families are embedding equivalents.*

Proof Sketch: It is easy to see that $P(m)$ can be embedded in $S(m)$ with simultaneous $O(1)$ dilation and expansion. We leave out the details from the extended abstract.

The injective function f, mapping the vertices of $S(m)$ to $P(m)$ is the identity function. It is easily seen that the exchange edges of $S(m)$ are routed by single edges in $P(m)$. Now, let i be even with $2 \le i \le m/2 - 2$. i is adjacent to $2 \cdot i$ by a perfect shuffle edge in $S(m)$. $f(i)$ is adjacent to $f(2 \cdot i)$ by an even tree edge in $P(m)$, as required. Next let i be even, with $2 \cdot i \ge m/2$. i is adjacent by a perfect shuffle edge to $2 \cdot i - m + 1$ in $S(m)$. In $P(m)$, $f(i)$ is adjacent to $f(i+1)$ by an exchange edge, and $f(i+1)$ is in turn adjacent to $f(2 \cdot i - m + 1)$ by an odd tree edge. The proof for the case where i is odd is similar. ∎

Let the subgraph of the paired tree induced by the even tree edges be called the *even tree* and the subgraph induced by the odd tree edges be called the *odd tree*. Both these trees are complete binary trees. Let $V' = \{v_1, v_2, \ldots, v_r\}$ be a set of vertices at the same level, say l', of the even tree (resp., the odd tree). V' is said to *subtend* subtree in the even tree (resp., the odd tree) if the set of descendants at level l', with level measured in the even tree (resp., the odd tree), of their common ancestor v_0 is exactly V'. In this case, the subgraph of the even tree (resp., the odd tree) induced by v_0 and its descendants is also a complete binary tree. Let V_e be a set of vertices in the even tree. Let V_o be the corresponding set of vertices in the odd tree given by $V_o = \{i : (i-1) \in V_e\}$. Then, V_e and V_o are called an *exchange pair*. Let us now make certain elementary observations. Their proofs are simple and are omitted.

Observation 1. The exchange pair of a set of nonleaf vertices in the even tree (resp., the odd tree) is a set of leaf vertices in the odd tree (resp., the even tree).

Observation 2. In $P(m)$, there is an exchange edge between $m-2$ and $m-1$. The exchange pair of the vertex set $\{(m-4), (m-6), \ldots, (m-2^{h+1})\}$ is the set of vertices at the first h levels of the odd tree.

Observation 3. In $P(m)$, vertices $(m-2), (m-4), \ldots, (m-2^{h+1})$ subtend a tree in the even tree.

Observation 4. Let V_e be a set of nonleaf vertices at level l' in the even tree (resp., the odd tree), such that V_e subtends a tree. Let V_o be the exchange pair of V_e. Then V_o subtends a tree in the odd tree (resp., the even tree).

Theorem 1. $R(h)$ *is embeddable in* $P(m)$, *where* $m = 2^{2h+1}$, *with* $O(1)$ *dilation and* $O(2^{h+1})$ *expansion. Thus, the n-vertex butterfly is embeddable in a shuffle-exchange with $O(1)$ dilation and $O(n)$ expansion.*

Proof Sketch: We will embed $R(h)$ in $P(m)$ level by level. First, the leaves of $R(h)$ (the vertices at level h in $R(h)$) are mapped to the vertices $(m-2), (m-4), \ldots, (m-2^{h+1})$ in lexicographic order. Thus, vertex 1^h gets mapped to $m-2$, $1^{(h-1)}0$ gets mapped to vertex $m-4$, and so on, with vertex 0^h getting mapped to $m-2^{h+1}$. We now have to embed the leaves of the two copies of $R(h-1)$. By Observation 3, $\{(m-2),(m-4),\ldots,(m-2^{h+1})\}$ subtends a tree T in the even tree. We embed the leaves of one copy of $R(h-1)$ in the vertex subset V_e, consisting of the parents of $\{(m-2),(m-4),\ldots,(m-2^{h+1})\}$ in T (in the same lexicographic order!), and the leaves of the other copy of $R(h-1)$ in the exchange pair V_o of V_e. Thus, one copy of vertex 1^{h-1} is mapped to vertex $\frac{m}{2}-2$ and the other copy of vertex 1^{h-1} is mapped to vertex $\frac{m}{2}-1$. Note that by Observation 1, V_o is a subset of the leaves of the odd tree. It is now easily shown that edges connecting the leaves of $R(h)$ and the leaves of the first copy of $R(h-1)$ are routed by single edges, while the edges connecting the leaves of $R(h)$ and the leaves of the second copy of $R(h-1)$ are routed by paths of length two.

V_e subtends a tree in the even tree, as it consists of all vertices at a certain level in T. By Observation 4, V_o also subtends a tree in the odd tree. Thus, this process can be continued, with V_e and V_o playing the role of $\{(m-2),(m-4),\ldots,(m-2^{h+1})\}$, until the entire recursive tree $R(h)$ is embedded. We leave out the further details from the extended abstract. It should, however, be clear that every edge of $R(h)$ is routed in $P(m)$ by a path of length at most two.

We still have to show that the embedding scheme described above defines a injective function. We prove this as follows. The vertices of $P(m)$ that we use are reached in one of two ways. They are either

1. those in the tree T subtended by $\{(m-2),(m-4),\ldots,(m-2^{h+1})\}$, or

2. those in a tree subtended by the exchange pair of a set of vertices that belong to some tree created earlier.

The exchange pair of a set of vertices is unique, and so is the tree that it subtends. Thus, two trees created by the second process cannot have a vertex in common. The only way for two vertices of $R(h)$ to map to the same vertex of $P(m)$ is if a tree created by the second process gets mapped to vertices of T. Since we only find exchange pairs of nonleaf vertices, Observation 1 implies that we used some vertex that is in the exchange pair of $\{(m-2),(m-4),\ldots,(m-2^{h+1})\}$. By Observation 2, the exchange pair of $\{(m-2),(m-4),\ldots,(m-2^{h+1})\}$ consists of the vertex $m-1$ which is in neither tree, and the vertices in the first h levels of the odd tree. But $R(h)$ has only h levels. Further, by Observation 1, trees created in the second process have their leaves mapped to the leaves of the even/odd trees. So the only vertices used in the construction described above are in the levels $h+1, h+2, \ldots, 2 \cdot h$ of the even and the odd trees. Thus, the embedding function is injective, and the theorem is proved. ∎

4. The Generalized Shuffle-Exchange

In this section, we provide our results concerning the generalized shuffle-exchange. We show two results of interest concerning this network, namely, that $gS(m+2)$ embeds $S(m)$ with constant dilation and that it embeds the X-tree with simultaneous $O(1)$

dilation and expansion. We then provide an embedding equivalent of the generalized shuffle-exchange. We first need the following intermediate results. In the following, let m be an even integer, such that $m/2$ is odd.

Lemma 4. *Let i be an even integer, such that $i/2$ is odd. Then there is a path of length three between vertices $i + 1$ and $i + 3$ in $gS(m)$.*

Proof Sketch: Vertex $i + 1$ is adjacent to some vertex x by a perfect shuffle edge, such that $2 \cdot x \bmod (m - 1) = i + 1$. $i + 1$ is an odd integer, implying that $2 \cdot x > m - 1$. Thus, $2 \cdot x - m + 1 = i + 1$, giving us $x = (m/2 + i/2)$. By the hypothesis, $m/2$ and $i/2$ are both odd, thus making x even. x is adjacent to $x + 1 = (m/2 + i/2 + 1)$ by an exchange edge. $x + 1$ is adjacent by a perfect shuffle edge to $2 \cdot (x + 1) \bmod (m - 1)$, which is $(m + i + 2) \bmod (m - 1)$. Thus, $x + 1$ is adjacent to $i + 3$. Therefore, we have a path $(i + 1), (m/2 + i/2), (m/2 + i/2 + 1), (i + 3)$, connecting vertices $i + 1$ and $i + 3$. ∎

Corollary 1. *Let i be an even integer, such that $i/2$ is odd. Then there is a path of length five between vertices i and $i + 2$ in $gS(m)$.*

Proof: There are exchange edges between vertices i and $i + 1$ and vertices $i + 2$ and $i + 3$, respectively. By Lemma 1, we have a path of length three between $i + 1$ and $i + 3$. ∎

Corollary 2. *There is a path of length at most five between i and $i + 2$ in $gS(m)$, for any integer $0 \le i \le m - 3$.*

Proof: We only need to prove the corollary for the case where i is odd and $(i - 1)/2$ is even and for the case where i is even and $i/2$ is also even. If i is odd, then there are exchange edges between vertices i and $i - 1$ and vertices $i + 2$ and $i + 1$, respectively. $i - 1$ and $(i - 1)/2$ are adjacent by a perfect shuffle edge. Further, if $(i - 1)/2$ is even, then there is an exchange edge between $(i - 1)/2$ and $(i - 1)/2 + 1$. Clearly, $(i - 1)/2 + 1$ is adjacent to $i + 1$ by a perfect shuffle edge. Thus, $i - 1$ and $i + 1$ are connected by a path of length three, and i and $i + 2$ are connected by a path of length five. ∎

We are now ready to prove the main theorems of this section. In the following, let $m = 2^k + 2$, for some integer k.

Theorem 2. *$gS(m)$ embeds $S(m - 2)$ with $O(1)$ dilation.*

Proof: Let f, the injective function mapping the vertices of $S(m - 2)$ to $gS(m)$, be defined by $f(i) = i$.

It is easy to see that the exchange edges of $S(m - 2)$ are routed by single edges in $gS(m)$.

We now show paths of constant length in $gS(m)$ that route the perfect shuffle edges of $S(m - 2)$. First, let i be an integer such that $2 \cdot i < m - 3$. Then, vertex i is adjacent to vertex $2 \cdot i$ in $S(m - 2)$, and vertex $f(i)$ is adjacent to vertex $f(2 \cdot i)$ in $gS(m)$. Next, let $i = (m - 2)/2$. In this case, vertex i is adjacent to vertex $(m - 2) \bmod (m - 3) = 1$ in $S(m - 2)$. Since $(m - 2)/2$ is even, there is an exchange edge between vertices i and $i + 1$ in $gS(m)$. Vertex $i + 1$ is adjacent in $gS(m)$ to vertex $2 \cdot (i + 1) \bmod (m - 1) = 1$, thus providing a path of length two between $f(i)$ and $f(1)$ in $gS(m)$. Finally, let $2 \cdot i > m - 1$. In this case, vertex i is adjacent to vertex $(2 \cdot i - m + 3)$ in $S(m - 2)$. In $gS(m)$, vertex

i is adjacent to vertex $(2 \cdot i - m + 1)$ by a perfect shuffle edge. By Corollary 2, we have a path of length at most five between vertices $(2 \cdot i - m + 1)$ and $(2 \cdot i - m + 3)$, thus providing us with a path of length at most six between $f(i)$ and $f(2 \cdot i - m + 3)$. ∎

We now show that the generalized shuffle-exchange embeds the X-tree with simultaneous $O(1)$ dilation and expansion.

Theorem 3. *The X-tree $X(h)$ can be embedded in $gS(m)$, where $m = 2^{h+1} + 2$ with $O(1)$ dilation.*

Proof: Let the root vertex of the X-tree be labeled 1. This fixes the labels of all vertices in $X(h)$. We now define a injective function f mapping the vertices of $X(h)$ to the vertices of $gS(m)$ as follows. A vertex in $X(h)$ labeled $< \beta_1 \beta_2 \dots \beta_k >$, where $1 \leq k \leq h, \beta_i \in \{0, 1\}$, is mapped by f to vertex $2 \cdot i$ in $gS(m)$, where i is the decimal equivalent of the binary number $\beta_1 \beta_2 \dots \beta_k$. Thus, the root vertex, labeled 1 in $X(h)$, is mapped to vertex 2 in $gS(m)$, the two children of the root vertex, labeled 10 and 11, are mapped to vertices 4 and 6 respectively, and so on.

We will now specify the $X(h)$-edge-routing paths in $gS(m)$, and show that they are each of constant length. Let us first deal with the edges of the underlying complete binary tree $T(h)$. Let a typical vertex of $X(h)$ be labeled $< \beta_1 \beta_2 \dots \beta_k >$, with $1 \leq k < h$. Such a vertex is adjacent in $T(h)$ to vertices $< \beta_1 \beta_2 \dots \beta_k 0 >$ and $< \beta_1 \beta_2 \dots \beta_k 1 >$. Let i be the decimal equivalent of $\beta_1 \beta_2 \dots \beta_k$. Then, $< \beta_1 \beta_2 \dots \beta_k >$ gets mapped to vertex $2 \cdot i$ in $gS(m)$. Since the decimal equivalent of $\beta_1 \beta_2 \dots \beta_k 0$ and $\beta_1 \beta_2 \dots \beta_k 1$ are $2 \cdot i$ and $2 \cdot i + 1$, respectively, $< \beta_1 \beta_2 \dots \beta_k 0 >$ and $< \beta_1 \beta_2 \dots \beta_k 1 >$ get mapped to vertices $4 \cdot i$ and $4 \cdot i + 2$, respectively in $gS(m)$. Vertex $2 \cdot i$ is adjacent to vertex $4 \cdot i$ by a perfect shuffle edge in $gS(m)$, thus providing a path of length one between the images of $< \beta_1 \beta_2 \dots \beta_k >$ and $< \beta_1 \beta_2 \dots \beta_k 0 >$. Further, $2 \cdot i$ is adjacent to $2 \cdot i + 1$ by an exchange edge, which in turn is adjacent to vertex $4 \cdot i + 2$ by a perfect shuffle edge, thus providing a path of length two between the images of $< \beta_1 \beta_2 \dots \beta_k >$ and $< \beta_1 \beta_2 \dots \beta_k 1 >$.

The edges of the X-tree left out so far are those that connect the vertices of each level of the X-tree in a path, with the vertices in the standard lexicographic order. We will now exhibit paths of constant length that route these edges in $gS(m)$. Notice that consecutive vertices in any such path are labeled in such a way that the difference in their binary strings is exactly 1. Thus, if x and y are consecutive vertices in such a path, and if x gets mapped to vertex i in $gS(m)$, then y gets mapped either to vertex $i - 2$ or to vertex $i + 2$ in $gS(m)$. By Corollary 2, there is a path of length at most five between x and y, thus proving the theorem. ∎

It was shown in [2] that the X-tree requires $\Omega(\log \log |X(h)|)$ dilation in $S(m)$ when m is restricted by $m \leq c \cdot |X(h)|$, for some constant c. We thus have that while $S(m)$ embeds in $gS(m + 2)$ with simultaneous $O(1)$ dilation and expansion, there is no simultaneous $O(1)$ dilation and expansion embedding of $gS(m)$ in the shuffle-exchange.

We conclude this section by providing a class of graphs that are equivalent to $gS(m)$.

The m-vertex *paired X-tree* $XP(m)$ is defined on the vertex set $V_m = \{0, 1, \dots, m - 1\}$, where $m = 2^i + 2$ for some nonnegative integer i. The edges of $XP(m)$ are partitioned into five groups.

1. *Even tree edges.* Let i be even, with $2 \leq i \leq m/2 - 3$. Then i is adjacent to $2 \cdot i$ and $2 \cdot i + 2$. $m/2 - 1$ is also adjacent to $m - 2$.

2. *Odd tree edges.* Let i be odd, with $m - 3 \leq i \leq m/2 + 2$. Then, i is adjacent to $2 \cdot i - m + 1$ and $2 \cdot i - m - 1$. $m/2$ is also adjacent to 1.

3. *Exchange edges.* Let i be even. Then i is connected to $i + 1$.

4. *Even tree paths* The vertices at each level of the even tree are connected in a path, with vertices in ascending order.

5. *Odd tree paths* The vertices at each level of the odd tree are connected in a path, with vertices in ascending order.

The proof of the following theorem will be provided in the final version of this paper.

Theorem 4. *The generalized shuffle-exchange and the paired X-tree graph families are embedding equivalents.*

References

[1] S. BHATT, F. CHUNG, F. LEIGHTON, A. ROSENBERG. Optimal Simulations of Tree Machines. *Proc. IEEE Symp. on Found. of Comp. Science,* 1986.

[2] S. BHATT, F. CHUNG, J. HONG, F. T. LEIGHTON, A. L. ROSENBERG. Optimal Simulations by Butterfly Networks. *ACM Symp. on the Theory of Computing,* 1988.

[3] A. DESPAIN AND D. PATTERSON X-Tree - A Tree Structured Multiprocessor Architecture. *5th Symp. on Computer Architecture,* 1978

[4] J. HONG, K. MEHLHORN AND A. ROSENBERG. Cost Tradeoffs in Graph Embeddings, with Applications. *J. ACM,* 1983.

[5] R. KOCH, T. LEIGHTON, B. MAGGS, S. RAO AND A. ROSENBERG. Work-Preserving Emulations of Fixed-Connection Networks. *ACM Symposium on Theory of Computing,* 1989.

[6] B. MONIEN AND I.H. SUDBOROUGH. Simulating Binary Trees on Hypercubes. *AWOC,* 1988.

[7] F. PREPARATA AND J. VUILLEMIN. The Cube-Connected Cycles: a Versatile Network for Parallel Computation *Communications of the ACM,* 1981.

[8] A. ROSENBERG. GRAPH EMBEDDINGS 1988: Recent Breakthroughs, New Directions. *AWOC,* 1988.

[9] J.D. ULLMAN. Computational Aspects of VLSI *Computer Science Press,* 1984.

Weighted Random Assignments
with Application to Hashing

Andrew Chi-Chih Yao

Department of Computer Science
Princeton University
Princeton, New Jersey, USA

Abstract In this talk we will study the optimal solution to a class of random assignment problems with dependent weights. The result is then used to show that, in double hashing, the expected retrieval cost with respect to an optimal static hash table is $O(1)$ even if the table is full. This confirms a conjecture of Gonnet and Munro (*SIAM J. on Computing* 8 (1979), 463-478).

Scheduling File Transfers
under Port and Channel Constraints

Shin-ichi Nakano   and   Takao Nishizeki

Department of Information Engineering

Faculty of Engineering

Tohoku University

Sendai-shi 980, Japan

Abstract

The file transfer scheduling problem was introduced and studied by Coffman, Garey, Johnson and LaPaugh. The problem is to schedule transfers of a large collection of files between various nodes of a network under port constraint so as to minimize overall finishing time. This paper extends their model to include communication channel constraint in addition to port constraint. We formulate the problem with both port and channel constraints as a new type of edge-coloring of multigraphs, called an fg-edge-coloring, and give an efficient approximation algorithm with absolute worst-case ratio 3/2.

1. Introduction

In this paper we study a scheduling problem on a computer network, that of transferring a large collection of files between various nodes of a network. In particular, we are interested in how collections of such transfers can be scheduled so as to minimize the total time for the overall transfer process.

This problem was introduced by Coffman *et al* [CG]. Their model represents an instance of the problem as a weighted undirected multigraph $G = (V, E)$, called a *file transfer graph*. The vertices of G correspond to the nodes of the network that are computer centers. Each vertex $v \in V$ is labeled with a positive integer $f(v)$ that represents the number of communication ports at the node corresponding to v. It is assumed that each communication module may be used as a transmitter and as a receiver. Each edge $e \in E$ is labeled with a positive integer $L(e)$ that represents the amount of time needed for the transfer of a file corresponding to e (the file is transmitted between the nodes corresponding to the end vertices of e). The authors assume, in addition, that once the transfer of a file begins it continues without interruption. They show the general problem to be NP-complete, but obtain polynomial time algorithms for various restrictions on the graph G. They furthermore give an approximation algorithm for the general graph G which has absolute worst-case ratio 5/2. In the case where G is an odd cycle, Choi and Hakimi [CH] obtain a polynomial time algorithm for the file transfer problem.

This model was extended to include the possibility of forwarding in [W]. It includes the case when a computer center u wishes to send a file to v but no direct link exists, the file must be sent to one or more intermediaries who will then send it on to v. Several special cases of this problem, which were previously solvable by polynomial time algorithms, are shown to be NP-complete when forwarding is included.

In this paper we study the file transfer problem with port **and** channel constraints. In our model an instance of the problem consists of an undirected multigraph $G = (V, E)$ together with positive integers $f(v)$ and $g(vw)$; $f(v)$ is associated with vertex $v \in V$ and $g(vw)$ with a pair of vertices v and w. We call this graph an *fg-file transfer graph*. In an *fg*-file transfer graph, vertices correspond to the nodes of the network, and edges correspond to files to be transferred. The integer $f(v)$ for a vertex v is its **port constraint**, that is the number of communication ports available for the simultaneous file transfers at a computer v, while $g(vw)$ for the pair of vertices v and w is its **channel constraint**, that is the number of communication channels between v and w. Our model does not allow forwarding. Each file is transferred directly between the centers that are its endpoint. We also assume that each amount of time needed to transfer a file is equal. That is we treat only the case when all $L(e)$ are equal or when we can partition each file into subfiles which are of same length. Since the problem to minimize the total time for the overall transfer process is NP-complete, it is unlikely that there is a polynomial-time algorithm to solve the problem exactly.

We formulate the problem above as a new type of edge-coloring of G, called an *fg*-edge-coloring. An *fg-edge-coloring* of G is a coloring of the edges of G such that

(a) at most $f(v)$ edges incident to v are colored with a single color for each vertex $v \in V$; and

(b) at most $g(vw)$ multiple edges joining v and w are colored with a single color for each pair of vertices v and w.

Figure 1 depicts an *fg*-coloring of a multigraph. Our scheduling problem is formulated as a problem to find an *fg*-coloring of G with the minimum number of colors. The number is called the *fg-chromatic index* $\chi'_{fg}(G)$ of G, and corresponds to the minimum finishing time in an optimal schedule. Note that edges colored with the same color correspond to files that can be transferred simultaneously.

Denote by $d(v)$ the number of edges incident to vertex v, by $E(vw)$ the set of multiple edges joining vertices v and w, and let $p(vw) = |E(vw)|$.

We show that a Shannon type upper bound holds for any multigraph G:

$$\chi'_{fg}(G) \le \left\lfloor \frac{3}{2}\Delta_{fg} \right\rfloor$$

where

$$\Delta_{fg} = \max\{\Delta_f, \Delta_g\}, \Delta_f = \max_{v \in V} \left\lceil \frac{d(v)}{f(v)} \right\rceil \text{ and } \Delta_g = \max_{E(vw) \subset E} \left\lceil \frac{p(vw)}{g(vw)} \right\rceil.$$

Clearly Δ_{fg}, Δ_f and Δ_g are lower bounds on $\chi'_{fg}(G)$. We furthermore give an $O(|E|^2)$ time algorithm for finding an fg-coloring of G using at most $(3/2)\Delta_{fg}$ colors. Thus the algorithm has the *absolute* worst-case ratio of 3/2, which is better than the ratio 5/2 of Coffman *et al*[CG].

In a special case when $f(v) = 1$ for all $v \in V$ and $g(vw) = 1$ for all $vw \in E$, the scheduling problem above reduces to the ordinary edge-coloring problem of multigraphs. For the case Nishizeki and Kashiwagi [NK] gave an $O(|E|^2)$ approximation algorithm of *asymptotic* worst-case ratio 11/10. On the other hand, when there is no channel constraint, that is, $g(vw) \geq p(vw)$ for all $vw \in E$, the scheduling problem reduces to the so-called f-coloring problem of multigraphs. For this case Nakano, Nishizeki and Saito [NNS] gave an $O(|E|^2)$ approximation algorithm of *asymptotic* worst-case ratio 9/8.

2. Main Theorem

For $S \subset V$, $E(S)$ denotes the set of edges joining vertices in S. Then we have another lower bound $t(G)$ on $\chi'_{fg}(G)$:

$$\chi'_{fg}(G) \geq t(G) = \max_{S \subset V} |E(S)|$$

where S runs over all subsets of V consisting of exactly three vertices v, w and x such that $f(v) = f(w) = f(x) = 1$.

The following theorem is a main result of this paper which is an upper bound better than $\left\lfloor \frac{3}{2}\Delta_{fg} \right\rfloor$.

THEOREM 1. Every multigraph G satisfies

$$\chi'_{fg}(G) \leq u(G)$$

where

$$u(G) = \max\left\{ t(G), \Delta_g, \left\lfloor \frac{5\Delta_f + 2}{4} \right\rfloor, \max_{vw \in E, f(v) \geq 2} \left\lceil \frac{d(v) + p(vw) - 1}{f(v) + g(vw) - 1} \right\rceil \right\}.$$

We will give a sketchy proof of Theorem 1 in Section 3.

One may assume without loss of generality that $f(v) \leq d(v)$ for all $v \in V$ and $g(vw) \leq p(vw)$ for all $E(vw) \subset E$. In this case the following corollary holds true.

Corollary 1. If $f(v) \le d(v)$ for every $v \in V$ and $g(vw) \le p(vw)$ for every $E(vw) \subset E$, then

$$\chi'_{fg}(G) \le \left\lfloor \frac{3}{2}\Delta_{fg} \right\rfloor.$$

Proof. It suffices to prove $u(G) \le \left\lfloor \frac{3}{2}\Delta_{fg} \right\rfloor$. Obviously the following inequalities hold for the first two terms of $u(G)$:

$$t(G) \le \left\lfloor \frac{3}{2}\Delta_{fg} \right\rfloor \text{ and } \Delta_g \le \left\lfloor \frac{3}{2}\Delta_{fg} \right\rfloor.$$

Since one may assume $\Delta_f \ge 2$, for the third term of $u(G)$ we have:

$$\left\lfloor \frac{5\Delta_f + 2}{4} \right\rfloor \le \left\lfloor \frac{3}{2}\Delta_f \right\rfloor \le \left\lfloor \frac{3}{2}\Delta_{fg} \right\rfloor.$$

We finally consider the last term of $u(G)$. For any $E(vw) \subset E$

$$\frac{d(v) + p(vw)}{f(v) + g(vw)} \le \max\left\{ \frac{d(v)}{f(v)}, \frac{p(vw)}{g(vw)} \right\}.$$

In particular, the following equation holds for any $E(vw) \subset E$ with $f(v) \ge 2$:

$$\left(\frac{3}{2}\frac{d(v) + p(vw)}{f(v) + g(vw)} - \frac{1}{2} \right) - \frac{d(v) + p(vw) - 1}{f(v) + g(vw) - 1}$$

$$= \frac{\{(d(v) + p(vw)) - (f(v) + g(vw))\}\{(f(v) + g(vw)) - 3\}}{2(f(v) + g(vw))(f(v) + g(vw) - 1)} \ge 0.$$

Therefore we have

$$\max_{vw \in E, f(v) \ge 2} \left\lceil \frac{d(v) + p(vw) - 1}{f(v) + g(vw) - 1} \right\rceil \le \left\lceil \frac{3}{2}\max\{\Delta_f, \Delta_g\} - \frac{1}{2} \right\rceil = \left\lfloor \frac{3}{2}\Delta_{fg} \right\rfloor.$$

Q.E.D.

3. Sketchy Proof of Main Result

We will prove Theorem 1 by induction on the number of edges. Let $e \in E(vw)$ be an arbitrary edge of G. By the inductive hypothesis the graph $G - e$ obtained from G by deleting e can be fg-colored with $u(G)$ colors. Before presenting the sketchy proof, we give some notations and lemmas.

Let U be the set of $u(G)$ colors available for an fg-coloring of G. An edge colored with color $c \in U$ is called a c-edge. The number of c-edges incident to vertex v is denoted by $d(v, c)$, while the number of c-edge in $E(vw)$ is denoted by $p(vw, c)$. Define $m(v, c) = f(v) - d(v, c)$ and $m(vw, c) = g(vw) - p(vw, c)$. Then G is fg-colored if and only if every color $c \in U$ satisfies $m(v, c) \ge 0$ for every vertex $v \in V$ and $m(vw, c) \ge 0$ for every multiple edges

$E(vw) \subset E$. Color c is *available at* v if $m(v,c) \geq 1$. Similary color c is *available at* $E(vw)$ if $m(vw,c) \geq 1$. We define

$$M(v) = \{c \in U | m(v,c) \geq 1\}$$

and

$$M(vw) = \{c \in U | m(vw,c) \geq 1\}.$$

Thus $M(v)$ is the set of colors available at vertex v, while $M(vw)$ is the set of colors available at multiple edges $E(vw)$.

In the case of an ordinary coloring, if the ends of an uncolored edge vw have a common available color c, that is, $c \in M(v) \cap M(w)$, then the coloring of G proceeds with coloring edge vw by color c. It is not the case when fg-coloring a graph. For, if $m(vw,c) = 0$, then one cannot color an uncolored edge vw with a color $c \in M(v) \cap M(w)$. Thus an uncolored edge $e \in E(vw)$ can be colored with color c only if $c \in M(v) \cap M(w) \cap M(vw)$. Such a color c does not always exist in a partial fg-coloring of G. We however claim that any partial fg-coloring of G using $u(G)$ colors can always be altered so that there is a color $c \in M(v) \cap M(w) \cap M(vw)$ for the ends of an uncolored edge vw.

Switching an alternating path is one of the standard techniques of an ordinary coloring [G,HNS,NK,V1,V2]. We also use it with some modifications. A walk is used instead of a path. A *walk* W is a sequence of distinct edges $v_0 v_1, v_1 v_2, \cdots, v_{k-1} v_k$, where the vertices $v_0, v_1, \cdots, v_k$ are not necessarily distinct. The *length* of W is the number of edges in W. Vertex v_0 is the *start vertex* of W and v_k the *end vertex*. Walk W is called a *cycle* if $v_0 = v_k$. When the edges of a walk W are colored with two colors α and β alternately, *switching* W means to interchange the colors α and β of edges in W. We define an "$\alpha\beta$-alternating walk" so that its switch would preserve an fg-coloring of G, as follows.

Let $G(\alpha, \beta)$ be the subgraph of G induced by all α- and β-edges. Delete successively all pairs of edges of color α and β respectively joining the same two vertices, and let G^* be the resulting graph in which there no longer exists such a pair. Denote by $E^*(vw)$ the set of multiple edges joining vertices v and w in $G^*(\alpha, \beta)$. Obviously each set $E^*(vw)$ contains only α- or β-edges. An $\alpha\beta$-alternating walk $W = v_0 v_1, v_1 v_2, \cdots, v_{k-1} v_k$ is a walk of length one or more in $G^*(a,b)$ such that

(1) the edges in W are colored alternately with α and β (the ith edge e_i of W is colored β if i is an odd number; otherwise e is colored α);

(2) if W is not a cycle then $m(v_0, \alpha) \geq 1$, while if W is a cycle of odd length then $m(v_0, \alpha) \geq 2$; and

(3) if W is not a cycle and is of even length then $m(v_k, \beta) \geq 1$, while if W is not a cycle and is of odd length then $m(v_k, \alpha) \geq 1$.

Thus a cycle in $G^*(\alpha, \beta)$ is an $\alpha\beta$-alternating walk (cycle) whenever it has even length and its edges are colored α and β alternately. The following lemma holds.

Lemma 1. Let G be fg-colored, and let W be any $\alpha\beta$-alternating walk in G. Furthermore, let m' represent the function m with respect to the new coloring after switching W. Then the following (a)–(c) hold.

(a) Switching W preserves an fg-coloring of G, that is, every color c satisfies $m'(v, c) \geq 0$ for any $v \in V$ and $m'(vw, c) \geq 0$ for any $E(vw) \subset E$.

(b) For any $E(vw) \subset E$

$$m'(vw, \alpha) \geq \min\{m(vw, \alpha), m(vw, \beta)\}$$

and

$$m'(vw, \beta) \geq \min\{m(vw, \alpha), m(vw, \beta)\}.$$

(c) If W passes through an α-edge in $E(vw)$ then $m'(vw, \alpha) > 0$, while if W passes through a β-edge in $E(vw)$ then $m'(vw, \beta) > 0$.

We denote by $W(\alpha, \beta, v_0)$ an $\alpha\beta$-alternating walk which starts with vertex v_0 and is not a cycle of even length. Switching $W(\alpha, \beta, v_0)$ may change $m(v, \alpha)$ or $m(v, \beta)$ only if v is the start or end vertex. On the other hand, switching an $\alpha\beta$-alternating cycle of even length changes neither $m(v, \alpha)$ nor $m(v, \beta)$ for any $v \in V$.

Lemma 2.

(a) $\sum_{c \in U} m(x, c) \geq 1$ for all vertex $x \in V$.

(b) $\sum_{c \in U} m(v, c) \geq 2$.

(c) $\sum_{c \in U} m(vw, c) \geq 1$.

(d) There exists a color c such that $m(v, c) + m(vw, c) \geq 2$.

Proof. (a)(b)(c) Omitted. (d) If $f(v) = 1$ then by (b) there exists a color $c \in M(v)$. Since $f(v) = 1$, $p(vw, c) = 0$. Therefore $m(v, c) + m(vw, c) \geq 2$. If $f(v) \geq 2$ then by the definition of $u(G)$ we have

$$u(G) \geq \frac{d(v) + p(vw) - 1}{f(v) + g(vw) - 1}.$$

Therefore

$$\sum_{c \in U}\{m(v, c) + m(vw, c)\}$$
$$= u(G) \cdot f(v) - d(v) + 1 + u(G) \cdot g(vw) - p(vw) + 1 \geq u(G) + 1.$$

Thus there exists a color c such that

$$m(v, c) + m(vw, c) \geq 2.$$

Q.E.D.

We shall consider five cases, but will sketch the first two cases only.

Case 1: *There is a color $c \in U$ such that $c \in M(v) \cap M(w) \cap M(vw)$.*

In this case we can simply color e by c.

We may now assume that Case 1 does not hold. By Lemma 2 (d) there are colors $\alpha \in U$ and $\beta \in U$ such that

$$m(v, \alpha) + m(vw, \alpha) \geq 2$$

and

$$m(w, \beta) + m(vw, \beta) \geq 2.$$

Case 2: $m(v, \alpha) \geq 2$.

We separate this case into seven subcases.

Case 2.1: $m(vw, \alpha) \geq 1$ and $m(w, \beta) \geq 2$.

Clearly $\alpha \neq \beta$ and $m(w, \alpha) = 0$. We can show that there exists a walk $W(\beta, \alpha, w)$. If $m(vw, \beta) \geq 1$, then switching the walk $W(\beta, \alpha, w)$ will make $m(vw, \alpha) \geq 1$ by Lemma 1. Now $m(v, \alpha) \geq 1$ and $m(w, \alpha) \geq 1$, hence Case 1 would apply.

If $m(vw, \beta) = 0$, then we can prove that there exists a walk $W(\beta, \alpha, w)$ containing no edge in $E(vw)$. Switching the walk $W(\beta, \alpha, w)$ reduces this case to Case 1.

Case 2.2: $m(vw, \alpha) \geq 1$, $m(w, \beta) = 1$ and $m(vw, \beta) \geq 1$.

Since Case 1 did not apply, $\alpha \neq \beta$ and $m(w, \alpha) = 0$. Switch the walk $W(\alpha, \beta, v)$. If the walk $W(\alpha, \beta, v)$ did not end at w, then we could color edge e with color β. If the walk $W(\alpha, \beta, v)$ ended at w then we could color edge e with color α.

Other subcases and cases are omitted for lack of space.

4. Approximation Algorithm

The proof sketched in the preceding section yields a polynomial-time algorithm which fg-colors any given multigraph G with at most $u(G)$ colors. The absolute worst case ratio of the algorithm is no greater than $3/2$ since $u(G) \leq (3/2)\Delta_{fg} \leq (3/2)\chi'_{fg}(G)$.

The algorithm is iterative in a sense that it colors the edges of G one by one. One can easily know that each of cases repeats the switching of an alternating walk at most constant times. Using data structure mentioned in [HNS], one can switch an alternating walk in $O(|E|)$ time. Therefore the algorithm runs in $O(|E|^2)$ time. Furthermore it uses $O(|E|)$ storage space.

REFERENCES

[CH] H. A. Choi S. L. Hakimi, "Scheduling file transfers for trees and odd cycles," SIAM J. on Comput., 16, pp. 162-168 (1987).

[CG] E. G. Coffman, Jr, M. R. Garey, D. S. Johnson and A. S. LaPaugh: "Scheduling file transfers," SIAM J. Comput., 14, 3, pp. 744-780 (1985).

[GJ] M. R. Garey and D. S. Johnson: "Computers and Intractability: A Guide to the Theory of NP-Completeness," W. H. Freeman & Co., New York (1979).

[G] M. K. Goldberg: "Edge-colorings of multigraphs: recoloring techniques," Journal of Graph Theory, 8, 1, pp. 122-136 (1984).

[H] S. L. Hakimi: "Further results on a generalization of edge-coloring," in Graph Theory with Applications to Algorithms and Computer Science, ed. Y. Alavi et al., pp. 371-389, John Wiley & Sons, New York (1985).

[HK] S. L. Hakimi and O. Kariv: "On a generalization of edge-coloring in graphs," Journal of Graph Theory, 10, pp. 139-154 (1986).

[HNS] D. S. Hochbaum, T. Nishizeki and D. B. Shmoys: "A better than "best possible" algorithm to edge color multigraphs," Journal of Algorithms, 7, 1, pp. 79-104 (1986).

[H] I. J. Holyer: "The NP-completeness of edge colourings," SIAM J. Comput., 10, pp. 718-720 (1980).

[NNS] S. Nakano, T. Nishizeki and N. Saito "On the f-coloring of multigraphs," IEEE Trans., Circuit and Syst., CAS-35, 3, pp. 345-353 (1988).

[NK] T. Nishizeki and K. Kashiwagi: "On the 1.1 edge-coloring of multigraphs,", SIAM J. Disc. Math., 3, 3, (1990).

[S] C. E. Shannon: "A theorem on coloring the lines of a network," J. Math. Phys., 28, pp. 148-151 (1949).

[V1] V. G. Vizing: "On an estimate of the chromatic class of a p-graph," Discret Analiz, 3, pp. 25-30 (in Russian) (1964).

[V2] V. G. Vizing: "The chromatic class of a multigraph," Kibernetica(Kief), 3, pp.29-39 (1965); Cybernatics, 3, pp. 32-41 (1965).

[W] J. Whitehead "The complexity of file transfer scheduling with forwarding," SIAM J. on Comput., 19, 2, pp. 222-245 (1990).

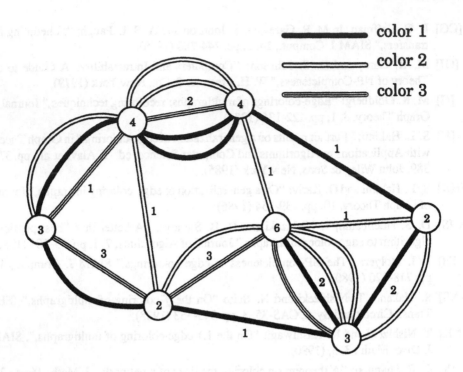

Fig. 1 An *fg*-coloring of a multigraph using three colors.

Substitution Decomposition on Chordal Graphs and Applications

Wen-Lian Hsu and Tze-Heng Ma

Institute of Information Science, Academia Sinica,

Taipei, Republic of China 11529

Abstract

In this paper, we present a linear time algorithm for substitution decomposition on chordal graphs. Based on this result, we develop a linear time algorithm for transitive orientation on chordal comparability graphs. Which reduces the complexity of chordal comparability recognition from $O(n^2)$ to $O(n + m)$. We also devise a simple linear time algorithm for interval graph recognition where no complicated data structure is involved.

1 Introduction

All graphs in this paper are simple and have no self-loops. Let $G = (V, E)$ be a graph. An undirected edge between vertices u and v is denoted by uv. A directed edge from u to v is written as (u, v). For undirected graphs, the neighborhood of a vertex v, $N(v)$, will denote the set $\{w \in V : vw \in E\}$. For a set S of vertices, $N(S) = \cup_{v \in S} N(v) \setminus S$. The *degree* of a vertex v, *deg(v)*, is the cardinality of $N(v)$. The *augmented neighborhood* of vertex v, $N^+(v)$, is $\{v\} \cup N(v)$.

A *chordal graph* is a graph with no induced subgraph isomorphic to a cycle C_k, $k \geq 4$. Chordal graphs have been studied extensively. They are also called *triangulated*, *rigid-circuit*, and *perfect elimination* graphs. There are several subclasses of chordal graphs which have gained a lot attention, e.g. interval graphs, split graphs, strongly chordal graphs, and chordal comparability graphs, which is the comparability graphs of *cycle-free* partial orders.

A *module* in an undirected graph $G = (V, E)$ is a set of vertices $S \subset V$ such that $N(S) = N(v) \setminus S \; \forall v \in S$. A module is *nontrivial* if $1 < |S| < |V|$. A *substitution decomposition* of a graph is to substitute a nontrivial module in the graph with a marker vertex and do the same recursively for the module and the substituted graph. A substitution decomposition is often represented by a tree; where each subtree represents a nontrivial module marked by its root. For general graphs, substitution decomposition takes $O(\min(n^2, m\alpha(m, n)))$ time [13] [14].

A vertex is *simplicial* if its neighborhood form a clique. A necessary and sufficient condition of a chordal graph is that it admits a *perfect elimination scheme*, which is a linear ordering of the vertices such that, for each vertex v, the neighbors of v, which are ordered before v, form a clique. A perfect elimination scheme of a chordal graph can be obtained by a *lexicographic ordering*, which can be carried out in linear time [12]. In the next section, we present an $O(n + m)$ algorithm for substitution decomposition on chordal graphs. Our algorithm uses a special ordering to force vertices in the same module to occur consecutively in this ordering.

In this paper, we call a graph *prime* if it does not contain a nontrivial module. Our linear time algorithm for substitution decomposition on chordal graphs enables us to find linear time algorithms for chordal comparability graph and interval graph recognition since a prime graph has the following properties: (i) if it is a comparability graph, there is a unique transitive orientation [6]; (ii) if it is an interval graph, there is a unique interval representation for the graph [8].

A directed graph $G = (V, E)$ is *transitive* if for all $u, v, w \in V$, $(u, v), (v, w) \in E \Rightarrow (u, w) \in E$. Since G has no self-loops, G must be acyclic. An undirected graph is a *comparability graph* if we can give each edge a direction such that the resultant directed graph is transitive. This process is called a *transitive orientation*. The complement of a comparability graph is called a *co-comparability graph*. The fastest algorithm [13] for recognizing a comparability graph involves two stages. First, the input graph is transitively oriented. This can be done in $O(n^2)$ time. Then we test whether this directed graph is transitive. The fastest algorithm for the testing takes time proportion to that of multiplying two $n \times n$ boolean matrices, which is currently $O(n^{2.376})$ [2].

A *chordal comparability graph* is a graph which is both chordal and comparable. Recently, an $O(n + m)$ algorithm is developed to test whether a directed chordal graph is transitive [11]. This brings the complexity of recognizing chordal comparability graphs down to $O(n^2)$. The new bottleneck is the transitive orientation of chordal graphs. In section 3, we present an algorithm which transitively orient a prime chordal comparability graph in $O(n + m)$ time. Combined with previous results, this yields an $O(n + m)$ algorithm for chordal comparability graph recognition.

A graph $G = (V, E)$ is called an *interval graph* if it is the intersection graph of a set of closed intervals on the real line. In other words, there is a one to one mapping between the vertices and intervals such that two vertices are adjacent iff their corresponding intervals overlap. Interval graphs are exactly chordal co-comparability graphs. This implies they can be recognized in polynomial time.

Booth and Lueker deviced an algorithm to recognize interval graphs in $O(n + m)$ time using a rather complicated data structure called a *PQ-tree* [1]. Korte and Möhring [7] simplified the operations on a PQ-tree by carrying out an incremental algorithm by a lexicographic ordering. In section 4, we present a linear time algorithm for recognizing prime interval graphs. Combined with the decomposition algorithm, this yields a linear time algorithm for interval graph recognition. We consider our algorithm to be much simpler than previous ones since there is no complicated data structure involved and the approach is intuitively appealing.

Limited by the physical length, all the proofs are omitted in this abstract. For the readers who want to verify the proofs for themselves, we advise them to make use of the geometric models whenever possible.

2 Substitution Decomposition on Chordal Graphs

Lexicographic ordering has been used to obtain perfect elimination schemes for chordal graphs in linear time [12]. A lexicographic ordering can be consider as a special kind of breadth first ordering. This ordering guarantees that if x is ordered before y and there is a vertex ordered before x which is a neighbor of one but not both of x, y, the first vertex added to the ordering with such property must be a neighbor of x. The linear time algorithm is built upon a partitioning procedure. One implementation of a lexicographic ordering looks like:

```
Lexicographic(G);
    S := {V};
    for i := 1 to n do
        begin
            v := the first element of the first set in S;
            remove v;
            π(i) := v;
            split each S_j ∈ S into N(v) ∩S_j and S_j−N(v);
            discard empty sets
        end
end Lexicographic;
```

The output ordering π is a *perfect elimination scheme* if and only if the input graph is chordal. Perfect elimination schemes play a central role in chordal graph recognition.

A *cardinality lexicographic ordering* is just a lexicographic ordering with the vertices sorted by their degrees prior to partitioning. The implication of this extra step is that when more than one vertex is eligible to be included to the ordering, we break the tie by choosing the vertex with maximal degree. Since we can count the degrees and bucket sort the vertices of a graph in linear time, cardinality lexicographic ordering can also be done in linear time.

Lemma 1 *Let S be a module in a chordal graph $G = (V, E)$. Either S is a clique or N(S) is a clique.*

If module S is a clique, every vertex in S has the same augmented neighborhood. Such a module is called a type I module. All type I module can be located in $O(n + m)$ time by partitioning the vertices using the augmented neighborhoods of all vertices. If there is a set with more than one vertex at the end of the partitioning process, it is a type I module.

Lemma 2 *If $N^+(u) \subset N^+(v)$, then $\pi^{-1}(u) > \pi^{-1}(v)$ in every cardinality lexicographic ordering.*

We call a module type II if it is connected but not type I. After all type I modules being removed, a cardinality lexicographic ordering will put the vertices in a type II module consecutively and the neighborhood of the module will be ordered before the module.

Lemma 3 *Let S be a connected module in a chordal graph G with no type I module. If π is a cardinality lexicographic ordering on G, then*
(i) $\pi^{-1}(u) < \pi^{-1}(v)$ $\forall u \in N(S), v \in S$
(ii) all vertices in S are ordered consecutively in π.

After getting the cardinality lexicographic ordering π, we scan the ordering from the last position. By lemma 3, if there is a type II module, the vertices in the module must be in consecutive positions with all neighbors ordered before the module. The algorithm to discover all type II modules in π tries to find the existence of such configurations. We use a "stack of stacks" to store scanned vertices. Each stack can be viewed as a candidate for a module. There are two conditions we have to enforce. First, no vertex in a stack has a neighbor beyond the bottom of the stack, because by lemma 3, all neighbors of a type II module will be ordered before the module. Secondly, the neighborhoods of vertices in the stack should agree outside the stack. Each time a new vertex v is scanned, we try to start a new stack. If there is an edge extended from the top stack down to a lower stack, all boundaries between these two stacks must be broken. With each stack, we store the size of the stack, the common neighborhood of vertices in the stack, and the minimum $\pi^{-1}(w)$ where w is the neighbor of some but not all vertices in the stack. For a stack to be eligible to be a module, w must be included in the stack. After v is processed, if the size of the top stack is greater than 1 and the neighborhoods of vertices in the top stack agree on all the unscanned vertices, we conclude this stack forms a module.

The modules can be found recursively. Each time a module is reported, we replace the top stack by a marker vertex, whose neighborhood is the AgreedSet of the stack. A pseudo-code implementation of the algorithm is listed below:

```
Modular_Decomp(G, π);
    i := 0;
    v := π(n);
    CreateStack(v);
    for j := n − 1 to 2 do
        begin
            v := π(j);
            CreateStack(v);
            let w be the neighbor of v with maximum π⁻¹(w)
            while π⁻¹(w) > STACK(i).bottom
                MergeTopTwoStacks;
            if STACK(i).size > 1 and STACK(i).MinDisagree ≥ j then
                report "vertices in STACK(i) form a module"
        end
end Modular_Decomp;


CreateStack(v);
    i := i+1;
    STACK(i).bottom := π⁻¹(v);
    STACK(i).MinDisagree := n;
    STACK(i).size := 1;
    STACK(i).AgreedSet := N(v)
end CreateStack;


MergeTopTwoStacks;
    S := STACK(i).AgreedSet ∩ STACK(i-1).AgreedSet;
    STACK(i-1).MinDisagree := min( STACK(i).MinDisagree,
        STACK(i-1).MinDisagree,
        (π⁻¹(v), v ∈ STACK(i).AgreedSet ∪ STACK(i-1).AgreedSet, v ∉ S) );
    STACK(i-1).AgreedSet := S;
    STACK(i-1).size := STACK(i).size + STACK(i-1).size;
    i := i-1
end MergeTopTwoStacks;
```

Theorem 1 *Chordal_Modular_Decomp correctly finds all type II modules in a chordal graph in $O(n + m)$ time.*

After we replace every type I and II module by a marker vertex, all possible modules left are those independent sets with the same neighborhoods. These modules, call them type III modules, can be found in linear time by partitioning the vertex set using their neighborhoods. In conclusion, we have the following theorem:

Theorem 2 *The substitution decomposition on a chordal graph can be carried out in $O(n + m)$ time.*

The decomposition tree can be easily constructed along with the decomposition process. Whenever a module S is reported, we replace S by a marker v with the same neighborhood as that of S. For the decomposition tree, create a tree rooted at v whose children are those vertices in S.

3   Transitive Orientation on Chordal Graphs

In this section, we provide a linear time algorithm to transitively orient a chordal graph which admits a unique transitive orientation. Combined with the result in last section, we can thus recognize chordal comparability graphs in linear time.

We call $P = (X, R)$ a *partially ordered set (poset)* where X is a set and R is an irreflexive transitive binary relation on X. We say x *dominates* y if $(x, y) \in R$. If (u, v) or $(v, u) \in R$, we say that u, v are comparable. A poset $P = (X, R)$ can be viewed as a transitive graph $G = (V, E)$ by taking X as the vertex set V, R as the edge set E. Chordal comparability graphs, when transitively oriented, become a class of poset called *cycle-free* partial orders. For more characteristics on cycle-free posets, see [3] [11].

A partial order can be expressed by its diagram, which is an undirected graph of minimum number of edges where there is an upward path from a to b if and only if (a, b) exists. A *chain* of a diagram is a path whose vertices are pairwise comparable. The chains of a partial order correspond to the cliques of its comparability graph. Throughout this section, the readers should keep in mind an imaginary diagram, which is the diagram of a unique orientation of the input graph. Our algorithm is much easier understood on this diagram.

We call a vertex w in a diagram *high* (resp. *low*) if it is dominated by (resp. dominates) a pair of incomparable vertices x, y. In a chordal comparability graph, a nonsimplicial vertex must be either high or low, but not both; since if w is made high by u, v, low by x, y, then u, v, x, y induce a cycle of length 4. According to this definition, it can be observed that on a chain in the diagram of a cycle-free poset, the high (resp. low) vertices are above (resp. below) all simplicial and low (resp. high) vertices. We say vertex x is higher (resp. lower) than y when x, y are in the same chain and x is above (resp. below) y.

One of the most widely used characterizations of chordal graphs is that the maximal cliques of a chordal graph can be connected to form a tree T such that for each vertex v, the subgraph induced on T by the maximal cliques containing v is connected [5]. Our algorithm applies a partitioning technique on the clique tree structure for the input chordal graph. At the end, all nonsimplicial vertices are marked either high or low and a topological sort for the target diagram is generated to provide the basis for a transitive orientation.

We first outline the algorithm. For briefness, all chains and cliques mentioned hereafter are assumed maximal unless otherwise stated. The basic idea is to take a confirmed high (or low) vertex x in a certain chain and try to force the common vertices in a neighboring chain to be low (or high). Formally, we claim the following:

Lemma 4 *Let C_i and C_j be two chains with intersection S. If x is highest (resp. lowest) in C_i, $x \notin S$, then every vertex in S must be low (resp. high).*

To take advantage of the property of lemma 4, we need to carry out our partitioning by an order where "extreme" vertices are considered first. This requirement can be easily met by the following observation.

Lemma 5 *For two adjacent high (resp. low) vertices x, y, $N(x) \supset N(y)$ iff x is higher (resp. lower) than y.*

For an input graph G, we create its clique tree representation. We also record the intersection of two adjacent maximal cliques on the edge connecting these cliques. These informations can be obtained in linear time. A queue for vertices is installed to store our candidates for partitioning. Initially, we choose the vertex x with maximum degree as our first candidate. x cannot be simplicial; otherwise, the whole graph is a clique. We can mark x high or low arbitrarily. We start our traversal from a clique in the clique tree containing x. Whenever we pass an edge, if x is recorded in that edge, we take x out of that edge and go on. Otherwise, we encounter the situation of lemma 4. We take every vertex recorded in this edge out; sort them by their degrees; mark them opposite to x; put them into the queue and backtrack. To avoid checking an edge while not doing anything, we remove the pointer leading to an edge when everything recorded in that edge is removed. This will not affect the correctness of the algorithm since our subsequent traversal is not going to pass beyond the edge anyway.

A topological sort is constructed along the process by growing a list of size n from both ends. Whenever a vertex enters the queue, if it is marked high, put it on the highest possible position of the list. Otherwise, put it on the lowest possible position. The rest of the vertices (simplicial) are put in the middle of the list by any order. We can then orient an edge from the endpoint with a lower position to the endpoint with a higher position on the list. If the input graph is transitive, this orientation yields a partial order whose diagram is exactly characterized by the high-low relations generated by our algorithm. We are now ready to state the main theorem of this section:

Theorem 3 *A prime chordal graph, $G = (V, E)$, can be transitively oriented in $O(n + m)$ time.*

The correctness of our algorithm relies on the fact that we never do anything that is not required. At the end, all nonsimplicial vertices will be marked since otherwise, there exists a module. The time complexity is analyzed by an amortized argument [15], we can "charge" our operations on $O(n + m)$ items where each item receives constant charges.

We can incorporate the orientation algorithm with the substitution decomposition algorithm for chordal graphs and obtain the following corollary. With the linear time algorithm for transitive verification for chordal graphs [10], we now have linear time algorithm for chordal comparability graph recognition.

Corollary 1 *Chordal comparability graphs can be transitively oriented in $O(n + m)$ time.*

4 Interval Graph Recognition

Interval grapgs has been a very useful model for many applications [9] [4]. The fastest algorithm to recognize interval graphs relies on the following property:

Theorem 4 *A graph G is an interval graph iff its maximal cliques can be linearly ordered such that, for each vertex v, the maximal cliques containing v occur consecutively.*

This linear ordering of the maximal cliques actually admits a clique tree which is a path. Unfortunately, although we can generate a clique tree for an interval graph in linear time, this clique tree is not necessarily a path. Booth and Lueker [1] created a data structure called a *PQ-tree* to capture the consecutive property of a set of intervals. This data structure yields a linear time algorithm to recognize interval graphs. However, their algorithm is quite involved. Korte and Möhring devised a simpler algorithm to recognize interval graphs which also runs in linear time. They observed that if the input vertices follow a lexicographic ordering, the operations on the PQ-tree can be simplified.

Hsu [8] proved that an interval graph has a unique maximal clique arrangement if the graph is prime. In this section, we provide a linear time algorithm to find the linear clique arrangement of a prime interval graph. Together with the linear time substitution decomposition algorithm, we have yet another linear time algorithm for interval graph recognition. Our algorithm uses only partitioning and lexicographic ordering; no special data structure is required. We consider it to be much simpler than previous algorithms. However, we want to remind the readers that Booth and Lucker's algorithm is much more flexible in the sense that it can test the consecutive 1's property of a boolean matrix. Korte and Möhring's algorithm, while must take the input in the form of a graph, does maintain the flexibility that the input is processed in an incremental fashion. Our algorithm only operates in an "off-line" fashion. The readers are encouraged to make comparisons among these algorithms.

Our algorithm is based on a graph partitioning idea. Initially, we obtain the clique tree representation of the input chordal graph as we did in last section. We partition the maximal cliques into two sets such that there is a linear clique arrangement (if G is an interval graph) where all the cliques in one set are at the left of the cliques in the other set. We then further refine our partition by the following observation:

Lemma 6 *Let A and B be two sets of maximal cliques where A is at the left of B. Suppose vertex v is shared by $C_A, C_B, C_A \in A, C_B \in B$. If X is a set of maximal cliques at the right of A, all maximal cliques in X containing v must be at the left of those not containing v. Symmetrically, if Y is a set of maximal cliques at the left of B, all maximal cliques in Y containing v must be at the right of those not containing v.*

The proof to this lemma, which is omitted here, can be observed from the geometric model of an interval graph. To start the partitioning process, we need an initial partition which admits a feasible linear clique arrangement if the input graph is interval. The following lemma provides an easy way to find such a configuration.

Lemma 7 *For any lexicographic ordering on a prime interval graph G, the maximal clique containing the last vertex must be leftmost or rightmost on the linear clique arrangement for G.*

The clique tree structure provides enough information for us to partition a prime interval graph into a unique linear maximal clique arrangement. We now present a detailed description of the algorithm:

```
procedure LinearMaxCliqueArrangement(G)
    find a clique tree of G;
    partition the last maximal clique of a lexicographic ordering
        from the rest of the maximal cliques;
    move the edges in the clique tree crossing these two sets of
        maximal cliques into CrossingEdge;
    Partitioning
end LinearMaxCliqueArrangement;


procedure Partitioning
    while CrossingEdge ≠ ∅ do begin
        pick edge C_i C_j from CrossingEdge;
        for each unprocessed v ∈ C_i ∩ C_j do
            begin
                for all sets S_i in P, partition S_i into
                    cliques containing v, S_i', and cliques not containing v, S_i'';
                if S_i' ≠ ∅ and S_i'' ≠ ∅ then do
                    begin
                        replace S_i by S_i', S_i'' according to lemma 6;
                        move the edges in the clique tree leading
                            out of S_i'into CrossingEdge
                    end
            end
    end {while}
end Partitioning;
```

We now present the main theorem of this section.

Theorem 5 *Given a prime interval graph G, a linear maximal clique arrangement of G can be obtained in $O(m + n)$ time.*

The proof to this theorem is very similar in principle to that we used in last section. First, we argue that the initial configuration and the subsequent partitionings are correct. Then we claim that after the algorithm ends, if the input prime graph is interval, we will have a unique partitioning since otherwise, there exists a module. The complexity of this algorithm is again based on an amortized analysis.

An interval model for the input graph G can be constructed from the linear clique arrangement. It's easy to test in linear time whether this model is consistant with G. Therefore, we have a linear time algorithm to recognize prime interval graphs. We can incorporate this algorithm with the substitution decomposition algorithm for chordal graphs to get a linear time algorithm for recognizing interval graphs. Whenever a module S is identified, it must be prime. We test whether S is an interval graph. If all the modules (including the last prime graph representing the input graph G) are interval graphs, G is an interval graph. Since we can always replace the interval of a marker vertex v by the interval model of the module it represents and still have an interval model, we have thus proved the following corollary.

Corollary 2 *An interval graph can be recognized in linear time.*

References

[1] K. S. Booth and G. S. Lueker, "Testing for the Consecutive Ones Property, Interval Graphs, and Graph Planarity Using PQ-tree Algorithms," *J. Comput. Syst. Sci.*, v.13, 1976, pp. 335-379.

[2] D. Coppersmith and S. Winograd, "Matrix Multiplication via Arithmetic Progressions," *Proceedings of the 19th Annual Symposium on the Theory of Computation*, 1987, pp. 1-6.

[3] D. Duffus, I. Rival, and P. Winkler, "Minimizing Setups for Cycle-free Ordered Sets," *Proc. of the American Math. Soc.*, v.85, 1982, pp. 509-513.

[4] P. C. Fishburn, *Interval Orders and Interval Graphs*, Wiley, New York, 1985.

[5] F. Gavril, "The Intersection Graphs of Subtrees in Trees are Exactly the Chordal Graphs," *J. Combin. Theory B*, v.16, 1974, pp. 47-56.

[6] M. C. Golumbic, *Algorithmic Graph Theory and Perfect Graphs*, Academic Press, New York, 1980.

[7] N. Korte and R. H. Möhring, "An Incremental Linear-Time Algorithm for Recognizing Interval Graphs," *SIAM J. Computing*, v.18, 1989, pp. 68-81.

[8] W. L. Hsu, *The Recognition and Isomorphism Problems for Circular-arc Graphs*, preprint, 1989.

[9] C. G. Lekkerkerker and J. Boland, "Representation of a Finite Graph by a Set of Intervals on the Real Line," *Fund. Math.*, v.51, 1962, pp. 45-64.

[10] J. H. Muller and J. Spinrad, "Incremental Modular Decomposition," *Journal of the ACM*, v.36, 1989, pp. 1-19.

[11] T. H. Ma and J. Spinrad, "Cycle-free Partial Orders and Chordal Comparability Graphs," *Order*, to appear.

[12] D. J. Rose, R. E. Tarjan, and G. S. Lueker, "Algorithmic Aspects of Vertex Elimination of Graphs," *SIAM J. Comput.*, v.5, 1976, pp. 266-283.

[13] J. Spinrad, "On Comparability and Permutation Graphs," *SIAM J. Comput.*, v.14, 1985, pp. 658-670.

[14] J. Spinrad, "P_4 Trees and Substitution Decomposition," *Discrete Applied Math.*, to appear, 1989.

[15] R. E. Tarjan, "Amortized Computational Complexity," *SIAM J. Alg. Disc. Meth.*, v.6, 1985, pp. 306-318.

Mixed-Searching and Proper-Path-Width

Atsushi TAKAHASHI, Shuichi UENO, and Yoji KAJITANI

Department of Electrical and Electronic Engineering
Tokyo Institute of Technology, Tokyo, 152 Japan

1   Introduction

This paper considers a new version of searching game, called mixed-searching, which is a natural common generalization of the edge-searching and node-searching extensively studied so far. We establish a relationship between the mixed-search number of a simple graph G and the proper-path-width of G introduced by the authors in [17]. Complexity results are also shown.

The *searching game* was introduced by Breisch [4] and Parsons [11]. In the searching game, an undirected graph G is considered as a system of tunnels. Initially, all edges of G are contaminated by a gas. An edge is *cleared* by some operations on G. A cleared edge is *recontaminated* if there is a path from an uncleared edge to the cleared edge without any searchers on its vertices or edges.

In the *edge-searching*, the original version of searching game, an edge is cleared by sliding a searcher along the edge. A search is a sequence of operations of placing a searcher on a vertex, deleting a searcher from a vertex, or sliding a searcher along an edge. The object of edge-searching is to clear all edges by a search. We call such a search an *edge-search*. An edge-search is *optimal* if the maximum number of searchers on G at any point is as small as possible. This number is called the *edge-search number* of G, and denoted by $es(G)$. LaPaugh proved that there exists an optimal edge-search without recontamination of cleared edges [8]. Megiddo, Hakimi, Garey, Johnson, and Papadimitriou showed that the problem of computing $es(G)$ is NP-hard for general graphs but can be solved in linear time for trees [9].

The *node-searching*, a slightly different version of searching game, was introduced by Kirousis and Papadimitriou [7]. In the node-searching, an edge is cleared by placing searchers at both its ends simultaneously. A *node-search* is a sequence of operations of placing a searcher on a vertex or deleting a searcher from a vertex so that all edges of G are simultaneously clear after the last stage. A node-search is optimal if the maximum number of searchers on G at any point is as small as possible. This number is called the *node-search number* of G, and denoted by $ns(G)$. Kirousis and Papadimitriou proved the following results: (1) There exists an optimal node-search without recontamination of cleared edges; (2) The problem of computing $ns(G)$ is NP-hard for general graphs; (3) $ns(G) - 1 \leq es(G) \leq ns(G) + 1$ [7].

The path-width of a graph was introduced by Robertson and Seymour [12]. Let G be a graph and $\mathcal{X} = (X_1, X_2, \ldots, X_r)$ be a sequence of subsets of $V(G)$. The *width* of $\mathcal{X}$ is $\max_{1 \le i \le r} |X_i| - 1$. $\mathcal{X}$ is called a *path-decomposition* of G if the following conditions are satisfied: (i) For any distinct i and j, $X_i \not\subseteq X_j$; (ii) $\bigcup_{1 \le i \le r} X_i = V(G)$; (iii) For any edge $(u, v) \in E(G)$, there exists an i such that $u, v \in X_i$; (iv) For all l, m, and n with $1 \le l \le m \le n \le r$, $X_l \cap X_n \subseteq X_m$. The *path-width* of G, denoted by $pw(G)$, is the minimum width over all path-decompositions of G. The unexpected equality $ns(G) = pw(G) + 1$ was mentioned by Möhring [10], and implicitly by Kirousis and Papadimitriou [6]. This provides a linear time algorithm to compute $ns(G)$ for trees [10, 15].

The *mixed-searching*, which is a natural common generalization of the edge-searching and node-searching, was introduced by Bienstock and Seymour [2], and independently by the authors [18]. In the mixed-searching, an edge is cleared by placing searchers at both its ends simultaneously or by sliding a searcher along the edge. A *mixed-search* is a sequence of operations of placing a searcher on a vertex, deleting a searcher from a vertex, or sliding a searcher along an edge so that all edges of G are simultaneously clear after the last stage. A mixed-search is optimal if the maximum number of searchers on G at any point is as small as possible. This number is called the *mixed-search number* of G, and denoted by $ms(G)$. Bienstock and Seymour proved that there exists an optimal mixed-search without recontamination of cleared edges [2]. This was proved independently by the authors in [18]. Bienstock and Seymour characterized the mixed-search number of a graph with minimum degree at least two by means of the concept of crusade, which is a sequence of sets of edges.

The proper-path-width of a graph was introduced by the authors in [17]. A path-decomposition $(X_1, X_2, \ldots, X_r)$ of G is called a *proper-path-decomposition* of G if $|X_l \cap X_n| \le |X_m| - 2$ holds for any l, m, and n ($1 \le l < m < n \le r$). The *proper-path-width* of G, denoted by $ppw(G)$, is the minimum width over all proper-path-decompositions of G.

In this paper, we prove that the problem of computing $ppw(G)$ is NP-hard for general graphs but can be solved in linear time for trees. We characterize the mixed-search number of a simple graph by means of the proper-path-width. That is, we establish the equality $ms(G) = ppw(G)$, which means that the problem of computing $ms(G)$ is NP-hard for general graphs but can be solved in linear time for trees.

2 Proper-Path-Width

Graphs we consider are nontrivial and connected, but may have loops and multiple edges unless otherwise specified. Let G be a graph, and $V(G)$ and $E(G)$ denote the vertex set and edge set of G, respectively.

Definition 1 ([17]) *Let $\mathcal{X} = (X_1, X_2, \ldots, X_r)$ be a sequence of subsets of $V(G)$. The width of $\mathcal{X}$ is $\max_{1 \le i \le r} |X_i| - 1$. $\mathcal{X}$ is called a* proper-path-decomposition *of G if the following conditions are satisfied: (i) For any distinct i and j, $X_i \not\subseteq X_j$; (ii) $\bigcup_{1 \le i \le r} X_i = V(G)$; (iii) For any edge $(u, v) \in E(G)$, there exists an i such that $u, v \in X_i$; (iv) For all l, m, and n with $1 \le l \le m \le n \le r$, $X_l \cap X_n \subseteq X_m$; (v) For all l, m, and n with $1 \le l < m < n \le r$, $|X_l \cap X_n| \le |X_m| - 2$. The* proper-path-width *of G, denoted by*

$ppw(G)$, is the minimum width over all proper-path-decompositions of G. If $\mathcal{X}$ satisfies (i), (ii), (iii), and (iv), $\mathcal{X}$ is called a path-decomposition of G. The path-width of G, denoted by $pw(G)$, is the minimum width over all path-decompositions of G.

Notice that $pw(G) \leq ppw(G) \leq pw(G) + 1$ for any graph G. It is not difficult to see the following lemma.

Lemma 1 (1) $\mathcal{X}$ satisfies condition (iv) in Definition 1 if and only if each vertex of G appears in consecutive X_i's.

(2) A path-decomposition $\mathcal{X}$ satisfies condition (v) in Definition 1 if and only if $|X_{i-1} \cap X_{i+1}| \leq |X_i| - 2$ holds for any i with $1 < i < r$.

A (proper-)path-decomposition with width k is called a k-(proper-)path-decomposition. A k-(proper-)path-decomposition $(X_1, X_2, \ldots, X_r)$ is said to be *full* if $|X_i| = k + 1$ $(1 \leq i \leq r)$ and $|X_j \cap X_{j+1}| = k$ $(1 \leq j \leq r - 1)$.

Lemma 2 If a graph G has a k-path-decomposition $\mathcal{X} = (X_1, X_2, \ldots, X_r)$ such that

$$(*) \quad |X_{i-1} \cap X_{i+1}| \leq k - 1 \ (1 < i < r),$$

then G has a full k-proper-path-decomposition.

Proof: Let $\mathcal{X} = (X_1, X_2, \ldots, X_r)$ be a k-path-decomposition of G satisfying $(*)$ such that $\sum_{i=1}^{r}(|X_i| - k)$ is maximum. We shall show that $\mathcal{X}$ is a full k-proper-path-decomposition of G. In the following, $X_j = \phi$ if $j \leq 0$ or $j > r$.

Assume that $|X_i| \leq k$ for some i $(2 \leq i \leq r)$. If $|X_{i-2} \cap X_i| = k-1$, let $v \in X_{i-1} - X_{i-2}$. Since $X_{i-1} \cap X_i = X_{i-2} \cap X_i$, $v \notin X_i$. If $|X_{i-2} \cap X_i| < k-1$, let $v \in X_{i-1} - X_i$. In either case, we have $v \notin X_i$ and $|X_{i-2} \cap (X_i \cup \{v\})| \leq k-1$. Since $v \notin X_{i+2}$, $|(X_i \cup \{v\}) \cap X_{i+2}| \leq k-1$. Thus, the sequence $\mathcal{X}' = (X_1, X_2, \ldots, X_{i-1}, X_i \cup \{v\}, X_{i+1}, \ldots, X_r)$ satisfies condition $(*)$ and conditions (ii), (iii), and (iv) in Definition 1. Assume that $X_j \subseteq X_i \cup \{v\}$ for some $j(\neq i)$. Since $v \notin \bigcup_{i+1 \leq p \leq r} X_p$, $j < i$. Thus $j = i - 1$ since $X_j = X_j \cap (X_i \cup \{v\}) \subseteq X_{i-1}$. Therefore, $(X_1, X_2, \ldots, X_{i-2}, X_i \cup \{v\}, X_{i+1}, \ldots, X_r)$ is a k-path-decomposition of G satisfying condition $(*)$. But this is contradicting to the choice of $\mathcal{X}$ since $|X_{i-1}| \leq k-1$. Thus $\mathcal{X}'$ is a k-path-decomposition of G. But again this is contradicting to the choice of $\mathcal{X}$. Thus $|X_i| = k + 1$ for any i $(2 \leq i \leq r)$. Since $(X_r, \ldots, X_1)$ is also a path-decomposition of G, $|X_i| = k + 1$ for any i $(1 \leq i \leq r)$.

Assume next that $|X_i \cap X_{i+1}| \leq k - 1$ for some i $(1 \leq i \leq r - 1)$. If $|X_{i-1} \cap X_{i+1}| = k - 1$, let $v \in X_i - X_{i-1}$. If $|X_{i-1} \cap X_{i+1}| < k - 1$, let $v \in X_i - X_{i+1}$. In either case, we have $v \notin X_{i+1}$ and $|X_{i-1} \cap (X_{i+1} \cup \{v\})| \leq k - 1$. If $|X_{i+1} \cap X_{i+2}| = k$, let $u \in (X_{i+1} \cap X_{i+2}) - X_i$. Note that $(X_{i+1} \cap X_{i+2}) - X_i \neq \phi$ since $|X_{i+1} \cap X_{i+2}| = k > k - 1 \geq |X_i \cap X_{i+1}|$. If $|X_{i+1} \cap X_{i+2}| < k$, let $u \in X_{i+1} - X_i$. In either case, we have $|(X_{i+1} - \{u\}) \cap X_{i+2}| \leq k - 1$. Since $v \notin \bigcup_{i+1 \leq j \leq r} X_j$ and $u \notin \bigcup_{1 \leq j \leq i} X_j$, the sequence $(X_1, \ldots, X_i, (X_{i+1} \cup \{v\}) - \{u\}, X_{i+1}, \ldots, X_r)$ is a k-path-decomposition of G satisfying condition $(*)$, contradicting the choice of $\mathcal{X}$. Thus $|X_i \cap X_{i+1}| = k$ for any i $(1 \leq i \leq r-1)$.

Thus, $\mathcal{X}$ is a full k-path-decomposition of G satisfying $(*)$, and so a full k-proper-path-decomposition of G by Lemma 1(2) since $|X_{i-1} \cap X_{i+1}| \leq k - 1 = |X_i| - 2$ $(1 < i < r)$. □

Lemma 3 *For any graph G with $ppw(G) = k$, there exists a full k-path-decomposition of G.*

Proof: A k-proper-path-decomposition $(X_1, X_2, \ldots, X_r)$ of G is a k-path-decomposition satisfying condition (*) in Lemma 2. Thus we obtain the lemma from Lemma 2. $\square$

A graph obtained from connected graphs H_1, H_2, and H_3 by the following construction is called a *star-composition* of H_1, H_2, and H_3: (i) Choose a vertex $v_i \in V(H_i)$ for $i = 1, 2,$ and 3; (ii) Let v be a new vertex not in $H_1, H_2,$ or H_3; (iii) Connect v to v_i by an edge (v, v_i) for $i = 1, 2,$ and 3. We define the family Ω_k of trees recursively as follows: (i) $\Omega_1 = \{K_{1,3}\}$; (ii) If Ω_k is defined, a tree T is in Ω_{k+1} if and only if T is a star-composition of (not necessarily distinct) three trees in Ω_k. A graph H is a *minor* of G if H is isomorphic to a graph obtained from a subgraph of G by contracting edges.

The following were proved by the authors in [17], in which Theorem B was used to prove Theorem A.

Theorem A ([17]) *For any tree T and an integer k ($k \geq 1$), $ppw(T) \leq k$ if and only if T contains no tree in Ω_k as a minor.*

Corollary A ([17]) *(1) The number of vertices of a tree in Ω_k is $\frac{3^{k+1}-1}{2}$ ($k \geq 1$). (2) $|\Omega_k| \geq k!^2$ ($k \geq 1$).*

Theorem B ([17]) *For any tree T and an integer k ($k \geq 1$), $ppw(T) \geq k+1$ if and only if T has a vertex v such that T/v has at least three connected components with proper-path-width k or more, where T/v is the graph obtained from T by deleting v.*

A k-*clique* of a graph G is a complete subgraph of G with k vertices. For a positive integer k, k-*trees* are defined recursively as follows: (i) The complete graph with k vertices is a k-tree; (ii) Given a k-tree Q with n vertices ($n \geq k$), a graph obtained from Q by adding a new vertex adjacent to the vertices of a k-clique of Q is a k-tree with $n + 1$ vertices. A k-tree Q is called a k-*path* if $|V(Q)| \leq k + 1$ or Q has exactly two vertices of degree k. A *partial k-path* is a subgraph of a k-path.

Theorem 1 *For any simple graph G and an integer k ($k \geq 1$), $ppw(G) \leq k$ if and only if G is a partial k-path.*

Proof: Suppose that $ppw(G) = h \leq k$. There exists a full h-proper-path-decomposition $\mathcal{X} = (X_1, X_2, \ldots, X_r)$ of G by Lemma 3. If $r = 1$ then G is a subgraph of a complete graph on $h + 1$ vertices, and so we conclude that G is a partial h-path. Thus we assume that $r \geq 2$. We construct a h-path H from $\mathcal{X}$ as follows:

(i) Let v_1 be a vertex in $X_1 \cap X_2$. Define that Q_1 is the complete graph on $X_1 - \{v_1\}$.

(ii) Define that Q_2 is the h-path obtained from Q_1 by adding v_1 and the edges connecting v_1 and the vertices in $X_1 - \{v_1\}$.

(iii) Given Q_i and the vertex $v_i \in X_i - X_{i-1}$ ($2 \leq i \leq r$), define that Q_{i+1} is the h-path obtained from Q_i by adding v_i and the edges connecting v_i and the vertices in $X_i - \{v_i\}$.

(iv) Define $H = Q_{r+1}$.

From the definition of full h-proper-path-decomposition, v_i $(2 \leq i \leq r)$ in (iii) is uniquely determined. Notice that $v_i \in X_{i+1}$ $(2 \leq i \leq r-1)$, for otherwise $|X_{i-1} \cap X_{i+1}| = h$. Since H is a h-tree and only the vertex in $X_2 - X_1$ and v_r have degree h, H is a h-path. Furthermore, we have $V(H) = V(G)$ and $E(H) \supseteq E(G)$ from the definitions of proper-path-decomposition and Q_i. Thus G is a partial h-path, and so a partial k-path.

Conversely, suppose, without loss of generality, that G is a partial h-path $(h \leq k)$ with n $(n > h)$ vertices and H is a h-path such that $V(H) = V(G)$ and $E(H) \supseteq E(G)$. It is well-known that H can be obtained as follows:

(i) Define that $Q_1 = R_1$ is the complete graph with h vertices.

(ii) Given Q_i, R_i, and a new vertex v_i $(1 \leq i \leq n - h)$, define that Q_{i+1} is the h-path obtained from Q_i by adding v_i and the edges connecting v_i and the vertices of R_i, and R_{i+1} is a h-clique of Q_{i+1} that contains v_i.

(iii) Define $H = Q_{n-h+1}$.

We define $X_i = V(R_i) \cup \{v_i\}$ $(1 \leq i \leq n-h)$ and $\mathcal{X} = (X_1, X_2, \ldots, X_{n-h})$. It is easy to see that $|X_i| = h + 1$ for any i, $\bigcup_{1 \leq i \leq n-h} X_i = V(H)$, and each vertex appears in consecutive X_i's. Thus $\mathcal{X}$ satisfies conditions (ii) and (iv) in Definition 1, and the width of $\mathcal{X}$ is h. Since $v_i \in X_i - X_{i-1}$ and $\phi \neq V(R_{i-1}) - V(R_i) \subseteq X_{i-1} - X_i$, $X_i \not\subseteq X_{i-1}$ and $X_{i-1} \not\subseteq X_i$ for any i. Thus $X_i \not\subseteq X_j$ for any distinct i and j, for otherwise $X_i = X_i \cap X_j \subseteq X_{i+1}$ $(i < j)$ or $X_i = X_i \cap X_j \subseteq X_{i-1}$ $(i > j)$. Hence $\mathcal{X}$ satisfies condition (i) in Definition 1. Since each edge of H connects v_i and a vertex in $V(R_i)$ for some i or connects vertices in $V(R_1)$, both ends of each edge of H is contained in some X_i. Thus $\mathcal{X}$ satisfies condition (iii) in Definition 1. Since $V(R_{i+1}) = X_i \cap X_{i+1}$, $|X_i \cap X_{i+1}| = |V(R_{i+1})| = h$ for any i with $1 \leq i < n - h$. Since $X_{i+1} - X_{i-1} = \{v_i, v_{i+1}\}$, $|X_{i-1} \cap X_{i+1}| = h - 1 = |X_i| - 2$ $(1 < i < n - h)$. Thus the sequence $\mathcal{X}$ is a full h-proper-path-decomposition of H from Lemma 1(2). Therefore, we have that $ppw(G) \leq ppw(H) \leq h \leq k$. $\square$

Arnborg, Corneil, and Proskurowski proved that the problem of deciding, given a graph G and an integer k, whether G is a partial k-path is NP-complete [1]. Thus we immediately have the following by Theorem 1.

Theorem 2 *The problem of computing $ppw(G)$ is NP-hard.*

It should be noted that Theorem A together with Robertson and Seymour's results on graph minors [13, 14] provides $O(n^2)$ algorithm to decide, given a tree T on n vertices, whether $ppw(T) \leq k$ for any fixed integer k, although it is not practical even if we could solve MINOR CONTAINMENT (see [5], for example) efficiently, because $|\Omega_k| \geq k!^2$ as is shown in Corollary A(2).

We show a practical algorithm to compute $ppw(T)$ for trees T based on Theorem B, and prove the following.

Theorem 3 *For any tree T, the problem of computing $ppw(T)$ is solvable in linear time.*

Proof: Our algorithm to compute $ppw(T)$ is shown in Fig. 1. The outline of the algorithm is as follows.

For any tree T with a vertex $v \in V(T)$ as the root, we define the path-vector $\overline{pv}(v,T) = (p_v, c_v, S_v)$. p_v describes the proper-path-width of T. c_v and S_v describe the condition of T as follows: If there exists $u \in V(T) - \{v\}$ such that T/u has two connected components with proper-path-width p_v and without v, then $c_v = 3$ and S_v is the path-vector of the connected component of T/u containing v; Otherwise, c_v is the number of the connected components of T/v with proper-path-width p_v and $S_v = nul$. Note that for any vertex u the number of connected components of T/u with proper-path-width p_v is at most two from Theorem B. Notice also that if there exists u such that T/u has two connected components with proper-path-width p_v and without v then u is uniquely determined. If there is no such u then the number of connected components of T/w with proper-path-width p_v and without v is not more than the number of connected components of T/v with proper-path-width p_v.

Suppose that a tree T_0 rooted at s is obtained from tree T_1 rooted at s and tree T_2 rooted at t by adding an edge (s,t). Based on Theorem B, Procedure MERGE recursively calculates the path-vector $\overline{pv}(s,T_0)$ of T_0 from the path-vector $\overline{pv}(s,T_1) = (p_s, c_s, S_s)$ of T_1 and the path-vector $\overline{pv}(t,T_2) = (p_t, c_t, S_t)$ of T_2.

Procedure DFS computes the path-vector of a maximal subtree rooted at s in T from the path-vectors of maximal subtrees rooted at children of s in T by using Procedure MERGE. Procedure MAIN obtains the proper-path-width of T from the path-vector of T obtained by Procedure DFS. The algorithm starts with the isolated vertices obtained from T by deleting all edges in T and obtains path-vector of T.

Procedure MERGE calculates the path-vector of the join of two subtrees T_1 and T_2 in $O(p)$ time where $p = \max(ppw(T_1), ppw(T_2))$. Note that the time complexity of Procedure MERGE is $O(1)$ except for recursive calls. Since the larger proper-path-width of two merged trees is reduced by at least one whenever Procedure MERGE is recursively called, the number of recursive calls is at most p.

From Corollary A(1), we have $p = O(\log n)$ where $n = |V(T)|$. Since Procedure MERGE is called at most once for any vertex, the time complexity of the algorithm is essentially $O(n \log n)$. By a careful use of pointers, same technique as used in [9], Procedure MERGE calculates the path-vector in $O(q)$ time where $q = \min(ppw(T_1), ppw(T_2))$. Thus we can prove that the time complexity of the algorithm is $O(n)$ time (See [9] for the proof). $\square$

We should mention that for any tree T with n vertices and $ppw(T) = k$, we can construct in $O(n \log n)$ time a k-proper-path-decomposition of T by a slight modification of the algorithm shown in Fig 1.

3 Mixed-Searching

In the *mixed-searching game*, a graph G is considered as a system of tunnels. Initially, all edges are contaminated by a gas. An edge is *cleared* by placing searchers at both its ends simultaneously or by sliding a searcher along the edge. A cleared edge is *recontaminated*

```
Procedure MERGE( pv(s,T₁),pv(t,T₂) )
{ input:pv(s,T₁),pv(t,T₂) }
{ output:pv(s,Tb) }
{ pv(tmp) = (p_tmp, c_tmp, s_tmp) }

 1. if p_s > p_t then
       if c_s ≤ 2 then
          pv(s,Tb) = pv(s,T₁);
    else
          pv(tmp) = MERGE( S_s, pv(t,T₂) );
          if p_s = p_tmp then
             pv(s,Tb) = (p_s + 1,0,nul);
       else
             pv(s,Tb) = (p_s,3,pv(tmp));
          endif
    endif

 2. if p_s = p_t then
       if c_s ≥ 2 or c_t ≥ 2 then
          pv(s,Tb) = (p_s + 1,0,nul);
       else if c_s = 0 then
          pv(s,Tb) = (p_s,1,nul);
       else if c_s = 1 then
          pv(s,Tb) = (p_s,2,nul);
    endif

 3. if p_s < p_t then
       if c_t ≤ 1 then
          pv(s,Tb) = (p_t,1,nul);
       else if c_t = 2 then
          pv(s,Tb) = (p_t,3,pv(s,T₁));

    else if c_t = 3 then
          pv(tmp) = MERGE( pv(s,T₁), S_t );
          if p_t = p_tmp then
             pv(s,Tb) = (p_t + 1,0,nul);
       else
             pv(s,Tb) = (p_t,3,pv(tmp));
          endif
    endif

 4. return( pv(s,Tb) );
    end

Procedure DFS( s ).
{ input: a vertex s }
{ output: the path-vector of the maximal subtree
                                     rooted at s }
 1. pv(s) = (1,0,nul);
 2. for all children t of s in T do
       pv(t) = DFS( t );
       pv(s) = MERGE( pv(s),pv(t) );
    endfor
 3. return( pv(s) );
    end

Procedure MAIN( T, r )
{ input: a tree T with a vertex r as the root }
{ output: proper-path-width ppw(T) }
 1. (p_r,c_r,S_r) = DFS( r );
 2. return( p_r );
    end
```

Figure 1: The algorithm to compute $ppw(T)$

if there is a path from an uncleared edge to the cleared edge without any searchers on its vertices or edges.

Definition 2 *A search is a sequence of the following operations: (a) placing a new searcher on a vertex; (b) deleting a searcher from a vertex;, (c) sliding a searcher on a vertex along an incident edge and placing the searcher on the other end; (d) sliding a searcher on a vertex along an incident edge; (e) sliding a new searcher along an edge and placing the searcher on its end; (f) sliding a new searcher along an edge.*

The object of mixed-searching game is to clear all edges by a search. We call such a search a *mixed-search*. A mixed-search is *optimal* if the maximum number of searchers on G at any point is as small as possible. This number is called the *mixed-search number* of G, and denoted by $ms(G)$.

We first show a relation to the edge-searching and node-searching: for any graph G, $es(G)-1 \leq ms(G) \leq es(G)$ and $ns(G)-1 \leq ms(G) \leq ns(G)$. The edge-search and node-search are special cases of the mixed-search by definition. Thus we have $ms(G) \leq es(G)$ and $ms(G) \leq ns(G)$. Using at most one more searcher to traverse an edge that is cleared by placing searchers at both its ends, we can convert any mixed-search to an edge-search. Thus $es(G) \leq ms(G) + 1$. Similarly, using at most one more searcher to clear an edge that is cleared by sliding a searcher along the edge, we can convert any mixed-search to a node-search. Thus $ns(G) \leq ms(G) + 1$. All four cases are possible as shown in Fig. 2.

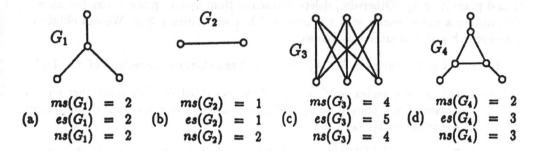

$$
\begin{array}{cccc}
ms(G_1) = 2 & ms(G_2) = 1 & ms(G_3) = 4 & ms(G_4) = 2 \\
\text{(a)} \quad es(G_1) = 2 & \text{(b)} \quad es(G_2) = 1 & \text{(c)} \quad es(G_3) = 5 & \text{(d)} \quad es(G_4) = 3 \\
ns(G_1) = 2 & ns(G_2) = 2 & ns(G_3) = 4 & ns(G_4) = 3
\end{array}
$$

Figure 2: Search numbers of graphs

A *crusade* in G, introduced by Bienstock and Seymour [2], is a sequence $(C_1, C_2, \ldots, C_r)$ of subsets of $E(G)$, such that $C_1 = \phi$, $C_r = E(G)$, and $|C_i - C_{i-1}| \leq 1$ for $1 \leq i \leq r$. The crusade uses at most k searchers if the number of vertices which are ends of an edge in C_i and also of an edge in $E(G) - C_i$ is at most k for $1 \leq i \leq r$. Bienstock and Seymour proved the following theorem.

Theorem C ([2]) *For any graph G with minimum degree at least two, $ms(G) \leq k$ if and only if there exists a crusade in G using at most k searchers.*

Moreover, they proved the following theorem by using the crusade.

Theorem D ([2]) *For any graph, there exists an optimal mixed-search without recontamination of cleared edges.*

This was proved independently by the authors in [18] using an optimal node-search without recontamination of cleared edges.

We obtain the following corollary from Theorem D.

Corollary 1 *For any graph G, there exists an optimal mixed-search without recontamination of cleared edges such that it is a sequence of operations (a), (b), or (c) of Definition 2, and satisfying the following two conditions:*

(i) every vertex is visited exactly once by a searcher,

(ii) every edge is visited at most once by a searcher.

A mixed-search described above is said to be *simple*.

Bienstock and Seymour characterized the mixed-search number of a graph with minimum degree at least two by the concept of crusade as shown in Theorem C. In the following, we characterize the mixed-search number of a simple graph by the proper-path-width.

Theorem 4 *For any simple graph G, $ms(G) = ppw(G)$.*

Proof: Suppose that $ppw(G) = k$ and $\mathcal{X} = (X_1, X_2, \ldots, X_r)$ is a full k-proper-path-decomposition of G. If $r = 1$ then let v_1 and u_1 be distinct vertices in X_1 and place k searcher on the vertices of $X_1 - \{v_1\}$. If $(u_1, v_1) \in E(G)$, slide a searcher on u_1 to v_1 and place it on v_1. Otherwise, delete a searcher from u_1 and place a searcher on v_1. This defines a mixed-search with k searchers. Thus we assume $r \geq 2$. We can obtain a mixed-search with k searchers as follows:

Step 1: Let v_1 be a vertex in $X_1 \cap X_2$. Place the k searchers on the vertices of $X_1 - \{v_1\}$.

Step 2: Let u_1 be a vertex in $X_1 - X_2$. If $(u_1, v_1) \in E(G)$, slide a searcher on u_1 toward v_1 and place it on v_1. Otherwise, delete a searcher from u_1 and place a searcher on v_1. Let $i = 1$.

Step 3: Repeat Step 3 while $i \leq r - 2$. Let $i = i + 1$. Let u_i be a vertex in $X_i - X_{i+1}$ and v_i be a vertex in $X_i - X_{i-1}$. If $(u_i, v_i) \in E(G)$, slide a searcher on u_i toward v_i and place it on v_i. Otherwise, delete a searcher from u_i and place a searcher on v_i.

Step 4: Let u_r be a vertex in $X_{r-1} \cap X_r$, and v_r be a vertex in $X_r - X_{r-1}$. If $(u_r, v_r) \in E(G)$, slide a searcher on u_r toward v_r and place it on v_r. Otherwise, delete a searcher from u_r and place a searcher on v_r.

From the definition of full k-proper-path-decomposition, both u_i $(1 \leq i \leq r - 1)$ and v_i $(2 \leq i \leq r)$ are uniquely determined. It should be noted that $((X_i - \{v_i\}) - \{u_i\}) \cup \{v_i\} = X_i \cap X_{i+1} = X_{i+1} - \{v_{i+1}\}$ and $u_{i+1} \in X_{i+1} - \{v_{i+1}\}$ for $1 \leq i \leq r - 1$. An edge with both its ends in $X_i - \{v_i\}$ $(1 \leq i < r)$ is cleared since the vertices in $X_i - \{v_i\}$ have searchers simultaneously in Step 1, 2, or 3. Also, an edge with both its ends in $X_r - \{u_r\}$ is cleared since the vertices in $X_r - \{u_r\}$ have searchers simultaneously in Step 4. Since G is simple, there exists at most one edge connecting u_i and v_i $(1 \leq i \leq r)$, and each edge (u_i, v_i), if exists, is cleared by sliding a searcher along the edge. Thus all edges are cleared at least once. Suppose that all edges connecting the vertices in $\bigcup_{1 \leq j \leq i-1} X_j$ are clear and k searchers are placed on the vertices in $X_i - \{v_i\}$. Since $u_i \notin \bigcup_{i+1 \leq j \leq r} X_j$, all edges incident to u_i except for (u_i, v_i), if exists, are clear when a searcher on u_i is deleted or slid from u_i. Thus, when the searcher is placed on v_i all edges in $\bigcup_{1 \leq j \leq i} X_j$ are clear and k searchers are placed on the vertices in $X_{i+1} - \{v_{i+1}\}$. Thus by induction no edge is recontaminated. Thus the search above is indeed a mixed-search with at most $ppw(G)$ searchers, and we have $ms(G) \leq ppw(G)$.

Conversely, suppose that we have a simple mixed-search S with k searchers. For the i-th operation of S, we define X_i as follows:

(1) When a searcher is placed on (deleted from) a vertex, we define X_i as the set of vertices having searchers.

(2) When a searcher is slid from u to v, we define X_i as the set consisting of u, v, and the vertices having searchers.

Let $\mathcal{X} = (X_1, X_2, \ldots, X_s)$ be the resulting sequence of sets of vertices. Since both ends of an edge which is cleared in the i-th operation are contained in X_i, all edges are contained in some X_i. Since S is simple, $\bigcup_{1 \leq i \leq s} X_i = V(G)$ and each vertex of G appears in consecutive X_i's. By the definition of X_i, $|X_i| \leq k + 1$ for any i. Let $\mathcal{X}' = (X_1', X_2', \ldots, X_r')$ be a maximal subsequence of $\mathcal{X}$ such that $X_i' \not\subseteq X_j'$ for any distinct i and j. Notice that $\mathcal{X}'$ satisfies the conditions (i), (ii), (iii), and (iv) in Definition 1. We shall show that k-path-decomposition $\mathcal{X}'$ satisfies condition (∗) in Lemma 2. If one of X_{i-1}', X_i', and X_{i+1}' is defined by (1), it is easy to see that $|X_{i-1}' \cap X_{i+1}'| \leq k - 1$. If all X_{i-1}', X_i', and X_{i+1}' are defined by (2), then $|X_i'| \leq k + 1$ and there exist distinct u and v in X_i' such that $u \notin X_{i+1}'$, and $v \notin X_{i-1}'$. Thus we have $|X_{i-1}' \cap X_{i+1}'| \leq k - 1$. Therefore, $\mathcal{X}'$ satisfies condition (∗) in Lemma 2, and there exists a full k-proper-path-decomposition of G by Lemma 2. Thus $ppw(G) \leq ms(G)$. □

It should be noted that Theorems A and 4 provide a structural characterization of trees T with $ms(T) \leq k$.

From Theorems 2, 3, and 4, we have the following complexity results on $ms(G)$.

Theorem 5 *The problem of computing $ms(G)$ is NP-hard for general graphs but can be solved in linear time for trees.*

We conclude with the following remarks:

1. Notice that Theorem 4 does not hold for multiple graphs. If G is the graph consisting of two parallel edges, $ppw(G) = 1$, and $ms(G) = 2$. However we can prove that $ppw(G) \leq ms(G) \leq ppw(G) + 1$ for any multiple graph G.

2. Bodlaender and Kloks showed an $O(n \log^2 n)$ time algorithm to decide whether $pw(G) \leq k$ for any graph G and a fixed integer k [3]. We can modify their algorithm to decide whether $ppw(G) \leq k$ for any graph G and a fixed integer k.

3. A relation between the mixed-searching and another searching game, called virus-searching [16], was mentioned in [18].

References

[1] S. Arnborg, D. G. Corneil, and A. Proskurowski, Complexity of finding embeddings in a k-tree, *SIAM J. Alg. Disc. Meth.*, 8(2), pp. 277–284, April 1987.

[2] D. Bienstock and P. Seymour, Monotonicity in graph searching; *Journal of Algorithms*, 12(2), pp. 239–245, 1991.

[3] H. L. Bodlaender and T. Kloks, Better algorithm for the pathwidth and treewidth of graphs, 1991, manuscript.

[4] R. L. Breisch, An intuitive approach to speleotopology, *Southwestern Cavers* (published by the Southwestern Region of the National Speleological Society), 6(5), pp. 72–78, 1967.

[5] D. S. Johnson, The NP-completeness column: an ongoing guide, *Journal of Algorithms*, 8, pp. 285–303, 1987.

[6] L. M. Kirousis and C. H. Papadimitriou, Interval graphs and searching, *Discrete Mathematics*, 55, pp. 181–184, 1985.

[7] L. M. Kirousis and C. H. Papadimitriou, Searching and pebbling, *Theoretical Computer Science*, 47, pp. 205–218, 1986.

[8] A. LaPaugh, *Recontamination does not help to search a graph*, Technical Report, Electrical Engineering and Computer Science Department, Princeton University, 1983.

[9] M. Megiddo, S. L. Hakimi, M. R. Garey, D. S. Johnson, and C. H. Papadimitriou, The complexity of searching a graph, *Journal of the Association for Computing Machinery*, 35(1), pp. 18–44, January 1988.

[10] R. H. Möhring, Graph problems related to gate matrix layout and PLA folding, in G. Tinhofer, E. Mayr, H. Noltemeier, and M. Syslo, editors, *Computational Graph Theory*, pp. 17–51, Springer-Verlag, Wien New York, 1990.

[11] T. D. Parsons, Pursuit-evasion in a graph, in Y. Alavi and D. Lick, editors, *Theory and Applications of Graphs*, pp. 426–441, Springer-Verlag, Berlin, 1976.

[12] N. Robertson and P. D. Seymour, Graph minors. I. Excluding a forest, *Journal of Combinatorial Theory*, Series B(35), pp. 39–61, 1983.

[13] N. Robertson and P. D. Seymour, Graph minors. XIII. The disjoint paths problem, 1986, preprint.

[14] N. Robertson and P. D. Seymour, Graph minors. XVI. Wagner's conjecture, 1987, preprint.

[15] P. Scheffler, A linear algorithm for the pathwidth of trees, in R. Bodendiek and R. Henn, editors, *Topics in Combinatorics and Graph Theory*, pp. 613–620, Physica-Verlag, Heidelberg, 1990.

[16] S. Shinoda, On some problems of graphs — including Kajitani's conjecture and its solution —, in *Proc. of 2nd Karuizawa Workshop on Circuits and Systems*, pp. 414–418, 1989, in Japanese.

[17] A. Takahashi, S. Ueno, and Y. Kajitani, Minimal acyclic forbidden minors for the family of graphs with bounded path-width, to appear in *Annals of discrete mathematics (Proceedings of 2nd Japan conference on graph theory and combinatorics, 1990)*. Also: *SIGAL* 91–19–3, IPSJ, 1991.

[18] A. Takahashi, S. Ueno, and Y. Kajitani, *Mixed-searching and proper-path-width*, Technical Report SIGAL 91–22–7, IPSJ, 1991.

Short Wire Routing in Convex Grids*

Frank Wagner   Barbara Wolfers

Institut für Informatik, Fachbereich Mathematik, Freie Universität Berlin,
Arnimallee 2–6, W1000 Berlin 33, Germany

Abstract

Knock-knee routing of two-terminal nets is mathematically modelled and successfully solved
in a lot of settings by multicommodity-flow techniques. The primary goal of routing, the fast
construction of a solution whenever it exists at all, is thus well understood and controlled.

A major drawback of these provably good algorithms is that they fail to reach any of the
secondary optimization criteria, wire length and number of bends. We identify a major reason
for this and suggest an empirically very successful way to overcome this fundamental weakness
in case of a rather general shape of the routing region, the convex grids.

The resulting new algorithm shares the good properties of the old one, it solves whenever it
is possible and is as efficient.

1 Introduction

One of the central steps in all existing systems for the layout of VLSI circuits is the solution of the
routing problem. Given a so-called routing *region*, *terminals* are to be connected in order to realize
given *nets*. The routing region is a finite subgraph of the infinite rectangular grid, a terminal is a
vertex of the routing region, a net, here two-terminal net, is a pair of two terminals.

The solution of the routing problem is a *layout*, i.e. each net is realized by a path through
the routing region connecting the terminals. In the application context of VLSI design the paths
correspond to wires between modules which are placed on a chip. Two paths have to be edge disjoint
(*knock-knee mode*). The *wiring* is done using a small number of *layers*, each wire *segment* (used edge
of the grid graph) is assigned to a layer such that conflicts between wires are avoided.

In general, as shown by [Kaufmann and Maley], it is $\mathcal{NP}$-complete to decide if there is a solution
for a given routing problem. On the other hand there are efficient routing procedures for large classes
of routing problems which include a major part of the problems of practical interest.

The routing problem can be considered as an integer *multicommodity-flow* problem on planar graphs. [Okamura and Seymour] describe implicitely an efficient solution procedure. This
is the core of a whole class of routing algorithms by [Frank], [Kaufmann and Mehlhorn],
[Nishizeki, Saito and Suzuki] and [Becker and Mehlhorn].

All of these find a solution, if there is one, but they tend to ignore and thus miss the secondary
optimization criteria, the total wire length, the length of the longest wire and the number of knock-
knees (corresponds to the number of vias and layers needed in the wiring phase).

Only in the case of *channels*, where the routing region is a rectangle and every net has one
terminal on the upper and one on the lower border, there are algorithms reaching (partially) both
goals mentioned above:

[Preparata and Lipski] and [Kuchem, Wagner and Wagner] bound the number of knock-knees and
as a consequence the number of layers needed, [Sarrafzadeh] and [Formann, Wagner and Wagner] the
wire lengths.

*Part of this research was done while the authors were with Lehrstuhl für angewandte Mathematik insbesondere
Informatik, Rheinisch-Westfälische Technische Hochschule Aachen

The algorithm presented in this paper works for the rather general case of *convex* routing regions, as described by [Nishizeki, Saito and Suzuki], i.e. between every pair of vertices of the routing region there is a connecting path with at most one bend; in particular there are no (nontrivial) holes. The terminals are allowed to be located anywhere on the border of the routing region. There is a slight difference to the rules for channels (at most one terminal per boundary point) concerning the usage of a boundary point for terminals of different nets:

Convex corners may be used by at most two terminals, concave corners by no terminal at all. See Figures 5,6 and 7 for typical examples of convex routing problems.

The paper is organized as follows:

In Section 2 we will give a simplified description of the algorithm by [Nishizeki, Saito and Suzuki]. In Section 3 we identify a major reason for its missing of the secondary optimization criteria and suggest an empirically very successful way to overcome this weakness.

2   An algorithm for convex grids

Firstly we need some terminology: The routing region is mathematically a planar graph $G = (V, E)$. If X is an arbitrary subset of the vertex set V, the *capacity*, *cap(X)*, is the number of edges with one endpoint in X and one in $V \backslash X$, the *density*, *dens(X)*, is the number of nets with one terminal in X and one in $V \backslash X$; the *free capacity*, *fcap(X)*, is defined as $cap(X) - dens(X)$.

[Okamura and Seymour] prove that the routing problem in planar graphs with all the terminals lying on the border (infinite face) is solvable if

$$(*) \qquad\qquad \forall X \subseteq V : fcap(X) \text{ is nonnegative and even.}$$

Trivially the condition $fcap(X) \geq 0$ is a necessary condition for the solvability of the routing problem. It is usually called the *cut condition*.

The algorithm of [Nishizeki, Saito and Suzuki] for convex grids can be traced back to the central theorem of [Okamura and Seymour]. It can be shown that for convex grids (*) is equivalent to the condition

$$(**) \qquad \begin{array}{lll} \forall v \in V & \text{lying on the border:} & fcap(\{v\}) \text{ is even} \\ \text{and } \forall X \subseteq V & \text{which are straight:} & fcap(X) \text{ is nonnegative.} \end{array}$$

A subset X is *straight* iff V can be divided into X and $V \backslash X$ by a straight horizontal or vertical line.

In addition they show that every solvable problem can be modified by adding so-called *dummy nets* such that every vertex on the border fulfills (**). The proposed procedure for the addition of dummy nets is not completely correct. [Lai and Sprague] fixed this mistake.

Their algorithm is very efficient, its running time is bounded by a linear function in the *area* (number of vertices) of the routing region.

After ensuring condition (**) the algorithm runs in principle as follows: The grid is scanned vertex by vertex in a row-wise fashion. For a certain vertex lying on a convex *corner*, i.e. a grid vertex of degree two, the algorithm checks if the two edges leaving the corner are needed by any solution path at all. If yes it determines for which solution path the edges have to be reserved. Then the corner and the two edges are deleted and the terminals are rearranged such that (**) continues to hold for the modified smaller routing problem. If the two edges were not needed, the addition of a dummy net is necessary to guarantee the condition. The solution of the given routing problem is then a combination of the two possibly reserved edges and the solution of the smaller subproblem. Scanning the whole grid we inductively get a layout.

The correct version of the routing algorithm sketched above finds the disjoint connecting paths whenever the problem is solvable at all. Unfortunately, the solutions in general have the crucial disadvantage that the proposed wires have a unnecessarily large length and a huge number of avoidable knock-knees occur.

Now a less informal description of the algorithm:

First some notations

UL:	leftmost point on the uppermost line
UR:	rightmost point on the uppermost line
L:	the set of all nodes not to the right of UL
U:	the set of all nodes on the uppermost line
BUL:	the point directly below UL
NUL:	the next point to the right of UL

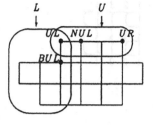

Algorithm ROUTING

(1)   if there are boundary points with odd free capacity then
(2)   add appropriate dummy nets
(3)   if $fcap(X) < 0$ for some straight X then
(4)   stop (routing problem not solvable)
(5)   while there is more than one horizontal line do
(6)   if UL has 2 terminals then
(7)   LEFT_CORNER
(8)   else if UR has 2 terminals then
(9)   RIGHT_CORNER
(10)   else if $fcap(L) > 0$ then
(11)   DUMMY
(12)   else find an APPROPRIATE_NET
(13)   delete UL resp. UR (and the two incident edges)
(14) for all nets with terminals on the last line
(15)   connect the terminals of a net by the sequence of the edges between them

Description of the subroutines:

Assume that the upper (shortest) line consists of more than two nodes (the other case is handled analogously).

LEFT_CORNER: UL has the two terminals i_1 and j_1 of nets $i = \{i_1, i_2\}$ and $j = \{j_1, j_2\}$. Assume that UL, i_2 and j_2 appear clockwise on the boundary in this order. The edges of node UL are reserved for the i-path and j-path.
   (1) $i_1 := $ NUL
   (2) $j_1 := $ BUL

RIGHT_CORNER: mirrorsymmetric to LEFT_CORNER

DUMMY:
   No solution path has to use UL. To preserve the evenness condition create a dummy net $d = \{d_1, d_2\}$.
   (1) $d_1 := $ NUL
   (2) $d_2 := $ BUL

APPROPRIATE_NET: $fcap(L)=0$ hence it follows that UL and the incident edges have to be used by an appropriate net which has to be determined. This net must have one terminal in L and one in $V \backslash L$. If there is such a net which in addition has a terminal on one of the two upper rows choose one of them arbitrarily; if not choose the first terminal i_1 of such a net which appears clockwise on the boundary after UL, i. e. the smallest in the sense of $\prec$. Reserve the two edges incident to UL and replace net i by two nets i' and i'' as follows:
   (1) $i'_1 := i_1$
   (2) $i'_2 := $ NUL
   (3) $i''_1 := i_2$
   (4) $i''_2 := $ BUL

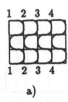
a)

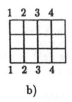
b)

Figure 1:

Figures 1a) and 1b) describe a trivial example exhibiting the typically constructed detours and an optimal solution in terms of wire length and number of knock-knees. The suboptimal wire lengths are in this case more than twice the minimum! In addition they have a lot of bends thereby producing a lot of knock-knees. This is for a lot of technological reasons a property which is not desired at all.

3 Improving the layout quality

A main reason for this behaviour is the handling of the dummy nets. The routing procedure does not distinguish between dummy and "real" nets. It produces a correct layout for both of them and wipes the dummy nets away leaving the paths realizing the real nets. For this reason it often happens that a real wire has a lot of detours caused by dummy nets (c.f. Figures 1a) and 1b)).

This happens especially if the routing region is much larger than necessary, that means if the problem instance is relatively easy and the lengths of the wires could be very small. In this case typically a lot of dummy nets have to be added to fulfill the evenness condition (line 2 of ROUTING). Additional dummy nets are introduced in procedure DUMMY (line 11 of ROUTING). This happens whenever there is a lot of free capacity.

Our algorithm avoids this behaviour from the beginning by systematically suppressing the dummy nets as much as possible. We utilize the implicit nondeterminism of the algorithm so that in case of choice between a dummy and a real net the real net has always priority as long as condition (∗∗) can be guaranteed.

(i) A simple but very effective modification can be done in LEFT_CORNER and RIGHT_CORNER. We describe the method by looking at LEFT_CORNER (RIGHT_CORNER may be handled mirrorsymmetricly). There are 2 terminals i_1 and j_1 at the upper left corner UL. In LEFT_CORNER we were sure that UL, i_2, j_2 appeared clockwise in this order on the boundary (denoted by $i_2 \preceq j_2$ in the following). If both, i and j, are real nets or both are dummy nets we do not change LEFT_CORNER.

But assume that there is a terminal k_1 of a real net k and a terminal d_1 of a dummy net d at UL. The two edges {UL,BUL} and {UL,NUL} have to be reserved for the wires of k and d. The decision which edges should belong to which wire is made by us only in dependence on the position of k_2, the other terminal of the real net. We decide as follows:

(1) **if** k_2 lies in L **then**
    k moves down and d to the right
(2) **else if** k_2 lies on the upper line **then**
    d moves down and k to the right
(3)   **else** move the terminals as in the original algorithm

We have to ensure that (∗∗) holds on:

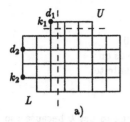

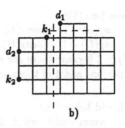

Figure 2: The interesting case: $d_2 \succ k_2$

(1): (k_2 lies in L)

The wire of net k should go down and not right because k_2 lies not to the right of UL but below UL.

$d_2 \preceq k_2$: We perform the same assignments as in the original LEFT_CORNER (which produces a correct solution).

$d_2 \succ k_2$: We have to show that $fcap'(X) \geq 0$ for all straight X and $fcap'(\{v\})$ even for all boundary nodes ($fcap'(X)$ is the modified $fcap(X)$ (after execution of LEFT_CORNER with our modification)).

$fcap'(X) \geq 0$: $fcap'(X) \neq fcap(X) \Rightarrow X = L$ or $X = U$

$$fcap'(L) = cap'(L) - dens'(L)$$
$$= cap(L) - 1 - (dens(L) + 1) = fcap(L) - 2.$$

But also in the original version of LEFT_CORNER we had in this case $fcap'(L) = fcap(L) - 2$. It follows $fcap'(L) \geq 0$.

$fcap'(U) = cap(U) - 1 - (dens(U) - 1) = fcap(U) \geq 0$.

$fcap'(\{v\})$ even: This is easy to see because $fcap'(\{v\})$ remains unchanged except $v = $ NUL and $v = BUL$.

$$fcap'(\{NUL\}) = cap(\{NUL\}) - 1 - (dens(\{NUL\}) + 1)$$
$$= fcap(\{NUL\}) - 2$$
$$fcap'(\{BUL\}) = fcap(\{BUL\}) - 2$$

It follows that the problem with the above modification remains solvable.

(2): (k_2 on the upper line)

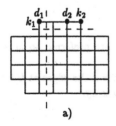

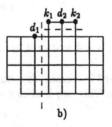

Figure 3: The interesting case: $d_2 \prec k_2$

In the original procedure LEFT_CORNER the wire of net k would use edge $\{UL, BUL\}$ if $d_2 \prec k_2$. That means it would detour because k_1 and k_2 both lie on the upper line. Instead of this it should use the direct connection with edge $\{UL, NUL\}$, that means k_1 moves to NUL, d_1 to BUL. With this modification the problem remains solvable:

$fcap'(\{v\})$ even (see (1))

$$fcap'(L) = cap(L) - 1 - (dens(L) - 1) = fcap(L) \geq 0$$
$$fcap'(U) = cap(U) - 1 - (dens(U) + 1) = fcap(U) - 2 \geq 0$$

$fcap(U) \geq 2$ because there are at most $cap(U) + 2$ terminals on the upper line and 4 of them, k_1, k_2, d_1, d_2, cannot contribute to $dens(U)$.

(3): (k_2 right of UL and below UL)

In this case we cannot decide which direction is the best referring to net k because the algorithm only makes local decisions. To ascertain solvability the algorithm proceeds like the original LEFT_CORNER.

The modification of the algorithm has the effect that the wires run as long as possible in the direction of their goal.

(ii) If the problem is "hard" that means there are many nets in a small region, for example if the algorithm proceeds to the end of the execution, subroutine APPROPRIATE_NET is very often executed since there is nearly no free capacity anymore.

If there is an appropriate net with a terminal on one of the two upper rows (a net in TOP) we do not just choose one of them arbitrarily. Instead we distinguish between three different types of such nets:

TOP1: one terminal t (the right) of the net $\{s, t\}$ is on the uppermost line, the net is a real net and there is no real net with both terminals on the uppermost line not to the right of t.

TOP2: the net is a dummy net.

TOP3: all other such nets.

Subroutine APPROPRIATE_NET could choose any net in TOP but it is more useful to give nets in TOP1 priority to nets in TOP2 and those priority to nets in TOP3.

TOP1: A net in TOP1 may be chosen without being harmful to the solution because a shortest connection of s and t may use the edges at UL; the second part of the condition avoids the enforcing of the detour of another net.

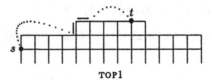

TOP1

TOP2: Because it does not matter if dummy nets detour or not they are preferred to nets in TOP3.

TOP3: If a net in TOP3 is choosen, one can easily see that it has to detour if the edges at UL are reserved to it. This may lead to a wire which is longer than necessary.

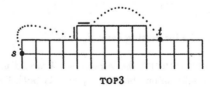

TOP3

The additional amount of running time of (i) is bounded by a constant number of additional comparisons and assignments in LEFT_CORNER and RIGHT_CORNER and no additional space is used.

In (ii) TOP1, TOP2 and TOP3 have to be built and updated. To guarantee the same (linear) time bound we instead organize these nets in the three data structures TOP1', TOP2 and TOP3'. TOP3' contains only those nets from TOP3 that have no terminal on the uppermost line, TOP1' all other real nets. Both primed datastructures are doubly linked lists of nets with pointers to the corresponding grid points of the terminals. They are built before the execution of a row and updated in constant time per step. The order of the list TOP3' is according to the position of the first of a nets two terminals with respect to $\prec$. In TOP1' we need an ordering of the nets according to the position of its terminals on the uppermost line because we look for the net $\{s_0, t_0\}$ in TOP1' with the leftmost terminal on the upper line. For this terminal t_0 we test if there is no net with both terminals to the left of it on the upper line. If there is no then net $\{s_0, t_0\}$ is in TOP1 else there is no net in TOP1 and $\{s_0, t_0\}$ is in TOP3. This can be done in constant time if we use arrays of length l (where l is the length of the upper line) and some pointers to store the nets with both terminals on the upper line and TOP1'. Note that nets may appear and disappear in TOP1, TOP2 and TOP3 after an execution of APPROPRIATE_NET or the other subroutines. But this modifications can be done in constant time. The required place for the data structures is $\mathcal{O}(n)$ where n is the number of vertices (an execution of a line with length l needs $\mathcal{O}(l)$ place).

Our algorithm produces in linear time a layout with short and relatively straight wires whenever it exists.

Comparing the algorithms on a lot of examples of different type and size, it turned out that the layouts of the new algorithm are of much higher quality in various aspects. Our algorithm produced for every example (cf. Figures 1, 5, 6 and 7) much shorter wires and dramatically less knock-knees in comparison with the algorithm of [Nishizeki, Saito and Suzuki]. In no case at all a deterioration of the layout properties was observed. Often the result is optimal in terms of wire length and knock-knees. The algorithm finds a solution if there is one. But in contrast to the algorithm of Nishizeki et. al., which tends fills the whole routing region with wires although it may not be necessary, the solution of our algorithm indicates if there is wasted room in the routing region. This information could be used to move the modules together in a new placement procedure and to save place.

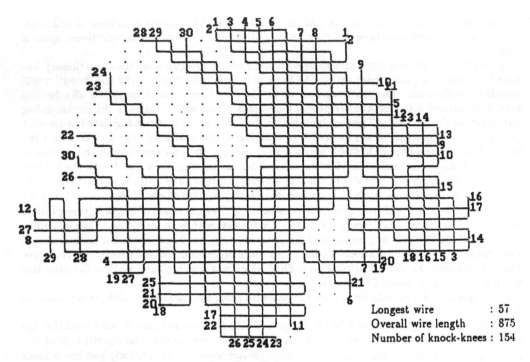

Figure 5a)

Longest wire   : 57
Overall wire length   : 875
Number of knock-knees : 154

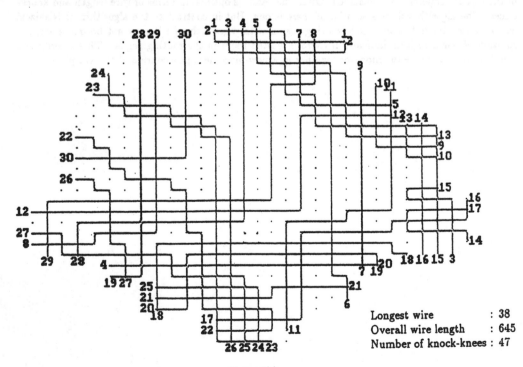

Longest wire   : 38
Overall wire length   : 645
Number of knock-knees : 47

Figure 5b)

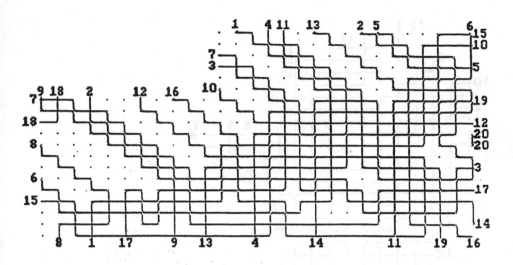

Figure 6a)

Longest wire   : 52
Overall wire length   : 598
Number of knock-knees : 92

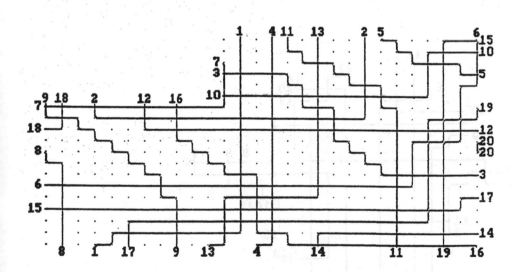

Figure 6b)

Longest wire   : 42
Overall wire length   : 410
Number of knock-knees : 5

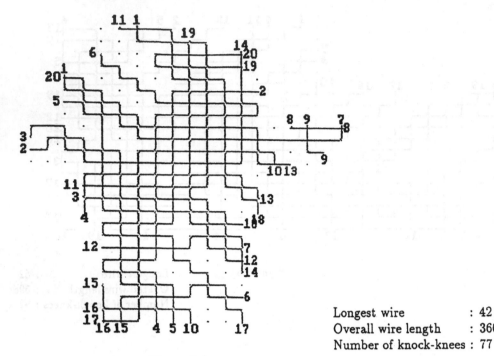

Longest wire : 42
Overall wire length : 360
Number of knock-knees : 77

Figure 7a)

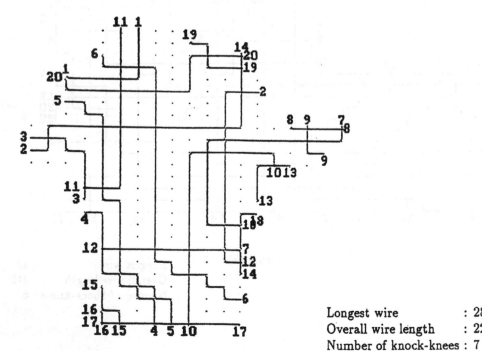

Longest wire : 28
Overall wire length : 228
Number of knock-knees : 7

Figure 7b)

References

[Brady and Brown] M. L. Brady, D. J. Brown: *VLSI-Routing: Four Layers Suffice*, 245-257, in: F. P. Preparata (Ed.), Advances in Computing Research: VLSI Theory, JAI Press, London, England (1984)

[Becker and Mehlhorn] M. Becker, K. Mehlhorn: *Algorithms for Routing in Planar Graphs*, Acta Informatica 23 (1986) 163-176

[Formann, Wagner and Wagner] M. Formann, D. Wagner, F. Wagner: *Routing through a Dense Channel with Minimum Total Wire Length*, Proceedings of the 2nd ACM-SIAM Symposium on Discrete Algorithms (SODA'91), (1991) 475-482

[Frank] A. Frank: *Disjoint Paths in a Rectilinear Grid*, Combinatorica 2 (1982) 361-371

[Kaufmann and Maley] M. Kaufmann, M. Maley: *Parity conditions in homotopic knock-knee routing*, to appear in Algorithmica

[Kaufmann and Mehlhorn] M. Kaufmann, K. Mehlhorn: *Routing through a Generalized Switchbox*, Journal of Algorithms 7 (1986) 510-531

[Kuchem, Wagner and Wagner] R. Kuchem, D. Wagner, F. Wagner: *Area-Optimal Three Layer Channel Routing*, Proceedings of the 30th IEEE Symposium on the Foundations of Computer Science (FOCS'89) (1989) 506-511

[Lai and Sprague] T.-L. Lai, A. Sprague: *On the Routability of a Convex Grid*, Journal of Algorithms 8 (1987) 372-384

[Nishizeki, Saito and Suzuki] T. Nishizeki, N. Saito, K. Suzuki: *A Linear-Time Routing Algorithm for Convex Grids*, IEEE Transactions on Computer-Aided Design 4 (1985) 68-75

[Okamura and Seymour] H. Okamura, P. D. Seymour: *Multicommodity-Flows in Planar Graphs*, Journal of Combinatorial Theory, Series B 31 (1981) 75-81

[Preparata and Lipski] F. P. Preparata, W. Lipski, Jr.: *Optimal three-layer channel routing*, IEEE Transactions on Computers, C-33 (1984) 427-437

[Sarrafzadeh] M. Sarrafzadeh: *Channel routing with provably short wires*, IEEE Transactions on Circuits and Systems, CAS-34 (1987) 1133-1135

A New Approach to Knock-Knee Channel Routing*

Dorothea Wagner

Fachbereich Mathematik

Technische Universität Berlin

Straße des 17. Juni 136, W-1000 Berlin 12, Germany

e-mail: dorothea@combi.math.tu-berlin.de

Abstract

We present a new channel routing algorithm in the knock-knee mode that produces for dense problems area-optimal layouts with minimum total wire length and $\mathcal{O}(n)$ bends (n number of nets), where the total number of bends is at most $d - 2$ (d density) more than the minimum. The running time is $\mathcal{O}(n)$. It thus improves the algorithm in [3] that determines area-optimal layouts with minimum total wire length in $\mathcal{O}(n^2)$ time, where the number of bends is $\Omega(n^2)$. Moreover, this is the first area-optimal layout algorithm with linear running time. The approach we use is completely different from all previously known algorithms. It is based on the notion of cycle graphs introduced in this paper.

1 Introduction

Channel routing is a key problem in the design of VLSI-circuits.

An instance consists of a rectangular grid-graph, the *channel* and a set of nets. A *net* is a pair of *terminals*, one on the upper and one on the lower boundary of the grid. These have to be connected. A solution *(layout)* is given by edge-disjoint paths *(wires)* that connect corresponding terminals within the channel. There are essentially two different routing models, *knock-knee* [2]-[10] and [12], and *Manhattan* [1],[11] and [13]. In this paper we consider the knock-knee model, where two wires may cross or both bend at a vertex, i.e. form a knock-knee. (In Manhattan routing knock-knees are not allowed.) Now, to avoid physical contacts between different wires, the edges of the paths are assigned to different layers, such that no two wires share a grid vertex in the same layer *(wiring)*. Connections between distinct layers (called *vias*) are placed on grid points.

The most important optimization criteria for the design of layout algorithms are the *layout area*, the *number of layers* and *vias* used for the wiring, the *length of the wires*, the *number of bends* and the *time complexity* of the algorithms.

---

*supported by the DFG under grant Mö 446/1-3.

In [4] and [9] algorithms that construct layouts of minimum area are presented, but the wirability for these layouts is not considered. It is known, that each layout is wirable in four layers [2], while it is $\mathcal{NP}$-complete to decide if a layout is wirable in three layers [7]. Only very restricted layouts are wirable in two layers [8], i.e. in general wirability in three layers is the best one can expect for a given layout. In [5] and [10] layout algorithms are given that always produce three-layer wirable layouts, but the minimum area is not achieved. Both, minimum area and wirability in three layers is guaranteed for the layout algorithm presented in [6]. The most efficient layout algorithms known up to now run in time $\mathcal{O}(n \log n)$ [6], [10] respectively $\mathcal{O}(n)$ [9], where the linear running time in the second case strongly depends on the fact, that the layouts produced by this algorithm are not area-optimal.

The length of the wires used in a layout were considered in [3] and [12]. The algorithm presented in [3] is the first that guarantees minimum total wire length for dense problems, i.e. all boundary points are occupied by terminals. The time complexity is $\mathcal{O}(n^2)$.

In this paper, we also consider the number of bends contained in a layout for a dense problem. We present a new algorithm that constructs a layout of minimum area with minimum total wire length and $\mathcal{O}(n)$ bends, where the number of additional bends in the layout is at most $d - 2$ (with d density of the problem, which is equal to the number of tracks). The time complexity is $\mathcal{O}(n)$, which is an substantial improvement of the result in [3] as well.

The paper is organized as follows: In Section 2 we review the main definitions and results used for our approach, in Section 3 the notion of the cycle-structure of a problem is introduced, that leads to a lower bound for the number of bends contained in a layout. The new algorithm is presented in Section 4. Section 5 contains a conclusion and a discussion of possible extensions of our approach.

2   Preliminaries

A *channel* of *width* w and *spread* s is a rectilinear grid determined by horizontal lines j, $0 \leq j \leq w + 1$ (called *tracks*) and vertical lines i, $1 \leq i \leq s$. A (*two-terminal*) *net* is given by a pair of integers $(t(l), l), 1 \leq t(l), l \leq s$, where $t(l)$ stands for the position of the *input terminal* on track 0, and l stands for the position of the *exit terminal* on track $w + 1$. A net $(t(l), l)$ is a *left* net (resp. *right* net) if $t(l) > l$ (resp. $t(l) < l$) and a *straight* net if $t(l) = l$. (In the following we often name the nets after the vertical line the exit terminal is placed on, i.e. $l := (t(l), l)$.) A *channel routing problem* (CRP) is a collection of nets, where no two nets share an input or exit terminal. A solution of a given CRP, called a *layout*, is a collection of n edge-disjoint paths (*wires*) through the grid (not using track 0 or $w + 1$) that connect the input and the exit terminal. A subpath (of length two) consisting of a horizontal edge followed by a vertical edge, or vice-versa, is called a *bend*. Obviously, a trivial CRP consisting only of straight nets is solvable without bends and with minimum total wire length by routing the nets vertically down from the input to the exit terminal. So, in the following we assume that the CRP contains at least one non-straight net.

The segment between two vertical lines i and $i + 1$ is called *column* $\overrightarrow{i}$ (resp. $\overleftarrow{i + 1}$). The (*local*) *density* of a column $\overrightarrow{i}$ is the number of nets that have to cross column $\overrightarrow{i}$ i.e.

$d\left(\vec{i}\right) := |\{(t(l), l)|t(l) \leq i < l \text{ or } l \leq i < t(l)\}|$. The *(global) density* of a CRP is the maximum of all local densities. Obviously, the density of the CRP is a lower bound for the channel width needed for the layout.

As stated in the introduction we assume that all grid points on the upper and lower track of the channel are occupied by terminals. From [4] we know that in this case a CRP of density d is solvable, iff the channel has width not smaller than d and has at least one additional vertical line to the right (left) of the rightmost (leftmost) terminal. So, if we have a CRP with n nets and density d, we assume that the channel has width d and spread $n+1$ to achieve area-optimal layouts. W.l.o.g. we may assume that the additional free vertical line lies to the left of the leftmost terminal.

For the design of a layout algorithm that guarantees minimum total wire length and the minimum number of bends, we need lower bounds for these parameters. A trivial lower bound for the total wire length used in a layout is the sum of the Manhattan-distances of the nets, i.e. the sum of vertical and horizontal edges between the input and the exit terminals of the nets. But, to avoid vertical conflicts between different wires, some wires are forced to *detour*. Clearly, the lower bound for the number of bends contained in a layout depends on the number of forced detours as well. A characterization of the detour-enforcing subproblems of a CRP is given in [3]. We briefly review these results.

There exist two types of detour-enforcing subproblems for a dense CRP, the *density valleys* and the *autonomous intervals*. An interval $[a, b]$ (of vertical lines) is a *density valley* iff there is a column to the left of a and a column to the right of b whose local densities are greater than that of column $\vec{i}$ for all $i \in [a, b-1]$. The interval $[a, b]$ is a *maximal density valley* iff $[a, b]$ is maximal with this property with respect to inclusion. An interval $[a, b]$ is called *autonomous* (*AI*) iff no net leaves $[a, b]$, i.e. $i \in [a, b] \Leftrightarrow t(i) \in [a, b]$, and neither $(t(a), a)$ nor $(t(b), b)$ are straight nets. To give a correct lower bound for the number and length of enforced detours, we may only consider *AI*'s enforcing extra detours, i.e. additional to those enforced by density valleys. So, an *AI* $[a, b]$ is called *bad* (for short *BAI*) if neither column $\overleftarrow{a}$ nor $\overrightarrow{b}$ are density valleys. For a *BAI* $[a, b]$ with l resp. r straight nets directly to the left resp. right, we call a the *good* end of the *BAI* if $l \leq r$, else we call b the good end.

Now, the number of detour edges contained in an area-optimal layout with minimum total wire length for a dense CRP is the number of horizontal edges needed to fill up the density valleys, plus the horizontal edges caused by the *BAI*'s (which are $2(\min\{l, r\} + 1)$ for a *BAI*.) Obviously, all *BAI*'s contain all columns of maximal density, and for two *BAI*'s $[a, b], [a', b']$ we have $a < a' < b' < b$ or $a' < a < b < b'$.

3 The Cycle Structure of a CRP

3.1 Lower bound for the number of bends

Clearly, every wire realizing a non-straight net contains an even number of bends, at least two. Now, bend-enforcing substructures are the cycles of a dense CRP. A sequence of nets $\langle (t(l_1), l_1), \ldots, (t(l_k), l_k) \rangle$ is a *cycle* iff $l_i = t(l_{i+1})$ for $1 \leq i < k$, and $l_k = t(l_1)$. Obviously, each cycle enforces two additional bends. Since the cycles of a CRP induce a partition of the set of nets, we have

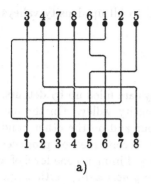 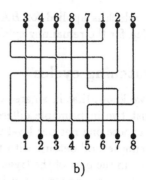

a)   b)

Figure 1:

Lemma 1 *A layout for a dense CRP with m non-straight nets consisting of c cycles contains at least $2m + 2c$ bends.*

A good strategy to avoid bends is to lay out each net entirely on one track, i.e. to perform the layout cycle-wise. If the CRP consists of exactly one cycle, a layout with minimum total wire length and minimum number of bends is easily constructed by first routing the net with the leftmost exit terminal on the uppermost track to the free vertical line to its left, and then routing the nets according to the cycle structure of the problem. Thereby, subsequent nets of the same type i.e. both left resp. right nets are routed on the same track, while a right net (left net) following a left net (right net) is on the track below. See Figure 1a).

But if the CRP consists of two or more cycles, the canonical procedure enforces unnecessarily long detours as to be seen in Figure 1b). Moreover, if the CRP contains a density valley this procedure possibly does not even achieve minimum area. See Figure 2a). Beside the cycles the detour enforcing subproblems of a CRP, i.e. *BAI*'s and density valleys induce additional bends. Therefore, to minimize the number of bends we must simultaneously minimize the number of detours in a layout.

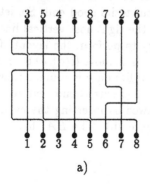 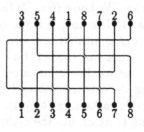

a)   b)

Figure 2:

We introduce a formal concept to perform from the cycles of a CRP one *cycle arrangement* that induces an area-optimal layout with minimum total wire length and nearly minimum number of bends. This concept is based on three operations on cycles:

inserting, which is applied for *BAI*'s, *patching*, which is used to fill up density valleys and *melting* for the remaining cyles.

3.2 Patching cycles

From [3] we know, that in an area-optimal layout density valleys are filled up by detours. There exist two in principle different methods to fill up a density valley. On the one hand, to choose one net per level that forms exactly one detour and on the other hand by producing several detours with the same net. Obviously, the second method produces more bends in the part of the layout related to the density valley. Filling up one level of a density valley corresponds to routing a detour between a *density decreasing* vertical line (i.e. a line l with $d(\overleftarrow{l}) > d(\overrightarrow{l})$) to the left and a *density increasing* vertical line (i.e. a line l with $d(\overleftarrow{l}) < d(\overrightarrow{l})$) to the right. (If $d(\overleftarrow{l}) = d(\overrightarrow{l})$, we call l *density preserving*.) There are four candidate nets for the detour, the two nets whose input resp. exit terminal lies on the density decreasing vertical line, and the two nets whose input resp. exit terminal lies on the density increasing vertical line.

The method used here to fill up a maximal density valley $[a, b]$ is to perform one detour per level as follows. For each density decreasing vertical line $i \in [a, b]$ the leftmost density increasing line $j \in [a, b]$ with $i < j$ and $d(\overleftarrow{i}) = d(\overrightarrow{j})$ is chosen (which clearly is unique). Then, according to the order the nets of the CRP are layed out, either net $(t(j), j)$ is routed until vertical line i just after the left net whose input terminal lies on i is layed out (on the same track), or the net $(t(l), l)$ with $t(l) = i$ is routed until vertical line j, just after the right net, whose input terminal lies on j is layed out. This procedure corresponds to a new arrangement of subsequences of the cycles "involved" in the density valley, which may be viewed as a *patching* of cycles. See Figure 2b). The density decreasing line here is 4, the density increasing line is 6, and net 6 performs the necessary detour.

3.3 Inserting cycles

Consider a CRP containing a *BAI*. Since a *BAI* is a CRP itself, it consists of one or more cycles. We know from [3] that each *BAI* enforces a detour to its good end. So, let [a,b] be a *BAI*, w.l.o.g. a is the good end of $[a, b]$. If $c < a$ is the rightmost vertical line to the left of a with $(t(c), c)$ is a non-straight net, our strategy to perform the layout of the cycle containing the net $(t(a), a)$ (which is clearly a left net) is to move $(t(a), a)$ to the left until vertical line c just after the net whose input terminal lies on c is routed, thus producing the detour enforced by $[a, b]$ simultanously with the two extra bends enforced by the cycle containing $(t(a), a)$. This procedure may be viewed as inserting the cycle containing $(t(a), a)$ into the cycle containing $(t(c), c)$. See Figure 3b). The *BAI* here is [4, 6], its good end is 7.

3.4 Melting cycles

As we have mentioned in 3.1, performing the layout for a CRP cycle-wise does not always lead to minimum total wire length. To overcome this problem we use the following idea: Consider two cycles C_1, C_2 such that C_1 contains a left net (resp. right net) $(t(l_1), l_1)$ and C_2 contains a left net (resp. right net) $(t(l_2), l_2)$ with $l_2 < t(l_1) < t(l_2)$ (resp. $l_2 >$

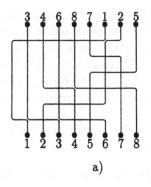

a)

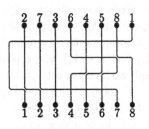

b)

Figure 3:

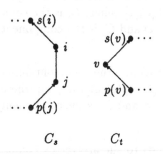

C_s $\qquad$ C_t

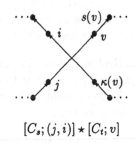

$[C_s; (j,i)] \star [C_t; v]$

Figure 4:

$t(l_1) > t(l_2))$. A layout of C_1 and C_2 may be performed in the following fashion. The nets of C_1 are routed according to their order until $(t(l_1), l_1)$ is layed out (where the first net of C_1 that is moved has to perform a detour). Then $(t(l_2), l_2)$ is layed out to vertical line $t(l_1)$ on the same track and followed by the remaining nets of C_2 according to their order until the net, whose input terminal lies on vertical line l_2 is moved. Now, $(t(l_2), l_2)$ may be moved again to its exit terminal, thereby forming two additional bends. Then the net of C_1, whose exit terminal lies on vertical line $t(l_1)$ is routed followed by the remaining nets of C_1. This procedure may be viewed as a melting of the cycles C_1 and C_2. A layout for the example of Figure 1b) after melting is shown in Figure 3a). Here $C_1 = \langle 1, 6, 3 \rangle, C_2 = \langle 2, 7, 5, 8, 4 \rangle$, the melting is performed with net 2.

These three operations on cycles patching, inserting and melting are based on the idea to interrupt the layout of a cycle at an appropriate vertical line and to perform the layout of a second cycle. For the layout of the second cycle one net is touched twice, i.e. layed out in two parts on different tracks, thus performing a detour in case of inserting or patching, respectively a staircase in case of melting. This procedure can be interpreted as a graph theoretic operation on the so-called cycle graph of a CRP.

3.5 The cycle graph of a CRP

Consider the cycles of a CRP. The *cycle graph* $G = (V, A)$ of the CRP is a directed graph whose vertex set $V := \{1, \ldots, n\}$ corresponds to the nets of the CRP, and for $1 \leq i, j \leq n$ $(j, i) \in A$ iff $t(j) = i$ for the nets $(t(i), i), (t(j), j)$. If $(j, i) \in A$, we call

$p(i) := j$ the *predecessor* of i and $s(j) := i$ the *successor* of j. Obviously, the components of G correspond to the cycles of the CRP. We define an *arc-vertex combination* of an arc (j, i) of a component $C_s = (V_s, A_s)$ (i.e. $(j, i) \in A_s$) and a vertex v of a component $C_t = (V_t, A_t)$ (i.e. $v \in V_t$) where possibly $C_s = C_t$ as follows. $[C_s; (j, i)] \star [C_t; v]$ is the directed graph with vertex set $V_s \cup V_t \cup \{\kappa(v)\}$ and the arc set $A_s \setminus \{(j, i)\} \cup A_t \setminus \{(p(v), v)\} \cup \{(j, v), (\kappa(v), v), (p(v), \kappa(v))\}$. See Figure 4. Notice that $[C_s; (j, i)] \star [C_t; v]$ consists of one component if $C_s \neq C_t$, and of two components if $C_s = C_t$.

Now, the operations patching, inserting and melting all three correspond to an arc-vertex combination of the corresponding components of the cycle graph.

1. Patching of the cycle containing the net $(t(i), i)$ for a density decreasing vertical line i and the cycle containing the net $(t(j), j)$ for the density increasing vertical line j, with $i < j$, corresponds to the following arc-vertex combination in the cycle graph: $[C_i; (p(i), i)] \star [C_j; j]$, respectively $[C_j; (p(j), j)] \star [C_i; i]$, where C_i is the component containing vertex i, which is related to net $(t(i), i)$ and C_j is the component containing vertex j, which is related to net $(t(j), j)$.

2. For a BAI $[a, b]$, inserting the cycle containing net $(t(a), a)$ into the cycle containing net $(t(c), c)$, corresponds to the arc-vertex combination $[C_c; (p(c), c)] \star [C_a; a]$, where C_c is the component of the cycle graph containing vertex c, and C_a is the component containing vertex a.

3. Melting cycles C_1 and C_2 as described in 3.4 corresponds to the arc-vertex combination $[C_{l_1}; (l_1, s(l_1))] \star [C_{l_2}; l_2]$, where C_{l_1} is the component related to C_1 and C_{l_2} the component related to C_2.

Notice, that the arc-vertex combinations corresponding to inserting and melting are always combinations of two different components, while patching might also correspond to the arc-vertex combination of one component.

4   The Algorithm

Informally, the algorithm determines from the cycle structure of the CRP a prescription for the order the nets are routed in the layout. In a preprocessing step the cycles of the CRP, the maximal density valleys and the BAI's are determined. For the density valleys the corresponding patching operations are performed. Then for each BAI separately the appropriate melting operations are applied. In a third procedure the BAI's are inserted recursively, and the nets performing detours or stairs are fixed. The procedure ends with one cycle arrangement that induces the order the nets are routed where exactly those nets appear more than once, which form a detour or a stair, i.e. produce additional bends.

For the formal description of the main part of the algorithm we use the cycle graph of the CRP. Assume that after the preprocessing step each vertex i knows its corresponding net $(t(i), i)$ and the status of the vertical line i (i.e. if i is density decreasing, increasing or preserving). In addition the values $d(\overleftarrow{i})$ and $d(\overrightarrow{i})$ are stored, if i is contained in a density valley, if i is the rightmost (resp. leftmost) non-straight vertical line to the left (resp. right) of the good end of a BAI and, if this is the case, the corresponding good end j. In the later case i is called the detour end of the BAI. C_i denote the component of the cycle graph containing i (at that state of the algorithm). Let $[a_1, b_1], ..., [a_s, b_s]$ be the BAI's, where $a_1 < a_2 < ... < a_s$ and $b_s < b_{s-1} < ... < b_1$; g_i denote the good end of $[a_i, b_i]$, k_i the detour end.

(* procedure patching*)
for $i = 1$ to n do   1
   if i is contained in a density valley and i is density decreasing then   2
     $j :=$ minimum of all $k > i$, k is density increasing and $d(\overleftarrow{i}) = d(\overrightarrow{k})$ 3
     apply $[C_i; (p(i), i)] \star [C_j; j]$ 4

(* procedure melting*)
for $i = 1$ to s do   5
$f := g_i$ 6
   repeat   7
     $j := s(f)$ 8
     if $f \in \{1, ..., n\}$ then (*f is not a "new" vertex $\kappa(v)$*)   9
      if $t(f) < f$ then   10
       $l :=$ maximum of all $t(k)$ not in C_f with $a_i \leq t(k) < t(f) < k \leq b_i$ 11
      else (*$t(f) > f$*)   12
       $l :=$ minimum of all $t(k)$ not in C_f with $a_i \leq k < t(f) < t(k) \leq b_i$ 13
      if l exists then $j := k$ (where $l = t(k)$)   14
       apply $C_f := [C_f; (f, s(f))] \star [C_j; j]$; $\kappa(j) := j$ 15
     $f := j$ 16
   until $f = g_i$ 17

(*procedure settle κ)
$f := 1$ 18
   repeat label f "visited"   19
     if $f = \kappa(l)$ then   20
      if all origins of l are "visited" then   21
       $\kappa(l) := l$ 22
      else $\kappa(l) := p(l)$ 23
     (*where $p(l)$ is the predecessor of the second l in case l appears twice *)
     $f := s(f)$ 24
   until $f = 1$ 25

(*procedure inserting*)
for $i = 1$ to n do   26
   $f := k_i$ 27
   (*where k_i is the last k_i from g_{i-1} in case k_i appears more than once *)
   apply $C_1 := [C_1; (p(f), f)] \star [C_{g_i}; g_i]$; $\kappa(g_i) := g_i$ 28

At the end of the algorithm C_1 induces the cycle arrangement that gives the order the nets have to be routed in the layout beginning with net $(t(1), 1)$ to be moved to the free vertical line to the left of the channel. Straight nets do not appear in this cycle arrangement, they are routed vertically down from the input to the exit terminal. The procedure settle κ, must be performed, since the direction of a detour filling a density valley depends on the order the nets appear in the resulting cycle arrangement, which is not yet fixed at the state of the algorithm when the corresponding patching operation is performed.

A layout determined by the algorithm is shown in Figure 5a) in comparison to a layout determined by the algorithm of [3] in Figure 5b).

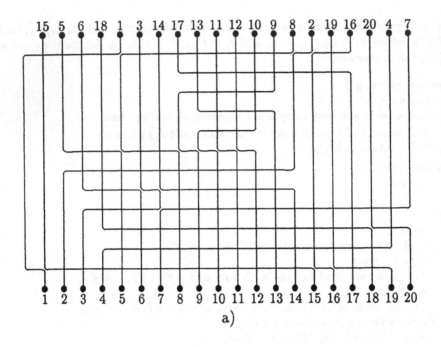

a)

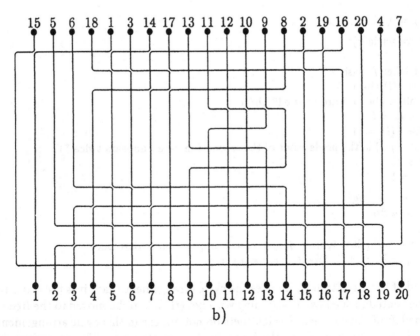

b)

Figure 5:

Theorem 1 *The algorithm determines for a dense CRP of density d an area-optimal layout with minimum total wire length and $\mathcal{O}(n)$ bends, where the number of bends additional to the minimum number is $d - 2$. The time complexity is $\mathcal{O}(n)$.*

The proof of the correctness of theorem 1 follows from a careful analysis of the algorithm and is contained in the full paper. Notice, that additional bends may only occur through meltings. In the worst case, a CRP consisting of one cycle is broken into $d/2$ cycles by the patching process thus enforcing $2 \cdot (d/2 - 1)$ additional bends by $d/2 - 1$ melting operations. The total number of bends is $\mathcal{O}(n)$ since there exist less than n BAI's to be inserted or cycles to be melt.

Its easy to see that the cycle graph, the density valleys and the related values for each vertical line are computable in $\mathcal{O}(n)$ time. To determine the BAI's in $\mathcal{O}(n)$ time we use the fact that they form a chain by interval containment. Candidate intervals for BAI's are all intervals $[a, b]$ with $d(\vec{a}) = d(\overleftarrow{b})$, $d(\vec{c}) < d(\vec{a})$ for $1 \leq c < a$, and $d(\overleftarrow{c}) < d(\overleftarrow{b})$ for $b < c \leq n$, which are computable in $\mathcal{O}(n)$ time. The test if such an interval is autonomous, can be done in $\mathcal{O}(n)$ time by traversing an array of length n, where in the i-th component the net $(t(i), i)$ is stored, at most twice. The good ends and detour ends of the BAI's are also easily determined in linear time.

The patching phase and the inserting phase both run in $\mathcal{O}(n)$ time. For the melting phase we have to realize that the determinations in line 10 to 12 need only $\mathcal{O}(n)$ in total. Consider e.g. the determination of the minimum of all $t(k)$ not in C_f with $k < t(f) < t(k)$ for one f. We have to scan an array of length n storing the net $(t(i), i)$ in component $t(i)$, from $t(f)$ to the first $t(k)$ with $t(k)$ not in C_f and $k < t(f)$. Since at that state of the algorithm the intervals related to the components C_i all overlap, we never have to check the array between $t(f)$ and $t(k)$ again for one of the following minimum determinations.

5 Conclusion

In this paper we introduced a completely new approach. All channel routing algorithms known before proceed track by track and on each track column by column performing the layout depending on the local state of the column. Our algorithm in contrast uses a global approach based on the new introduced cycle structure of the CRP. The advantage is its optimal running time and the nearly optimal number of bends it produces. As a consequence the number of knock-knees is small as well and the possible combinations of knock-knees per vertical line is restricted. Since the knock-knees contained in a layout are a fundamental property for its wirability [8] our approach is probably useful to find area-optimal layouts with minimum total wire length that are also wirable in only three (instead of four) layers.

Acknowledgement The author wants to thank Frank Wagner, Stefan Felsner for fruitful discussions, and especially Karsten Weihe for implementing the algorithm and helpful comments.

References

[1] B.S. Baker, S.N. Bhatt, T. Leighton. An approximation algorithm for Manhattan routing. *Advances in Computing Research, Vol. 2, VLSI Theory (ed. F.P. Preparata) JAI Press Inc. (1984) 205-229*

[2] M.L. Brady, D.J. Brown. VLSI routing: Four layers suffice. *Advances in Computing Research, Vol. 2, VLSI Theory (ed. F.P. Preparata) JAI Press Inc. (1984) 245-257*

[3] M. Formann, D. Wagner, F. Wagner. Routing through a dense channel with minimum total wire length. *Proc. of the 2nd Ann. ACM-SIAM Symposium on Discrete Algorithms (1991) 475-482*

[4] A. Frank. Disjoint paths in a rectilinear grid. *Combinatorica 2 (4) (1982)361-371*

[5] T. Gonzales, S. Zheng. Simple Three-layer channel routing algorithms. *Proc. of AWOC 88 (ed. J.H. Reif), LNCS 319 (1988) 237-246*

[6] R. Kuchem, D. Wagner, F. Wagner. Area-optimal three-layer channel routing. *Proc. of the 30th Ann. Symposium on Foundations of Computer Science (1989) 506-511*

[7] W. Lipski, Jr. On the structure of three-layer wirable layouts. *Advances in Computing Research, Vol. 2, VLSI Theory (ed. F.P. Preparata) JAI Press Inc. (1984) 231-243*

[8] W. Lipski, Jr., F.P. Preparata. A unified approach to layout wirability. *Mathematical Systems Theory, Vol. 19 (1987) 189-203*

[9] K. Mehlhorn, F.P. Preparata. Routing through a rectangle. *J. ACM, Vol. 33, No. 1 (1986) 60-85*

[10] F.P. Preparata, W. Lipski, Jr. Optimal three-layer channel routing. *IEEE Trans. on Computers, C-33 (5) (1984) 427-437*

[11] R.L. Rivest, C.M. Fiduccia. A greedy channel router. *Proc. 19th Design Automation Conference (1982) 418-424*

[12] M. Sarrafzadeh. Channel routing with provably short wires. *IEEE Trans. on Circuits and Systems, CAS-34 (9) (1987) 1133-1135*

[13] T. Szymanski. Dogleg channel routing is NP-complete. *IEEE Trans. on Computer-Aided Design of Integrated Circuits and Systems, CAD-4 (1) (1985) 31-41*

Circuit Partitioning Algorithms:
Graph Model versus Geometry Model *

Tetsuo Asano

Osaka Electro-Communication University, Japan

Takeshi Tokuyama

IBM Research, Tokyo Research Laboratory

One of the most important problems in VLSI layout design is a circuit partitioning problem, for which a number of algorithms have been presented. Given a set of modules together with a net list, the problem here is to find an optimal partition of the module set into two so that the areas occupied by modules are comparable in the two sides and the number of interconnections between two sides is minimized. For this problem this paper compares the method based on graph representation with the one of solving the bipartition problem after mapping modules into points in the plane.

1 Introduction

A number of methods have been proposed for determining placement of modules in VLSI chip. One of the most popular methods is "Mincut Placement Algorithm" proposed by Lauther in 1980 [La80]. The basic strategy of this algorithm is so-called divide-and-conquer. That is, a set of modules is divided into two parts so that the number of interconnections between different sides is minimized under the constraint that the total areas are comparable in the two sides. Then, an optimal placement of modules is determined in each side in a recursive manner. Finally, the interconnections between the two sides are completed.

This Mincut Algorithm seems to be very promising if we can find an optimal partition of modules. Unfortunately the problem seems to be intractable [GJS76]. This is mainly because of exponentially many different partitions. Two different approaches may be considered. In one approach we rely on heuristic search toward an approximated goal. In this case, for example, maximization of the minimum connectivity (strength of connection) among modules in the same part is one such approximated goal. The other approach is to find a partition that is best in a restricted search space.

The above described problem of partitioning a set of modules into two parts so as to minimize the number of nets between different sides has usually been formulated using a graph representation. In one such representation vertices correspond to modules and an edge between two vertices has a weight proportional to the connectivity between the two corresponding modules. Then, an approximated goal is to find a partition that maximizes the minimum connectivity between modules in the same side.

The second approach we present is totally different. We first compute connectivities among modules as before. Then, we map modules into points in the plane in such a way that the distance between any two points is anti-proportional to the connectivity between the corresponding modules as much as possible. Thus, two tightly connected modules should be placed close to each other. Furthermore, we put a restriction on partitions, that is, we only consider partitions by straight lines (called linear partitions).

---

*This work was partially supported by Grant in Aid for Scientific Research of the Ministry of Education, Science and Cultures of Japan

A combinatorial observation tells us the fact that there are only $O(n^2)$ different linear partitions, which allows us to examine all possible linear partitions in polynomial time. One more nice thing about it is that such thorough examination can be carried out in an efficient way using Geometric Transform that maps points into lines and lines into points and Topological Walk Algorithm for searching in the transformed plane.

We could find a partition of a point set to minimize the larger diameter of the resulting sets [ABKY88]. In this case it is known that there is an optimal partition which is linear. This makes it somewhat reasonable for us to consider linear partitions alone.

2 Problem Formulation

In this section we prepare several terminologies and notations.

Let S be a set of modules $m_1, m_2, \cdots, m_n$. A net N_j is specified as a set of modules to be interconnected. Although it is usually given as a set of terminals there is no need here to know which terminals to be interconnected. We assume that there is no net interconnecting terminals of only one module. Area of a module m_i is denoted by $area(m_i)$. For a set A of modules the sum of the areas of the constituent modules is denoted by $area(A)$.

Now we are ready to define our problem.

[Partitioning Problem]
[INSTANCE] (1) A set of modules $S = \{m_1, m_2, \cdots, m_n\}$.
(2) Area $area(m_i)$ of each module m_i, $i = 1, 2, \cdots, n$.
(3) A set of nets (net list).
(4) A real number r, $0 \leq r < 1$, which is allowance on the relative difference of areas of two parts.
[QUESTION]
Is there a partition (S_1, S_2) of S such that the number of nets interconnecting modules in both sides, that is, $|\{N_j | N_j \cap S_1 \neq \phi \text{ and } N_j \cap S_2 \neq \phi\}|$, is minimized under the constraint

$$\frac{|area(S_1) - area(S_2)|}{area(S)} \leq r,$$

or equivalently

$$\frac{1-r}{2} area(S) \leq area(S_1) \leq \frac{1+r}{2} area(S).$$

The above condition is referred to as the **area constraint**.

Unfortunately, the problem is known to be NP-complete (see [GJS76]), which means that there seems to be no efficient algorithm for finding an optimal solution for this problem. Since we need an efficient (polynomial-time) algorithm, we must either try to achieve an approximated goal or by some heuristic algorithm or find an optimal solution in a restricted space. Several different approximated goals may be considered. Many of them are concerned about the notion of connectivity or dissimilarity between modules. For two modules their connectivity is high if there are many nets interconnecting them. The dissimilarity is anti-proportional to connectivity. Then, approximated goals may be maximizing the minimum connectivity in each side, maximizing the sum of connectivities between modules in the same side, minimizing the maximum dissimilarity in each side, and so on.

[Example] An example of a circuit is shown in Figure 1. It consists of 17 modules/pads denoted by a through q and 24 nets. Area of each module is showm below.

	a	b	c	d	e	f	g	h	i	j	k	l	m	n	o	p	q
area	2	2	2	2	2	2	25	20	6	9	16	12	9	6	16	8	25

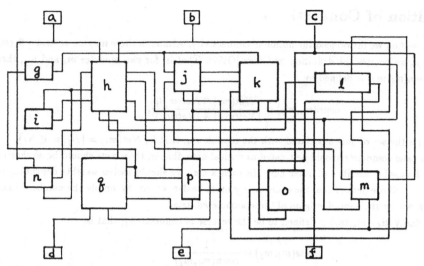

$Net(1) = \{a, h, n\}, Net(2) = \{b, h, m\}, Net(3) = \{c, m, o\}, Net(4) = \{d, q\}, Net(5) = \{e, j, p\},$
$Net(6) = \{f, k\}, Net(7) = \{g, j, k\}, Net(8) = \{g, p, q\}, Net(9) = \{h, i\}, Net(10) = \{h, i, n\},$
$Net(11) = \{h, l\}, Net(12) = \{h, l, m\}, Net(13) = \{h, m\}, Net(14) = \{h, m\}, Net(15) = \{h, n\},$
$Net(16) = \{i, q\}, Net(17) = \{j, k\}, Net(18) = \{j, k\}, Net(19) = \{j, k, p\}, Net(20) = \{j, p\},$
$Net(21) = \{k, l, p\}, Net(22) = \{l, m, o\}, Net(23) = \{l, o\}, Net(24) = \{p, q\}$

Figure 1: An example with 17 modules/pads and 24 nets.

The net list is given as a set of nets expressed as sets of modules to be interconnected. Positions of terminals are not interested here. An optimal bipartition of the module set when the relative difference of the total areas in two sides is within 10% is shown by a wavy line in Figure 2. In the partition the sum of area of modules in the upper part is 75 while that for the lower part is 70.

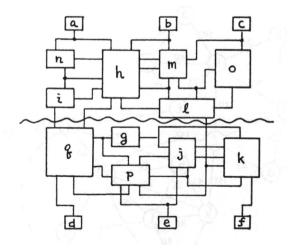

Figure 2: An optimal bipartition when $r \leq 0.1$.

3 Definition of Connectivity

In the previous section we introduced the notion of connectivity. Although there may be several different ways to define it, we borrow the definition by Otten [Ot82]. That is, for two modules m_i and m_j, their connectivity $conn(m_i, m_j)$ is defined by

$$conn(m_i, m_j) = \frac{|Net(m_i) \cap Net(m_j)|}{|Net(m_i)| + |Net(m_j)|},$$

where $Net(m_i)$ is the set of nets associated with the module m_i, that is, $Net(m_i) = \{N_j | m_i \in N_j\}$.

In practical cases some nets should be treated as critical nets, that is, those nets should be as shortest as possible with possible sacrifice of other nets. In such a case we should assign weights depending on their importance to those nets. Then, the connectivity must be defined not by simply the number of nets associated with but by the sum of weights of those associated nets.

The dissimilarity $diss(m_i, m_j)$ is defined to be the reverse of connectivity, that is,

$$diss(m_i, m_j) = \frac{1}{conn(m_i, m_j)}.$$

Here the dissimilarity between modules which have no common nets is defined to be positive infinite.

Based on the dissimilarity information we can define a weighted graph such that vertices correspond to modules and edge weights are dissimilarities between modules, respectively. It is a complete graph in which many edges may have infinite weights. Here note that only direct relationship between modules is considered in the above definition. In other words, the dissimilarity between two modules takes a finite value if and only if there is any net directly interconnecting them. Therefore, even if module m_1 is tightly connected with module m_2 and m_2 is tightly connected with m_3, the dissimilarity between m_1 and m_3 becomes infinite if there is no net interconnecting m_1 and m_3. In order to take such transitive relationship into accounts, the weight of an edge (m_i, m_j) is modified to be the length of the shortest path (the sum of the weights on the least-weight path) between m_i and m_j. By $d(m_i, m_j)$ we denote the weight of the edge (m_i, m_j), that is, the length of the shortest path between them in the dissimilarity graph. The resulting weighted graph is referred to as the transitive dissimilarity graph.

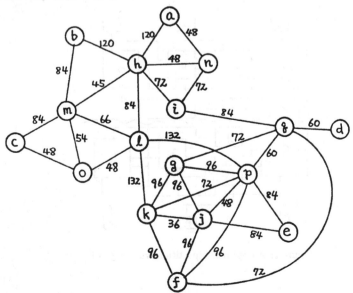

Figure 3: Original dissimilarity graph for the example given in Figure 1 (edges with infinite weights are omitted).

Figure 3 shows the original dissimilarity graph for the example given in Figure 1. The transitive dissimilarity graph for this graph is illustrated in Figure 4 in the form of a matrix.

	a	b	c	d	e	f	g	h	i	j	k	l	m	n	o	p	q
a:	0	216	225	264	348	396	276	96	120	312	312	180	141	48	195	264	204
b:	216	0	168	336	366	366	348	120	192	318	282	150	84	168	138	282	276
c:	225	168	0	345	312	312	324	129	201	264	228	96	84	177	48	228	285
d:	264	336	345	0	204	276	132	216	144	168	192	252	261	216	300	120	60
e:	348	366	312	204	0	204	180	300	228	84	120	216	282	300	264	84	144
f:	396	366	312	276	204	0	180	300	300	120	84	216	282	348	264	156	216
g:	276	348	324	132	180	180	0	228	156	96	96	228	273	228	276	96	72
h:	96	120	129	216	300	300	228	0	72	252	216	84	45	48	99	216	156
i:	120	192	201	144	228	300	156	72	0	192	216	156	117	72	171	144	84
j:	312	318	264	168	84	120	96	252	192	0	36	168	234	264	216	48	108
k:	312	282	228	192	120	84	96	216	216	36	0	132	198	264	180	72	132
l:	180	150	96	252	216	216	228	84	156	168	132	0	66	132	48	132	192
m:	141	84	84	261	282	282	273	45	117	234	198	66	0	93	54	198	201
n:	48	168	177	216	300	348	228	48	72	264	264	132	93	0	147	216	156
o:	195	138	48	300	264	264	276	99	171	216	180	48	54	147	0	180	240
p:	264	282	228	120	84	156	96	216	144	48	72	132	198	216	180	0	60
q:	204	276	285	60	144	216	72	156	84	108	132	192	201	156	240	60	0

Figure 4: Transitive dissimilarity graph in the matrix form.

4 Approach Based on Graph Model

In this section we present an algorithm for achieving an approximated goal for the Circuit Partitioning Problem using the transitive dissimilarity graph defined in the previous section. Several different approximated goals could be considered. Typical ones among them would be as follows.

(1) $\sum_{m_i \in S_1, m_j \in S_2} d(m_i, m_j) \to \min$,

(2) $\max(diam(S_1), diam(S_2)) \to \min$, and

(3) $diam(S_1) + diam(S_2) \to \min$,

where

$$diam(S_k) = \max\{d(m_i, m_j)|m_i, m_j \in S_k\}$$

is the largest edge weight within S_k and is called the diameter of S_k.

Several results are known for achieving the above goals in the case where the area constraint is not taken into accounts. The first problem of minimizing the sum of weights within the resulting subsets seems to be intractable. For the second problem the first efficient algorithm was presented by Avis [Av86], which was later improved from $O(e \log n)$ time to $O(e \log^* n)$ time by Monma and Suri [MS88], where e is the number of edges. The third goal is more hard to achieve. Hansen and Jaumard [HJ87] presented an $O(n^3 \log n)$ time algorithm, which was later improved to $O(n \cdot e \log n)$ time again by Monma and Suri [MS88].

As for the area constraint, Avis [Av86] presents an algorithm for finding a balanced bipartition, that is, a partition into two modules sets of equal cardinality, that minimizes the larger diameter. A key observation is that there is a bipartition for which the resulting larger diameter is less than a threshold t if the graph defined by those edges of the transitive dissimilarity graph which have weights greater than or equal to t is bipartite. It can be determined by dynamic programming whether such a balanced bipartition exists. Here, we generalize the Avis' algorithm so as to accommodate the area constraint.

This generalization can be easily achieved if area of each module is an integer or there is a common divisor for the areas so that each area can be represented as an integer and the sum of areas is bounded by some small constant independent of n.

[Lemma] Given a transitive dissimilarity graph and a threshold t, it can be determined in $O(n \cdot area(S))$ time whether there is a bipartition satisfying the area constraint such that the larger diameter is at most t.

Based on this lemma, we obtain the following algorithm.

[Algorithm Based on Graph Model]

(input) A transitive dissimilarity graph G and a real number r to represent the area constraint.

(Step 1) Sort the transitive dissimilarities in the ascending order. Let the sorted list be $t_1, t_2, \cdots, t_q, q = n(n-1)/2$.

(Step 2) We perform a binary search on the sorted list. At each iteration, for t_i we check whether there is a bipartition of larger diameter $< t_i$ that satisfies the area constraint. Then, the largest such t_i gives us the solution.

The above algorithm runs in $O(n^2 \log n + n \cdot area(S) \log n)$ time since the first step can be done in $O(n^2 \log n)$ time and in the second step binary searches are iterated $O(\log n)$ times and each iteration (dynamic programming) takes $O(n \cdot area(S))$ time. A similar approximation scheme can be found in [HS76] where an $\epsilon-$approximation is discussed.

5 Algorithms Based on Geometry Model

In the graph model described in the previous section we first defined the transitive dissimilarity graph and then tried to find a bipartition of a vertex set so as to minimize the larger diameter of the resulting vertex sets while the area constraint is satisfied. In the geometry model to be described in this section we map the vertices of the transitive dissimilarity graph into points in the plane so that squared distance between any two points are closest possible to the transitive dissimilarity between two corresponding vertices (the weight of the edge between those corresponding vertices in the transitive dissimilarity graph).

The most fundamental problem in the geometry model is thus how to map vertices of the transitive dissimilarity graph into points in the plane. We could rely on the principal coordinate analysis to be explained below.

Suppose we have mapped the vertices into points in the plane. Then, the approximated goal considered in the previous section is to find a bipartition of the point set so as to minimize the larger diameter of the resulting sets. An algorithm for finding such an optimal bipartition of a point set is presented by Asano, et al. [ABKY88]. The algorithm runs in $O(n \log n)$ time. Note that any algorithm based on the graph model described in the previous section requires at least time proportional to the number of edges, which is usually $O(n^2)$. One important observation here is that for any point set there is an optimal bipartition by a straight line (such a partition is called a linear partition). This observation suggests us that restriction to linear partitions may not be so harmful, that is, even if we restrict partitions only to linear ones, we may have an optimal partition in many cases.

The following natural question comes up.

[Problem G1] Given a set of points corresponding to modules in the plane, find a **linear partition** of S into two disjoint subsets that minimizes the number of nets interconnecting modules corresponding to points separated by the partition.

What happens if we take the area constraint into accounts.

[Problem G2] Solve the Problem G1 under some area constraint determined by a parameter r, $0 < r < 1$.

Moreover, in practical cases some modules (or pads) should be placed in predefined sides. Then, the problem becomes still more complicated.

[Problem G3] Solve the Problem G2 when some modules must be placed in predetermined sides.

We shall show that all of the above problems can be solved efficiently by Topological Walk Algorithm on arrangement of lines devised by Asano et al. [AGT91].

5.1 Principal Coordinate Analysis

Given a matrix representing squared distances among n items, we can embed those n items into points in the n-dimensional space to realize the distance matrix. Here, we want to embed those items into points in the plane. For this purpose we use a so-called principal coordinate analysis, which will be described below. Let $W = (w_{ij})$ be the matrix representing the transitive dissimilarities among modules. Then, we define a matrix $A = (a_{ij})$ by

$$a_{ij} = -\tfrac{1}{2}(w_{ij} - w_{i*} - w_{*j} + w_{**}),$$

where $w_{i*} = w_{*i} = \frac{1}{n}\sum_{j=1}^{n} w_{ij}$ and $w_{**} = \frac{1}{n^2}\sum_{i=1}^{n}\sum_{j=1}^{n} w_{ij}$.

Since A is a real symmetric matrix, it can be diagonalized:

$$A = V\Lambda V',$$

where Λ is a diagonal matrix concisting of A's eigenvalues and $V = (v_1, v_2, \cdots, v_n)$ is a matrix whose columns are the corresponding eigenvectors.

Then, choose two largest eigenvalues λ_1 and λ_2 and let v_1 and v_2 be their corresponding eigenvectors, respectively. Then, v_1 and v_2 are chosen as the two axes. More exactly, the vector X

$$X = [\sqrt{\lambda_1}v_1, \sqrt{\lambda_2}v_2]$$

specifies the x- and y-coordinates of the corresponding points.

Figure 5 shows the resulting set of points when we applied the principal coordinate analysis to the circuit shown before.

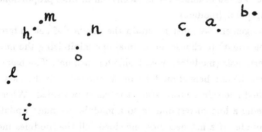

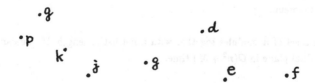

Figure 5: The resulting set of points by applying the principal coordinate analysis to the circuit in Figure 1.

5.2   Problem G1: Optimal Linear Partition

The key idea behind our algorithm to be presented is the duality transform between points and lines. We map a point $p = (a, b)$ into the line $T_p : y = ax + b$ in the dual plane and a line $L : y = kx + d$ into a point $T_L : (-k, d)$. One important property of the duality transform is that it keeps the vertical relationship between points and lines. That is, if a point p lies above (below, respectively) a line L then the corresponding line T_p passes above (below, resp.) the corresponding point T_L. We first map a set of points in the plane into a set of lines in the dual plane. Then, those lines partition the plane into many small regions (called cells).

Figure 6 shows the set of lines corresponding to the set of points in Figure 5. Here note that each cell can be characterized by vertical position with respect to each line. In the figure, the large cell denoted as R_1, for example, corresponds to the partition $\{a, b, c, h, i, m, n, o\}/\{d, e, f, g, j, k, p, q\}$ since it lies below the lines a, b, c, h, i, m, n, o and above the lines d, e, f, g, j, k, p, q. By the property of the duality transform an arbitrary point in this cell corresponds to a line in the original plane that passes above the points a, b, c, h, i, m, n, o and below the points d, e, f, g, j, k, p, q. Therefore, if we examine all the cells in the dual plane, it means we have examined all possible linear partitions of the original point set.

Topological Walk is an efficient algorithm for visiting all the cells in an arrangement of lines in such a way that any two cells visited consecutively are adjacent. Here it should be noted that Topological Sweep[HG86] is not appropriate for the purpose here because it can enumerate all the intersections in the arrangement but there are jumps between intersections visited consecutively.

We start a walk from a certain cell. First of all we evaluate the partition corresponding to the starting cell. When we move from one cell to an adjacent cell in the arrangement, the partition corresponding to the latter cell differs by just one point (module) from that of the former partition. For example, when we walk from the cell R_1 to its adjacent cell R_2 (see Figure 6) by crossing the line i, the point i is moved from one side to the other. Therefore, we can evaluate the adjacent cell in time proportional to the number of nets associated with the moving point (module).

During the walk on the arrangement we must maintain the number of nets interconnecting different sides (referred to as the crossing count) to choose an optimal one minimizing the number. For this purpose we maintain numbers of terminals (modules) in each side for each net. The balancing factor of a net is defined to be the ratio between the numbers of modules in the two sides. If the balancing factor of a net is zero or infinite, the net is called one-sided. Otherwise, it is called two-sided. When we move from one cell to its adjacent cell by crossing a line corresponding to a module, we must update balancing factors of all nets connected to the module. If a net becomes one-sided (all the modules incident with the net belong to the same side) by this movement, we decrement the crossing count. On the other hand, if a net which was one-sided becomes two-sided we increment the count. Thus, the total time for updating the balancing factors and crossing count is $O(M)$, where M is the length of the net list. Since there are $O(n^2)$ cells in the arrangement of n lines and all the cells are visited in time $O(n^2)$ time by Topological Walk, the time required to implementing the exhaustive search on the arrangement is $O(n^2 + M)$. Formally, we have the following theorem.

[Theorem 1]   Given a set of n modules together with a net list of length M, we can find an optimal linear partition in the dual plane in $O(n^2 + M)$ time.

5.3   Area Constraint

It is very easy to take the area constraint into accounts. Since we examine all possible linear partitions, we just choose an optimal one among those partitions satisfying the area constraint. As was stated before, cells are visited in such a way that two consecutive cells are adjacent, and one module moves from one side to the other at each movement. Therefore, once we evaluate the area of each side at the starting cell, only constant time is required to check the area constraint each time. Thus, the total time needed remains just the same as that in the case without considering the area constraint.

5.4 Side-Fixed Modules

There may be several side-fixed modules in practical cases, that is, some modules (or pads) must be placed in predefined sides. For simplicity of the argument, let A be a set of modules that must be placed above a separating line and B be that of modules that must be placed below it. Then, we do not need to examine any cell above any line corresponding to a module in A or below any line corresponding to a module in B. So, the search space is restricted to the cells below any line of A and above any line of B, which results in the convex region bounded from above the lines of A and bounded from below the lines of B. If there are K intersections in the convex region consisting of s edges, Topological Walk visits all the cells in the convex region in $O(K + n \log(n \cdot s))$ time. In our case s is smaller than n, and thus the overall time complexity is expressed as $O(K + M + n \log n)$.

6 Conclusions

In this paper we have presented two different approaches toward circuit partitioning problem. One approach is based on a graph defined by connectivities among modules. The other one is based on geometric transformation. Since we have applications to VLSI layout design in mind, we need to check the effectiveness of these approaches by experiments, which was left as future subjects.

Acknowledgment   The first author would like to thank Dr. R. S. Tsay at IBM Watson Research Center for introducing the problem. The authors would also like to thank Dr. J. Matousek for his valuable discussion. This work was partially supported by Grant in Aid for Scientific Research of the Ministry of Education, Science and Cultures of Japan.

Reference
[Av86] David Avis: "Diameter Partitioning, Discrete and Computational Geometry," 1, pp.265-276 (1986).
[ABKY88] T. Asano, B. Bhattacharya, J. M. Keil, and F. F. Yao: "Clustering Algorithms Based on Minimum and Maximum Spanning Trees", Proc. 4th Ann. ACM Symp. Computational Geometry, Urbana-Champaign, pp.252-257, 1988.
[AGT91] T. Asano, L.J. Guibas and T. Tokuyama: "Walking on an Arrangement Topologicaly," to appear in Proc. 7th Ann. ACM Symp. Computational Geometry, North Conway, 1991.
[GJ79] M.R. Garey and D.S. Johnson: "Computers and Intractability," Freeman, New York, 1979.
[GJS76] M.R. Garey, D.S. Johnson and L. Stockmeyer: "Some Simplified NP-complete Graph Problems," Theor. Comput. Sci. 1, pp.237-267, 1976.
[HG86] H. Edelsbrunner and L. J. Guibas: "Topologically Sweeping an Arrangement", Proc. 18th ACM Symp. on Theory of Computing, pp.389-403, 1986.
[HJ87] P. Hansen and B. Jaumard: "Minimum sum of diameters clustering", Journal of Classification, pp.215-226, 1987.
[HS76] E. Horowitz and S. Sahni: "Exact and Approximate Algorithms for Scheduling Nonidentical Processors", J. of ACM, vol.23, pp.317-327, 1976.
[La80] U. Lauther: "A Min-cut Placement Algorithm for General Cells Assemblies Based on a Graph Representation", J. Digital Systems, vol.4, pp.21-34, 1980.
[MS88] C. Monma and S. Suri: "Partitioning Points and Graphs to Minimize the Maximum or Sum of Diameters", Proc. of the 6th International Conference on the Theory and Applications of Graphs, Kalamazoo, 1988.
[Ot82] R.H.J.M. Otten: "Automatic Floorplan Design", Proc. 19th Design Automation Conf., pp.261-267, 1982.

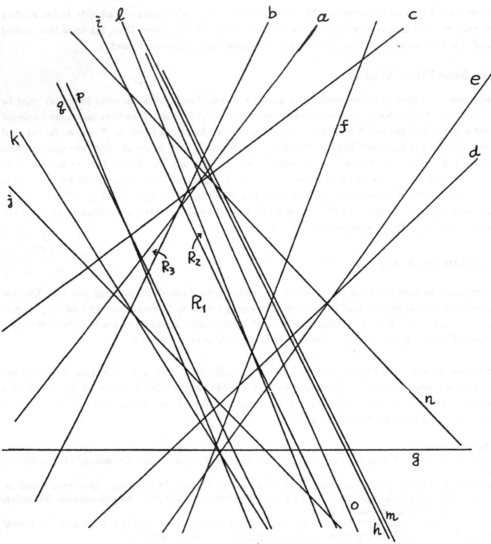

$R_1 : \{a, b, c, h, i, \ell, m, n, o\} / \{d, e, f, g, j, k, p, g\}$

$R_2 : \{a, b, c, h, \ell, m, n, o\} / \{d, e, f, g, i, j, k, p, g\}$

$R_3 : \{a, c, h, i, \ell, m, n, o\} / \{b, d, e, f, g, j, k, p, g\}$

Figure 6: Arrangement of lines in the dual plane which correspond to points in Figure 5.

Identifying 2-Monotonic Positive Boolean Functions in Polynomial Time

E. BOROS*, P.L. HAMMER*, T. IBARAKI† and K. KAWAKAMI†

Abstract

We consider to identify an unknown Boolean function f by asking an oracle the functional values $f(a)$ for a selected set of test vectors $a \in \{0,1\}^n$. If f is known to be a positive function of n variables, the algorithm by Gainanov can achieve the goal by issuing $O(mn)$ queries, where $m = |\min T(f)| + |\max F(f)|$ and $\min T(f)$ (resp. $\max F(f)$) denotes the set of minimal true vectors (resp. maximal false vectors) of f. However, it is not known whether this whole task including the generation of test vectors can be carried out in polynomial time in n and m or not. To partially answer this question, we propose here two algorithms that, given an unknown positive function f of n variables, decide whether f is 2-monotonic or not, and if f is 2-monotonic, output sets $\min T(f)$ and $\max F(f)$. The first algorithm uses $O(nm^2 + n^2m)$ time and $O(nm)$ queries while the second one uses $O(n^3m)$ time and $O(n^3m)$ queries.

1 Introduction

We investigate in this paper the problem of identifying an unknown Boolean function f by asking an oracle the functional values $f(a)$ for a selected set of test vectors $a \in \{0,1\}^n$, and propose two polynomial time algorithms when f is known to belong to a subclass of positive (i.e., monotone) Boolean functions. To be more specific, for a given unknown positive function f of n variables, we present two algorithms that decide whether f is 2-monotonic or not, and if f is 2-monotonic, output the set of minimal true vectors $\min T(f)$ and the set of maximal false vectors $\max F(f)$ (see Section 2 for the definitions of these terminologies). The first algorithm uses $O(nm^2 + n^2m)$ time and $O(nm)$ queries, while the second one uses $O(n^3m)$ time and $O(n^3m)$ queries, where

$$m = |\min T(f)| + |\max F(f)|. \tag{1}$$

(Throughout this paper, the stated computation time does not include the time spent on oracle to answer the given queries.) The class of 2-monotonic positive functions is important in practice as it properly includes the class of positive threshold functions (e.g., [8]).

The problem of identifying Boolean functions arises in various settings of theory and practice. A first application is the testing of logic circuits, in which a selected set of input vectors are applied to the circuit to make sure that it satisfies the given specification. This is essentially the identification the Boolean function realized by the circuit. The minimization of the number of test vectors and the time to generate them is important to make the test procedure efficient.

*RUTCOR - Rutgers Center for Operations Research, Bush Campus, Rutgers University, New Brunswick, NJ 08903, USA (Netaddress: boros@dimacs.rutgers.edu)

†Department of Applied Mathematics and Physics, Faculty of Engineering, Kyoto University, Kyoto, Japan 606 (Netaddress: ibaraki@kuamp.kyoto-u.ac.jp)

Another example is found in the process of forming a concept from partially observed data [4], in which hypotheses for the functional form of a hidden Boolean function are generated assuming that the class the function belongs to is known. The number of test vectors in this case is equal to the number of experiments to be conducted before convergence.

Probably the most rigorous mathematical basis for our problem is provided by the recent development of computational learning theory. The problem discussed in this paper is an example of exact learning (see, e.g., [1]), in which it is allowed to ask membership queries (i.e., whether a given a is a true vector or a false vector of f) to the teacher (or oracle) but no other types of queries. Our result shows that a polynomial time exact learning in this category is possible for the class of 2-monotonic positive Boolean functions.

If no a priori information about f is available, it is obvious that the Boolean function f cannot be identified usless the values $f(a)$ for all 2^n vectors $a \in \{0,1\}^n$ are tested. Therefore the problem becomes interesting only when some knowledge about f, e.g., the class of functions it belongs to, is at hand. An important class in the above second and third problem settings is the class of positive (i.e., monotone) Boolean functions. A positive function f can be characterized by the set of minimal true vectors $\min T(f)$ and the set of maximal false vectors $\max F(f)$. Therefore, we mean by identification of f to compute sets $\min T(f)$ and $\max F(f)$. As it is known that the above m can become as large as

$$\binom{n}{\lfloor \frac{n}{2} \rfloor} + \binom{n}{\lfloor \frac{n}{2} + 1 \rfloor},$$

the time complexity should be measured by both n and m. The paper by Gainanov [5] (also see [1]) contains an algorithm that identifies a positive function f of n variables by issuing $O(mn)$ queries. This means that the identification can be done in polynomial time if equivalence queries (i.e., whether the target function f is equivalent to the function f' currently being learned) are allowed in addition to the membership queries. When only membership queries are allowed, however, it is not known yet whether it can be done in polynomial time or not, in spite of long intensive studies on this topic (e.g., [7,11,5,6]).

Our result shows that it can be done in polynomial time if additional knowledge that the function f is 2-monotonic is available. It makes use of the above result of Gainanov, and a new simple characterization of a 2-monotonic positive function by the sets of minimal true vectors and maximal false vectors (the same task can also be done by using the recent result ([2,3,10,9]) that dualization of a regular Boolean function (a special case of a 2-monotonic positive function) can be done in $O(mn)$ time).

2 Definitions and basic properties

A Boolean function f of n variables is a mapping $f:\{0,1\}^n \to \{0,1\}$. We usually refer a Boolean function as a function and a Boolean vector $a \in \{0,1\}^n$ as a vector. A vector a is called a true (resp. false) vector if $f(a) = 1$ (resp. 0). The set of true (resp. false) vectors of a function f is denoted by $T(f)$ (resp. $F(f)$). A function f is called positive (or monotone) if $a \leq b$ always implies $f(a) \leq f(b)$, where $\leq$ is understood in the numerical sense. The set of minimal vectors in $T(f)$ is denoted by $\min T(f)$, and similarly the set of maximal vectors in $F(f)$ by $\max F(f)$. A positive function f is completely characterized by one of the $\min T(f)$ and $\max F(f)$, since f is for example defined by

$$f(a) = \begin{cases} 1, & \text{if } a \geq b \text{ for some } b \in \min T(f), \\ 0, & \text{otherwise.} \end{cases}$$

It is known in Boolean algebra that another characterization of a positive function f is that f has a disjunctive form in which all literals appear uncomplemented. In this case, each prime implicant of f corresponds one-to-one to a minimal true vector of f.

The *dual* f^d of a function f is defined by

$$f^d(x) = \bar{f}(\bar{x}),$$

where $\bar{f}$ (resp. $\bar{x}$) denotes the complement of f (resp. x). The Boolean expression of f^d is obtained from that of f by exchanging $\wedge$ (and) and $\vee$ (or), as well as constants 1 and 0.

Let f and g be functions of n variables (not necessarily positive). If any $a \in \{0,1\}^n$ satisfying $g(a) = 1$ also satisfies $f(a) = 1$, then we denote $g \subseteq f$. If $g \subseteq f$ and there exists a vector a satisfying $g(a) = 0$ and $f(a) = 1$, we denote $g \subset f$.

An assignment A of binary values 0 or 1 to k variables $x_{i_1}, x_{i_2}, \ldots, x_{i_k}$ out of all n variables is called a *k-assignment*, and is denoted by

$$A = \{x_{i_1} \leftarrow a_1, x_{i_2} \leftarrow a_2, \ldots, x_{i_k} \leftarrow a_k\}, \tag{2}$$

where each of $a_1, \ldots, a_k$ is either 1 or 0. Let the complement of A, denoted by $\bar{A}$, represent the assignment obtained from A by complementing all the 1's and 0's of A. When a function $f(x)$ of n variables and a k-assignment A are given,

$$f_A(x) = f(x; x_{i_1} \leftarrow a_1, x_{i_2} \leftarrow a_2, \ldots, x_{i_k} \leftarrow a_k)$$

denotes the function of $(n-k)$ variables obtained by fixing variables $x_{i_1}, x_{i_2}, \ldots, x_{i_k}$ as specified by A.

Let f be a Boolean function of n variables. If either $f_A \subseteq f_{\bar{A}}$ or $f_A \supseteq f_{\bar{A}}$ holds for every k-assignment A, then f is said to be *k-comparable*. If a function f is k-comparable for every k such that $1 \le k \le m$, then f is said to be *m-monotonic*. (For more detailed discussion on these topics, see [8] for example.) In particular, f is 1-monotonic if $f_{(x_i \leftarrow 1)} \supseteq f_{(x_i \leftarrow 0)}$ or $f_{(x_i \leftarrow 1)} \subseteq f_{(x_i \leftarrow 0)}$ holds for any $i \in \{1, 2, \ldots, n\}$. It can be shown that f is positive if and only if f is 1-monotonic and $f_{(x_i \leftarrow 1)} \supseteq f_{(x_i \leftarrow 0)}$ holds for every i.

Now consider a 2-assignment $A = \{x_i \leftarrow 1, x_j \leftarrow 0\}$. If $f_A \supseteq f_{\bar{A}}$ (resp. $f_A \supset f_{\bar{A}}$) holds, this is denoted as $x_i \succeq_f x_j$ (resp. $x_i \succ_f x_j$). Variables x_i and x_j are said to be *comparable* if either $x_i \succeq_f x_j$ or $x_i \preceq_f x_j$ holds. When $x_i \succeq_f x_j$ and $x_i \preceq_f x_j$ hold simultaneously, it is denoted as $x_i \approx_f x_j$. If f is 2-monotonic, this binary relation $\succeq_f$ over the set of variables is known to be a total preorder [8]. A 2-monotonic positive function f of n variables is called *regular* if

$$x_1 \succeq_f x_2 \succeq_f x_3 \succeq_f \ldots \succeq_f x_n. \tag{3}$$

Any 2-monotonic positive function becomes regular by permuting variables.

3 Outline of the algorithms

We present an outline of our algorithms that decide, for a given unknown positive function f, whether f is 2-monotonic or not, and if f is 2-monotonic, output $\min T(f)$ and $\max F(f)$, by evaluating $f(a)$ for some selected set of vectors a. The details of each step and the analysis of its time complexity will be given in the subsequent sections.

In each of our algorithms, we maintain two subsets of vectors MT and MF, where it is known that

$$MT \subseteq \min T(f) \quad \text{and} \quad MF \subseteq \max F(f) \tag{4}$$

hold. Call a vector a *unknown* if

$$a \not\succeq b \text{ for any } b \in MT, \quad \text{and } a \not\preceq b \text{ for any } b \in MF, \tag{5}$$

as $f(a)$ for such a cannot be deduced from the knowledge of MT and MF.

In each iteration, we test whether the two positive functions g_1 and g_0 defined by

$$
\begin{aligned}
\min T(g_1) &= MT \\
\min T(g_0) &= CMF \equiv \{\bar{a} | a \in MF\}
\end{aligned}
\tag{6}
$$

satisfy the following conditions, where $\bar{a}$ denotes the complement of a.

(a) Both g_1 and g_0 are 2-monotonic.

(b) The orders of variables for g_1 and g_0 coincide, i.e., $x_i \preceq_{g_1} x_j$ if and only if $x_i \preceq_{g_0} x_j$ for any i and j.

If some of these conditions are violated, there are the following two outcomes.

(i) f is concluded not being 2-monotonic. In this case, our algorithms halt at this point.

(ii) An unknown vector a is found for the current MT and MF. In this case, a is modified into a vector c such that

$$c \in (\min T(f) \cup \max F(f)) \setminus (MT \cup MF). \tag{7}$$

Then MT or MF is augmented with c, and the algorithms proceed to the next iteration.

On the other hand, if g_1 and g_0 satisfy the above conditions (a) and (b), we test whether the current MT and MF satisfy

$$MT = \min T(f) \text{ and } MF = \max F(f). \tag{8}$$

The following outcomes are possible.

(iii) Condition (8) holds. Then $g_1 = f$ and f is identified (also $g_0 = g_1^d = f^d$ holds). Our algorithms halt here.

(iv) Condition (8) does not hold. Then an unknown vector a is found and the algorithms proceed as in the above (ii).

Initially MT and MF are appropriately prepared, and then the above procedure is repeated until it halts in (i) or (iii). The key points here are how to execute the following steps in polynomial time.

1. Prepare the initial sets MT and MF.

2. Check if conditions (a) and (b) hold for the current g_1 and g_0 of (6), and if condition (a) or (b) is violated, conclude that f is not 2-monotonic or provide an unknown vector a.

3. Check if the termination condition (8) holds or not, and, if not, provide an unknown vector a.

4. Given an unknown vector a, modify it into a vector c satisfying (7).

These points will be separately discussed in the subsequent sections. The contribution of this paper mainly consists in providing polynomial time algorithms for executing 2 and 3.

4 Construction of a minimal true vector or a maximal false vector of f

Given an unknown vector a of (5) for a positive function f, Gainanov [5] gives an algorithm that finds a vector c satisfying (7). It proceeds as follows.

Assume that an unknown vector a satisfies $f(a) = 1$, and a contains k elements of 1, $a_{j_1}, a_{j_2}, \ldots, a_{j_k}$. Let e_i denote the unit vector whose components are all 0 except at the i-th component. Then the sequence $f(a^i), i = 1, 2, \ldots, k$, with

$$
\begin{aligned}
a^0 &= a, \\
a^i &= a \oplus f(a^1)e_{j_1} \oplus \ldots \oplus f(a^{i-1})e_{j_{i-1}} \oplus e_{j_i}, \quad i = 1, 2, \ldots, k,
\end{aligned}
$$

provides a minimal true vector $c \in \min T(f) \setminus MT$ by $c = a^p$, where p is the maximum index such that $f(a^p) = 1$.

The case of $f(a) = 0$ is similarly treated, and a vector $c \in \max F(f)$ is found. In either case, at most $n + 1$ vectors are tested and the time required to generate all a^i is $O(n)$ (since at most two components of a^i is modified to obtain a^{i+1}).

5 Algorithm with $O(nm^2 + n^2m)$ time and $O(nm)$ queries

5.1 Initialization

If $MT = \emptyset$ and $MF = \emptyset$, any a is an unknown vector. It is convenient to start with $a^1 = (111 \ldots 1)$. If $f(a^1) = 0$, then the positivity of f implies that f is constantly 0 (i.e., f is identified). Therefore assume that $f(a^1) = 1$ and obtain a vector c^1 of (7) by the algorithm described in Section 4. Similar procedure is then applied to $a^0 = (000 \ldots 0)$. If $f(a^0) = 1$, then f is constantly 1 (i.e., f is identified), or else a vector c^0 of (7) is obtained. Our algorithm initializes MT and MF as

$$
MT := \{c^1\} \quad \text{and} \quad MF := \{c^0\}.
$$

5.2 Checking the 2-monotonicity of g_i

As g_1 and g_0 of (6) can be treated in the parallel manner, we refer either of g_1 and g_0 as g and the corresponding MT and CMF as M in this subsection. That is, $\min T(g) = M$ holds. The algorithm described below decides if g is 2-monotonic or not, and if g is 2-monotonic, it also computes the $\preceq_g$-order over the n variables.

The general strategy is to find pairs of vectors a and b such that

$$
a \in M, \quad g(b) = 0, \tag{9}
$$

and

$$
\begin{aligned}
&a_i = 1, \ a_j = 0, \ b_i = 0, \ b_j = 1, \\
&a_k = b_k \ \text{for } k \neq i, j. \tag{10}
\end{aligned}
$$

For this pair a and b to exist, first note that

$$
a_i = 1 \quad \text{and} \quad a_j = 0 \tag{11}
$$

must hold. Also there must *not* be any $d \in M$ such that

$$
d_i = 0, \ d_j = 1, \ \text{and} \ d_k \leq a_k \ \text{for } k \neq i, j, \tag{12}
$$

since such d implies $g(b) = 1$ for the b obtained from a by complementing components i and j (hence condition (9) is violated).

Lemma 1 Let g and M be defined as above.

(i) $x_i \preceq_g x_j$ holds if and only if there is no pair a and b satisfying (9) and (10).

(ii) There is a pair of vectors a and b satisfying (9) and (10) if and only if there is some $a \in M$ satisfying (11), for which no $d \in M$ satisfies condition (12). $\square$

Note that the existence of i and j of (11)(12) for $a, d \in M$ is equivalent to

$$d_j = 1, a_j = 0 \text{ hold for exactly one } j,$$
$$d_k \leq a_k \text{ for all } k \neq j, \tag{13}$$

since $d_k < a_k$ must hold for some k as any vector in M is maximal, which is then regarded as i.

Now to test conditions (11)(12), we prepare an $n \times n$ matrix $P(a)$ for each $a \in M$. Initially it is defined by

$$P_{ij}(a) = \begin{cases} 1 & \text{if } a_i = 1 \text{ and } a_j = 0 \\ 0 & \text{otherwise.} \end{cases} \tag{14}$$

Upon termination, it has the following property:

$$P_{ij}(a) = \begin{cases} 1 & \text{if } a \text{ satisfies (11) and there is no } d \in M \text{ satisfying (12)} \\ 0 & \text{otherwise.} \end{cases} \tag{15}$$

From these $P(a)$, we then obtain an $n \times n$ matrix P by

$$P = \sum_{a \in M} P(a). \tag{16}$$

Lemma 2 For a positive function g of (6),

(i) $x_i \succ_g x_j$ if and only if $P_{ij} > 0$ and $P_{ji} = 0$,

(ii) $x_i \approx_g x_j$ if and only if $P_{ij} = 0$ and $P_{ji} = 0$,

(iii) x_i and x_j are not comparable (then g is not 2-monotonic) if and only if $P_{ij} > 0$ and $P_{ji} > 0$. $\square$

Now we consider how to compute these $P(a)$ and P. Initially we start with P with all $P_{ij} = 0$ for $M = \emptyset$. Assume that $P(a)$ for all $a \in M$ and P have been computed for the current M, and c is going to be added to M. In this iteration, the following steps are performed before M is augmented as $M := M \cup \{c\}$. The total computation time is $O(n|M| + n^2)$.

Step 1. Prepare initial $P(c)$ by (14).

Step 2. Compare c with each $e \in M$ to see if condition (13) holds. If c and e can be regarded as a and d in (13), respectively, then let $P_{ij}(c) := 0$ if i and j satisfy (11) and (12). On the other hand, if c and e are regarded as d and a, respectively, then apply similar operations to $P(e)$.

Step 3. Update P to reflect the modifications in Step 2.

Based on this P, g can be tested whether it is 2-monotonic or not by applying Lemma 2. If it is 2-monotonic, the $\preceq_g$-order can be computed from P by Lemma 2.

5.3   When g is not 2-monotonic

Assuming that the g tested in the previous section is not 2-monotonic, we show here that either it can be concluded that f is not 2-monotonic or an unknown vector is obtained (as discussed in (i) and (ii) of Section 3). Let $P_{ij} > 0$ and $P_{ji} > 0$ (see Lemma 2) be obtained as a result of $P_{ij}(a) > 0$ and $P_{ji}(b) > 0$ for $a, b \in M$, in the procedure of the previous subsection. This means that $a_i = 1$, $a_j = 0$, $b_i = 0$, $b_j = 1$. Then define a' (resp. b') by complementing components i and j of a (resp. b). We consider the cases of $g = g_1$ and $g = g_0$ separately.

If $g = g_1$ (hence $M = MT$), then the above a' and b' do not belong to MT, since otherwise a does not satisfy condition (ii) of Lemma 1.

(i) If $f(a') = f(b') = 0$, then we can conclude that f is not 2-monotonic. This is because $x_i \npreceq_f x_j$ follows from $f(a) = 1$ and $f(a') = 0$, and $x_j \npreceq_f x_i$ from $f(b) = 1$ and $f(b') = 0$.

(ii) If at least one of $f(a')$ and $f(b')$, say $f(a')$, is 1, then $a' \notin MT \cup MF$, i.e., a' is an unknown vector.

On the other hand, if $g = g_0$ (hence $M = CMF$), we consider $\bar{a}, \bar{b}, \bar{a}', \bar{b}'$, where $\bar{a}$ denotes the complement of a, etc. By definition (6) of g_0, we have the following.

(i′) If $f(\bar{a}') = f(\bar{b}') = 1$, then f is not 2-monotonic.

(ii′) If at least one of $f(\bar{a}')$ and $f(\bar{b}')$, say $f(\bar{a}')$, is 0, then $\bar{a}'$ is an unknown vector.

5.4   When variable orders of g_1 and g_0 do not coincide

Assume that g_1 and g_0 of (6) are both 2-monotonic but the orders of variables $\preceq_{g_1}$ and $\preceq_{g_0}$ do not coincide. We show in this case that an unknown vector can be identified for use in (ii) of Section 3.

For simplicity, let $x_i \succ_{g_1} x_j$ but $x_i \approx_{g_0} x_j$ or $x_i \prec_{g_0} x_j$. Then, as discussed in Subsection 5.2, there is a vector $a \in MT$ with

$$P_{ij}(a) > 0 \quad \text{and} \quad a_i = 1, a_j = 0, \tag{17}$$

and vector a' obtained from a by complementing a_i and a_j does not belong to MT. This a' does not belong to MF either, i.e., it is an unknown vector. For if $a' \in MF$, then $\bar{a}' \in CMF$ and $\bar{a} \notin CMF$, implying that $P_{ij}(\bar{a}') > 0$ for g_0, a contradiction to the assumption that $x_i \approx_{g_0} x_j$ or $x_i \prec_{g_0} x_j$.

Similarly, if $x_i \succ_{g_0} x_j$ but $x_i \approx_{g_1} x_j$ or $x_i \prec_{g_1} x_j$, there is a vector $a \in CMF$ satisfying (17). Define a' similarly, and then $\bar{a}'$ is an unknown vector.

5.5   Checking if $g_1 = f$

Assume now that both g_1 and g_0 are 2-monotonic, and the orders $\preceq_{g_1}$ and $\preceq_{g_0}$ coincide. We show how to test condition (8) of Section 3, and how to obtain an unknown vector if (8) does not hold (i.e., steps (iii) and (iv) of Section 3). For simplicity of discussion, we assume in this subsection that

$$x_1 \succeq_{g_i} x_2 \succeq_{g_i} \cdots \succeq_{g_i} x_n, \quad \text{for } i = 1, 2, \tag{18}$$

that is, g_1 and g_0 are regular. Let

$$\begin{aligned} T &= \{a \mid \exists a' \in MT \text{ such that } a' \le a\} \\ F &= \{b \mid \exists b' \in MF \text{ such that } b' \ge b\}. \end{aligned} \tag{19}$$

By definition of g_1 and g_0, we have $T \cap F = \emptyset$. This T is *left-shift stable*, i.e.,

$$a \in T \text{ implies } a + e_i - e_j \in T \text{ for any } i < j \text{ such that } a_i = 0 \text{ and } a_j = 1, \tag{20}$$

where e_i is the unit vector with its i-th component being 1, and F is *right-shift stable*, i.e.,

$$b \in F \text{ implies } b - e_i + e_j \in F \text{ for any } i < j \text{ such that } b_i = 1 \text{ and } b_j = 0. \tag{21}$$

Lemma 3 Let MT, MF and T, F be defined as above. If MT and MF further satisfy the properties that

(i) $a - e_j \in F$ for any $a \in MT$ and $a_j = 1$, and

(ii) $b + e_j \in T$ for any $b \in MF$ and $b_j = 0$,

then there is no unknown vector, i.e., $MT = \min T(f)$ and $MF = \max F(f)$ (hence $g_1 = f$). $\square$

The conditions (i) and (ii) in the lemma can be tested as follows. For a given $a \in MT$, compute

$$J_b(a) = \{j \mid a_j > b_j\}, \quad b \in MF. \tag{22}$$

If $J_b(a) = \{j\}$, then $a - e_j \leq b$ holds. Therefore condition (i) holds for a if

$$J(a) \equiv \bigcup_{|J_b(a)|=1} J_b(a) \tag{23}$$

satisfies condition

$$J(a) = \{j \mid a_j = 1\}. \tag{24}$$

On the other hand, if $J(a) \subset \{j \mid a_j = 1\}$, then those defined by

$$a - e_i, \quad i \in \{j \mid a_j = 1\} \backslash J(a) \tag{25}$$

are all unknown vectors, which can then be used in (iv) of Section 3. Since $J_b(a)$ can be computed in $O(n)$ time for each $b \in MF$, condition (23) can be tested in $O(n|MF|)$ time for each $a \in MT$. Similar argument can then be applied to condition (ii) for $b \in MF$.

Remark. Although we did not employ in our algorithm, the condition $MT = \min T(f)$ can also be checked by utilizing polynomial time algorithms for dualizing a regular function ([2,3,9,10]). Since $\max F(g_1) = \{\bar{a} \mid a \in \min T(g_1^d)\}$, compute $\max F(g_1)$ by applying such a dualization algorithm to g_1. Then $MT = \min T(f)$ holds if and only if $\max F(g_1) = MF$. The time required for dualization is $O(n|MT| + n|MF|) = O(nm)$ ([2,10]), and the time for checking $\max F(g_1) = MF$ can be $O(n|MF|)$ if we sort both sets lexicographically in $O(n|MF|)$ time and then compare them. $\square$

5.6 Description of the algorithm

program IDENTIFY-1

    begin

        Initialize MT and MF as described in Subsection 5.1 $\{f = 0 \text{ or } f = 1 \text{ may be concluded}$ here. All vectors in $MT \cup MF$ are unscanned.$\}$;

        repeat

            while g_1 or g_0 is not 2-monotonic, or orders $\preceq_{g_1}$ and $\preceq_{g_0}$ do not coincide

            $\{g_1(\text{resp. } g_0) \text{ is defined by } \min T(g_1) = MT(\text{resp. } \min T(g_0) = \{\bar{a} | a \in MF\})$

            and 2-monotonicity of g_i as well as the $\preceq_{g_i}$ order is checked as described

in Subsection 5.2} **do**
  begin {See Subsections 5.2, 5.3 and 5.4}
    if f is concluded not to be 2-monotonic {see (i) or (i') of
    Subsection 5.3} **then halt else**
      begin
        Using an unknown vector a found as described in (ii) or (ii')
        of Subsection 5.3, or in Subsection 5.4, obtain a vector c of (7) as
        described in Section 4;
        $MT := MT \cup \{c\}$ or $MF := MF \cup \{c\}$ depending upon $f(c) = 1$
        or $f(c) = 0$, respectively {The added c is unscanned. Although
        not explicitly stated, matrices $P(a)$ for $a \in M$ and P are updated
        as described in Subsection 5.2.}
      end
  end;
  repeat {Check if (8) holds}
    if there is an unscanned vector $a \in MT$ or $b \in MF$ **then** choose such
    a vector and test if condition (i) or (ii) of Lemma 3 holds, respectively
    {as described in Subsection 5.5}
  until new unknown vectors are found by (25) of Subsection 5.5, or there is
  no unscanned vector in $MT \cup MF$;
  if unknown true vectors $a^1, a^2, \ldots, a^k$ and unknown false vectors $b^1, b^2, \ldots, b^h$
  have been found in the above repeat-loop $\{k + h > 0\}$ **then**
    begin
      Obtain minimal true vectors $c^1, c^2, \ldots, c^k$ from $a^1, a^2, \ldots, a^k$
      as described in Section 4;
      $MT := MT \cup \{c^1, c^2, \ldots, c^k\}$;
      Obtain maximal false vectors $d^1, d^2, \ldots, d^h$ from $b^1, b^2, \ldots, b^h$
      as described in Section 4;
      $MF := MF \cup \{d^1, d^2, \ldots, d^h\}$
    end
  until $\min T(f) = MT$ {this holds if no unknown vector is found in the
  above inner repeat-loop};
  Halt {f is identified, i.e., $\min T(f) = MT$ and $\max F(f) = MF$}
end.

We now analyze the time complexity. Each time the while-loop or the if-block after the inner repeat-loop is executed, MT or MF is augmented by new vectors. Thus these are updated at most $m = |\min T(f)| + |\max F(f)|$ times.

Note first that checking the 2-monotonicity of g_1 and computing its order $\preceq_{g_1}$ is done by maintaining matrices $P(a)$ and P for MT, as explained in Subsection 5.2. The data of $P(a)$ and P are updated only when MT is augmented with new vectors. Since $O(n|MT| + n^2)$ time is required for each vector in MT, as noted in Subsection 5.2, total time here is $O(n|MT|^2 + n^2|MT|)$. Similar argument can also be applied to g_0 and MF, and total time required is $O(n|MF|^2 + n^2|MF|)$. The sum of these two is bounded from above by $O(nm^2 + n^2m)$. The conditions (i) and (ii) of Lemma 3 are tested in the inner repeat-loop. Since $O(n|MF|)$ (resp. $O(n|MT|)$) time is required for each $a \in MT$ (resp. $b \in MF$) as explained in Subsection 5.5, total time here is $O(n|MT||MF|) = O(nm^2)$. Finally, whenever an unknown vector is found, a new vector c in $MT \cup MF$ is computed by the algorithm of Gainanov in Section 4. Since each execution requires $O(n)$ time, total time here is $O(nm)$. Summing the above terms, we see that the time complexity of IDENTIFY-1 is

$$O(nm^2 + n^2 m).$$

The next theorem includes also the bound on the number of queries.

Theorem 1 Given an unknown positive function f of n variables, algorithm IDENTIFY-1 decides whether f is 2-monotonic or not, and if f is 2-monotonic, it outputs $\min T(f)$ and $\max F(f)$. The time required is $O(nm^2 + n^2 m)$ and the number of queries to oracle is $O(nm)$, where $m = |\min T(f)| + |\max F(f)|$. □

6 Algorithm with $O(n^3 m)$ time and $O(n^3 m)$ queries

6.1 Reducing time complexity of IDENTIFY-1

Since $m \gg n$ can be expected in usual cases, the most time consuming portions of IDENTIFY-1 (i.e., of $O(nm^2)$) are the computation of matrices $P(a)$ and P for $M = MT$ and CMF (Subsection 5.2) and the test for $g_1 = f$ (Subsection 5.5). We show here that this time bound can be reduced at the cost of increasing the number of queries.

We first consider the computation of $P(a)$ and P. For simplicity, consider the case of $M = MT$ and $g = g_1$, since the case of $M = CMF$ is similar. Instead of conditions (9)(10) of Subsection 5.2, we look for pairs of vectors a and b such that

$$
\begin{aligned}
& a \in MT, \\
& a_i = 1,\ a_j = 0, \\
& b \text{ is obtained from } a \text{ by complementing } a_i \text{ and } a_j, \\
& f(b) = 0,
\end{aligned}
\tag{26}
$$

and compute $n \times n$ matrix $P(a), a \in MT$, newly defined here by

$$
P_{ij}(a) = \begin{cases} 1 & \text{if } a \text{ and } b \text{ satisfy (26)} \\ 0 & \text{otherwise.} \end{cases}
\tag{27}
$$

P is then defined by (16). Since the condition $g(b) = 0$ in (9) is replaced by $f(b) = 0$ here, $P(a)$ is now independent of other members of MT. Therefore, once it is computed at the time of generating $a \in MT$, there is no need for further updates.

The computation of $P_{ij}(a)$ for $a \in MT$ is done by generating vectors b obtained by complementing a_i and a_j of a, for all pairs of $a_i = 1$ and $a_j = 0$, and then asking the oracle if $f(b) = 0$ holds or not. Since there are $O(n^2)$ vectors b, the time and the number of queries required to construct $P(a)$ is $O(n^2)$ for each $a \in MT$.

If $f(b) = 1$ holds for a vector b generated as above, then there are two cases: $b \in T$ and $b \notin T$ (T is defined in (19)). In the latter case, b is concluded to be an unknown vector and can be used as in (ii) of Section 3. However, checking if $b \in T$ is not very cheap and appears to require $O(n|MT|)$ time for each b (i.e., simply compare b with every $c \in MT$ to see if $b \geq c$ holds). This task can be reduced to $O(n)$ time as follows if we allow to use additional queries. For a vector b with $f(b) = 1$, apply the procedure of Section 4 to obtain a vector $c \in \min T(f)$ in $O(n)$ time, by using $O(n)$ queries. Then whether $c \in MT$ or not can be answered in $O(n)$ time, assuming that all the vectors a in MT are stored in a binary tree of $O(nm)$ size, constructed according as $a_i = 0$ or 1, for $i = 1, 2, \ldots, n$. Obviously $b \in T$ if and only if $c \in MT$; if $b \notin T$, the vector c is added to MT in our algorithm (i.e., (ii) of Section 3).

The time and the number of queries required for all the b generated from a vector $a \in MT$ are therefore $O(n^3)$.

When $P_{ij}(a)$ for all $a \in MT \cup MF$ for the current sets MT and MF are constructed, and there are no new vector b that should be added to MT or MF, it is not difficult to show that Lemma 2 of Subsection 5.2 holds also for the matrix P constructed in this section.

The time bound and the number of queries for checking conditions (i) and (ii) of Lemma 3 can also be improved to $O(n^2)$ for each $a \in MT$ and $b \in MF$, respectively, by a similar technique.

6.2 Description of the algorithm

The identification algorithm with the above modifications is almost the same as IDENTIFY-1 given in Subsection 5.6. The while-loop in IDENTIFY-1 is now modified as follows. The resulting algorithm is called IDENTIFY-2.

> **while** g_1 or g_0 is not 2-monotonic, or orders $\preceq_{g_1}$ and $\preceq_{g_0}$ do not coincide **do**
> > **begin**
> > > **if** f is concluded not to be 2-monotonic **then halt else**
> > > > **begin**
> > > > > Using an unknown vector a found as described in (ii) or (ii′)
> > > > > of Subsection 5.3, or in Subsection 5.4, obtain a vector c of (7) as
> > > > > described in Section 4;
> > > > > $MT := MT \cup \{c\}$ or $MF := MF \cup \{c\}$ depending upon $f(c) = 1$
> > > > > or $f(c) = 0$, respectively {the added c is unscanned}
> > > > **end;**
> > > **while** there is $a \in MT \cup MF$ for which $P(a)$ is not computed **do**
> > > > **begin**
> > > > > Choose such a vector $a \in MT \cup MF$ and compute $P(a)$
> > > > > {as described in Subsection 6.1};
> > > > > **if** new vectors $c \in MT \cup MF$ are generated while computing $P(a)$
> > > > > {as described in Subsection 6.1} **then** add such c to MT or MF
> > > > > depending upon $f(c) = 1$ or $f(c) = 0$, respectively
> > > > **end**
> > **end**

The analysis of the time and the number of queries proceeds in a manner similar to Subsection 5.6. The time required for each $P(a), a \in MT \cup MF$, is $O(n^2)$ as discussed in Subsection 6.1, and hence

$$O(n^2|\min T(f)| + n^2|\max F(f)|) = O(n^2 m)$$

for all such a. The number of queries in this part is also $O(n^2 m)$. To check if $b \in T$ (or $b \in F$) for a vector b generated from such a (and also to compute new vectors c to be added to MT or MF), $O(n^3)$ time and queries are necessary for each $a \in MT \cup MF$. Therefore total time and total number of queries in this part are both

$$O(n^3|\min F(f)| + n^3|\max F(f)|) = O(n^3 m).$$

The time and the number of queries to check condition (i) or (ii) of Lemma 3 is $O(n^2)$ for each vector in $MT \cup MF$ as described in Subsection 6.1. Total time and number of queries in this part is therefore $O(n^2 m)$. Other computation is minor, and can be treated as in Subsection 5.6.

Theorem 2 Given an unknown positive function f of n variables, algorithm IDENTIFY-2 decides whether f is 2-monotonic or not, and if f is 2-monotonic, it outputs $\min T(f)$ and $\max F(f)$. The time required is $O(n^3 m)$ and the number of queries to oracle is $O(n^3 m)$. $\square$

Acknowledgement The discussion with Prof. Y. Crama of University of Limburg was very beneficial. This research was partially supported by the Ministry of Education, Science and Culture of Japan under a Scientific Grant-in-Aid.

References

[1] D. Angluin, Queries and concept learning, *Machine Learning*, 2 (1988), 319-342.

[2] P. Bertolazzi and A. Sassano, An $O(mn)$ time algorithm for regular set-covering problems, *Theoretical Computer Science*, 54 (1987), 237-247.

[3] Y. Crama, Dualization of regular Boolean functions, *Discrete Applied Mathematics*, 16 (1987), 79-85.

[4] Y. Crama, P. L. Hammer and T. Ibaraki, Cause–effect relationships and partially defined boolean functions, *Annals of Operations Research*, 16 (1988), 299-326.

[5] D. N. Gainanov, On one criterion of the optimality of an algorithm for evaluating monotonic Boolean functions, *U.S.S.R. Computational Mathematics and Mathematical Physics*, 24 (1984), 176–181.

[6] A. V. Genkin and P. N. Dubner, Aggregation algorithm for finding the informative features, *Automation and Remote Control*, (1988), 81–86.

[7] J. Hansel, On the number of monotonic Boolean functions of n variables, *Cybernetics Collection*, 5 (1968), 53-58.

[8] S. Muroga, *Threshold Logic and Its Applications*, John Wiley and Sons, 1971.

[9] U. N. Peled and B. Simeone, Polynomial-time algorithms for regular set-covering and threshold synthesis, *Discrete Applied Mathematics*, 12 (1985), 57–69.

[10] U. N. Peled and B. Simeone, An $O(nm)$-time algorithm for computing the dual of a regular Boolean function, Technical Report, University Illinois at Chicago (1990).

[11] N. A. Sokolov, On the optimal evaluation of monotonic Boolean functions, *U.S.S.R. Computational Mathematics and Mathematical Physics*, 22 (1979), 207–220.

An Average Case Analysis of Monien and Speckenmeyer's Mechanical Theorem Proving Algorithm

*T. H. Hu , *C. Y. Tang, and **R. C. T. Lee

*   T. H. Hu and C. Y. Tang are with the Institute of Computer Science, National Tsing Hua University, Hsinchu 30043, Taiwan, Republic of China.

**   R. C. T. Lee is with the National Tsing Hua University, Hsinchu 30043, Taiwan, and Academia Sinica Taipei, Taiwan, Republic of China.

Abstract : In this paper, we shall give an average case analysis of a mechanical theorem proving algorithm based upon branching techniques for solving the k–satisfiability problem. The branching algorithm is a modified version of Monien and Speckenmeyer's branching algorithm [Monien and Speckenmeyer 1985]. Monien and Speckenmeyer's branching algorithm has a worst case time complexity which is strictly better than 2^n [Monien and Speckenmeyer 1985]. Based upon the probability distribution model that given r clauses, each clause is randomly chosen from the set of all k–literal clauses over n variables and each clause is chosen independently with others, we can show that our branching algorithm runs in exponential expected time under the condition that $\lim_{r,n \to \infty} \frac{t}{n} \to \infty$ and k is a constant.

Section 1. Introduction

The problem whether a boolean formula F in conjunctive normal form is satisfiable is known as the satisfiability problem. If the number of literals in each clause of a formula is restricted by k, the satisfiability problem for such formulas is called the k–satisfiability problem.

The satisfiability problem and the k–satisfiability problem, $k \geq 3$, are well known NP–complete problems [Cook 1971, karp 1972, Garey and Johnson 1979]. In artificial intelligence, solving the satisfiability problem is called mechanical theorem proving. There are many algorithms published for solving the satisfiability problem. Some of these algorithms are based upon backtracking techniques [Bitner and Reingold 1975, Purdom 1983], some are based upon the Davis–Putnam method [Davis and Putnam 1960, Goldberg, Purdom and Brown 1982, Franco and Paul 1983], and some are based upon the resolution principle [Robinson 1965, Chang 1970, Chang and Slagle 1971, Galil 1977, Tseitin 1968].

The NP–completeness of the satisfiability problem and the k–satisfiability problem, $k \geq 3$, indicates that there probably do not exist algorithms which solve these problems in polynomial time in worst cases. Much has been done in studying the behavior of algorithms to solve the satisfiability problem and the k–satisfiability problem. Some of this work is dedicated to the average case analysis of different versions of the Davis–Putnam method [Goldberg 1979, Franco

1986, Purdom and Brown 1981, 1985] and one of this work deals with the average case analysis of the resolution principle [Hu, Tang, and Lee 1991].

Some algorithms are shown to have polynomial behavior [Goldberg 1979, Purdom 1983, Purdom and Brown 1981, 1983, 1985, Iwama 1989, Hu, Tang, and Lee 1991], while in [Franco and Paul 1983], for some class of distributions a weaker form of Davis–Putnam method is shown to require exponential time with probability 1.

In this paper, we shall analyze an algorithm based upon branching techniques for solving the k–satisfiability problem. This branching algorithm is a modified version of Monien and Speckenmeyer's branching algorithm [Monien and Speckenmeyer 1985]. Monien and Speckenmeyer's branching algorithm has a worst case time complexity which is strictly better than 2^n [Monien and Speckenmeyer 1985]. In this paper, we shall give an average case analysis of our algorithm used to solve the k–satisfiability problem. Based upon the probability distribution that given r clauses, each clause is randomly chosen from the set of all k–literal clauses over n variables and each clause is chosen independently with others, we can show that our branching algorithm runs in exponential expected time under the condition that $\lim_{n,r\to\infty} \frac{r}{n} \to \infty$, and k is a constant. In Section 2, we shall describe our branching algorithm. In Section 3 and Section 4, an average case analysis of our branching algorithm will be shown.

Section 2. The Branching Algorithm

Let $X_1, X_2, ..., X_n$ denote boolean variables whose values are either true or false. Let $\neg X_i$ denote the negation of X_i. A literal is either a variable or its negation, denoted by X_i and $\neg X_i$ respectively. The complement of a literal X ($\neg X$) is $\neg X$ (X). A clause is a disjunction of literals, denoted by $L_1 \vee L_2 \vee ... \vee L_v$, such that no variable appears more than once and the complementary pair does not appear in a clause. A formula is a finite set of clauses in conjunctive form. The satisfiability problem is to determine if a formula is true for any truth values assigned to the variables. The k–satisfiability problem is a satisfiability problem where each clause has exactly k literals in the given formula.

The satisfiability problem is a famous NP–complete problem and so is the k–satisfiability problem when k is greater than or equal to three. Monien and Speckenmeyer [Monien and Speckenmeyer 1985] proposed a mechanical theorem proving algorithm based upon a branching technique. Our branching algorithm is its modified version. Since our algorithm is for solving general cases of the satisfiability problem, It also works for the k–satisfiability problem because the k–satisfiability problem is a special form of the satisfiability problem. In this paper, we shall give an average case analysis of our algorithm for solving the k–satisfiability problem. First of all, we shall describe the branching algorithm concept in the following.

Definition 1 [Monien and Speckenmeyer 1985] : Given a set S of clauses over the set V of variables $\{X_1, X_2, ..., X_n\}$, a truth assignment T is a function which gives the truth values to the variables of V. A partial truth assignment T' is a function which gives the truth values to the variables of a subset V' of V. A partial truth assignment T' is called an *autark* with respect to S if and only if, for each clause C in S, C is either made true by T' or none of variables in C is assigned any truth value by T'.

Let $S(n,r)$ denote the set of all conjunctive normal form formulas in which each formula contains at most n variables and at most r clauses, where n and r are integers.

Given a formula $F \in S(n,r)$, consider any clause $C = \{L_1, L_2,..., L_v\} \in F$. The following v partial truth assignments

$$A_1: \quad L_1 = True,$$
$$A_2: \quad L_1 = False, L_2 = True,$$
$$\vdots$$
$$A_v: \quad L_1 = L_2 = \cdots = L_{v-1} = False, L_v = True,$$

will all satisfy the clause C. For these v partial truth assignments, we have the following rules :

(1) Autark Rule : If there exists an A_i, $1 \leq i \leq v$, such that A_i is an autark, then delete all clauses which are satisfied by A_i. Monien and Speckenmeyer [Monien and Speckenmeyer 1985] proved that the remaining set F' of clauses is satisfiable if and only if F is satisfiable.

(2) Branching Rule : If none of A_1, A_2, ..., and A_v is an autark, then we construct v sub-formulas, $F_1, F_2, ...,$ and F_v by the following processes :

(i) If a clause is made true by A_i, then delete it from F.

(ii) If a literal in a clause $C \in F$ is made false by A_i, then delete it from C.

(iii) The remaining set of clauses is called F_i.

Again, Monien and Speckenmeyer [Monien and Speckenmeyer 1985] proved that F is unsatisfiable if and only if each F_i, $1 \leq i \leq v$, is unsatisfiable.

Now, we formally present the branching algorithm as follows :

The Branching Algorithm

Input : A set F of clauses over the variable set $\{X_1, X_2,..., X_n\}$

Output : Whether F is satisfiable or not

Step 1.   if $F = \phi$ then return F satisfiable;

Step 2.   if $\square \in F$ then return F unsatisfiable;

Step 3.   choose any clause $\{L_1, L_2, ..., L_v\} \in F$ with length v;

Step 4.   set $A_1: \quad L_1 = True,$
$$A_2: \quad L_1 = False, L_2 = True,$$
$$\vdots$$
$$A_v: \quad L_1 = L_2 = \cdots = L_{v-1} = False, L_v = True,$$

Step 5.   if there exists an autark A_i, $1 \leq i \leq v$,

then apply the Autark Rule to produce F' and recursively apply the Branching Algorithm to F'.

else apply the Branching Rule to produce $F_1, F_2, ..., F_v$ and recursively apply the Branching Algorithm to $F_1, F_2, ..., F_v$.

In the above algorithm, if none of the A_i's is an autark, then there will be v branches. If any A_i is an autark, then there is only one branch.

There is a slight difference between our branching algorithm and Monien and Speckenmeyer's branching algorithm [Monien and Speckenmeyer 1985]. In Step 3 of each iteration, our algorithm chooses any clause in F while their algorithm uses a shortest clause selection heuristic.

Two simple examples are given to describe the above algorithm as follows :

Example 1 :

Let us consider the following set of clauses:

$$
S: \quad
\begin{cases}
X_1 \vee & \neg X_2 \vee & X_3 & & (1) \\
X_1 \vee & X_2 \vee & & X_4 & (2) \\
& & X_3 \vee & X_4 & (3) \\
& X_2 \vee & & \neg X_4 & (4) \\
& & \neg X_3 \vee & \neg X_4 & (5)
\end{cases}
$$

At first, we choose any clause in S. Suppose that Clause (4) is chosen. Based upon Clause (4), we shall have two partial truth assignments : $(X_2=\text{True})$ and $(X_2=\text{False}, X_4=\text{False})$. In this step, we first check whether $(X_2=\text{True})$ or $(X_2=\text{False}, X_4=\text{False})$ is an autark or not. Since both of them are non–autark, we use the Branching Rule. Two sets S_1 and S_2 are constructed, where S_1 corresponds to the partial truth assignment $(X_2=\text{True})$ and S_1 corresponds to the partial truth assignment $(X_2=\text{False}, X_4=\text{False})$.

$$
\begin{aligned}
S_1: &\quad
\begin{cases}
X_1 \vee & X_3 & & (1)' \\
& X_3 \vee & X_4 & (3)' \\
& \neg X_3 \vee & \neg X_4 & (5)' \\
X_1 & & & (2)'' \\
& X_3 & & (3)''
\end{cases} \\
S_2: &
\end{aligned}
$$

With the above processing, we have a corresponding tree structure as follows :

$$
\begin{array}{ccc}
& S & \\
\swarrow & & \searrow \\
S_1' & & S_2
\end{array}
$$

Now, we need to apply the Branching Algorithm to S_1 and S_2. Suppose that S_1 is tested first. In S_1, each clause has length two. We may suppose that Clause (1)' is chosen. Then we are given two partial truth assignments : $(X_1=\text{True})$ and $(X_1=\text{False}, X_3=\text{True})$. The partial truth assignment $(X_1=\text{True})$ is an autark. By applying the Autark Rule, S_1 becomes

$$
S_{11}: \quad
\begin{cases}
X_3 \vee & X_4 & (3)' \\
\neg X_3 \vee & \neg X_4 & (5)'
\end{cases}
$$

by deleting all clauses which are made true by the partial truth assignment $(X_1=\text{True})$ from S_1.

Now, we apply the branching technique to S_{11}. We choose Clause (3)' and then have two partial truth assignments $(X_3=\text{True})$ and $(X_3=\text{False}, X_4=\text{True})$. Among these two partial truth assignments, (X_3=False, X_4=True) is an autark. Because of this autark, S_{11} becomes an empty set. We thus have found a truth assignment (X_1=True, X_2=True, X_3=False, X_4=True) satisfying the set S of clauses and return "S is satisfiable". The tree structure corresponding to the above processing as follows :

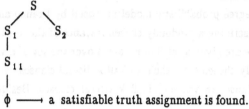

$\phi \longrightarrow$ a satisfiable truth assignment is found.

Example 2 :

Let us consider the following set of clauses :

$$
S : \begin{cases}
X_1 \vee X_2 & \text{(1)} \\
X_1 \vee \neg X_2 & \text{(2)} \\
\neg X_1 \vee X_2 & \text{(3)} \\
\neg X_1 \vee \neg X_2 & \text{(4)}
\end{cases}
$$

In this example, we can see that each clause is with length two. Suppose that we choose Clause (1) and two partial truth assignments, (X_1=True) and (X_1=False, X_2=True), are created. These two partial truth assignments are both non-autarks. Therefore, by the Branching Rule, two sets S_1 and S_2 are constructed by the partial truth assignments (X_1=True) and (X_1=False, X_2=True). S_1 and S_2 are as follows :

$$
S_1 : \begin{cases}
X_2 & \text{(3)'} \\
\neg X_2 & \text{(4)'}
\end{cases}
$$
$$
S_2 : \begin{cases}
\square & \text{(2)"}
\end{cases}
$$

where $\square$ denotes an empty clause. The corresponding tree structure is as follows :

$$
\begin{array}{ccc}
 & S & \\
\swarrow & & \searrow \\
S_1 & & S_2 = \{\square\}
\end{array}
$$

We then need to apply the Branching Algorithm to S_1 and S_2. Suppose that S_1 is processed first, and Clause (3)' is chosen. The corresponding partial truth assignment is (X_2=True). Because (X_2=True) is not an autark, we need to apply the Branching Rule. We can see that only one branch is found, which is $S_{11}=\{\square\}$. Thus, S_{11} is unsatisfiable. The corresponding tree structure becomes

$$
\begin{array}{ccc}
 & S & \\
\swarrow & & \searrow \\
S_1 & & S_2 = \{\square\} \\
| & & \\
S_{11} = \{\square\} & &
\end{array}
$$

Then, we traverse to Set S_2 and find that S_2 contains an empty clause. We finally go back to S and no more branch can be found. The branching algorithm stops and returns "S is unsatisfiable".

Section 3. A Glimpse of the Analysis of the Branching Algorithm

In this paper, we shall give an average case analysis of the Branching Algorithm for solving the k–satisfiability problem.

The constant degree probabilistic model proposed by Franco and Paul [Franco and Paul 1983] is suitable for describing a randomly chosen instance of the k–satisfiability problem. It is defined as follows : We are given a set S of r clauses over the set of n variables $\{X_1, X_2, ..., X_n\}$. Each clause is randomly chosen from the set of all k–literal clauses over $\{X_1, X_2,..., X_n\}$. There are $\binom{n}{k} \cdot 2^k$ different clauses in the set of all k–literal clauses. Besides, each clause is chosen independently with others and all clauses are chosen with the same probability.

Essentially, our average case analysis of the algorithm goes as follows :

(1) We use the size of the proof tree as the measurement of the time–complexity of the Branching Algorithm.

(2) Each node corresponds to a set of clauses. For an unsatisfiable case, each terminating node corresponds to an empty clause.

(3) We can prove that for a randomly chosen set of clauses, if $\lim_{n,r \to \infty} \frac{r}{n} \to \infty$, n is large and k is a constant, for each node with level less than $\left\lceil \frac{\sqrt{n}}{k} \right\rceil$ of the proof tree, the expected number of clauses corresponding to this node is $\mathcal{O}(r)$ and the expected length of clauses corresponding to this node is still k. In other words, for nodes of the proof tree above $\left\lceil \frac{\sqrt{n}}{k} \right\rceil$, it is unlikely that it contains empty clauses because an empty clause has length zero.

(4) In other words, if the input is unsatisfiable, then the depth of the proof tree is expected to be larger than $\left\lceil \frac{\sqrt{n}}{k} \right\rceil$.

(5) We further prove that for each node of the proof tree with level less than $\left\lceil \frac{\sqrt{n}}{k} \right\rceil$, its expected number of branches is k.

(6) Thus, if the input is unsatisfiable, then the expected size of the proof tree will be proved to be larger than $k^{\left\lceil \frac{\sqrt{n}}{k} \right\rceil}$ if certain conditions are satisfied.

(7) In [Franco and Paul 1983], it was proved that the probability that a randomly chosen set of clauses, under the constant degree model, is unsatisfiable approaches to 1. Therfore, we conclude that the average case time–complexity of the Branching Algorithm is exponential under the condition that $\lim_{n,r \to \infty} \frac{r}{n} \to \infty$, n is large and k is a constant.

Section 4. The Analysis of the Branching Algorithm

In this section, we shall discuss an average case analysis of the Branching Algorithm for solving the k–satisfiability problem. By the inferences of Section 3, we shall show that the expected time complexity of the Branching Algorithm for a randomly chosen instance is exponential under the constant degree model. In the following, we assume that we are given a conjunctive normal form formula F in which each clause C is randomly chosen from the set of all k–literal clauses. If a clause $C \in F$, then C has k literals and we denote $C = \{L_1, L_2,..., L_k\}$, where $L_i, 1 \leq i \leq k$, is a literal.

Note that only at the very beginning our data set is a k–literal data set. As soon as a branching is made, it is possible that clauses may have different lengths. For instance, consider the following example :

$$
S: \begin{cases}
X_1 v & X_2 v & X_3 & (1) \\
\neg X_1 v & & X_3 v & X_4 & (2) \\
\neg X_1 v & \neg X_2 v & & X_4 & (3) \\
& & \neg X_3 v & \neg X_4 v & X_5 & (4)
\end{cases}
$$

The above data set is a 3–literal data set. Suppose that we use the first clause to construct the following partial truth assignments :

A_1: $X_1 :=$ True,

A_2: $X_1 :=$ False, $X_2 :=$ True,

A_3: $X_1 :=$ False, $X_2 :=$ False, $X_3 :=$ True.

Since none of them is an autark, we shall have the following sets of clauses, as illustrated in Figure 4–1.

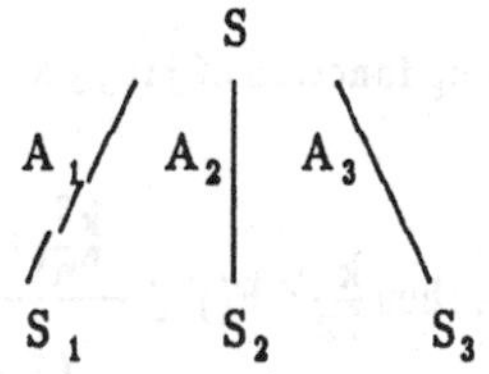

Figure 4–1

In Figure 4–1, S_1, S_2 and S_3 are as follows :

$$
S_1: \begin{cases}
X_3 v & X_4 & (2)' \\
\neg X_3 v & \neg X_4 v & X_5 & (4)'
\end{cases}
$$

$$
S_2: \begin{cases}
X_4 & (3)'' \\
\neg X_4 v & X_5 & (4)''
\end{cases}
$$

$$
S_3: \begin{cases}
\neg X_4 v & X_5 & (4)'''
\end{cases}
$$

It can be seen that for both S_1 and S_2, the lengths of clauses within them vary.

In the following, we shall define several terms and prove several lemmas. Note that if a term or a theorem is based upon k–literal, then it is used only for discussing the initial step of the algorithm.

In the algorithm, we always randomly pick a clause. And, initially, we use the following k partial truth assignments based upon C :

$$A_1: \quad L_1 := \text{True},$$
$$A_2: \quad L_1 := \text{False}, L_2 := \text{True},$$
$$\vdots$$
$$A_k: \quad L_1 := L_2 := \cdots := L_{k-1} := \text{False}, L_k := \text{True}.$$

which all satisfy C.

It may appear that each A_i has a special form . However, note that each A_i is constructed out of a randomly chosen clause C. Therefore, A_i can be considered as a randomly chosen partial truth assignment with i variables. In the rest of the paper, we shall treat A_i as a randomly generated partial truth assignment so far as all other clauses are concerned. Of course, it should be noted that all A_i's satisfy clause C from which it is constructed.

In the following, let $P_{aut}(A_i)$ be the probability that an arbitrary partial truth assignment A_i is an autark with respect to a randomly chosen set S of r k–literal clauses under the constant degree model over n variables.

<u>Lemma 1 :</u> $P_{aut}(A_i) = (1 - \dfrac{\sum\limits_{s=1}^{i} \binom{i}{s}\binom{n-i}{k-s}2^{-s}}{\binom{n}{k}})^r$.

<u>Lemma 2 :</u> $\sum\limits_{s=1}^{j} \binom{j}{s}\binom{n-j}{k-s}2^{-s}$ is an increasing function of j if $j \le k \le \lceil n^{\frac{1}{2}} - 1 \rceil$.

<u>Lemma 3 :</u> If $F(j) = \dfrac{\sum\limits_{s=1}^{j} \binom{j}{s}\binom{n-j}{k-s}2^{-s}}{\binom{n}{k}}$, then $\dfrac{k}{2n} \le F(j) \le \dfrac{\frac{k^2}{2n}(1 - (\frac{k^2}{2n})^k)}{(1 - \frac{k^2}{2n})}$ for $1 \le j \le k \le \lceil n^{\frac{1}{2}} - 1 \rceil$.

<u>Lemma 4 :</u> $P_{aut}(A_i) \le e^{-\frac{rk}{2n}}$ for $1 \le i \le k \le \lceil n^{\frac{1}{2}} - 1 \rceil$.

<u>Theorem 1 :</u> Given initially a set S of r k–literal clauses and a randomly selected clause C, the probability that none of the partial truth assignments A_1, A_2, ..., A_k constructed out of C is an autark approaches to one if $\lim\limits_{n \to \infty} \dfrac{rk}{2n} \to \infty$ and k is a constant.

Thus, in the root node, we can expect that the number of branches to be k. In the following, we shall derive the expected number of branches in al other nodes of the proof tree.

If a clause C is not satisfied by a partial truth assignment, then we say that the clause C survives from the partial truth assignment. And we call two partial truth assignments distinct if no variable is contained in both of these two partial truth assignments.

For a proof tree in the Branching Algorithm, each node corresponds to a set of clauses and each edge corresponds to a partial truth assignment. We shall have the following lemma.

<u>Lemma 5 :</u> Let N_1 and N_2 be two nodes of a proof tree where N_1 is the ancestor node of N_2. Let E be the edge linking N_1 and N_2. Let N_1 correspond to a set S of clauses containing a variables X_1, X_2, ..., X_a and E corresponds to a partial truth assignment A_i containing variables X_1, X_2, ..., X_b, where $b \leq a$. Then the set S' of clauses corresponding to N_2 may only contain variables X_{b+1}, X_{b+2}, ..., X_a.

Lemma 5 implies that every two partial truth assignments corresponding to two edges are distinct in any path from the root node to a node of the proof tree. That is, all partial truth assignments corresponding to the edges of such path in the proof tree are mutually distinct.

<u>Lemma 6 :</u> Given t mutually distinct partial truth assignments At_1, At_2, ..., At_t where each At_m corresponds to v_m variables, $1 \leq m \leq t$, a clause C which is randomly chosen from the set of all k–literal clauses survives from these t partial truth assignments with probability

$$1 - \frac{\sum_{s=1}^{u} \binom{v'}{s}\binom{n-v'}{k-s}2^{k-s}}{\binom{n}{k} \cdot 2^k}, \text{ where } v' = \sum_{m=1}^{t} v_m \text{ and } u = \min(k,v').$$

<u>Lemma 7 :</u> In each node with level less than $\left\lceil \frac{\sqrt{n}}{k} \right\rceil$ in the proof tree, the corresponding expected number of clauses is $\mathcal{O}(r)$ under the condition that k is a constant and n is large.

<u>Lemma 8 :</u> The expected length of a clause which survives in each node with level less than $\left\lceil \frac{\sqrt{n}}{k} \right\rceil$ is k if $\lim_{n,r \to \infty} \frac{r}{n} \to \infty$ and k is a constant.

<u>Theorem 2 :</u> In each branched node with level less than $\left\lceil \frac{\sqrt{n}}{k} \right\rceil$ in the proof tree, the expected number of branches is k with probability tending to 1 if n is large, k is constant, and $\lim_{r,n \to \infty} \frac{r}{n} \to \infty$.

<u>Theorem 3 :</u> The expected size of the proof tree is greater than $k^{\lceil \sqrt{n}/k \rceil}$ with probability 1 if $\lim_{r,n \to \infty} \frac{r}{n} \to \infty$ and k is a constant.

Our result can now be summarized in Theorem 4.

<u>Theorem 4 :</u> If $\lim_{n \to \infty} \frac{r}{n} \to \infty$, k is a constant, and $k \geq 3$, the Branching Algorithm runs in exponential expected time.

Section 5. Concluding Remarks

In this paper, we showed an average case analysis of a mechanical theorem proving algorithm based upon branching techniques for solving the k–satisfiability problem. The branching algorithm was a modified version of Monien and Speckenmeyer's branching algorithm [Monien and Speckenmeyer 1985]. Our analysis showed that based upon the probabilistic model that given r clauses, each clause is randomly chosen from the set of all k–literal clauses over n variables and each clause is chosen independently with others, the branching algorithm runs in exponential expected time under the condition that $\lim_{r,n\to\infty} \frac{r}{n} \to \infty$ and k is a constant.

References

[Bitner and Reingold 1975]   Bitner, J. R. and Reingold, E. M. Backtracking Programming Techniques, Communications of the Association for Computing Machinery, vol.18, no.11, 1975, pp.651–665.

[Chang and Lee 1973] Chang, C. L. and Lee, R. C. T., Symbolic Logic and Mechanical Theorem Proving, Academic Press, New York, 1973.

[Chang 1970]   Chang, C. L., The Unit Proof and the Input Proof in Theorem Proving, Journal of the Association for Computing Machinery, vol. 17, no.4, pp.698–707.

[Chang and Slagle 1971]   Chang, C. L. and Slagle, J. R., Completeness of Linear Resolution for Theories with Equality, Journal of the Association for Computing Machinery, vol. 18, no.1, pp.126–136.

[Cook 1971]   Cook, S. A., The Complexity of Theorem–Proving Procedures, Proceeding Third ACM Symposium on Theory of Computing, 1971, pp.151–158

[Davis and Putnam 1960]   Davis, M. and Putnam, H., A Computing Procedure for Quantification Theory, Journal of the Association for Computing Machinery, vol. 7, no.7, 1960, pp.201–215.

[Franco 1986]   Franco, J., On the Probabilistic Performance of Algorithms for the Satisfiability Problem, Information Processing Letters, vol.23, no.2, 1986, pp.103–106.

[Franco and Paul 1983]   Franco, J. and Paul, M., Probabilistic Analysis of the Davis – Putnam Procedure for Solving the Satisfiability Problem, Discrete Applied Mathematics, vol.5, no.1, 1983, pp.77–87.

[Galil 1977]   Galil, Z., On the Complexity of Regular Resolution and the Davis – Putnam Procedure, Theoretical Computer Science, vol.4, 1977, pp.23–46.

[Garey and Johnson 1979]   Garey, M. R. and Johnson, D. S., Computer and Intractability: A Guide to the Theory of NP–completeness, Freeman, San Francisco, 1979.

[Goldberg 1979]   Goldberg, A., Average Case Complexity of the Satisfiability Problem, Proceeding of Fourth Workshop on Automated Deduction, 1979, pp.1–6.

[Goldberg, Purdom, and Brown 1982]   Goldberg, P. W., Purdom, P. W., and Brown, C. A., Average Time Analyses of Simplified Davis – Putnam Procedures, Information Processing Letters, vol.15, no.2, 1982, pp.72–75.

[Hu, Tang, and Lee 1991]   Hu, T. H., Tang, C. Y., and Lee, R. C. T., An Average Case Analysis of a Resolution Principle Algorithm in Mechanical Theorem Proving, the Annals of Mathematics and Artificial Intelligence, 1991, to appear.

[Iwama 1989]   Iwama, K., CNF Satisfiability Test by Counting and Polynomial Average Time, SIAM Journal on Computing, vol.18, no.2, 1989, pp.385–391.

[Karp 1972]   Karp, R. M., Reducibility among Combinatorial Problem, in R. E. Miller and J. W. Thatcher (eds.), Complexity of Computer Computations, Plenum Press, New York, 1972, pp.85–103.

126

[Monien and Speckenmeyer 1985]   Monien, B. and Speckenmeyer, E., Solving
    Satisfiability in Less Than 2^n Steps, Discrete Applied Mathematics, vol.10, no.3, 1985,
    pp.287–295.
[Purdom 1983]   Purdom, P. W., Search Rearrangement Backtracking and
    Polynomial Average Time, Artificial Intelligence, vol.21, no.1, 1983, pp.117–133.
[Purdom and Brown 1980]   Purdom, D. W. and Brown, C. A., An Analysis of
    Backtracking with Search Rearrangement, Indiana University Computer Science,
    Technique Report, No.89, 1980.
[Purdom and Brown 1981]   Purdom, D. W. and Brown C. A., Polynomial
    Average – Time Satisfiability Problems, Indiana University Computer Science Report,
    No.118, Bloomington, IN.
[Purdom and Brown 1983]   Purdom, D. W. and Brown, C. A., An Analysis of
    Backtracking with Search Rearrangement, SIAM Journal on Computing, vol.12, no.4,
    1983, pp.717–733.
[Purdom and Brown 1985]   Purdom, P. W. and Brown, C. A., The Pure Literal
    Rule and Polynomial Average Time, SIAM Journal on Computing, vol.14, no.4, 1985,
    pp.943–953.
[Robinson 1965]   Robinson, J. A., Machine Oriented Logic Based on the
    Resolution Principle, Journal of the Association for Computing Machinery, vol.12, no.1,
    pp.23–41.
[Tseitin 1968]   Tseitin, G.S., On the Complexity of Derivations in the
    Propositional Calculus, in: A.O. Slisenko ed., Structures in Constructive Mathematics and
    Mathematical Logic, Part II (translated from Russian), 1968, pp.115–125.

An On-line Algorithm for Navigating in Unknown Terrain

Kwong-fai Chan   Tak Wah Lam

Department of Computer Science
University of Hong Kong, Hong Kong

Abstract

Suppose that a robot is required to traverse a terrain spred with impenetrable rectangular obstacles. The robot has no information of the terrain in advance, but it can see and move in any direction. In this paper we construct an on-line algorithm for the robot to determine an obstacle-free path to its destination dynamically. Our algorithm guarantees that the ratio of the distance traversed to the length of the shortest path is optimized.

1   Introduction

With a motivation from motion-planning in robotics, the problem of finding the shortest path for a point object to traverse a terrain with disjoint obstacles has been studied intensively. Most previous works assumed that information of the terrain is fully available in advance, so a complete plan could possibly be made beforehand [1, 8, 11, 12]. This paper, however, focuses on cases where information of the terrain such as location and size of obstacles must be acquired dynamically [9, 5, 13, 14].

The Traversal Problem: In this paper we consider the following problem. There are possibly unlimited number of disjoint impenetrable opaque obstacles spreading over a boundless two-dimensional terrain. An obstacle is a rectangle whose size, location, and orientation is arbitrary. A robot, assumed to be a point object, is initially at a point S and has no knowledge of obstacles on the terrain. The robot can see and traverse in any direction, yet the robot doesn't know the existence of an obstacle unless the obstacle has been within its field of vision. The robot is given a target point T and is required to move to T.

Note that the robot in this problem requires an on-line algorithm [15, 10], in the sense that at any time the algorithm determines the path of the robot by making reference only to obstacles that the robot has seen so far. In other words, the path of the robot is adjusted dynamically. Our primary concern of such algorithm is the total distance traversed by the robot.

Measurement of Performance: To evaluate the performance of an on-line algorithm, we analyze its competitiveness. That is, we compare the performance of an on-line algorithm against the performance of an optimal off-line algorithm which has complete knowledge of the future. Competitive analysis was first introduced by Sleator and Targan [15] for evaluating different heuristics for maintaining a search list, and was then widely recognized to be better than the traditional worst-case and average-case analysis for studying a variety of on-line problems [2, 4, 6, 7, 10, 13].

Let $\mathcal{A}$ be any on-line algorithm for the traversal problem. For any problem instance I, let $D_A(I)$ be the distance traversed by a robot that uses algorithm $\mathcal{A}$, let $D(I)$ be the length of the shortest obstacle-free path. Following the literature, we say that $\mathcal{A}$ is c-competitive if there is a constant a such that for all problem instance I, $D_A(I) \leq cD(I)+a$. The competitiveness coefficient of $\mathcal{A}$ is the infimum of c such that $\mathcal{A}$ is c-competitive.

In some cases, we might be only able to show an on-line algorithm $\mathcal{A}$ to be asymptotically c-competitive, in the sense that there is a sublinear function f (i.e., $\lim_{x \to \infty} \frac{f(x)}{x} = 0$) such that for any problem instance I, $D_A(I) \leq cD(I) + f(D(I))$. Nevertheless, we can prove that $\inf\{c \mid \mathcal{A}$ is c-competitive$\} = \inf\{c \mid \mathcal{A}$ is asymptotic c-competitive$\}$. This means that the competitiveness coefficient of an on-line algorithm is determined by its asymptotic behavior.

Related Works: A similar but more restrictive problem was considered in [9, 5], where it was assumed that the robot can perceive an obstacle only when the robot bumps into it. In this case, it is easy to construct an adverse problem instance in which the robot is blocked by all obstacles, no matter how intelligent the robot is. Thus, the problem becomes less challenging. Papadimitriou and Yannakakis [13] were the first to study the problem considered in this paper. In particular, they showed that if obstacles are arbitrarily thin rectangles (i.e., of unbounded aspect ratio*), then there is no on-line algorithm that has a bounded competitiveness coefficient. When all obstacles were unit-size squares parallel to the horizontal axis, they could find an asymptotic $1\frac{1}{2}$-competitive algorithm, however. In this case, they also proved that no algorithm has a competitiveness coefficient less than $1\frac{1}{2}$, though their proof relies on the assumption that the robot always walks along-side obstacles. Recently Chan [3] has shown that if this assumption is taken out, a 1.21-competitive on-line algorithm can be constructed for the adverse problem instance used in that proof. In fact, the robot can move arbitrarily and may sometimes move backward in order to widen its view.

Summary of Results: As we are not interested in cases where the distance between S and T, denoted $d(S,T)$, is too small in comparing with the size of obstacles, we assume that there is a sublinear real function g such that the length of any obstacle's edge is bounded by $g(d(S,T))$. In this paper we construct an asymptotic $(\frac{r}{2} + 1)$-competitive on-line algorithm for the traversal problem in which every obstacle has an aspect ratio bounded by some constant r. The algorithm is optimal as we can prove that no on-line algorithm has a competitiveness coefficient less than $\frac{r}{2} + 1$. These results imply that the complexity of the traversal problem is characterized by the aspect ratio rather than the actual size of obstacles.

The remainder of the paper is organized as follows. First, a simple algorithm is presented to introduce some of the ideas. It is also used in the final algorithm. Then, the optimal algorithm as well as the analysis is given. Finally, the lower bound is shown.

Definitions and Notations: Consider any two points p, q on the terrain. Let $d(p,q)$ denote the Euclidean distance between p and q. If p is apart from the destination T, then pT-axis is defined as a vector pointing from p to T. To measure how much a traversal from a point q_1 to another point q_2 has advanced in the direction parallel to pT-axis, a ratio is defined as follows. Let t be the actual distance traversed, and let q_1' (q_2') be the projection of q_1 (q_2) on pT-axis. Then the pT-advancement ratio is equal to

---

*The aspect ratio of a rectangle is the ratio of its longer side to its shorter side.

$d(q_1', q_2')/t$. Intuitively, a small pT-advancement ratio means the robot has little progress in the direction parallel to the pT-axis and is not desirable.

2 A Simple Algorithm

The following algorithm, called *Along-ST-Axis*, guides the robot to traverse from S to T. The idea is based on the optimal algorithm by Papadimitriou and Yannakakis [13] for problem instances with horizontal square obstacles.

Traverse along ST-axis towards T. When an obstacle is met, traverse along-side the obstacle until ST-axis is reached again. See figure 1. Note that there are always two different paths to traverse along-side the obstacle; pick the one with which at the first instance in moving away from ST-axis, the change of the direction of traversal is no more than $\frac{\pi}{2}$. After reaching ST-axis again, continue to traverse along ST-axis.

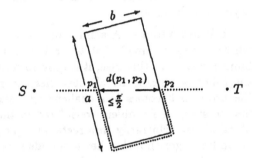

Figure 1: *Along-ST-Axis.*

The performance of *Along-ST-Axis*, though far from optimal, is $(2r+1)$-competitive.

Proposition 1 Applying the algorithm *Along-ST-Axis*, the robot traverses a distance of at most $(2r+1)d(S,T)$.

Proof: It is sufficient to show that on average the ST-advancement ratio is at least $1/(2r+1)$. When the robot traverses along ST-axis, the ratio is always 1. Consider the situation when the robot circumvents an obstacle. Let p_1 be the first point at which the robot meets the obstacle, and let p_2 be the point at which it leaves the obstacle. Let a (b) be the length of the first edge (the second edge) touched by the robot. (See figure 1) In circumventing the obstacle, the robot achieves an ST-advancement ratio of at least $d(p_1, p_2)/(2a+b)$. As $d(p_1, p_2) \geq b$ and $\frac{a}{b} \leq r$, the ratio is at least by $1/(2r+1)$. $\quad\square$

3 Circumventing an Obstacle

Consider that the robot meets an obstacle at a point p and its way to the destination T is blocked. We say that the robot circumvents the obstacle if the robot walks along-side the obstacle to a point q such that the line segment joining q and T no longer intersects with the obstacle. Notice that the point q is not unique. In this section we show that if p is not too close to T, the robot must be able to circumvent the obstacle in such a way that the pT-advancement ratio is at least $1/(\frac{r}{2}+1)$. Intuitively, the ratio guarantees that the robot is getting closer to T by a factor of roughly $1/(\frac{r}{2}+1)$ of the distance just traversed.

Suppose the robot meets a rectangular obstacle α at a point p. If p is one of the four corners of the obstacle, the robot can see two edges of α at the same time. One of the two edges must make an angle of at most $\frac{\pi}{4}$ with pT-axis. Traversing along-side that edge up to the end, the robot can circumvent α with a pT-advancement ratio of at least $\cos\frac{\pi}{4}$ which is greater than $1/(\frac{r}{2}+1)$.

Next, we study the cases where p is not a corner of α. Consider the orientation of α with reference to pT-axis. There are two possible cases as shown in figure 2. As the

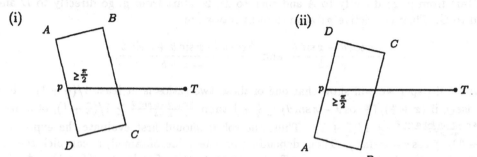

Figure 2: Two different orientations of α with respect to pT-axis.

two cases are symmetric, only case (i) is considered here. Assume the corners of α are labeled as shown in figure 2(i). Let θ be the acute angle between the edge from AD and any line perpendicular to pT-axis. Let $d(A, D) = a$, $d(A, B) = b$, and $d(A, p) = x$. Note that the robot at p can find out θ, a, and x but not b. Depending on different values of θ, a, and x, the robot makes different moves.

(a) $\frac{\pi}{4} < \theta < \frac{\pi}{2}$: Consider the traversal which starts from p and goes directly to D. Unless the robot is very close to T (i.e., $d(p,T) < \sqrt{2}a$), the robot must have circumvented α. The pT-advancement ratio is again at least $\cos\frac{\pi}{4}$ which is greater than $\frac{r}{2}+1$. ♠

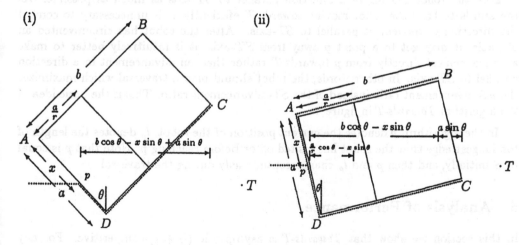

Figure 3: Circumventing an obstacle.

(b) $\theta \le \frac{\pi}{4}$ and $\theta \ge \arctan\frac{a}{rx}$: See figure 3(i). Consider the traversal which starts from p, goes directly to D and then to C, the pT-advancement ratio is

$$\frac{b\cos\theta - x\sin\theta + a\sin\theta}{a - x + b}.$$

Through some careful analysis, we can prove that this ratio is strictly greater than $1/(\frac{r}{2}+1)$. Details will be given in the full paper. ♠

(c) $\theta \leq \frac{\pi}{4}$ and $\theta < \arctan\frac{a}{rx}$: See figure 3(ii). Consider the following two traversals: 1. start from p, go directly to A and then to B; 2. start from p, go directly to D and then to C. Their respective pT-advancement ratios are

$$\frac{b\cos\theta - x\sin\theta}{x+b} \quad \text{and} \quad \frac{b\cos\theta - x\sin\theta + a\sin\theta}{a - x + b}.$$

In the full paper we will show that one of these two ratios is at least $1/(\frac{r}{2}+1)$. More precisely, if $(x + \frac{a}{r})/(\frac{a}{r}\cos\theta - x\sin\theta) \leq \frac{r}{2}+1$ then $\frac{b\cos\theta - x\sin\theta}{x+b} \geq 1/(\frac{r}{2}+1)$; otherwise $\frac{b\cos\theta - x\sin\theta + a\sin\theta}{a-x+b} \geq 1/(\frac{r}{2}+1)$. Thus, the robot should first evaluate the expression $(x + \frac{a}{r})/(\frac{a}{r}\cos\theta - x\sin\theta)$. Then, depending on the value obtained, it can pick the appropriate path which guarantees a pT-advancement ratio of at least $1/(\frac{r}{2}+1)$. ♠

In summary, the robot can always circumvent the obstacle with a pT-advancement ratio of at least $1/(\frac{r}{2}+1)$. As a matter of fact, in some situations such as the one in figure 3(i), though the robot has already reached a point at which the obstacle no longer blocks it from T, it still continues to walk along-side the obstacle in order to achieve a large enough pT-advancement ratio.

4 An Asymptotic $(\frac{r}{2}+1)$-Competitive Algorithm

The basic idea of *Along-ST-Axis* and Papadimitriou and Yannakakis's algorithm [13] is to keep the robot moving in a direction parallel to ST-axis as much as possible. As the aim is to have the robot moving towards T efficiently, it is unnecessary to compel the direction of movement parallel to ST-axis. After the robot has circumvented an obstacle, it may get to a point p away from ST-axis, it is intuitively better to make an advancement directly from p towards T rather than an advancement in a direction parallel to ST-axis. In other words, the robot should plan a traversal which maximizes the pT-advancement ratio instead of the ST-advancement ratio. This is the basic idea of the algorithm *Towards-T* in figure 4.

In the algorithm, p denotes the current position of the robot; l_p denotes the length of the longest edge that the robot has visited on or before reaching p. Note that p is equal to S initially, and then p and l_p change continuously during the traversal.

5 Analysis of Performance

In this section we show that *Towards-T* is asymptotic $(\frac{r}{2}+1)$-competitive. For any problem instance, let P be the point at which the robot enters Phase 2 of the algorithm; let l be the value l_P, i.e., the length of the longest edge visited by the robot during the traversal from S to P. The performance of each phase is analyzed separately as

---

Phase 1

1. Traverse straightly towards T until $d(p,T) \leq (r^2 + 6r)l_p$ or an obstacle is met.

2. If $d(p,T) \leq (r^2 + 6r)l_p$ then goto Phase 2; if an obstacle is met, circumvent the obstacle according to the rules specified in section 3 and then go back to Step 1.

Phase 2

1. Consider the current position p of the robot as a new starting point and use the algorithm *Along-ST-Axis* to traverse from p to T.

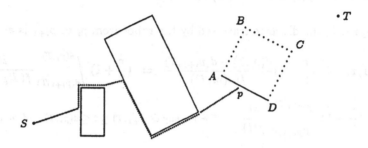

Figure 4: Algorithm *Towards-T*.

follows: Phase 1—the total distance traversed from S to P is at most $(\frac{r}{2} + 1)d(S,T) + O(l \log \frac{d(S,T)}{l})$; Phase 2—the distance traversed from P to T is at most $(2r+1)(r^2+6r)l$. Note that any feasible path from S to T is of length at least $d(S,T)$.

Proposition 2 *From S to P, the robot traverses a distance of at most*

$$(\frac{r}{2} + 1)d(S,T) + (\frac{r}{2} + 1)^2 l \log_e \frac{d(S,T)}{l}.$$

Proof: Suppose the robot has circumvented a sequence of obstacles $\alpha_1, \alpha_2, \cdots, \alpha_m$ in Phase 1. Define a sequence of points $p_0, p_1, p_2, \cdots, p_k$ on the terrain such that $p_0 = S$; for $1 \leq i \leq m$, p_{2i-1} is the point at which the robot reaches α_i and p_{2i} is the point at which it has just circumvented α_i. Clearly, p_{2m} is the point at which the robot leaves the last obstacle in Phase 1. Define $k = 2m$ if $p_{2m} = P$; otherwise define $k = 2m + 1$ and $p_k = P$.

For $0 \leq i < k$, let p'_{i+1} be the projection of p_{i+1} on $p_i T$-axis, and let x_i denote the distance $d(p_i, p'_{i+1})$. Intuitively, x_i measures how much the robot in the traversal from p_i to p_{i+1} has advanced in the direction of $p_i T$-axis. In the following we show that the distance traversed by the robot from p_i to p_{i+1} is at most $(\frac{r}{2} + 1)x_i$, and prove that x_i itself can be bounded by a factor of how much the robot has shortened its distance from T (i.e., $d(p_i, T) - d(p_{i+1}, T)$). Since the sum of $d(p_i, T) - d(p_{i+1}, T)$ over all i in the range $[0, k-1]$ is exactly $d(p_0, T) - d(p_k, T)$, this gives us a way to bound the sum of all x_i and the overall distance traversed.

For any even i, the robot moves strictly from p_i towards T and stops at p_{i+1}. Thus, p_{i+1} is on $p_i T$-axis and the distance traversed from p_i to p_{i+1} is exactly x_i. For any odd i, the robot circumvents an obstacle in traversing from p_i to p_{i+1}. By the results in section 3, the $p_i T$-advancement ratio is at least $1/(\frac{r}{2}+1)$. The distance traversed from p_i to p_{i+1} is at most $(\frac{r}{2}+1)x_i$. Moving from p_i to p_{i+1}, the robot shortens its distance from T by $d(p_i, T) - d(p_{i+1}, T)$. Intuitively, if p_i is far away from T, x_i can be approximated by $d(p_i, T) - d(p_{i+1}, T)$. Their actual relationship is stated in Lemma 3. The proof will be provided in the full paper.

Lemma 3 For $0 \leq i < k$, $x_i \leq (d(p_i, T) - d(p_{i+1}, T))/f(d(p_i, T))$, where f is a real function defined as $f(t) = 1 - (\frac{r}{2}+1)l/(t-2l)$.[†]

For $0 \leq i < k$, the distance traversed by the robot from p_i to p_{i+1} is at most

$$(\frac{r}{2}+1)x_i \ \leq \ (\frac{r}{2}+1)\frac{d(p_i, T) - d(p_{i+1}, T)}{f(d(p_i, T))} \ = \ (\frac{r}{2}+1)\int_{d(p_{i+1}, T)}^{d(p_i, T)} \frac{dt}{f(d(p_i, T))}$$

$$\leq \ (\frac{r}{2}+1)\int_{d(p_{i+1}, T)}^{d(p_i, T)} \frac{dt}{f(t)}. \quad \%\% \text{ for any } d(p_{i+1}, T) \leq t \leq d(p_i, T), f(d(p_i, T)) \geq f(t).$$

The total distance traveled from S ($= p_0$) to P ($= p_k$) is bounded by

$$\sum_{0 \leq i < k} (\frac{r}{2}+1)\int_{d(p_{i+1}, T)}^{d(p_i, T)} \frac{dt}{f(t)} \ = \ (\frac{r}{2}+1)\int_{d(P, T)}^{d(S, T)} \frac{dt}{f(t)}.$$

Now we need a lower bound of $d(P, T)$. Step 1 of *Towards-T* seems to suggest $d(P, T) = (r^2 + 6r)l$, but this isn't correct for some cases. A more careful analysis shows that $d(P, T)$ is indeed at least $(r+4)l$. Therefore,

$$(\frac{r}{2}+1)\int_{d(P, T)}^{d(S, T)} \frac{dt}{f(t)} \ \leq \ (\frac{r}{2}+1)\int_{(r+4)l}^{d(S, T)} \frac{dt}{f(t)} \ \leq \ (\frac{r}{2}+1)d(S, T)+(\frac{r}{2}+1)^2 l\log_e(\frac{d(S, T)}{l}). \quad \square$$

Proposition 4 From P to T, the robot traverses a distance of at most $(2r+1)(r^2 + 6r)l$.

Proof: P, the starting point of Phase 2, is at a distance of at most $(r^2 + 6r)l$ from T. By proposition 1, the distance traversed is at most $(2r+1)(r^2 + 6r)l$. $\square$

In conclusion, the total distance traversed in the two phases is bounded by

$$(\frac{r}{2}+1)d(S, T) + (\frac{r}{2}+1)^2 l\log_e \frac{d(S, T)}{l} + (2r+1)(r^2 + 6r)l.$$

The $(\frac{r}{2}+1)$ competitiveness coefficient of *Towards-T* follows from the fact that for some sublinear real function h, $(\frac{r}{2}+1)^2 l\log_e \frac{d(S, T)}{l} + (2r+1)(r^2 + 6r)l$ can be bounded by $h(d(S, T))$.

[†]Note that for $t \geq (\frac{r}{2}+3)l$, $f(t)$ is between 0 and 1 and is strictly increasing.

6   A Lower Bound

In this section we show that any on-line algorithm for the traversal problem has a competitiveness coefficient of at least $\frac{r}{2} + 1$. This implies that the algorithm *Towards-T* has an optimal competitiveness coefficient.

Let A be any on-line algorithm for the traversal problem. In the following we construct an adverse problem instance I_A against which the robot using A would perform poorly. Most obstacles in I_A are fixed in advance, but some are defined gradually as the robot moves towards T and their locations depend on the route traversed by the robot. To ensure I_A to be well defined, new obstacles are defined only in an area that has not been within the view of the robot so far. From the viewpoint of the robot, all obstacles are stationary.

In I_A all obstacles are rectangles of size exactly $r \times 1$. Let δ be any positive number less than $\frac{1}{5}$, and let n be an integer greater than 1. The starting point S and the destination T are at a distance of $2n(1 + \delta)$ apart. Obstacles are arranged into $2n + 1$ columns as shown in figure 5. Each obstacle is at a distance of δ from others. There are infinite

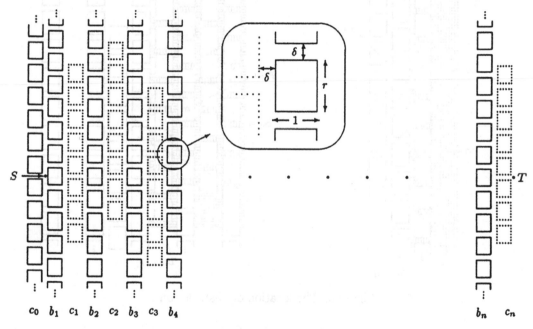

Figure 5: adverse problem instance I_A.

obstacles on column c_0 and columns $b_1, b_2, \cdots, b_n$. Their positions are fixed in advance, and their function is to block the view of the robot. On columns $c_1, c_2, \cdots, c_n$, there are, however, only eight consecutive obstacles. For any $1 \leq i \leq n$, the locations of obstacles on c_i are defined inductively according to the trajectory of the robot.

Base case, $i = 1$: Let R_1 be the shaded region in figure 6(i). As R_1 encloses S but not T, the robot must cross the boundary of R_1 eventually. Let p_1 be the point where the robot reaches the boundary of R_1 for the first time. Note that p_1 is somewhere between

c_0 and b_1. During the traversal from S to p_1, the robot is unable to see anything to the right of b_1. Let α be the obstacle on b_1 closest to p_1. Then the eight consecutive obstacles on c_1 are positioned in such a way that α is next to their middle. See figure 6(i).

Inductive step, $1 < i \leq n$: Assume that p_{i-1} and the obstacles on c_{i-1} have been defined. Let R_i be the shaded region in figure 6(ii), which is defined with respect to the obstacles on c_{i-1}. Clearly, R_i encloses p_{i-1} but not T. Suppose the robot, after its first arrival at p_{i-1}, traverses to the boundary of R_i for the first time at a point p_i. Note that p_i may not lie between c_{i-1} and b_i. Up to this moment, the robot has never got a chance to see anything beyond b_i. Among all obstacles on b_i, let α be the one closest to p_i. Then the eight consecutive obstacles on c_{2i} are positioned in such a way that α is next to their middle. See figure 6(ii).

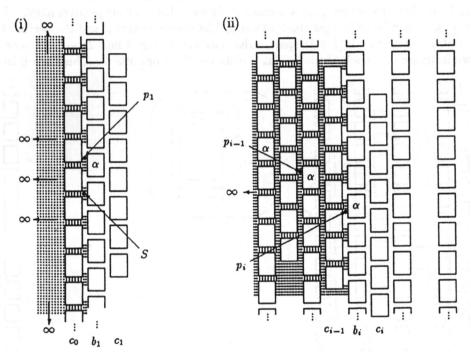

Figure 6: The location of obstacles on c_i.

Proposition 5 *For any $1 \leq i < n$, the length of the shortest route from p_i to p_{i+1} is at least $r + 2 - \delta$.*

The shortest route from p_n to T is also at a distance of at least $r + 2 - \delta$. Thus, for the problem instance I_A, the algorithm A would guide the robot to traverse a distance of at least $(r + 2 - \delta)n$. Next, we show a good off-line solution for I_A.

Proposition 6 *For some integer $k \leq \sqrt{n}$, there is a horizontal line which intersects at most $\sqrt{n}$ obstacles on I_A and is at a distance of $(8k + \frac{1}{2})(r + \delta)$ away from the ST-axis.*

By proposition 6, we can construct a feasible route of length at most $2n(1+\delta)+51r\sqrt{n}$ for I_A.

Theorem 7 *Let A be any on-line algorithm for the traversal problem. If A is c-competitive, then c is at least $\frac{r}{2}+1$.*

Corollary 8 *Any on-line algorithm for the traversal problem has a competitiveness coefficient of at least $\frac{r}{2}+1$.*

References

[1] T. Asano, T. Asano, L. Guibas, J. Hershberger and H. Imai, Visibility of Disjoint Polygons, *Algorithmica*, 1, 1986, 49-63.

[2] M. Bern, D.H. Greene, A. Raghunathan, and M. Sudan, Online Algorithms for Locating Checkpoints, *Proceedings of the 22nd Annual ACM Symposium on Theory of Computing*, 1990, 359-368.

[3] K.F. Chan, Two Results in Algorithm Design, *Manuscript*.

[4] R. Cole, and A. Raghunathan, Online Algorithms for Finger Searching, *Proceedings of 31st Annual Symposium on Foundations of Computer Science*, 1990, 480-489.

[5] P. Eades, X. Lin and N.C. Wormald, Performance Guarantees For Motion Planning With Temporal Uncertainty, *Manuscript*.

[6] A. Fiat, Y. Tabani, Y. Ravid, Competitive k-Server Algorithms, *Proceedings of 31st Annual Symposium on Foundations of Computer Science*, 1990, 454-463.

[7] A.R. Karlin, M.S. Manasse, L. Rudolph, and D.D. Sleator, Competitive Snoopy Caching, *Algorithmica*, 3, 1988, 79-119.

[8] D.T. Lee, T.H. Chen and C.D. Yang, Shortest Rectilinear Paths Among Weighted Obstacles, *Proceedings of the 6th Annual ACM Symposium on Computational Geometry*, 1990, 301-310.

[9] V.J. Lumelsky and A.A. Stepanov, Path-Planning Strategies For a Point Mobile Automaton Moving Amidst Obstacles of Arbitrary Shape, *Algorithmica*, 2, 1987, 403-430.

[10] M.S. Manasse, L.A. McGeoch, and D.D. Sleator, Competitive Algorithms for On-line Algorithms, *Proceedings of the 20th Annual ACM Symposium on Theory of Computing*, 1988, 322-333.

[11] J.S.B. Mitchell, Planning Shortest Paths, Ph.D. Dissertation, Dept. of Oper. Res., Stanford University, 1986.

[12] J.S.B. Mitchell and C.H. Papadimitriou, The Weighted Region Problem: Finding shortest Paths Through a Weighted Planar Subdivision, *Journal of the Association for Computing Machinery*, 38(1), 1991, 18-73.

[13] C.H. Papadimitriou and M. Yannakakis, Shortest Paths Without a Map, *Proceedings of ICALP 1989*, 610-620.

[14] J.F. Schwartz and M. Sharir, Algorithmic Motion Planning in Robotics, *Handbook of Theoretical Computer Science*, 1990, MIT Press, 391-430.

[15] D.D. Sleator and R.E. Tarjan, Amortized Efficiency of List Update and Paging Rules, *Communications of the ACM*, 28(2), 1985, 202-208.

On Maintaining the Width and Diameter of a Planar Point-set Online[1]

Ravi Janardan

Department of Computer Science, University of Minnesota
Minneapolis, MN 55455, U.S.A.

Abstract. Efficient online algorithms are presented for maintaining the (almost-exact) width and diameter of a dynamic planar point-set, S. Let n be the number of points currently in S, let W and D denote the width and diameter of S, respectively, and let α and β be positive, integer-valued parameters. The algorithm for the width problem uses $O(\alpha n)$ space, supports updates in $O(\alpha \log^2 n)$ time, and reports in $O(\alpha \log^2 n)$ time an approximation, $\hat{W}$, to the width such that $\hat{W}/W \leq \sqrt{1 + \tan^2 \frac{\pi}{4\alpha}}$. The algorithm for the diameter problem uses $O(\beta n)$ space, supports updates in $O(\beta \log n)$ time, and reports in $O(\beta)$ time an approximation, $\hat{D}$, to the diameter such that $\hat{D}/D \geq \sin(\frac{\beta}{\beta+1} \frac{\pi}{2})$. Thus, for instance, even for α as small as 5, $\hat{W}/W \leq 1.01$, and for β as small as 11, $\hat{D}/D \geq .99$. All bounds stated are worst-case. Both algorithms, but especially the one for the diameter problem, use well-understood data structures and should be simple to implement. The diameter result yields a fast implementation of the greedy heuristic for maximum-weight Euclidean matching and an efficient online algorithm to maintain approximate convex hulls in the plane.

1   Introduction

Let S be a finite set of n points in the plane. Two attributes that are used often to measure the spread of the points of S are its width and diameter. The *width*, W, of $S

[1] Research supported in part by a Graduate School Faculty Summer Research Fellowship and by the Army High Performance Computing Research Center, both at the University of Minnesota. E-mail address: janardan@cs.umn.edu.

is the distance between the closest pair of parallel lines that support S (i.e., parallel lines that enclose S). The *diameter*, D, of S is the maximum Euclidean distance between two points of S. The width of S can be computed in $O(n \log n)$ time and $O(n)$ space and has applications to collision-avoidance problems and to the approximation of polygonal curves [HT85]. The diameter of S can also be computed in $O(n \log n)$ time and $O(n)$ space and is used extensively in clustering applications [PS88].

The algorithms cited above are optimal but assume that S is fixed and does not change over time. In many applications, however, S is modified through insertion and deletion of points and from time to time it is necessary to determine the current width or diameter. Moreover, the updates and queries are performed online, i.e., they are not known beforehand and each operation must be performed before the next one can be seen. This motivates the problem of designing online width- and diameter-maintenance algorithms that are more efficient than doing a straightforward $O(n \log n)$-time recomputation.

Previous work on these problems is as follows. For the width problem, Agarwal and Sharir [AS91] have recently considered a restricted, offline version of the problem: Given a sequence of n updates on S and a real number $K > 0$, does the width of S ever become less than or equal to K during the sequence? In [AS91] an $O(n \log^3 n)$-time algorithm is presented for this problem and the existence of an online algorithm is posed as an open problem. For the diameter problem, an online algorithm with update time $O(\sqrt{n} \log n)$ is sketched in [Sup90]. Dobkin and Suri [DS89] have given a solution which uses $O(n)$ space, $O(\log^2 n)$ amortized update time, and $O(1)$ query time. However, their result holds only in the semi-online model, in which insertions are fully online but the deletion time of each point must be announced at the time it is inserted. In [DS89], an online algorithm with $O(n)$ update time is also attributed to Overmars [Ove81].

In this paper we present efficient, online algorithms for maintaining the width and diameter of S. Although our algorithms do not provide the exact width or diameter, they provide approximations that can be made arbitrarily close to optimal. As noted by Bentley, Faust, and Preparata [BFP82], such solutions are quite appropriate in, for example, statistical applications, where the input data is not exact anyway, but is known to a certain precision. The following theorems summarize our main results:

Theorem 1 [WIDTH] *Let S be a finite set of n points in the plane. Let W be the width of S and let $\alpha \geq 1$ be an integer-valued parameter. There is an online algorithm for maintaining S under insertions, deletions, and width queries. The algorithm uses $O(\alpha n)$ space, supports updates in $O(\alpha \log^2 n)$ time, and reports in $O(\alpha \log^2 n)$ time a pair of parallel supporting lines of S with distance $\hat{W}$, such that $\hat{W}/W \leq \sqrt{1 + \tan^2 \frac{\pi}{4\alpha}}$. (Thus, for instance, for*

$\alpha = 5$, $\hat{W}/W \leq 1.01$ and for $\alpha = 16$, $\hat{W}/W \leq 1.001$.) All bounds are worst-case. $\square$

Theorem 2 [DIAMETER] Let S be a finite set of n points in the plane. Let D be the diameter of S and let $\beta \geq 1$ be an integer-valued parameter. There is an online algorithm for maintaining S under insertions, deletions, and diameter queries. The algorithm uses $O(\beta n)$ space, supports updates in $O(\beta \log n)$ time, and reports in $O(\beta)$ time a pair of points of S at distance $\hat{D}$, such that $\hat{D}/D \geq \sin(\frac{\beta}{\beta+1}\frac{\pi}{2})$. (Thus, for instance, for $\beta = 11$, $\hat{D}/D \geq .99$ and for $\beta = 35$, $\hat{D}/D \geq .999$.) All bounds are worst-case. $\square$

Both algorithms use well-understood data structures. In particular, the algorithm for the diameter problem uses only balanced binary search trees and so should be quite easy to implement. From the diameter result, we get two additional results: (1) An efficient implemention of the greedy heuristic for maximum-weight Euclidean matching in the plane, which is faster than a previous algorithm [Sup90] by a $\sqrt{n}$-factor, but at the expense of a slightly inferior matching and (2) An efficient online algorithm for maintaining an approximation to the convex hull of a dynamic point-set in the plane. See Theorems 3 and 4 in Section 3.

For lack of space, some proofs and details are omitted here but can be found in the full paper [Jan91].

2 The dynamic width problem

As noted in [AS91] the width problem does not appear to be decomposable in the usual ways [Ove81, DS89] and so known dynamization techniques cannot be applied. Another approach is to consider a characterization of width based on the convex hull CH(S) of S. Let e be any edge of CH(S) and v any vertex. Let ℓ_e be the supporting line of e and ℓ_v the line through v that is parallel to ℓ_e. The pair (e, v) is called an *antipodal pair* of CH(S) if ℓ_e and ℓ_v support CH(S). It is well-known [HT85] that the width of S is the minimum distance between lines ℓ_e and ℓ_v, taken over all antipodal pairs (e, v) of CH(S). Thus one could maintain CH(S) using the Overmars-van Leeuwen algorithm [OvL81] and search it for the antipodal pair which admits parallel supporting lines with the smallest distance. Unfortunately, there seems to be no efficient way of conducting this search.

We use geometric duality to solve the problem in the dual space where the interpretation of distance between parallel lines has a nice monotonicity property, which permits efficient searching. However, in the dual space, the distance between parallel lines can get distorted considerably. We show how to ameliorate this problem by maintaining a family of coordinate systems so that in at least one system the distortion is small.

2.1 Interpreting width in the dual space

Let $\mathcal{F}$ be the following duality transform: If p is the point (a, b), then $\mathcal{F}(p)$ is the line $y = ax + b$ and if ℓ is the (non-vertical) line $y = mx + c$, then $\mathcal{F}(\ell)$ is the point $(-m, c)$. (Thus parallel lines are mapped to points with the same abscissa.) Map the points of S to lines in the dual space and orient each line so that it points upwards (rightwards if it is horizontal). Let L (resp. R) be the boundary of the intersection of the halfplanes that lie to the left (resp. right) of the lines. Let U (resp. D) be the upward-convex (resp. downward-convex) chain of $CH(S)$ lying between its leftmost and rightmost vertices, both inclusive. It is well-known that each vertex of U maps to the supporting line of an edge of L. In particular, the supporting line of an edge (u, v) of U is the vertex of L formed by the intersection of $\mathcal{F}(u)$ and $\mathcal{F}(v)$. Analogous remarks apply to D and R. (See Figure 1.)

Let v be any vertex of $CH(S)$, say on D. Let e_1 and e_2 be the edges incident with v and let ℓ_1 and ℓ_2 be their supporting lines. Consider a line ℓ that rotates about v, always supporting $CH(S)$. If v is not the leftmost or rightmost vertex of $CH(S)$, then in the dual space, the set of allowable lines ℓ maps to the line segment $\mathcal{F}(v) \cap R$, with endpoints $\mathcal{F}(\ell_1)$ and $\mathcal{F}(\ell_2)$. (For example, in Figure 1, if $v = 7$, then $\mathcal{F}(7) \cap R$ is the segment labeled $7'$.) If v is the leftmost or rightmost vertex of $CH(S)$, then the set of allowable lines ℓ maps to the rays $\mathcal{F}(v) \cap R$ and $\mathcal{F}(v) \cap L$, originating at $\mathcal{F}(\ell_1)$ and $\mathcal{F}(\ell_2)$, respectively (assuming that $e_1 \in D$ and $e_2 \in U$). Thus, in Figure 1, if $v = 5$, then the rays are the ones labeled $5'$.

Let (e, v) be an antipodal pair of $CH(S)$ and wlog let $e \in U$. Since U is convex, $v \in D$. Let ℓ_e and ℓ_v respectively be the parallel lines through e and v that support $CH(S)$. Then $\mathcal{F}(\ell_v)$ is the point at which the vertical line through the point $\mathcal{F}(\ell_e)$ on L intersects R. Similarly if $e \in D$ and $v \in U$. In Figure 1, if $e = (3, 4)$, then $v = 7$ and $\mathcal{F}(\ell_7)$ is the point on R that lies vertically below the common endpoint of the edges $3'$ and $4'$.

2.2 The searching strategy

For any vertex p on L, let the distance of p to R, denoted by $\rho(p, R)$, be the length of the vertical line segment which originates at p and ends at a point on R. The following lemma is the basis of our searching algorithm.

Lemma 1 *Among the vertices of L, let p_0 be a vertex for which $\rho(p_0, R)$ is smallest. Index the vertices on L that are to the right of p_0 as $p_1, p_2, \ldots$, in order from p_0. Then there is an index $k \geq 0$ such that $\rho(p_i, R) = \rho(p_0, R)$, $0 \leq i \leq k$, and $\rho(p_i, R) > \rho(p_{i-1}, R)$ for $i > k$. (A similar statement applies to vertices on L that are to the left of p_0 and to vertices on R.)*

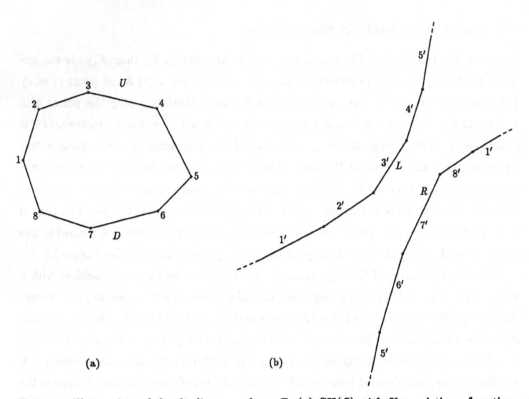

Figure 1: Illustration of the duality transform $\mathcal{F}$. (a) CH(S) with U consisting of vertices 1–5 and D consisting of vertices 5–8, 1. (b) The duals L of U and R of D. Note that for each vertex i of CH(S), we have labeled with i' the edge(s) supported by the line $\mathcal{F}(i)$. (To avoid clutter, the supporting lines themselves are not shown.)

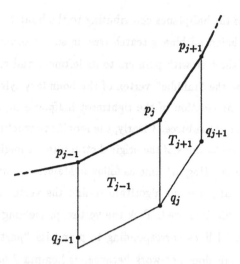

Figure 2: Proving monotonicity of ρ in Lemma 1.

Proof   Suppose that the lemma is false. Then there is a $j \geq 1$ such that $\rho(p_{j-1}, R) <$ $\rho(p_j, R)$ and $\rho(p_j, R) \geq \rho(p_{j+1}, R)$. Let q_{j-1}, q_j, and q_{j+1} be the points at which the vertical lines through p_{j-1}, p_j, and p_{j+1}, respectively, meet R. Thus the segments $p_{j-1}q_{j-1}, p_jq_j$, and $p_{j+1}q_{j+1}$ have lengths $\rho(p_{j-1}, R)$, $\rho(p_j, R)$, and $\rho(p_{j+1}, R)$, respectively. Consider Figure 2, which shows two parallelograms, T_{j-1} and T_{j+1}, where two of the sides of T_{j-1} are $p_{j-1}p_j$ and p_jq_j and two of the sides of T_{j+1} are p_jp_{j+1} and p_jq_j. Since $\rho(p_{j-1}, R) < \rho(p_j, R)$, q_{j-1} must lie strictly above the lower-left corner of T_{j-1} and since $\rho(p_{j+1}, R) \leq \rho(p_j, R)$, q_{j+1} cannot lie below the lower-right corner of T_{j+1}. (The figure shows q_{j+1} at this corner, its lowest possible position.) It follows that the portion of R from q_{j-1} to q_{j+1} cannot belong to the convex boundary of the intersection of right halfplanes — a contradiction. $\square$

Lemma 1 suggests that p_0 can be found by binary search: Let p_{mid} be the middle vertex of L and let p_l (resp. p_r) be its neighboring vertex to the left (resp. right). Compute $\rho(p_{mid}, R)$, $\rho(p_l, R)$, and $\rho(p_r, R)$ by binary search on R. If $\rho(p_{mid}, R) \leq \rho(p_l, R)$ and $\rho(p_{mid}, R) \leq \rho(p_r, R)$ then set p_0 equal to p_{mid} and stop. Otherwise, if $\rho(p_{mid}, R) > \rho(p_l, R)$, then eliminate the portion of L to the right of p_{mid}; otherwise, eliminate the portion of L to the left of p_{mid}. If the vertices of L and R are each stored in order at the leaves of a balanced binary search tree and if $|L| + |R| = n$, then the search takes $O(\log^2 n)$ time.

In [OvL81], Overmars and van Leeuwen show how the intersection of n halfplanes can be maintained in a balanced binary search tree in $O(n)$ space and $O(\log^2 n)$ update time. Their structure makes the halfplanes contributing to the boundary of the intersection available at the leaves of a balanced binary search tree, in sorted order of their slopes. By augmenting each node u of the tree with pointers to its leftmost and rightmost leaves, we can compute in constant time the "middle" vertex of the boundary lying in u's subtree, i.e., the vertex formed by the intersection of the rightmost halfplane in u's left subtree and the leftmost halfplane in u's right subtree. Clearly, the modified structure can still be maintained within the time and space bounds of the original structure. It might now appear that we can report the exact width in $O(\log^2 n)$ time as follows: Maintain L and R in separate instances of the above structure and use our algorithm to find the vertex on L minimizing the distance to R. Repeat this for R as well. Pick the vertex, p, yielding the smaller distance and report the primal parallel lines corresponding to p and its "partner" on the other convex chain. Unfortunately, this does not work because, as Lemma 2 below states, $\mathcal{F}$ does not preserve the distance between parallel lines. Thus the answer produced can be made arbitrarily bad by choosing the slope of the optimal supporting lines suitably large.

Lemma 2 *Let g and h be parallel lines in the plane, with slope m. The distance between the points $\mathcal{F}(g)$ and $\mathcal{F}(h)$ is $\sqrt{1+m^2}$ times the distance between g and h.* $\square$

2.3   The approximate-width maintenance algorithm

We alleviate the above problem as follows: Let $\alpha \geq 1$ be an integer parameter. We maintain a family of coordinate systems, $\mathcal{C}_i = (X_i, Y_i)$, $0 \leq i \leq \alpha$, with positive x- and y-axes X_i and Y_i, respectively, where X_0 is horizontal and for $i = 1, 2, \ldots, \alpha$, X_i makes an angle of $\frac{\pi}{2\alpha}$ with X_{i-1}. The lemma below implies that the optimal pair of supporting lines has "small" slope in one of the $\mathcal{C}_i$.

Lemma 3 *Let $\mathcal{C}_i$, $0 \leq i \leq \alpha$, be a family of coordinate systems as described above. For any line ℓ in the plane, there is a coordinate system in which the absolute value of the slope of ℓ is at most $\tan \frac{\pi}{4\alpha}$.*

Proof   Let θ_0 be the angle (in radians) made by ℓ with X_0. It can be verified that for any i, $0 \leq i \leq \alpha$, the angle, θ_i, that ℓ makes with X_i is either $\theta_0 - \frac{i\pi}{2\alpha}$ or $\pi + \theta_0 - \frac{i\pi}{2\alpha}$. Thus, $|m_i| = |\tan(\theta_0 - \frac{i\pi}{2\alpha})|$. Suppose that $\theta_0 \in [0, \frac{\pi}{2}]$. Thus there is a j, $1 \leq j \leq \alpha$, such that $\theta_0 \in [(j-1)\frac{\pi}{2\alpha}, j\frac{\pi}{2\alpha}]$. If $\theta_0 \leq (j - \frac{1}{2})\frac{\pi}{2\alpha}$, then $\theta_0 - (j-1)\frac{\pi}{2\alpha} \leq \frac{\pi}{4\alpha}$, and so $|m_{j-1}| = |\tan(\theta_0 - (j-1)\frac{\pi}{2\alpha})| \leq \tan \frac{\pi}{4\alpha}$. If $\theta_0 \geq (j - \frac{1}{2})\frac{\pi}{2\alpha}$, then $j\frac{\pi}{2\alpha} - \theta_0 \leq \frac{\pi}{4\alpha}$ and so $|m_j| = |\tan -(j\frac{\pi}{2\alpha} - \theta_0)| = |\tan(j\frac{\pi}{2\alpha} - \theta_0)| \leq \tan \frac{\pi}{4\alpha}$. Similarly if $\theta_0 \in [\frac{\pi}{2}, \pi]$. $\square$

The update and query algorithms are now as follows: Assume that the points of S are specified in $\mathcal{C}_0$. Thus a point $p = (x, y)$ corresponds to the point $p_i = (x \cos \frac{i\pi}{2\alpha} +$

$y \sin \frac{i\pi}{2\alpha}, y \cos \frac{i\pi}{2\alpha} - x \sin \frac{i\pi}{2\alpha})$ in C_i, $0 \leq i \leq \alpha$. For each C_i, $0 \leq i \leq \alpha$, we maintain two instances of the Overmars–van Leeuwen structure, one for the intersection, L_i, of the left halfplanes and for the intersection, R_i, of the right halfplanes. To insert (resp. delete) a point $p = (x, y)$ in S, for each i, $0 \leq i \leq \alpha$, we compute p_i and insert (resp. delete) the halfplane to the left (resp. right) of $\mathcal{F}(p_i)$ into the structure for L_i (resp. R_i). The total time is $O(\alpha \log^2 n)$, where n is the size of S currently. To answer a width query, for each i, $0 \leq i \leq \alpha$, we use the search algorithm based on Lemma 1 to find the vertex of L_i (resp. R_i) for which the distance to R_i (resp. L_i) is minimum and note this vertex and its partner point on R_i (resp. L_i). Among the $2(\alpha+1)$ point pairs thus found, we determine the pair for which the distance between the corresponding primal pair of parallel lines is minimum and report these lines. The query time is $O(\alpha \log^2 n)$. The following lemma provides an upper bound on the distance between the reported lines, which completes the proof of Theorem 1.

Lemma 4 *Let $\hat{W}$ be the distance between the lines reported by the query algorithm and let W be the width of S. Then $\hat{W}/W \leq \sqrt{1 + \tan^2 \frac{\pi}{4\alpha}}$.*

Proof Let ℓ and ℓ' be parallel supporting lines of S with distance W. By Lemma 3, there is a j, $0 \leq j \leq \alpha$, such that the slope m_j of these lines in C_j satisfies $|m_j| \leq \tan \frac{\pi}{4\alpha}$. Let d_j be the distance between the points $\mathcal{F}(\ell_j)$ and $\mathcal{F}(\ell'_j)$, where ℓ_j and ℓ'_j are the transformed versions of ℓ and ℓ' in C_j. Let d'_j be the smallest vertex-to-convex-chain distance found by the query algorithm in C_j, let m'_j be the slope of the lines in the primal space that correspond to the points with distance d'_j, and let W' be the distance between the lines.

We have $d_j = W\sqrt{1 + m_j^2}$ and $d'_j = W'\sqrt{1 + m'^2_j}$. Since $d'_j \leq d_j$, we have $W'\sqrt{1 + m'^2_j} \leq W\sqrt{1 + m_j^2}$, and so $W' \leq W\sqrt{1 + m_j^2} \leq W\sqrt{1 + \tan^2 \frac{\pi}{4\alpha}}$. The lemma now follows since the pair of lines reported has distance $\hat{W} \leq W'$. $\square$

3 The dynamic diameter problem

We now prove Theorem 2. Let $\beta \geq 1$ be an integer-valued parameter. It is well-known [PS88] that the diameter of S is realized by two vertices of $\text{CH}(S)$. Intuitively, our approach is to walk around $\text{CH}(S)$ and view it from β predetermined directions. For each direction, we determine the "leftmost" and "rightmost" vertices of $\text{CH}(S)$ as viewed from that direction, and report as the answer the farthest such pair found. In reality, we do not have to maintain $\text{CH}(S)$ dynamically; it is enough to maintain the leftmost and rightmost vertices of S for each direction.

Specifically, we maintain a family of coordinate systems, C_i, $0 \leq i \leq \beta$, with positive x- and y-axes X_i and Y_i, respectively, where X_0 is horizontal and for $i = 1, 2, \ldots, \beta$, X_i makes an angle of $\frac{\pi}{\beta+1}$ with X_{i-1}. The points of S are specified in C_0 and any point $p = (x, y)$ in C_0 corresponds to the point $p_i = (x \cos \frac{i\pi}{\beta+1} + y \sin \frac{i\pi}{\beta+1}, y \cos \frac{i\pi}{\beta+1} - x \sin \frac{i\pi}{\beta+1})$ in C_i, $0 \leq i \leq \beta$. For each i, we store the points of C_i at the leaves of a balanced binary search tree, T_i, in increasing order by X_i-coordinate. T_i uses $O(n)$ space, supports insertions and deletions in $O(\log n)$ time, and allows access to the minimum and maximum X_i-coordinates in $O(1)$ time, via explicitly-maintained pointers to the leftmost and rightmost leaves.

To insert (resp. delete) a point $p = (x, y)$, we insert (resp. delete) p_i in each T_i, $0 \leq i \leq \beta$. To answer a diameter query, we determine the Euclidean distance between the points with maximum and minimum X_i-coordinate in each T_i and report the pair that is farthest apart. The time and space bounds claimed in Theorem 2 are obvious. The following lemma proves the lower bound on the reported diameter claimed in Theorem 2.

Lemma 5 *Let $\hat{D}$ be the Euclidean distance between the points reported by the query algorithm and let D be the diameter of S. Then $\hat{D}/D \geq \sin(\frac{\beta}{\beta+1}\frac{\pi}{2})$.*

Proof For $i = 0, 1, \ldots, \beta$, draw lines parallel to Y_i through the points of S with the least and greatest X_i-coordinate. Each pair of lines encloses the points of S and so S is contained in the convex polygon, P, that is the intersection of all such lines. We refer to either of the two lines parallel to Y_i as ℓ_i. Let $L = \{\ell_i \mid 0 \leq i \leq \beta\}$.

Any vertex v of P is the intersection point of two or more lines of L. It can be shown that the lines of L through v can be ordered so that their subscripts are consecutive. This implies that for some i there are lines ℓ_i and ℓ_{i+1} through v such that the angle formed by the lines and contained (at least partially) inside P is $\theta = \pi - \frac{\pi}{\beta+1} = \frac{\beta}{\beta+1}\pi$.

Among all the points of the plane contained in P, the two farthest points, u and v, are vertices of P. The segment uv divides the angle $\theta = \frac{\beta}{\beta+1}\pi$ at v into angles θ_i and θ_{i+1}, where θ_i is bounded by uv and ℓ_i and θ_{i+1} by uv and ℓ_{i+1}. Wlog, let $\theta_i \geq \theta_{i+1}$, so that $\theta_i \geq \frac{\beta}{\beta+1}\frac{\pi}{2}$. (See Figure 3.) Let F be the Euclidean distance between u and v and let D' be the length of the perpendicular from u to ℓ_i. Also, let D_j denote the distance between the two lines ℓ_j, $0 \leq j \leq \beta$. We have $D'/F = \sin \theta_i = \sin(\frac{\beta}{\beta+1}\frac{\pi}{2})$. Now, u must be contained between the two lines ℓ_i, for otherwise u cannot belong to P. Thus $D_i \geq D'$. Furthermore, $\hat{D} \geq D_i$ and $D \leq F$. Thus $\hat{D}/D \geq \sin(\frac{\beta}{\beta+1}\frac{\pi}{2})$. $\square$

A similar approach can be used to compute the diameter of a static set of n points in the plane: For $0 \leq i \leq \beta$, transform the points into C_i, determine the Euclidean distance between the points with least and greatest X_i-coordinate, and pick the maximum of this

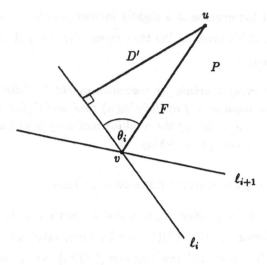

Figure 3: Angle θ_i is at least $\frac{\beta}{\beta+1}\frac{\pi}{2}$.

distance. The only data structure needed is a linked list.

Corollary 1 *An approximation to the diameter of a (static) set of n points in the plane can be computed in $O(\beta n)$ worst-case time and $O(n)$ space such that the ratio of the approximate diameter to the actual diameter is at least $\sin(\frac{\beta}{\beta+1}\frac{\pi}{2})$. Here $\beta \geq 1$ is an integer parameter.* $\Box$

$\Theta(n \log n)$ time is both necessary and sufficient to solve the problem exactly [PS88]. Very recently, in [AMS91], $O(n/\epsilon^{(d-1)/2})$-time algorithm has been given to compute the diameter of a d-dimensional point-set with a relative error of $1 - \epsilon$, for any $\epsilon > 0$.

3.1 A fast implementation of the greedy heuristic for maximum-weight Euclidean matching

The maximum-weight Euclidean matching problem requires a pairing of $2n$ points in the plane such that the weight of the matching, defined to be the sum of the Euclidean distances between the paired points, is maximized. A well-known heuristic for this problem, known as the *greedy heuristic* [Avi83], works by repeatedly pairing the farthest points in the current set and then deleting them. In [Avi83] it is shown that the weight of the resulting matching is at least one-half that of the optimal matching and in [Sup90] an $O(n\sqrt{n}\log n)$-time implementation of the heuristic is given. Using Theorem 2, we can obtain a faster implementation at the expense of a slightly inferior matching: We first insert the points into the structure of Theorem 2. We then repeatedly query it, pair the returned points, and then delete them.

Theorem 3 *The greedy heuristic for maximum-weight Euclidean matching on 2n points in the plane can be implemented in $O(\beta n \log n)$ time and $O(\beta n)$ space, where $\beta \geq 1$ is an integer parameter. The weight of the resulting matching is at least $\frac{1}{2}\sin(\frac{\beta}{\beta+1}\frac{\pi}{2})$ times the weight of the maximum-weight matching.* $\square$

3.2 Maintaining an approximate convex hull

The dynamic convex hull problem requires that a point-set, S, be maintained online under insertions and deletions so that $CH(S)$ can be enumerated quickly. The best algorithm known is due to Overmars and van Leeuwen [OvL81], which uses $O(n)$ space, supports updates in $O(\log^2 n)$ time, and enumerates $CH(S)$ in time proportional to its size. Finding a more efficient solution has been a long-standing open problem. Here we present a simple, efficient approach to maintain a certain approximation to $CH(S)$.

$CH(S)$ is contained in the convex polygon P, defined in the proof of Theorem 2. We simply use P as an approximation to $CH(S)$. An update in S is handled in $O(\beta \log n)$ time as before. P has at most $2(\beta+1)$ vertices which can be enumerated in order in $O(\beta)$ time by computing the intersections between successive lines in the sequence $\ell_0, \ell_1, \ell_2, \ldots, \ell_\beta, \ell_0$. P is an approximation to $CH(S)$ in the sense described in Theorem 4. In [BFP82], Bentley, Faust, and Preparata show that this notion of approximation is natural for some applications.

Theorem 4 *Let S be a dynamic set of n points in the plane and let $CH(S)$ be its convex hull. Let $\beta \geq 1$ be an integer parameter. There is a convex polygon P of size at most $2(\beta+1)$ which contains $CH(S)$ and approximates it in the following sense: For any point p on P, let $d(p)$ be the distance to the closest point on $CH(S)$ and let d^* be the maximum of $d(p)$ taken over all points p on P. Let D be the diameter of S. Then $d^*/D \leq \frac{1}{2}(1/\tan(\frac{\beta}{\beta+1}\frac{\pi}{2}))$. (Thus, for instance, for $\beta = 7$, $d^*/D \leq 0.1$ and for $\beta = 78$, $d^*/D \leq .01$.) P can be maintained in $O(\beta n)$ space and $O(\beta \log n)$ update time, and can be listed out in $O(\beta)$ time.*

Proof The space and time bounds are clear. We prove the bound on the approximation as follows. Clearly, d^* is realized at a vertex, v, of P. Assuming that $P \neq CH(S)$, we have that $v \notin S$ and so v must be the intersection point of exactly two lines ℓ_i and ℓ_{i+1}. Also, the angle between these two lines and lying inside P is $\frac{\beta}{\beta+1}\pi$. Let a and b be the points of S lying on ℓ_i and ℓ_{i+1}, respectively. To maximize the distance of v to $CH(S)$, we assume that the portion of $CH(S)$ from a to b is a straight-line segment, ab. The closest point to v on $CH(S)$ is no farther than the intersection point, q, of the perpendicular

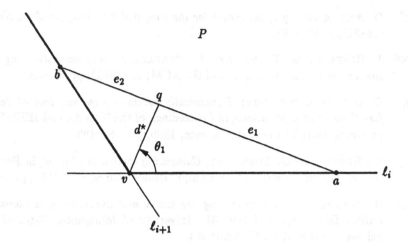

Figure 4: Illustration for the bound on the approximation in the proof of Theorem 4

from p to ab. (See Figure 4.) Let e_1 and e_2 be the lengths of the segments aq and bq, respectively and wlog let $e_2/e_1 = r \leq 1$. Since $d^*/e_1 = 1/\tan\theta_1$ and $D \geq (r+1)e_1$, we have $d^*/D \leq 1/((r+1)\tan\theta_1) = 1/(\tan(\frac{\beta}{\beta+1}\pi - \theta_1) + \tan\theta_1)$. The righthand side is maximized when $\theta_1 = \frac{\beta}{\beta+1}\frac{\pi}{2}$ and so $d^*/D \leq \frac{1}{2}(1/\tan(\frac{\beta}{\beta+1}\frac{\pi}{2}))$. $\square$

4 Conclusion

We have presented efficient algorithms for maintaining the width and diameter of a planar point set under online updates. From a practical standpoint, our algorithms are attractive because they are simple and yet provide almost-exact answers in linear space and low (i.e., log-squared or better) update and query time (when α and β are chosen as constants). From a theoretical viewpoint, however, it is still important to investigate whether efficient exact algorithms exist. We leave this as an open problem.

References

[AMS91] P. AGARWAL, J. MATOUŠEK, AND S. SURI, *Farthest neighbors, maximum spanning trees and related problems in higher dimensions*, in Proceedings of the 2nd Workshop on Data Structures and Algorithms, August 1991. To appear.

[AS91]   P. AGARWAL AND M. SHARIR, *Planar geometric location problems and maintaining the width of a planar set*, in Proceedings of the 2nd Annual ACM-SIAM Symposium on Discrete Algorithms, January 1991, pp. 449–458.

[Avi83]   D. AVIS, *A survey of heuristics for the weighted matching problem*, Networks, 13 (1983), pp. 475–493.

[BFP82]   J. BENTLEY, M. FAUST, AND F. PREPARATA, *Approximation algorithms for convex hulls*, Communications of the ACM, 25 (1982), pp. 64–68.

[DS89]   D. DOBKIN AND S. SURI, *Dynamically computing the maxima of decomposable functions, with applications*, in Proceedings of the 30th Annual IEEE Symposium on Foundations of Computer Science, 1989, pp. 488–493.

[HT85]   M. HOULE AND G. TOUSSAINT, *Computing the width of a set*, in Proceedings of the 1st Annual Symposium on Computational Geometry, 1985, pp. 1–7.

[Jan91]   R. JANARDAN, *On maintaining the width and diameter of a planar point-set online*, Tech. Report TR–91–31, University of Minnesota, Dept. of Computer Science, August 1991. (Submitted.).

[Ove81]   M. OVERMARS, *Dynamization of order-decomposable set problems*, Journal of Algorithms, 2 (1981), pp. 245–260.

[OvL81]   M. OVERMARS AND J. VAN LEEUWEN, *Maintenance of configurations in the plane*, Journal of Computer and System Sciences, 23 (1981), pp. 166–204.

[PS88]   F. PREPARATA AND M. SHAMOS, *Computational geometry – an introduction*, Springer–Verlag, 1988.

[Sup90]   K. SUPOWIT, *New techniques for some dynamic closest-point and farthest-point problems*, in Proceedings of the 1st Annual ACM-SIAM Symposium on Discrete Algorithms, January 1990, pp. 84–90.

Optimal Triangulations by Retriangulating

Herbert Edelsbrunner
University of Illinois at Champaign

ABSTRACT

Given a set S of n points in the plane, a triangulations that minimizes the maximum angle can be contructed in polynomial time by repeatedly deleting all edges that intersect some line segment and retriangulating the thus created polygonal regions. The same method can be used to compute a triangulation that maximizes the minimum triangle height. The currently most efficient implementation of this paradigm runs in time $O(n^2 \log n)$ and storage $O(n)$.

While the above method fails when the objective is to minimize the length of the longest edge, such an optimal triangulation can be computed as follows. First, construct the relative neighborhood graph of the points and then optimally triangulate the thus obtained polygonal regions. This idea leads to an algorithm that runs in $O(n^2)$ time and storage.

In the first case (minmax angle and maxmin height) the retriangulation is done algorithmically, and in the second case (minmax length) it is a proof technique. In particular, various retriangulation methods are used to prove that, indeed, there is a minmax length triangulation that contains the relative neighborhood graph as a subgraph and to show several additional properties necessary for the final algorithm.

Approximating Polygons and Subdivisions
with Minimum Link Paths

Leonidas J. Guibas   John E. Hershberger   Joseph S. B. Mitchell*
Stanford and DEC SRC   DEC SRC   Cornell University

Jack Scott Snoeyink†
Utrecht University

Abstract

We study several variations on one basic approach to the task of simplifying a plane polygon or subdivision: Fatten the given object and construct an approximation inside the fattened region. We investigate fattening by convolving the segments or vertices with disks and attempt to approximate objects with the minimum number of line segments, or with near the minimum, by using efficient greedy algorithms. We also discuss additional topological constraints such as simplicity.

1 Introduction

In the practical application of computers to graphics, image processing, and geographic information systems, great gains can be made by replacing complex geometric objects with simpler objects that capture the relevant features of the original. The need for simplification is most clearly seen in cartography. McMaster [20] lists ways that current methods and technology benefit from data simplification and reduction, including reduced storage space and faster vector operations, vector to raster conversion, and plotting. Improving computation and plotting capabilities does not always help; currently, the speed of data communication is often the bottleneck. Even manual cartography depends on simplification: boundaries must be simplified when drawing a map at a smaller scale or the map becomes unreadable because of the inconsequential information it presents.

The theme of our approach to the task of simplifying a plane path, polygon, or subdivision is: *Fatten the given object and construct an approximation inside the fattened region.* This theme has many variations. In this section, we consider some of them applied to a the problem that cartographers call *line simplification*; in section 2 we briefly survey the literature on this and related approximation problems.

A list of n points $p_1, p_2, \ldots, p_n$ defines a *polygonal chain* with line segments or *links* $\overline{p_i p_{i+1}}$. Given a polygonal chain C, the line simplification problem asks for a polygonal chain $\tilde{C}$ with fewer than n links that represents C well. If the criterion of representing C well is that every point of the approximation $\tilde{C}$ be within ε of a point of C, then the following fattening method could be used. Paint C with a circular brush of radius ε to obtain a fattened region. Then use a minimum link path algorithm to approximate C within the fattened region, as illustrated in figure 1a.

Mathematically, this fattening entails computing the *convolution* of a path, polygon, or subdivision S with a disk (or some other shape) to obtain a region $\mathcal{R}$ in the plane. Computing

---

*Partially supported by a grant from Hughes Research Laboratories, Malibu, CA, and by NSF Grant ECSE-8857642.

†Partially supported by the ESPRIT Basic Research Action No. 3075 (project ALCOM). On leave from from the Department of Computer Science of the University of British Columbia.

convolutions is not our primary focus, but an appendix to the full paper outlines several known methods [11, 18]. The given polygon or subdivision S then defines a homotopy class of curves that can be deformed to S without leaving the fattened region $\mathcal{R}$. We can then attempt to find a minimum link representative of the homotopy class. Section 3 makes the definitions for such a "homotopy method" more precise. Its four subsections contain the following results:

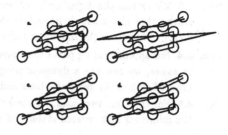

Figure 1: Fattening and approximating

Sec. 3.1 We briefly outline minimum link path algorithms developed in a previous paper [13] and apply them to approximate paths and polygons. These are greedy algorithms that, after the region $\mathcal{R}$ has been triangulated, find a path in time proportional to the number of triangles that the path passes through.

Sec. 3.2 In contrast, we show that the problem of computing a minimum link subdivision is NP-hard. The difficulties comes in optimal placement of vertices of degree three or more; if these are fixed, then we can find the optimum for each chain independently using a minimum link path algorithm.

Sec. 3.3 Returning to polygons, we show that the problem of finding a minimum link simple polygon, that is, one with no self-intersections, is also NP-hard.

Sec. 3.4 Given a region $\mathcal{R}$ with h holes, we show that we can find a simple polygon enclosing the holes with at most $O(h)$ links more than the minimum link polygon.

Returning to the line simplification problem, we can see some "features" of this fattening method that are undesirable in some applications. For example, convolution may create quite large regions where the original chain C was dense in the plane; vertices p_i in these regions can be quite far from the approximation $\tilde{C}$, even though every point of $\tilde{C}$ is close to C. Also, the convolution itself is difficult to compute robustly.

To address these types of problems, we consider fattening just the vertices p_i of the chain C by replacing each vertex with a disk of radius ε. We then require that our approximation "visit" each of these disks in order. This method, illustrated in figure 1b, would ensure that vertices of the chain C would be within ε of its minimum link approximation $\tilde{C}$. If we further restrict the path to turn only inside the vertex disks as shown in figure 1c, then $\tilde{C}$ would also remain within ε of the original chain C. An alternative shown in figure 1d, which is more in the spirit of the convolution approach and for which minimum link paths are easier to compute, is to convolve each link of C separately with a disk of radius ε, glue the resulting *tubes* at the vertex disks that they share, then compute a minimum link path in this region. Notice that turns are allowed in the tubes and not just the vertex disks, but also that the region formed is not planar—it overlaps itself at every angle—thus, this is different from the homotopy approach.

Section 4 generalizes this slightly to a problem we call *ordered stabbing*: given an ordered list of convex objects, find a polygonal chain that visits the objects in order. We have taken the name from Egyed and Wenger [7], who developed a linear-time greedy algorithm for computing a line stabbing disjoint objects in order, if such a line exists. As well as the three possible restrictions on vertices of the approximation (no restriction, in objects, or in tubes), we also study various definitions of visiting or stabbing order for intersecting objects in five subsections:

Sec. 4.1 We examine Egyed and Wenger's algorithm [7], which uses a Graham scan to compute an ordered stabbing line of disjoint objects, if one exists.

Sec. 4.2 We extend this algorithm to stab intersecting objects with a line under four definitions of visiting or stabbing order.

Sec. 4.3 We extend the definition of ordered stabbing to polygonal paths. Stabbing line algorithms then give a simple procedure for computing a path that is at most a multiplicative factor of two from the minimum link stabbing path.

Sec. 4.4 For approximating paths that have no restrictions on vertices or are restricted to turn in tubes, we can use a dynamic programming approach to compute the minimum link ordered stabber in quadratic time and linear space.

Sec. 4.5 For the same two cases, we show that a greedy algorithm can, in linear time, compute the minimum link ordered stabber of disjoint objects.

2 Previous results on approximation

Cartographers have a large catalog of algorithms for the line simplification problem and many measures by which to classify them [5, 19, 20, 21]. Their algorithms either seek a rough but quick reduction of the data or else an accurate but slow reduction. comparative tests, the Douglas-Peucker algorithm (also proposed by Ramer) [6, 26] produces the most satisfactory output, but its speed has been criticized. Its worst-case running time is quadratic in current implementations, but this can be improved to $O(n \log n)$ [14].

A common feature of these algorithms is that they use original data points as vertices of the approximation, even though they acknowledge that these vertices come with some error from a digitizer. This could be reasonable, except the volumes of data and slowness of accurate reduction algorithms lead to using two or more phases of approximation. In the process of reducing a stream of data obtained from a digitizer to the vectors to be plotted on a map, a cartographer may first cast out points until the remaining points are separated by at least ε, and then apply a more complex line simplification algorithm to reduce the data further for storage or display. Though the properties of the individual algorithms are characterized and classified, the properties of these heuristic combinations are not. Criteria much like our ε fattening [4, 25, 27] are then used *a posteriori* to test the quality of the resulting approximations.

Imai and Iri [15, 16, 17] and other researchers [2, 3, 7, 12, 22, 24, 29] have chosen mathematical criteria for the approximations and then sought efficient algorithms to find best approximations. The algorithms they have developed, however, have quadratic or greater running times—especially for those that use original data points as vertices of the approximation.

We remove the restriction that vertices of the approximation must be original data points in an attempt to find faster algorithms that fulfill mathematical specifications. Our goal is linear or $O(n \log n)$ algorithms that find the best approximation. Failing that, we may look for a slower algorithm or find a suboptimal approximation—we usually opt for the latter, especially if we can determine how close the approximation is to the optimal.

3 Homotopy classes and minimum link representatives

Suppose we have a region $\mathcal{R}$ of the plane and a path, polygon, or subdivision S contained in the region $\mathcal{R}$. If we bend and move the components of S, without leaving $\mathcal{R}$, we obtain other paths, polygons, or subdivisions that could be said to be equivalent to S by deformation within $\mathcal{R}$. The topological concept of *homotopy* formally captures this notion of deformation. We omit the formal definitions from this abstract; see a basic topology book such as [23].

3.1 Computing minimum link paths and polygons of a given homotopy type

In a companion paper [13], we investigate the problem of computing minimum link paths and closed curves of a given homotopy class in triangulated polygons. Thus, we merely state the theorem for this abstract.

Theorem 3.1 (From [13]) *A minimum link path α' that is homotopic to a chain α can be computed in time proportional to the number of links of α and the number of triangles intersecting*

α and α'. *A polygon α' that is homotopic to a polygon α and has the minimum number of links if α' is non-convex or at most one more than the minimum number if α' is convex can be computed in the same time.*

These paths and polygons are computed by a greedy procedure, following Suri [28] and Ghosh [9]. In brief, the idea is to start at some point and illuminate as much of the region $\mathcal{R}$ as possible; this is as far as one link can reach. Repeat the illumination from the appropriate boundary (determined by the homotopy class of α) of the lit area until the goal point is found.

3.2 The min-link subdivision problem is NP-complete

Given a subdivision S in a polygonal region, P, the min-link subdivision problem (MinLinkSub) asks for the polygonal subdivision S' homeomorphic to S in P that is composed of the minimum number of line segments. We can also look at the decision problem: Given S and P and an integer k, is there a polygonal subdivision S' with at most k segments that is homeomorphic to S in P? We use the decision problem to show that MinLinkSub is NP-complete.

Before we argue that the decision problem is NP-hard, we note that the planar case of a problem that Garey, Johnson and Stockmeyer [8] have called maximum 2-sat (Max2Sat) is NP-complete.

The general case of Max2Sat is: Given a set of variables V, an integer k, and disjunctive clauses $C_1, C_2, \ldots, C_p$, each containing one or two variables, determine if some truth assignment to the variables satisfies at least k clauses. The *variable graph* of an instance of Max2Sat is defined to be the graph $G = (V, E)$, with an edge $(u, v) \in E$ if and only if the variables u and v both appear (either negated or unnegated) in some clause C_i. An instance of Max2Sat is *planar* if its variable graph is planar.

Theorem 3.2 *Planar maximum 2-sat (Max2Sat) is NP-complete.*

In theorem 3.3 we prove that MinLinkSub is NP-hard by reduction from the restricted version of planar Max2Sat. That is, given an instance of planar Max2Sat, we construct an instance of MinLinkSub that has a solution if and only if the instance of Max2Sat has a solution. We will use a similar reduction to the minimum link simple polygon problem in subsection 3.3.

Rather than prove theorem 3.3, let us take an informal look at the gadgets for truth assignments and for unary and binary clauses that are used in the construction. We embed the variable graph of the 2-sat instance in the plane such that no edge is vertical, then we fatten each vertex to a disk and each edge to a rectangular strip and require that the subdivision lies within the resulting region. Within each disk we place *true* and *false* points, directly above and below the disk center, and force the vertex of the minimum-link subdivision to lie at one of these points by using appropriate gadgets.

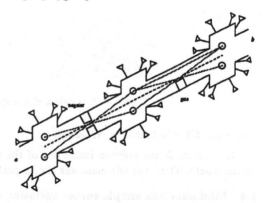

Figure 2: A negater and gate for $(\overline{a} \vee b)$

For the binary clauses on two variables, we divide the rectangular strip of the fattened edge joining the two variables into four strips. In each we form *negaters* for variables that need them and a *gate* to simulate an OR gate. Figure 2 illustrates a negater and gate combination for the clause $(\overline{a} \vee b)$—dashed lines are subdivision edges, solid lines are region boundaries, and grey lines are possible satisfying assignments.

Figure 3: An enforcer and its cone

For a unary clause, we add an *enforcer* pointing to the *true* point for a positive clause and the *false* point for a negative clause. Figure 3 illustrates an enforcer—dashed lines are subdivision edges and solid lines are region boundaries. The enforcer can be realized by four line segments if and only if the subdivision vertex lies in the shaded cone.

In a minimum link subdivision, each clause that is not satisfied requires one extra line segment. Thus, there is a number, k', such that k clauses of the instance of Max2Sat can be satisfied if and only if the instance of MinLinkSub uses at most k' line segments. In the full paper, Theorem 3.3 shows that this construction can be carried out.

Theorem 3.3 *MinLinkSub is NP-hard.*

Placement of vertices of degree at least three is the difficult part of MinLinkSub:

Theorem 3.4 *MinLinkSub is in NP.*

Proof: One can guess the vertices and compute shortest paths using the algorithm of the previous section. ∎

3.3 Minimum link simple polygons

In the full paper, we show that the problem of finding a minimum link simple polygon of a given homotopy type (MinLinkSP) is *NP*-hard by a reduction from planar Max2Sat. The reduction is much like the one used in that section: We embed an Euler tour of the variable graph as a simple closed curve in the plane and place obstacles so that graph vertices are pinned in place. Then we form toggle switches at each graph vertex and use enforcers to ensure that an approximate path can be interpreted as a truth assignment. Finally, we arrange negaters and gates so that an edge of the graph can be embedded using fewer links if the clause is satisfied. See figure 4. Other pictures and proofs are omitted from this abstract.

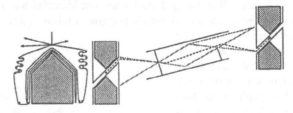

Figure 4: A toggle and a negater/gate combination

Theorem 3.5 *MinLinkSP is NP-complete.*

One can break the polygon inside one of the vertex gadgets and anchor its endpoints to obtain a path. Thus, the minimum link simple path problem is also *NP*-complete.

3.4 Minimum link simple curves enclosing all holes

The reduction in the previous section requires a linear number of obstacles both inside and outside the curve; whether one can efficiently find a minimum link simple curve in a polygon with h holes that encloses all the holes is an open question. We can find a simple curve that has only $O(h)$ more segments than the (non-simple) minimum link curve; this is independent of the number of segments of the minimum link curve. We identify $O(h)$ *junction triangles* of the triangulation—triangles that must be cut on all three sides by a path—and group the rest of the triangles into *corridors*. In each corridor we find the minimum link path.

Theorem 3.6 *In a polygon P of n vertices with h holes, one can, in $O(n)$ time, find a simple closed curve enclosing all the holes that has $O(h)$ segments more than the minimum link curve of the same homotopy class.*

The worst case for our procedure results in $10h - 12$ additional line segments. We have yet to find a polygon that requires more than $2h - 2$ additional segments to make a minimum link curve simple.

4 Ordered Stabbing

In this section, we study the *ordered stabbing* problem: Given an ordered sequence of n convex objects, $\mathcal{O} = \{O_1, O_2, \ldots, O_n\}$, find a polygonal chain, consisting of the minimum number of line segments, that *visits* the objects in order. This generalizes section 3.1, the computation of a minimum link path of a given homotopy type in a triangulated manifold, because in that section the task is to visit the sequence of triangles intersected by the Euclidean shortest path with the additional restriction that the stabbing path must remain in the manifold.

Different variants of the ordered stabbing problem arise from different definitions of "visiting" as well as from restrictions on the stabbed objects or stabbing path. Section 4.1 considers stabbing a sequence of disjoint objects with a line in linear time; the visiting order of the objects is naturally the order of their intersections with the line. Section 4.2 extends this algorithm to intersecting objects under four definitions of visiting order: entering objects in order, leaving in order, entering and leaving in order, or choosing one point from each object to visit in order. For the first three definitions, the algorithm still runs in linear time. For the last, if the n objects to be stabbed are disks or polygons of constant size, then the algorithm can be implemented to run in $O(n \log n)$ time.

Ordered stabbing with a polygonal path is more difficult than with a line. Section 4.3 does point out that the stabbing line algorithms give a simple way to compute a stabbing path that has at most twice the number of links needed—even when the path vertices are required to lie inside the given objects. Section 4.4 gives a dynamic programming method to compute the true minimum link stabbing path when the placement of vertices unrestricted or is restricted to lie in *tubes*—that is, inside the convex hull of two adjacent objects. The algorithm that it gives runs in quadratic time and linear space. Finally, with the same placement of vertices, section 4.5 shows that a greedy algorithm can, in linear time, compute a minimum link path that visits disjoint objects in the correct order.

4.1 Ordered stabbing of disjoint objects with a line

Egyed and Wenger [7] looked at the problem of stabbing disjoint objects in order with a line. They showed that the actual shape of the object mattered less than the ability to find inner and outer common tangents—if one assumed that computing these tangents took constant time, then one could find a line stabbing the objects in order by a simple Graham scan. We first reinvent (and simplify) their algorithm for stabbing disjoint objects with a line and then modify it in the next subsection to handle convex objects whose boundaries intersect in at most two points under four different definitions of *visiting order*.

If α is a direction, then let $-\alpha$ denote the reverse direction. We call an object $O \in \mathcal{O}$ a *support object* for direction α if there is a line ℓ_α in direction α such that O lies on and to the left of ℓ_α and every other object $O' \in \mathcal{O}$ contains a point on or to the right of ℓ_α. The line ℓ_α is called a *support line* for direction α and the point or points of $O \cap \ell_\alpha$ are called *support points*. We can observe the following connection between support lines and stabbing lines.

Observation 4.1 *The lines parallel to direction α that stab a set of objects $\mathcal{O}$ are the lines to the right of support line ℓ_α and to the left of support line $\ell_{-\alpha}$, if any.*

By analogy with the convex hull, we can define the *support hull* of a set of n objects as the circular list of support objects, ordered by the angles of their support lines. Repetitions are possible, as figure 5 shows, but if any two objects O and O' have at most two outer common tangents, then any subsequence of the list can have only two alternations between O and O'. Thus, the size of the list is at most $2n - 2$ by Davenport-Schinzel sequence bounds [1].

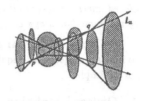

Figure 5: Support hull with limiting lines

A support line ℓ_α is a *limiting line* if its reverse $\ell_{-\alpha}$ is also a support line, as shown in figure 5. Limiting lines are analogous to inner common tangents. A limiting line ℓ_α hits two support points; we name them the *first contact* p and *second contact* q so that the vector $q - p$ has direction α. Similarly, we name the objects that contain these points the *first* and *second contact objects* for ℓ_α. We can distinguish two types of limiting lines: ℓ_α is a *counterclockwise (ccw) limiting line* if the first contact p is the support point for ℓ_α, as shown in figure 5, and a *clockwise (cw) limiting line* if the second contact q is the support point for ℓ_α.

Limiting lines are stabbers, as you can see from the figure, but rotating a ccw limiting line counterclockwise gives a line that is no longer a stabber. In our ordered stabbing problems, we will find at most one limiting line of each type; they will delimit the possible slopes for stabbing lines. The above and below portions of the support hull between these slopes limit the extent that a stabbing line can move up and down. Thus, the hulls and limiting lines give a linear size description of all possible stabbers. In the rest of this section, we show how to maintain this description under the assumption that basic operations such as computing the intersection of an object with a line and computing common tangents of two objects take constant time. We prove the following theorem.

Theorem 4.1 *Let $\mathcal{O} = \{O_1, O_2, O_3, \ldots\}$ be a sequence of disjoint convex objects. One can compute a line that stabs the longest possible prefix $O_1, O_2, \ldots, O_i$ in order using $O(i)$ time and space.*

Proof: We outline the idea; the full paper gives more complete pseudocode.

We can easily compute a description of all ordered stabbers for O_1 and O_2: Initialize the ccw limiting line t and the cw limiting line t' to the appropriate inner common tangents directed from O_1 toward O_2. Two portions of the support hull have slopes that fall between the slopes of t and t'; these portions are delimited by the contact points of t and t'. We name them the *above hull*, A, and the *below hull*, B, as shown. To represent A and B, we store the list of support objects in a *deque*—a doubly-ended queue—which we will maintain by a Graham scan procedure [10]. Initially, both deques contain O_1 followed by O_2.

We would like to add successive objects and maintain the description of ordered stabbers. Given the above and below hulls A and B and limiting lines t and t' after the first i objects, we want to add object O_{i+1}. We first define the *line-stabbing wedge* to be the region between t and t' that is left of object O_i— drawn shaded in figure 6. A point p in the line-stabbing wedge has the property that there is a line ℓ through p that visits the first i objects before visiting p. Thus, if O_{i+1} does not intersect the wedge, then no stabbing line visits the first $i + 1$ objects in order. If it does, then we update the limiting lines, which are ordered stabbing lines, and the portions of the support hull. ∎

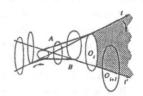

Figure 6: Updating t and the hulls

4.2 Ordered stabbing of intersecting objects with a line

In this section, we extend our algorithm to convex objects that intersect in at most two points and, thus, have at most two outer common tangents. There are several possible definitions when objects intersect; we consider four of them in this section. All four definitions will be equivalent to the natural definition if the objects are disjoint.

Given two points p and q on a directed line ℓ, we say that $p \prec q$ if the vector from p to q is in the direction of ℓ. Let the intersection $\ell \cap O_i$ have extreme points $a_i \prec b_i$. Given a sequence of objects $O_1, O_2, \ldots, O_n$ and a line ℓ such that the intersection $\ell \cap O_i$ has extreme points $a_i \prec b_i$, we say that ℓ *visits the objects in order* if

Def. 1: Line ℓ exits the objects in the correct order: For $i < j$, we have $b_i \prec b_j$.

Def. 2: Line ℓ enters the objects in the correct order: For $i < j$, we have $a_i \prec a_j$.

Def. 3: Line ℓ both enters and exits the objects in the correct order: For $i < j$, we have $a_i \prec a_j$ and $b_i \prec b_j$.

Def. 4: Line ℓ hits points $p_1, p_2, \ldots, p_n$, with $p_i \in \ell \cap O_i$, in the correct order: For $i < j$, the point $p_i \prec p_j$.

Definitions 1 and 2 could be considered equivalent: given an algorithm that computes stabbing lines for one definition we can compute stabbing lines for the other by just reversing the sequence of objects. We will, however, combine the algorithms for 1 and 2 to handle definition 3. Since the algorithms that compute stabbers without reversing the sequence are slightly different, we treat definitions 1 and 2 separately.

The rest of this section proves theorem 4.2.

Theorem 4.2 *Let* $\mathcal{O} = \{O_1, O_2, O_3, \ldots\}$ *be a sequence of convex objects whose boundaries intersect pairwise in two points. With any of definitions 1–3 for visiting order, one can compute a line that stabs the longest possible prefix* $O_1, O_2, \ldots, O_i$ *in order using* $O(i)$ *time and space. One can use definition 4 if the objects are polygons of constant size or equal radius circles; the time bound increases to* $O(i \log i)$.

Proof for Def. 1: Let us begin with definition 1: exiting the objects in the correct order. A way to view the result that we are trying to obtain is to imagine that the objects are painted on the plane in reverse order—starting with object O_n. An ordered stabbing line must exit a visible portion of the boundary of each object. We will not compute this "painting" because it could have quadratic complexity; it will, however, guide us in modifying our algorithm to add object O_{i+1} and update the description of the stabbers of the first i objects. In this abstract, we just outline how to maintain this description.

To add O_{i+1}, we must determine if any ordered stabbers of $O_1, \ldots, O_i$ exit O_{i+1} after O_i. As before, define the line-stabbing wedge to be the region between the limiting lines t and t' and right of O_i. Because O_i is exited last, no object O_j with $j < i$ intersects the wedge. Also as before, if O_{i+1} does not intersect the wedge then no stabbing line exists.

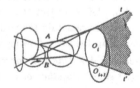

Figure 7: Updating t under def. 1

In our imaginary painting, O_{i+1} may be obscured by O_i; thus, we discard portions of O_{i+1} that lie outside the line-stabbing wedge. If what is left of O_{i+1} does not intersect the ccw limiting line t, then we must update the support hulls and the line t. Once the support hull A is updated, t moves clockwise until it comes to rest on the object or point that is last in A. This may cause objects to be removed from the front of B. The cw limiting line is adjusted in a similar fashion. ∎

Proof for Def. 2: Stabbing lines satisfying definition 2, entering the objects in the correct order, must hit the boundaries of objects in a "painting" that starts with O_1. They can be found by a similar algorithm.

Define the line-stabbing wedge to be the convex region bounded by the two limiting lines and not left of O_i. Following the painting model, discard portions of O_i that lie inside O_{i+1}. If the remaining portion of O_i no longer intersects the ccw (or cw) limiting line, or if O_{i+1} does not intersect the line, then we must update the support hull and limiting line as before. We again use a Graham scan to maintain support points and support objects in A and B with the key property that the support points or objects for the limiting lines are the first and last entries in A and B. ■

Proof for Def. 3: We can combine the two previous algorithms to find stabbing lines satisfying definition 3. Given the support hulls A and B and limiting lines after the first i objects we need to determine the ordered stabbing lines that enter and exit O_{i+1} after O_i. Unless O_{i+1} intersects the line-stabbing wedges of both definitions 1 and 2, there are no stabbing lines of the first $i + 1$ objects.

First, discard portions of O_i that lie in O_{i+1} and update the support hulls and limiting lines as under definition 2 if the remaining portion of O_i no longer intersects one of the limiting lines. Next, discard portions of O_{i+1} that lie in O_i and update according to definition 1 if necessary.

If objects O_1 and O_2 intersect, then the initial support hulls A and B are the upper and lower intersection points, respectively, of the boundaries of O_1 and O_2. The initial limiting lines are the two orientations of the line through the two intersection points. ■

Proof for Def. 4: The fourth definition is different from the others in that it involves choosing points rather than defining an order for intervals. There is an equivalent formulation in terms of intervals, however: no later interval may end before an earlier one begins.

Lemma 4.1 *Let $[a_i, b_i]$, for $i \in [1 \ldots n]$, be non-empty intervals of the real line. One can choose a set of points $\{p_1, p_2, \ldots, p_n\}$ with $p_i \in [a_i, b_i]$ and $p_i \leq p_j$ for all $1 \leq i < j \leq n$ if and only if there is no pair $j < k$ with $b_k < a_j$. Furthermore, the p_is can be chosen from the set $\{a_1, a_2, \ldots, a_n\}$.*

We are not be able to give a linear time algorithm for this definition of visiting order because the line-stabbing wedge has non-constant complexity. When our objects are constant size polygons or equal radius circles, however, we can maintain the wedge by an intersection algorithm that allows us to stab i objects in $O(i \log i)$ time.

As before, we want the line-stabbing wedge of the first i objects to be the locus of all points p that have a line that visits the i objects before visiting p. Assume that we have two limiting lines t and t' that define an angle of less than $180°$ and let W_j be the region between these lines and not left of object O_j. Define the line-stabbing wedge as the intersection $\bigcap_{j \leq i} W_j$.

We can maintain the wedge as n objects are added incrementally using $O(n \log n)$ total time, according the the following lemma.

Lemma 4.2 *One can incrementally form all wedges for a sequence of n convex polygons with $O(n)$ sides altogether or n unit radius circles in a total of $O(n \log n)$ time.*

This completes the proof of theorem 4.2.

4.3 Ordered stabbing with a polygonal chain

The problem of ordered stabbing with a polygonal chain instead of a line has its own complications and variations. We can make one simple observation in this section, however; the line stabbing algorithms give a simple means to find a stabbing chain that approximates the minimum link stabber within a multiplicative factor of two. First, though, we discuss the variations that arise by different definitions of visiting order and restrictions on vertex placement.

The first three definitions of visiting order of the previous section defined an pairwise ordering on the intervals in which objects intersected the stabbing line. Because a polygonal chain can intersect an object in several intervals, one can no longer arbitrarily compare pairs of intervals. We could apply the definitions of visiting order to the interval between the first and last intersection of the stabbing chain with an object, but this type of global criterion works against efficient greedy algorithms; when considering O_{i+1}, an algorithm must worry about not entering any later object rather than looking only at O_i and a couple of limiting lines. Instead, we choose a set of intervals $\{I_1, I_2, \ldots, I_n\}$, where I_j is a maximal connected interval of the intersection $\pi \cap O_j$ such that the desired property holds for those intervals. This solution, in the spirit of definition 4, still allows local algorithms.

Further variations arise from different restrictions on the vertices of the approximation. As mentioned in the introduction, we will concentrate on three, listed in order of decreasing restriction.

1. Turn in objects: Each vertex of the approximation must lie in one of the original objects.

2. Turn in tubes: Each vertex of the approximation must lie within a region bounded by two adjacent objects and their outer common tangents.

3. No restriction: The approximate path can turn anywhere.

Using the algorithms for ordered stabbing with a line, there is a simple method to find a stabbing path for any of these variations using at most twice the minimum number of links.

Theorem 4.3 *One can compute an ordered stabbing path with vertices inside objects $O_1, O_2, \ldots, O_n$ that has less than twice as many segments as the minimum link stabbing path.*

Figure 8 illustrates that a greedy approach, always attempting to stab as many objects as possible, can approach the factor of two from the minimum link path when path vertices must lie inside stabbed objects. The bound of theorem 4.3 is tight. This is in contrast to the algorithms for minimum link paths in simple polygons [9, 13, 28]

Figure 8: The greedy path (dotted) versus the minimum path (solid)

and the case in section 4.5, where greedy methods do obtain a minimum link stabbing path.

4.4 A dynamic programming approach

In this section, we solve several variations of the ordered stabbing problem by dynamic programming.

For each object O_i, we compute the length of the minimum link ordered stabbing chains that stab objects O_1 through O_i and also all possible final segments of minumum chains. Lemma 4.4 shows that these final segments have a constant size description. Theorem 4.4 shows that, for definitions 1, 2, and 4 of visiting order, we can compute the path length and final segments for O_i from the final segments for objects O_j, with $j < i$, by using the line stabbing algorithms of sections 4.1 and 4.2. Thus, for definitions 1 and 2, we obtain a minimum link stabbing paths in $O(n^2)$ time and linear space. For definition 4, the time increases to $O(n^2 \log n)$ and the objects must be constant-size polygons or equal radius circles. All three restrictions on turns or vertices are supported.

This should be compared to the general graph-based approach of Imai and Iri [17], which, in our terminology, would create a graph with an edge (j, k) if there is an ordered stabber from O_j through O_k and then search the graph for the shortest path. Our dynamic programming method shares the problem of a super-quadratic running time, but saves a factor of $O(n)$ in space by better organization of computation.

In sections 4.1 and 4.2 we formed line-stabbing wedges under visiting orders 1, 2, and 4, which were the locus of all points p such that some line stabbed the first i objects before stabbing p. We can generalize this definition to polygonal chains: the *chain-stabbing wedge* W_i of the first i

objects is the locus of all points p such that there is a minimum link chain that visits the first i objects and then visits p.

To help make this definition into an efficient computation, we show first that chain-stabbing wedges can enlarge only when the path gains an extra link.

Lemma 4.3 *If the chain-stabbing wedges W_i and W_{i+1} both have minimum stabbing paths with k-links, then $W_i \supseteq W_{i+1}$.*

Next, we show that chain-stabbing wedges really are wedge-like—that they are bounded by two rays.

Lemma 4.4 *The chain-stabbing wedge W_i is a region that contains a ray through every point and whose boundary consists of two rays and portions of obstacle boundaries that form a convex or concave curve depending on the definition of visiting order.*

Now, we prove the following theorem.

Theorem 4.4 *Under visiting order definitions 1, 2, or 4, one can compute the minimum link path visiting objects $O_1, O_2, \ldots, O_n$ in order that either has no restrictions on vertices or has vertices in or between adjacent objects. Space is $O(n)$. Under definitions 1 and 2 the time is $O(n^2)$. Under definition 4, the time increases to $O(n^2 \log n)$ and the objects must be constant-size polygons or equal radius circles.*

4.5   A linear-time greedy algorithm

We have seen in section 4.3 that a purely greedy approach—always attempting to stab to the greatest numbered object with each segment of the path—can be a factor of two from the minimum link path. The dynamic programming approach, by way of contrast, chooses $k-1$ links so as to allow the kth link to reach as far as possible. Unfortunately, the dynamic programming approach makes this choice for each object, thus spending up to a quadratic amount of work. We really only need to make this decision for the last object stabbed by each segment of a minimum link stabbing path. If we return to stabbing disjoint objects, as in section 4.1, we can determine these "last objects" and compute a minimum link stabbing path in linear time if the vertices of the path are unrestricted or are forced to lie in tubes between adjacent objects.

Theorem 4.5 *Given n disjoint objects $O_1, O_2, \ldots, O_n$, one can compute, in $O(n)$ time, the minimum link ordered stabbing path whose vertices either have no restrictions or lie in or between adjacent objects.*

5   Conclusions and open problems

We have examined minimum link approximations that lie in convolutions or are ordered stabbers as part of a basic approach to approximating paths, polygons, and subdivisions. We have developed some efficient algorithms and indicated that others are unlikely to ever be developed.

There are many avenues that we hope to explore further—the most important being practical studies of implementations of theoretically efficient approximation methods. A few of the many open questions that remain are: Is computing the minimum link simple polygon enclosing all holes *NP*-complete? What other restrictions on approximation can be handled in subquadratic time? For example, the vertices may be required to lie within some $\delta < \varepsilon$ of the original path. Can subquadratic time algorithms be developed for ordered stabbing with the other definitions of visiting order?

References

[1] P. K. Agarwal, M. Sharir, and P. Shor. Sharp upper and lower bounds on the length of general Davenport-Schinzel sequences. *J. Comb. Theory, A*, 52:228–274, 1989.

[2] H. Alt, J. Blömer, M. Godau, and H. Wagener. Approximation of convex polygons. In *Seventeenth ICALP*, number 443 in LNCS, pages 703–716. Springer-Verlag, 1990.

[3] R. Bellman. On the approximation of curves by line segments using dynamic programming. *CACM*, 4:284, 1961.

[4] M. Blakemore. Generalisation and error in spatial data bases. *Cartographica*, 21:131–139, 1984.

[5] B. Buttenfield. Treatment of the cartographic line. *Cartographica*, 22:1–26, 1985.

[6] D. H. Douglas and T. K. Peucker. Algorithms for the reduction of the number of points required to represent a line or its caricature. *The Canadian Cartographer*, 10(2):112–122, 1973.

[7] P. Egyed and R. Wenger. Ordered stabbing of pairwise disjoint convex sets in linear time. *Disc. App. Math.*, to appear.

[8] M. R. Garey, D. S. Johnson, and L. Stockmeyer. Some simplified NP-complete graph problems. *Theoretical Comp. Sci.*, 1:237–267, 1976.

[9] S. K. Ghosh. Computing the visibility polygon from a convex set and related problems. *J. Alg.*, 12:75–95, 1991.

[10] R. Graham. An efficient algorithm for determining the convex hull of a finite planar set. *Info. Proc. Let.*, 1:132–133, 1972.

[11] L. Guibas, L. Ramshaw, and J. Stolfi. A kinetic framework for computational geometry. In *Proc. 24th FOCS*, pages 100–111, 1983.

[12] S. L. Hakimi and E. F. Schmeichel. Fitting polygonal functions to a set of points in the plane. *CVGIP: Graph. Mod. Image Proc.*, 53(2):132–136, 1991.

[13] J. Hershberger and J. Snoeyink. Computing minimum length paths of a given homotopy class. In *WADS '91 Proceedings*, 1991.

[14] J. Hershberger and J. Snoeyink. An implementation of the Douglas-Peucker line simplification algorithm using at most $cn \log n$ operations. In preparation, 1991.

[15] H. Imai and M. Iri. Computational-geometric methods for polygonal approximations of a curve. *Comp. Vis. Graph. Image Proc.*, 36:31–41, 1986.

[16] H. Imai and M. Iri. An optimal algorithm for approximating a piecewise linear function. *J. Info. Proc.*, 9(3):159–162, 1986.

[17] H. Imai and M. Iri. Polygonal approximations of a curve—formulations and algorithms. In G. T. Toussaint, editor, *Computational Morphology*. North Holland, 1988.

[18] K. Kedem, R. Livne, J. Pach, and M. Sharir. On the union of Jordan regions and collision-free translational motion amidst polygonal obstacles. *Disc. & Comp. Geom.*, 1:59–71, 1986.

[19] R. B. McMaster. A statistical analysis of mathematical measures for linear simplification. *Amer. Cartog.*, 13:103–116, 1986.

[20] R. B. McMaster. Automated line generalization. *Cartographica*, 24(2):74–111, 1987.

[21] R. B. McMaster. The integration of simplification and smoothing algorithms in line generalization. *Cartographica*, 26(1):101–121, 1989.

[22] A. Melkman and J. O'Rourke. On polygonal chain approximation. In G. T. Toussaint, editor, *Computational Morphology*. North Holland, 1988.

[23] J. R. Munkres. *Topology: A First Course*. Prentice-Hall, Englewood Cliffs, N.J., 1975.

[24] J. O'Rourke. An on-line algorithm for fitting straight lines between data ranges. *CACM*, 24(9):574–578, Sept. 1981.

[25] J. Perkal. On the length of empirical curves. In *Discussion Paper 10, Michigan Inter-University Community of Mathematical Geographers*, University of Michigan, Ann Arbor, 1966.

[26] U. Ramer. An iterative procedure for the polygonal approximation of plane curves. *Comp. Vis. Graph. Image Proc.*, 1:244–256, 1972.

[27] A. Rosenfeld. Axial representation of shape. *Comp. Vis. Graph. Image Proc.*, 33:156–173, 1986.

[28] S. Suri. A linear time algorithm for minimum link paths inside a simple polygon. *Comp. Vis. Graph. Image Proc.*, 35:99–110, 1986.

[29] G. Toussaint. On the complexity of approximating polygonal curves in the plane. In *Proc. IASTED, International Symposium on Robotics and Automation*, Lugano, Switzerland, 1985.

An incremental algorithm for constructing shortest watchman routes

Xue-Hou TAN, Tomio HIRATA and Yasuyoshi INAGAKI
Faculty of Engineering, Nagoya University
Chikusa-ku, Nagoya 464, Japan

Abstract

The problem of finding the shortest watchman route in a simple polygon P through a point s on its boundary is considered. A route is a watchman route if every point inside P can be seen from at least one point along the route. We present an incremental algorithm that constructs the shortest watchman route in $O(n^3)$ time for a simple polygon with n edges. This improves the previous $O(n^4)$ bound.

1 Introduction

Stationing watchmen in an n-wall art gallery room so that every point in the interior of the room can be seen by at least one watchman is the well known "Art Gallery" problem. The room is a simple polygon and the watchmen are stationary points which can see any points connected to them by a line segment that entirely lies in the polygon. The goal is to minimize the number of watchmen. A survey of Art Gallery and related problems can be found in [6]. While the watchman route problem, posed by Chin and Ntafos [2], deals with finding the shortest route from a point x back to itself so that every point in the polygon can be seen from at least one point along the route.

An $O(n)$ algorithm for the shortest watchman route in a simple rectilinear polygon with n edges is given by Chin and Ntafos [2]. Let us first give a brief review of their algorithm. Given a rectilinear polygon P, we can identify a set of line segments, called essential cuts, inside the polygon so that any watchman route must visit them and any route which visits them is a watchman route. See Fig. 1a. Moving on an essential cut C can see either whole top (left) or whole bottom (right) piece of P incident to C. This piece is called the non-essential piece of C. Removing these non-essential pieces will not affect construction of the shortest watchman route. Let P' denote the resulting polygon. P' is then triangulated using the existing linear-time algorithm [1]. Fig. 1b shows a possible triangulation. Since the shortest watchman route should visit the essential cuts in the order in which they appear in the boundary of P', the essential cuts are then used as mirrors to roll-out the polygon P' in that order. Furthermore, a point x which lies on the shortest watchman route is computed. By now, the problem of finding the shortest watchman route is reduced to that of finding the shortest path from x to its image x' in the rolled-out polygon. The shortest path between two points can then be constructed using $O(n)$ algorithms [4]. Fig. 1c shows the rolled-out polygon as well as the shortest path which starts at point x and ends at its image x'. Finally, the shortest watchman route is obtained by folding back the shortest path. See Fig. 1d.

Chin and Ntafos also presented an $O(n^4)$ algorithm for constructing the shortest watchman route in a simple polygon through a starting point s specified on its boundary [3]. Similarly, a set of essential cuts can be defined inside the given polygon (see also Section 2). But now all non-essential pieces of essential cuts can not be removed in determining the shortest watchman route. Instead, an initial watchman route is first constructed in the polygon obtained by removing the non-essential pieces of some selected essential cuts. Then, by checking local optimality of the route at the selected essential cuts, the route goes through a sequence of adjustments. Each adjustment involves a change in the set of selected cuts and results in a shorter route. Finally, the shortest watchman route is obtained when no more adjustments can be made. The problem without a starting point remains open. (Usually, the starting point s can be easily selected so that it lies on the shortest watchman route.) Note that the algorithm for simple rectilinear polygons does not assume a starting point, i.e., it finds the shortest watchman route overall.

In this paper, we improve Chin and Ntafos' result for simple polygons. The intuition for our algorithm is as follows. The watchman, starting from s, visits the essential cuts one after another. When a new essential cut C is involved, we compute the shortest watchman path from s to a point on C so that all the essential cuts added by now are visited. As only one essential cut is added each time, it is possible to construct the new shortest watchman path from the previous one at small cost. Such an incremental method gives an $O(n^3)$ algorithm for constructing shortest watchman routes in simple polygons.

Section 2 of this paper gives a brief review of essential cuts and shortest watchman routes. Section 3 presents the incremental algorithm in detail. The time complexity of our algorithm is analyzed in Section 4. Discussion and application are finally given in Section 5.

2   Essential cuts and shortest watchman routes

Let P be an n-sided simple polygon with a point s on its boundary. We assume that P is given by the sequence of its vertices in the clockwise order from s. A vertex is *reflex* if the internal angle is greater than 180^0. P can be partitioned into two pieces by a "cut" that starts at a reflex vertex v and extends either edge incident to v until it first intersects the boundary. We say a cut is a *visibility cut* if it produces a *convex* angle ($< 180^0$) at v in the piece of P containing s. (Some reflex vertices may not contribute to any visibility cut.) Such a visibility cut "resolves" the reflexivity at v; in order to see the edge (or corner) incident to it, a watchman route needs to visit only one point on that cut (Fig. 2). Since the shortest path between s and a cut C need not go over C (as viewed from s), the piece of P containing s is called the essential piece of C. A cut C is described by a pair of points (l, r), where l (left endpoint) is the endpoint of C that is first visited in a clockwise scan of P and r (right endpoint) is the other endpoint. These points lie on the boundary of P and are not always vertices of P. The orientation of a cut C is supposed to be from l to r. Thus, s always falls to the right side of visibility cuts.

A watchman route must visit all visibility cuts so that each corner of P can be seen. But some of them are not important in determining the shortest watchman route. We say cut C_j **dominates** cut C_i if C_j appears between two endpoints of C_i in a clockwise scan of the boundary. Clearly, if C_j dominates C_i, any route that visits C_j will automatically visit C_i, i.e., C_i can be disregarded in determining the shortest watchman route. A cut is called an **essential cut** if it is not dominated by any other cuts. It is important to observe that any watchman route must visit these essential cuts and any route that visits them is a watchman route.

The essential cuts can be identified in $O(n)$ time by applying the clockwise scanning scheme. Let $C_1, C_2, \cdots, C_m$ be the sequence of essential cuts indexed in the clockwise order of their left endpoints. The set of essential cuts is then partitioned into cut corners. A **cut corner** is a subset of consecutive essential cuts $C_i, C_{i+1}, \cdots, C_j$ such that each C_k intersects with C_{k-1} ($i < k \leq j$), and C_i and C_j do not intersect with C_{i-1} and C_{j+1}, respectively. Clearly, any pair of the cuts derived from different cut corners can not intersect each other. This partition of essential cuts into cut corners takes $O(n)$ time.

Let us now review some known results on shortest watchman routes.

Lemma 1 *[3] The shortest watchman route in a simple polygon P through a point s on its boundary is unique.* □

In fact, Lemma 1 can be stated in a more general form: Given a source point s on P's boundary and a target point t on cut C_i, the shortest watchman path from s to t, which lies in the essential piece of C_i and visits $C_1, C_2, \cdots, C_{i-1}$, is unique.

Lemma 2 *[3] The shortest watchman route should visit the cut corners in the order in which they appear in the boundary of P.* □

Lemma 2 states that the shortest watchman route need not cross itself among cut corners. Lemma 2 also applies inside a cut corner, i.e., the shortest watchman route does not properly intersect itself. Convexity of the shortest watchman route in a cut corner is also showed in [3]. If a concave section occurs inside a cut corner, the section can be stretched out to obtain a shorter route. Concave sections can occur only among cut corners.

3   Incremental construction of shortest watchman routes

Let us first see what is the problem faced with us. Consider an intersection between two essential cuts. On one side of the intersection, one cut dominates the other cut while the opposite is true on the other side of the intersection. An intersection thus corresponds to a switch in (visibility) dominance between the intersecting cuts. The concept of essential cuts is then needed to be refined further. In a cut corner, an essential cut is intersected with at most $m - 1$ cuts and thus divided into at most $m - 1$ **fragments**. We can similarly define the dominance among fragments: fragment f dominates fragment g if any route that visits f also visits the cut to which g belongs. Each fragment thus has a unique set of essential cuts dominated by it. (Note that two fragments may dominate each other.) In order to formulate the watchman route problem, we define a set of fragments, called **watchman fragment set**: (i) (completeness) the union of dominances of the fragments is the whole set of essential cuts and (ii) (independence) no one is dominated by any other fragments. Thus, visiting these fragments in any order will give a watchman route and the optimum watchman route for the fragment set need not go over any fragments. With respect to a watchman fragment set, we distinguish a fragment as an *active* or *unactive* fragment according to whether it belongs to the fragment set or not. A cut is active if it contains an active fragment. Otherwise, it is unactive. Given a watchman fragment set, we can construct the corresponding (optimum) watchman route using the same approach as that is described in [2]. Specifically, the non-essential pieces of all active essential cuts are removed, the resulting polygon P' is triangulated and then rolled-out using the active fragments as mirrors. The optimum watchman route is then obtained by constructing the shortest path from s to its image s' in the rolled-out polygon. From these watchman routes, we can easily find the shortest one. A bit of thought will convince the reader that the number of possible watchman fragment sets can be high to exponential in the number of essential cuts.

We shall present below an $O(n^3)$ algorithm for the watchman route problem. Our algorithm proceeds in an incremental way. That is, the watchman, starting from s, visits the essential cuts one by one in order. When cut C_i is involved, we compute the intersections of C_i with C_1, C_2, $\cdots$, C_{i-1} and then construct the shortest watchman path P_i from s to some point s_i of C_i in the essential piece of C_i so that all the essential cuts with index less than i are visited along the path. Finally, the shortest watchman route is obtained when the watchman returns to s. For each cut C_i, a list is maintained to hold the intersections with the cuts added by now. The intersections are ordered from l_i to r_i. Obviously, these lists should be dynamically maintained and the fragments of cuts are also dynamically changed. It takes $O(n^2 log\, n)$ time and $O(n^2)$ space to maintain these lists.

Next, we describe how to choose a point on each cut so that the shortest watchman paths between s and them can be easily constructed. The chosen points are called "images". For a point p and a segment Q, p's image on Q is the point of Q which is closest to p. Images in a cut corner are defined as follows. Let C_i be the first (least indexed) cut in a cut corner. The left endpoint l_i is defined as the current starting image s_i. We then denote s_{i+1} as the image of s_i on cut C_{i+1}, s_{i+2} as the image of s_{i+1} on cut C_{i+2} and so on. In the case that s_k lies in the non-essential piece of C_{k+1}, i.e., the segment (s_{k-1}, s_k) intersects with cut C_{k+1}, the image s_{k+1} of s_k on C_{k+1} is undefined (since C_{k+1} has already be visited by any watchman path from s to s_k). The next image s_{k+2} of s_k on C_{k+2} is then considered. Observe that image s_i always lies to the right side of C_{i+1} (if s_{i+1} is defined) and thus the set of the images in a cut corner is that of the vertices of a convex polygon. The computation of images in polygon P takes linear time.

For simplicity, we assume that all of the cuts have the images defined on them. (In our incremental algorithm, a cut having no defined image is used only for computing cut intersections and can thus be handled together with the image-defined cut just after it.) For completeness, we set $s_{m+1} = s$. Now our watchman starts from s, visits in turn s_1, s_2, $\cdots$, s_m and finally returns to s_{m+1}. The shortest watchman path P_i, which starts at s and ends at s_i, visits all the cuts with index less than i.

Similar to the watchman route, we can describe a watchman path from s to s_i by a set of the fragments with index less than i. The index of a fragment is that of the cut to which it belongs. Because of the difference between watchman routes and watchman paths, we need to define the dominance of an image. We say image s_i dominates cut C_h if s and s_i lie to the different sides of C_h. Then the union of dominances of not only the fragment set but also the end point s_i is the set $\{C_1, C_2, \cdots, C_{i-1}\}$. (Note that C_i does not appear in the dominance set $\{C_1, C_2, \cdots, C_{i-1}\}$.) Thus, the watchman fragment set for the shortest watchman route is also computed incrementally.

Before we describe how the increment is done and what is its nature, let us consider how the shortest watchman paths can come in contact with the essential cuts. A shortest watchman path P_l makes a **reflection contact** with an essential cut C_i if the path comes into the cut at some point and then reflects on the cut and goes away from that point (Fig. 3a). There exists only one point in common between the path and the cut. The reflection is *perfect* when the incoming angle of the path with the cut is equal to the outgoing angle. A shortest watchman path P_l makes a **crossing contact** with an essential cut C_i if the path crosses the cut once or twice (Fig. 3b). There exist one common point or two common points between the path and the cut. The degenerate case of a crossing contact where the path crosses the cut twice but does not properly cross the cut is called a **tangential contact** (Fig. 3c). In this case, they share a line segment. (Tangential contacts are not degenerate cases of reflect contacts since a single point can not form a line segment.) When the shortest watchman path reflects on essential cuts, it must reflect perfectly unless the reflection contacts occur at cut intersections. For a watchman fragment set, the corresponding optimum path makes reflection contacts with the active cuts and crossing contacts with the unactive cuts.

At the initial step of our algorithm, we look for the shortest watchman path P_k with the greatest index k which is just the line segment (s, s_k). The path P_k can be simply found by connecting s to s_k in order until the segment (s, s_{k+1}) does not intersect all the cuts with index less than $k+1$. Clearly, this initial step takes $O(k^2)$ time. The watchman fragment set for P_k is initially set to be empty.

Assume inductively that P_l and thus the associated watchman fragment set are given. Now we shall describe how P_{l+1} is obtained when C_{l+1} is added. P_{l+1} should be constructed from P_l, rather than recomputed from scratch each time. The procedure consists of (1) finding an initial path P_{l+1}^0 which makes use of most of P_l and visits all of the cuts with index less than $l+1$ and (2) adjusting the current path P_{l+1}^m until it becomes optimum. The completeness and independence of the watchman fragment sets should be retained at both steps. The path obtained at the end is P_{l+1}. The associated watchman fragment set is also maintained. We shall describe these two steps in detail.

Finding an initial path P_{l+1}^0

Let us first give two lemmas concerning P_l, which will assure that we can find a reasonable path P_{l+1}^0, i.e., it is convex within each cut corner.

Lemma 3 P_l *lies in the essential piece of* C_{l+1}.

Proof: This can be simply proved by noticing that s and s_l lie in the essential piece of C_{l+1} and P_l can not intersect with C_{l+1}. □

Lemma 4 *In a cut corner, the projection of the last segment of P_l on C_l is behind s_l and the projection of the segment (s_l, s_{l+1}) is ahead of s_l.*

Proof: The first part directly follows from the fact that P_l consists of convex chains within cut corners. Recall that the endpoints l_l and r_l are to the left of the endpoints l_{l+1} and r_{l+1}

on the boundary of polygon P, respectively. It is also easy to see that the projection of the segment (s_l, s_{l+1}) is ahead of s_l. $\square$

Now we consider how to find the initial path P_{l+1}^0. When C_l and C_{l+1} are in the different cut corners, i.e., s_{l+1} is the starting image of the next cut corner, P_{l+1}^0 is simply found by adding the shortest path between s_l and s_{l+1} to P_l. The shortest path between two points in a simple polygon can be found by using the existing linear time algorithms [4]. The fragment of C_l containing s_l is then inserted into the current watchman fragment set. More studies are needed when C_l and C_{l+1} belong to the same cut corner. Consider the configurations between the last segment of P_l and s_{l+1}. With respect to cut C_l, s_{l+1} might be to the left or right of it. By virtue of Lemma 3 and Lemma 4, these two basic cases are shown in Fig. 4a-4b. If s is to the right of C_l (Fig. 4a), P_{l+1}^0 is simply formed by adding the segment (s_l, s_{l+1}) to P_l. The fragment of C_l containing s_l is inserted into the current watchman fragment set. If s_l is to the left of C_l (Fig. 4-b1,b2), we first look for the unactive cuts (with index less than l) which have s and s_{l+1} in the right side of them and are crossed exactly once by the last segment of P_l. Among these cuts, we select one whose intersection with P_l is nearest to s_l. If there exists such cut, P_{l+1}^0 is obtained by replacing the segment of P_l from the intersection to s_l with the segment from the intersection to s_{l+1} (Fig. 4-b1). The fragment of the selected cut containing that intersection is inserted into the watchman fragment set. Otherwise, P_{l+1}^0 is obtained by replacing the last segment of P_l with the segment which starts at the contact point of P_l with the last active cut and ends at s_{l+1} (Fig. 4-b2). There is no change in the watchman fragment set in this case. In the special case where the last segment of P_l overlaps with C_l (see Fig. 4c), P_{l+1}^0 is obtained by replacing the last segment of P_l with the segment which starts at the last reflect contact point of P_l and ends at s_{l+1}. The current watchman fragment set remains unchanged.

In the above way, we obtain an initial path P_{l+1}^0 that remains convex within each cut corner and visits all of the cuts with index less than $l + 1$. Suppose that the cut which has the point connected to s_{l+1} in P_{l+1}^0 has index l' (e.g., l' coincides with l in Fig. 4a). It is important to observe that the part of P_{l+1}^0 that is completely overlapped with P_l (from s to the point of $C_{l'}$ which is connected to s_{l+1}) has been already adjusted, and is optimal as it is. Clear, the operations done in this step do not affect the completeness and the independence of the watchman fragment set. This step also takes linear time.

Adjusting the current path P_{l+1}^m

We say a watchman path P is *adjustable* on a cut C if P makes a reflection contact with C and the incoming angle of P with C is not equal to the outgoing angle. The appropriate adjustment of the contact point on C results in a shorter path. The new path, of course, must be remained as a watchman path.

Since the incoming angle with respect to $C_{l'}$ is not equal to the outgoing angle, the path P_{l+1}^0 is adjustable on $C_{l'}$. Thus, P_{l+1}^0 is not optimal with respect to the current fragment set. Using the current active fragments, we optimize the path P_{l+1}^0 into the path P_{l+1}^1. Now P_{l+1}^1 is optimal with respect to the current watchman fragment set and taken as the starting path at this step. The shortest watchman path P_{l+1} is then found by adjusting the current path P_{l+1}^m, where $m \geq 1$. Each adjustment involves a change in the watchman fragment set and results in a shorter path. This suggests in principle a simple algorithm:

$m = 1$;
WHILE P_{l+1}^m is adjustable **DO**

Take an adjustable cut and update the watchman fragment set;
Construct the path P_{l+1}^{m+1} according to the new fragment set;
$m = m + 1$;
OD

An adjustment can only occur at the intersection of two essential cuts. As Chin and Ntafos showed in [3], there are three types of adjustments on an active cut C_i (Fig. 5). In Fig. 5, the incoming angle of P_{l+1}^m with C_i is assumed to be smaller than the outgoing angle. (The symmetric case is omitted). The bold and discontinuous segments in Fig. 5 stand for the active fragments before and after an adjustment, respectively. A possible next path P_{l+1}^{m+1} is also shown. Except for Fig. 5b, the indexs of the participating cuts are represented with respect to index i, i.e., $h < i < j < k$.

In Fig. 5a, P_{l+1}^m makes reflection contacts with both C_i and C_h at their intersection. The adjustment involves moving the contact point of C_i to the left. The next path, P_{l+1}^{m+1}, will make a reflection with C_i but a crossing contact with C_h. Thus, the current fragment of C_i is replaced by the next fragment and the fragment of C_h is deleted from the fragment set. We call this a **(-1)-adjustment** since the number of active fragments is decreased by 1. Note that we should not move the contact point of C_h to the left as that would leave C_i without a contact with the new path.

In Fig. 5b, P_{l+1}^m makes a reflection contact with C_i and a normal crossing contact with $C_{i'}$. The adjustment involves moving the contact point of C_i to the left, i.e., replacing the current fragment of C_i by the next fragment. The next path P_{l+1}^{m+1} still makes a crossing contact with $C_{i'}$. This is called a **0-adjustment**. Index i' can be smaller or greater than index i.

In Fig. 5c, P_{l+1}^m makes a reflection contact with C_i but a special crossing contact with C_j, i.e., the crossing contact with C_j has degenerated into a reflection or a tangential contact. From the definition of the watchman fragment set, index j can not be smaller than index i; otherwise we can not assure that C_j must be visited by the watchman path. The adjustment involves substituting the current fragment of C_i with the next fragment and inserting the fragment of C_j next to p_{ij} (the intersection of C_i and C_j) into the fragment set. For the tangential contact case, the next active cut C_k should be also considered (Fig. 5-c2). In order to shorten the path, the incoming angle of P_{l+1}^m with C_k must be greater than the outgoing angle. Thus, the current fragment of C_k should be also substituted by the next fragment. Furthermore, we can add the segment (p_{ij}, p_{jk}) of C_j, rather than a single fragment of C_j, to the watchman fragment set so that the following possible 0-adjustments on C_j can be eliminated. After the new path P_{l+1}^{m+1} is constructed, the segment (p_{ij}, p_{jk}) should be substituted back with the fragment of C_j which now contains the contact point of P_{l+1}^{m+1} with C_j. We called it a **(+1)-adjustment** since the number of active fragments is increased by 1. Depending on whether the incoming angle of P_{l+1}^m with C_j is greater (Fig. 5-c1,c2) or smaller (Fig. 5-c3) than the outgoing angle, the next path P_{l+1}^{m+1} will be shorter, or the same as P_{l+1}^m. In the latter case, a (-1)-adjustment on C_j follows. This (-1)-adjustment should be done immediately after the (+1)-adjustment so that the path can still be shortened. (Note that the tangential contacts for (+1)- adjustments were not considered in [3].)

Clearly, each adjustment retains the completeness and the independence of the current fragment set unchanged. Our algorithm starts from an initial path P_{l+1}^0, which is relatively close to the shortest watchman path P_{l+1}. Then, the current path goes through a sequence of adjustments. Since each adjustment results in a shorter path and the number of possible watchman fragment sets is finite, we can eventually obtain P_{l+1}.

4 Analysis of the algorithm

We define the direction of an adjustment on an active cut C as the shift direction of the reflection contact point on C.

Lemma 5 *The adjustments on an active cut are all in the same direction at the step of adjusting the path P_{l+1}^m.*

Proof: Since the part of P_{l+1}^0 which overlaps with P_l has been already adjusted, we can adjust P_{l+1}^0 only at $C_{l'}$. Suppose this adjustment makes the contact point of $C_{l'}$ move in the left direction (with respect to the watchman path). Consider the rolled-out version of the path P_{l+1}^0 (Fig. 6). The effect of moving the contact point on $C_{l'}$ to the left will make the following paths P_{l+1}^m ($m \geq 1$) adjustable on the other active cuts. But the adjustments made on these cuts (including $C_{l'}$) will be all in the same direction; otherwise the path can not be shortened. The path P_{l+1}^m will gradually move close to the shortest watchman path P_{l+1}, but can never go over P_{l+1} since P_{l+1}^m must retain convexity in the rolled-out polygon. We thus obtain that the adjustments on an active cut are all in the same direction at the step of adjusting the path P_{l+1}^m. $\Box$

Lemma 6 *The procedure of constructing P_{l+1} from P_l requires $O(l)$ adjustments.*

Proof: Consider how the active cuts and the unactive cuts with respect to P_l change when P_{l+1} is constructed. First, an active cut can become unactive during the procedure of constructing P_{l+1} from P_l, but it can never be active again. Once a cut becomes unactive because of a (-1)-adjustment, the following paths will remain the crossing contact with that cut by virtue of Lemma 5. Next, we show that once a cut becomes active, it can never be unactive again. The proof is by contradiction. After cut C_j becomes active because of a (+1)-adjustment on cut C_i, the adjustments on C_j will be, say, all in the right direction. Suppose that in the following process, there exists another active cut $C_{j'}$ on which a (-1)-adjustment makes C_j unactive. The following paths will move further right to the intersection $p_{j'j}$ along $C_{j'}$. On the other hand, if there exists such a cut $C_{j'}$, index j' must be smaller than index j and the intersection p_{ij} must be left to $C_{j'}$ (see Fig. 7). Then, the path just after C_j becoming active makes a crossing contact with $C_{j'}$. Since $C_{j'}$ is supposed to be active (while C_j remains active), the path can be then adjusted to the situation where it makes a tangential contact with the unactive cut $C_{j'}$ (like Fig. 5-c2), which requires a (+1)-adjustment on C_j to make $C_{j'}$ active. This (+1)-adjustment makes the contact point on $C_{j'}$ move to the left of the intersection $p_{j'j}$. The following paths will go further left along $C_{j'}$ according to Lemma 5. Therefore, the adjustments on $C_{j'}$ can not make C_j unactive. This contradicts with our assumption. In summary, there are $O(l)$ (+1)-adjustments and (-1)-adjustments during the procedure of constructing P_l from P_{l+1}. Consider 0-adjustments now. Let us charge the cost of a 0-adjustment to the unactive cut. From Lemma 5 and convexity of the watchman paths in cut corners, an unactive cut can take part in at most two 0-adjustments. Thus, the total number of 0-adjustments is also $O(l)$. This completes the proof. $\Box$

Theorem 1 *The time complexity of the incremental algorithm is $O(n^3)$.*

Proof: The incremental algorithm for constructing the shortest watchman route requires at most $O(n^2)$ adjustments. For each adjustment, we need to construct the corresponding optimum path for the new fragment set, which takes linear time. Thus, the time complexity of all adjustments is $O(n^3)$. This dominates the time complexities of the other steps in the algorithm. (The space requirement of our algorithm is $O(n^2)$.) □

Fig. 8 shows main steps of incremental construction of the shortest watchman route in a simple polygon P. The bold segments stand for the active fragments. In Fig. 8a, P_3, P_4 and P_5^0 are shown. P_3 is the watchman path with the greatest index which is just a line segment. In order to obtain P_4, C_1 becomes active. P_5^0 is the initial path and is not optimal with respect to the current fragment set. Fig. 8b gives the starting path P_5^1. A $(+1)$-adjustment on C_1 produces the next path P_5^2 or, exactly, P_5 (in order to speed up the adjusting procedure, two consecutive fragments of C_2 are used for this adjustment). See Fig. 8c. The shortest watchman route P_6 is shown in Fig. 8d. This route reflects perfectly on C_1, C_2, C_4 and C_5. The method for computing P_6 is shown in Fig. 9. A possible triangulation of the polygon P', which is obtained by removing the non-essential pieces of C_1, C_2, C_4 and C_5, is given in Fig. 9a, and the rolled-out polygon with respect to the active fragments and the shortest watchman route P_6 from s to s' are shown in Fig. 9b.

5 Discussion and application

We have presented an $O(n^3)$ algorithm for the watchman route problem in simple polygons. The time complexity of the algorithm is mainly determined by the number of adjustments. Our algorithm needs a total of $O(n^2)$ adjustments. Can this be further reduced? $O(n \log n)$ adjustments seem possible for a divide-and-conquer algorithm [7].

For simple rectilinear polygons, it is proved [2] that the overall shortest watchman route is unique (except for very special cases where finite shortest watchman routes exist). Thus, the watchman route problem for simple rectilinear polygons is solved overall. It is then natural to ask for the similar solution for simple polygons without specifying the starting point. However, the problem still remains open. Besides, finding shortest watchman routes in polygons with holes and in simple polyhedra is known to be NP-hard [2].

A similar problem, called the **robber route problem**, is introduced in [5]. The problem generalizes the watchman route problem in two sides. First, it requires to see a specified subset of edges of P instead of all of them. Second, the route should avoid to be seen from a given set of points inside P. A possible application of the robber route problem is in path planning where a low flying aircraft is required to photograph installations while avoiding detection. If a robber route exists, the problem can be reduced to the watchman route problem [5]. Using our algorithm, the robber route problem for simple polygons can be also solved in $O(n^3)$ time. This improves the previous $O(n^4)$ bound.

172

References

[1] B. Chazelle, Triangulating a simple polygon in linear time, *Proceedings, 31th Annu. IEEE Symp. Found. of Comput. Sci.*(1990), pp. 220-229.

[2] W.P.Chin and S.Ntafos, Optimum watchman routes, *Inform. Process. Lett.* 28, 39-44, 1988.

[3] W.P.Chin and S.Ntafos, Shortest watchman routes in a simple polygon, *Discrete Comput. Geometry* 6, 9-31, 1991.

[4] L.Guibas, J.Hershberger, D.Leven, M.Sharir and R.Tarjan, Linear time algorithms for visibility and shortest path problems inside simple polygons, in *Proc. 2nd ACM Symp. Comput. Geometry*(1987), pp.1-13.

[5] S.Ntafos, The robber route problems, *Inform. Process. Lett.* 34 (1990) 59-63.

[6] J.O'Rourke, *Art Gallery theorems and algorithms*, Oxford University Press, 1987.

[7] X.H.Tan, T.Hirata and Y.Inagaki, Constructing shortest watchman routes by divide-and-conquer, Manuscript in preparation.

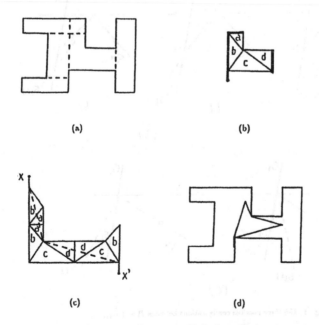

(a)

(b)

(c)

(d)

Fig. 1. Finding shortest watchman routes in simple rectilinear polygons

due to Chin and Ntafos's algorithm.

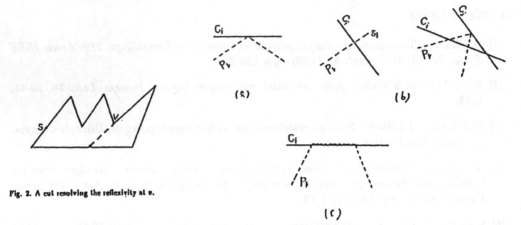

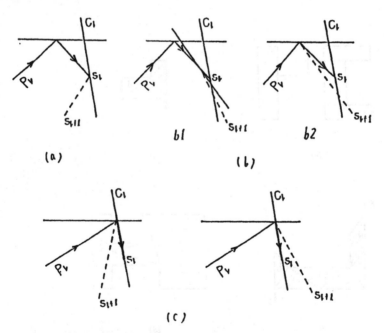

Fig. 2. A cut resolving the reflexiyity at v.

Fig. 3. The three basic contacts between P_i and C_i.

Fig. 4. The three possible configurations between P_i and s_{i+1}.

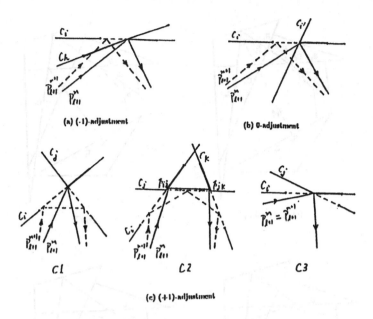

(a) (-1)-adjustment

(b) 0-adjustment

C1

C2

C3

(c) (+1)-adjustment

Fig. 5. Types of adjustments on an active cut C_i.

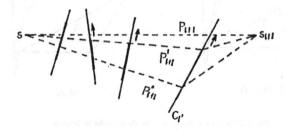

Fig. 6. The adjustments on a single cut are all in the same direction.

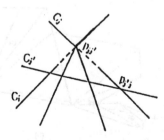

Fig. 7. Illustration for the proof of Lemma 6.

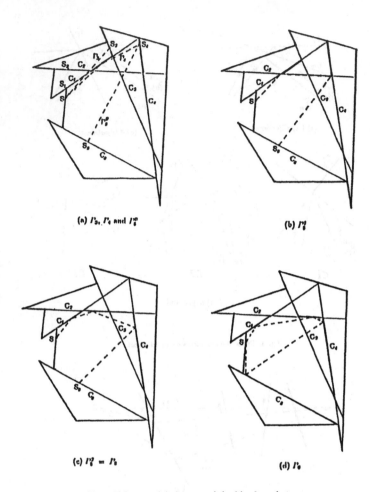

(a) P_3, P_4 and P_5^0

(b) P_5^1

(c) $P_5^2 = P_5$

(d) P_6

Fig. 8. Main steps of the incremental algorithm for an instance
of the watchman route problem.

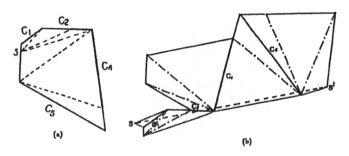

(a)

(b)

Fig. 9. The construction of P_6

On Hitting Grid Points in a Convex Polygon with Straight Lines *

H. S. Lee

Institute of Computer Science and Information Engineering,
National Chiao Tung University,
Hsinchu, Taiwan 30050,
Republic of China

R. C. Chang[†]

Institute of Computer and Information Science,
National Chiao Tung University,
Hsinchu, Taiwan 30050, Republic of China
E-mail: rcchang@twnctu01.bitnet

Abstract

We consider the following problem: Find a set of parallel straight lines with equal spacing to hit all m grid points in a closed region bounded by a convex polygon P with n vertices such that no grid points in a plane fall between two adjacent lines of these parallel lines and size of this set is minimal. We use continued fraction expansions to explore the combinatorial properties of this problem and propose an $O(n + log m)$ approximation algorithm which guarantees finite performance ratio.

1   Introduction

Hitting set problem is a well known NP-complete problem [1]. Hassin and Megiddo [3] recently proposed approximation algorithms with a finite performance ratio to solve some special cases of hitting set problem: Given m compact subsets of R^d, find a set of straight lines with minimum cardinality so that each of the given subsets is hit by at least one line and slopes of hitting lines are restricted to a finite set. Here, we restrict our attention to find a set of parallel straight lines with equal spacing to hit all m grid points in a closed region bounded by a convex polygon P with n vertices such that no grid points in a plane fall between two adjacent lines of these parallel lines and size of this set is minimal (Figure 1). By a grid point, we mean a point with integer coordinates.

---

*This research work was partially supported by the National Science Council of the Republic of China under grant No. NSC80-0408-E009-03.

[†]To whom all the correspondences should be sent.

This problem arises in structuring grid points in a convex polygon such that searching on grid points or enumeration of grid points can be done efficiently when collinear grid points possess some favorable properties [5]. Another possible application is to minimize number of bombers for an air raid against a cluster of fuel tanks scattered regularly in a convex region so each pass of a bomber over the area is a straight line along which fuel tanks are destroyed [3].

Let S_v denote set of parallel lines of equal spacing with direction v to cover all grid points in P such that no grid points in the plane fall between two adjacent lines in S_v. Let $|S_v|$ denote the cardinality of S_v. Direction of set of parallel lines obviously determines cardinality of the set. In this paper, we use continued fraction expansion [2,7,8] to explore combinatorial properties of size of set of parallel lines for all possible directions, proposing an approximation algorithm with finite performance ratio. Performance ratio is the ratio of the solution value delivered by the algorithm to the optimal solution value. We first classify possible directions by continued fraction expansions into three classes. We then derive various bounds of size of the set of parallel lines for each class, showing one specific direction u is quasi-optimal. That is, there exists a constant c such that $c|S_u| < |S_v|$ for any v. Size of the set of parallel lines with direction u is within constant ratio of the minimal. Assume convex polygon P with n vertices is contained in a U by U grid plane. Time complexity to find a quasi-optimal solution is $O(n + log U)$. If there are m grid points in P and m is roughly equal to the area of P, time complexity is $O(n + log m)$.

2 A Classification of Possible Directions of Parallel Lines

Since each line covering more than one grid point has a slope of rational number, direction of a set of parallel straight lines covering grid points in a convex polygon can be represented by a vector consisting of two relative prime integers. A vector consisting of two relative prime integers will be used to represent the direction of a set of parallel lines; such a vector is called direction vector. Euclidean norm of a vector is defined as length of a vector as usual.

Definition 2.1 ([10]) *The diameter of a convex polygon is the maximum distance between any two of its vertices. A diametral pair of a convex polygon is a pair of its vertices whose distance is exactly the diameter of the polygon (Figure 2).*

With loss of generality, we shall assume there is only one diametral pair. If there are more than one diametral pair, tie can be broken by choosing any diametral pair. Let α denote slope of line passing through the diametral pair. Without loss of generality, assume $0 \le \alpha \le 1$. $(1, \alpha)$ is then the vector of line passing through the diametral pair. Let θ_v denote angle between direction vector v and $(1, \alpha)$. Size of $|S_v|$ depends on two factors: θ_v and $\|v\|$, length of v. θ_v affects range to be covered by the parallel lines, and length of direction vector v determines distance between two adjacent parallel lines. θ_v and $\|v\|$, however, are not two independent factors. Size of S_v is proportional to θ_v and $\|v\|$. Not departing from the topics in this section, derivation of exact relationship between $|S_v|$ and v will be discussed in the next section. Distance between two adjacent parallel lines being in inverse proportion to length of its direction vector will only be shown here.

Lemma 2.1 *Let d_v denote the distance between two adjacent parallel lines in S_v, where v is a direction vector. Then $d_v = \frac{1}{\|v\|}$.*

Let octants of plane be defined as eight wedges bounded by lines $x = y$, $x = -y$, $x = 0$, and $y = 0$ (Figure 3). Each direction vector v in Octant 1 has one correspondent in each octant. Vector $(4, 1)$ is in Octant 1 for example. Its correspondents in each octant are $(1, 4)$, $(-1, 4)$, $(-4, 1)$, $(-4, -1)$, $(-1, -4)$, $(1, -4)$ and $(4, -1)$ in counterclockwise. Since each direction vector v not in Octant 1 has one correspondent v' of same length in Octant 1 and θ_v is greater than $\theta_{v'}$, will only pay attention to direction vectors in Octant 1.

There is an one-to-one correspondence between direction vectors in Octant 1 and irreducible fractions in the interval $[0, 1]$. The ascending sequence of all irreducible fractions in $[0, 1]$ with denominators less than or equal to d is called Farey series of order d [6]. " $\frac{0}{1}, \frac{1}{4}, \frac{1}{3}, \frac{1}{2}, \frac{2}{3}, \frac{3}{4}, \frac{1}{1}$ " is a Farey series of order 4, for example. There are approximately $\frac{3U^2}{\pi^2}$ fractions with denominators no greater than U in $[0, 1]$ (see Farey series in [4]). Grid size U is normally very large [9]. An exhaustive search for optimal direction would be inefficient. A more efficient search is then essential. Since lengths of direction vectors in Octant 1 are dominated by denominators of the fractions and denominators of series vary drastically, a straightforward binary search for optimal direction is impossible.

An efficient search scheme in conjunction with continued fraction expansions, which is as effective as binary search, is proposed. By continued fraction expansions, a subsequence of Farey series can be obtained; denominators in this subsequence will become greater as fractions become closer to α in either direction. In the next section, a direction corresponding to one fraction in this subsequence is a quasi-optimal direction will be shown in next section.

To classify directions in Octant 1, some results in number theory [6] will first be introduced. Since α, the slope of line through diametral pair, is a real number, by the method called continued fraction expansion [7], α can be uniquely written in the form

$$\alpha = a_0 + \cfrac{1}{a_1 + \cfrac{1}{a_2 + \cdots}}$$

where $a_0 = \lfloor \alpha \rfloor$ and $a_1, a_2, \ldots$ are positive integers. In fact, $a_1, a_2, \ldots$ can be defined by the recurrence

$$\alpha_0 = \alpha, a_0 = \lfloor \alpha \rfloor, \alpha_{k+1} = \frac{1}{\alpha_k - a_k}, a_{k+1} = \lfloor \alpha_{k+1} \rfloor.$$

Let the rational number

$$\frac{g_k}{h_k} = a_0 + \cfrac{1}{a_1 + \cfrac{1}{\ddots a_{k-1} + \frac{1}{a_k}}}.$$

$\frac{g_k}{h_k}$ is then called k-th convergent of α [7]. For example, if $\alpha = \frac{123}{638}$, by continued fraction expansions,

$$\frac{123}{638} = 0 + \cfrac{1}{5 + \cfrac{1}{5 + \cfrac{1}{2 + \cfrac{1}{1 + \frac{1}{7}}}}}.$$

Its convergents are then $\frac{g_0}{h_0} = \frac{0}{1}$, $\frac{g_1}{h_1} = \frac{1}{5}$, $\frac{g_2}{h_2} = \frac{5}{26}$, $\frac{g_3}{h_3} = \frac{11}{57}$, $\frac{g_4}{h_4} = \frac{16}{83}$, and $\frac{g_5}{h_5} = \frac{123}{638}$. Arranging these convergents in ascending order, we have a subsequence of Farey series of order greater than 638: $\frac{0}{1}, \frac{5}{26}, \frac{16}{83}, \frac{123}{638}, \frac{11}{57}, \frac{1}{5}$.

We have following properties about convergents of α :

Lemma 2.2 ([7])

1. $g_{k+1} = a_{k+1}g_k + g_{k-1}, h_{k+1} = a_{k+1}h_k + h_{k-1}$,

2. $g_{k+1}h_k - g_k h_{k+1} = (-1)^k$,

3. $|\alpha - \frac{g_k}{h_k}| < \frac{1}{h_k h_{k+1}}$.

Convergents of α satisfy the recurrence relation in Property 1. If $0 \leq \alpha \leq 1$, initial values in the recurrence relation would be $(h_{-1}, g_{-1}) = (0, 1)$ and $(h_0, g_0) = (1, 0)$. Let $g_{k,a} = a g_{k-1} + g_{k-2}$ and $h_{k,a} = a h_{k-1} + h_{k-2}$, where a is a integer. For $k \geq 1$, $\frac{g_{k,a}}{h_{k,a}}$ is defined to be a quasi-convergent of α if $1 \leq a \leq a_k - 1$ [6]. One property of quasi-convergents is stated as follows:

Lemma 2.3 ([6]) *If k is even, then $\frac{g_{k-2}}{h_{k-2}} < \frac{g_{k,1}}{h_{k,1}} < \cdots < \frac{g_{k,a_k-1}}{h_{k,a_k-1}} < \frac{g_k}{h_k} \leq \alpha < \frac{g_{k,a_k+1}}{h_{k,a_k+1}} < \frac{g_{k-1}}{h_{k-1}}$, while the inequality signs are reversed when k is odd.*

For example, if $\alpha = \frac{123}{638}$, quasi-convergents between $\frac{g_0}{h_0}$ and $\frac{g_2}{h_2}$ are: $\frac{g_{2,1}}{h_{2,1}} = \frac{g_0 + 1 \times g_1}{h_0 + 1 \times h_1} = \frac{1}{6}$, $\frac{g_{2,2}}{h_{2,2}} = \frac{g_0 + 2 \times g_1}{h_0 + 2 \times h_1} = \frac{2}{11}$, $\frac{g_{2,3}}{h_{2,3}} = \frac{g_0 + 3 \times g_1}{h_0 + 3 \times h_1} = \frac{3}{16}$, $\frac{g_{2,4}}{h_{2,4}} = \frac{g_0 + 4 \times g_1}{h_0 + 4 \times h_1} = \frac{4}{21}$. That is, $\frac{g_0}{h_0} = \frac{0}{1} < \frac{1}{6} < \frac{2}{11} < \frac{3}{16} < \frac{4}{21} < \frac{g_2}{h_2} = \frac{5}{26} < \frac{123}{638} < \frac{g_1}{h_1} = \frac{1}{5}$. Any quasi-convergent is a linear combination of two consecutive convergents. Some irreducible fractions with denominators less than 638 falling between two consecutive quasi-convergents also exist. For example, $\frac{3}{17}$ and $\frac{5}{28}$ fall between $\frac{1}{6}$ and $\frac{2}{11}$. These irreducible fractions can be easily obtained by following lemma:

Lemma 2.4 ([6]) *Let F_d denote the Farey series of order d. Two fundamental properties of F_d are:*

1. *$\frac{t_1}{s_1}$ and $\frac{t_2}{s_2}$ are adjacent in F_d if and only if $s_1 + s_2 > d$ and $|s_1 t_2 - s_2 t_1| = 1$.*

2. *If $\frac{t_1}{s_1}$ and $\frac{t_2}{s_2}$ are adjacent in F_{d-1} but separated in F_d, there will be only one intervening fraction in F_d, and it will be $\frac{t_1+t_2}{s_1+s_2}$.*

For example, $\frac{1}{6}$ and $\frac{2}{11}$ are adjacent in F_{11} and the irreducible fraction in-between with denominator less than or equal to 17 would be $\frac{1+2}{6+11} = \frac{3}{17}$.

Possible directions of parallel lines in Octant 1 can be classified as follows: Type 1 directions are those directions corresponding to convergents of α. Type 2 directions are those directions corresponding to quasi-convergents of α. Type 3 directions are those directions which correspond to irreducible fractions that are neither convergents nor quasi-convergents of α, that is, the irreducible fractions falling between quasi-convergents. For example, if $\alpha = \frac{123}{638}$, $(1, 0)$ and $(5, 1)$ are type 1 directions, $(6, 1)$ and $(11, 2)$ are type 2 directions, and $(17, 3)$ is a type 3 direction. Let $v_k = (h_k, g_k)$ and $v_{k,a} = (h_{k,a}, g_{k,a})$. For $k \geq 0$, v_k then denotes a type 1 direction; for $k \geq 1$ and $0 < a < a_k$, $v_{k,a}$ denotes a type 2 direction. Let v^* denote a direction of type 3. Note that by Lemma 2.2 and definition of quasi-convergent, v_k and $v_{k,a}$ can be expressed as linear combinations of v_{k-1} and v_{k-2}:

$$\begin{cases} v_k = a_k v_{k-1} + v_{k-2} \\ v_{k,a} = a v_{k-1} + v_{k-2}. \end{cases}$$

Also, if we treat v_k and $v_{k,a}$ as 3-space vectors, $\|v_k \times v_{k-1}\| = \|v_{k-1} \times v_{k-2}\| = \ldots = \|v_0 \times v_{-1}\| = 1$ and $\|v_{k,a} \times v_{k,a+1}\| = \|v_{k-1} \times v_{k-2}\| = 1$ for $k > 0$, where $\times$ denotes the corss product operation. In following section, bounds for size of S_v for directions of different types will be derived, showing one of v_k's is a quasi-optimal direction.

3 Bounds on Size of Parallel Lines

First, an estimation function $f(v)$ will be derived for $|S_v|$, where v is a direction vector. Estimation function is in proportion to length of v and angle between v and vector $(1, \alpha)$, which is the vector of line joining diametral pair. Upper and lower bounds of $f(v)$ for various types of directions are then investigated. One of the directions being a quasi-optimal direction will also be shown.

3.1 Derivation of $f(v)$

Width of a convex polygon is defined as follows (Figure 2). Let $\{p, q\}$ be a diametral pair of P. Draw two supporting lines parallel with line through p and q such that P is bounded by these two lines. Distance between these two lines is then said to be a width of P. In remained of this paper, l will be used to denote diameter of P; w to denote width of P.

Let $v^\perp$ be the direction orthogonal to v. Span of P in direction $v^\perp$, denoted as $S(P, v^\perp)$, is defined to be distance between two parallel supporting lines of P with direction v (Figure 4). Size of S_v is then equal to $\lfloor S(P, v^\perp)/d_v \rfloor$ or $\lfloor S(P, v^\perp)/d_v \rfloor + 1$. Let P' be the smallest rectangle containing P with two sides parallel with line joining diametral pair of P (Figure 5). In following, $S(P', v^\perp)$ will be shown as able to approximate $S(P, v^\perp)$.

Lemma 3.1

$$S(P, v^\perp) < S(P', v^\perp) < 5S(P, v^\perp)$$

Since θ_v is the angle between v and vector $(1, \alpha)$, and length and width of P' are l and w (Figure 6), we have $S(P', v^\perp) = l\sin\theta_v + w\cos\theta_v$. Let $L(v) = S(P', v^\perp)/d_v = (l\sin\theta_v + w\cos\theta_v)\|v\|$. From above discussions and Lemma 3.1, we know that the size of S_v can be approximated by $L(v)$. That is,

$$|S_v| - 1 < L(v) < 5|S_v| - 5. \tag{1}$$

Let $f(v) = (l\sin\theta_v + w)\|v\|$. We have following lemma.

Lemma 3.2

$$L(v) \leq f(v) \leq 2L(v)$$

Combining Lemma 3.2 and (1), we have the following theorem.

Theorem 3.1

$$|S_v| - 1 < f(v) < 10|S_v| - 10$$

$f(v)$ can be used as an approximation function for $|S_v|$. Observing $f(v)$, we find that f is proportional to length of direction vector v and θ_v, angle between direction vector and vector $(1, \alpha)$.

3.2 Bounds on $|S_v|$

Let i be the largest integer such that $h_i \leq min\{\sqrt{\frac{l}{w}}, l\}$. $\frac{g_i}{h_i}$ is the closest convergent to α with denominator less than or equal to $min\{\sqrt{\frac{l}{w}}, l\}$. In this subsection, bounds for $f(v_k)$ and bounds for $f(v_{k,a})$ are first derived, where v_k corresponds to convergent $\frac{g_k}{h_k}$ and $v_{k,a}$ corresponds to quasi-convergent $\frac{g_{k,a}}{h_{k,a}}$. Bound for $f(v^*)$ is then derived where v^* corresponds to an irreducible fraction which is neither a convergent nor a quasi-convergent of α. Bounds of $|S_v|$ for directions of different types are finally summarized, and v_i is shown to be a quasi-optimal direction.

Bounds of $f(v_k)$ for $k \geq 0$ are stated in the following lemma:

Lemma 3.3 *If $\sqrt{\frac{l}{w}} < l$ then*

1. $f(v_k) > \frac{1}{2\sqrt{2}}\sqrt{lw}$ for $k \neq i$ and $k \geq 0$,

2. $f(v_i) < 2\sqrt{2}\sqrt{lw}$;

otherwise, $f(v_i) < 2\sqrt{2}$.

When $\sqrt{\frac{l}{w}} < l$, lower bound of $f(v_{k,a})$ for $k \geq 1$ and $0 < a < a_k$ is stated in following lemma:

Lemma 3.4 *If $\sqrt{\frac{l}{w}} < l$ and $\frac{g_{k,a}}{h_{k,a}}$ is a quasi-convergent of α, then*

$$f(v_{k,a}) > \frac{1}{12\sqrt{2}}\sqrt{lw}.$$

Let $v^* = (s,t)$ where $\frac{t}{s}$ is an irreducible fraction in $[0,1]$ which is not a convergent nor quasi-convergent of α. Now we want to derive the lower bound for $f(v^*)$ when $\sqrt{\frac{l}{w}} < l$.

Lemma 3.5 *$f(v^*) > \frac{1}{12\sqrt{2}}\sqrt{lw}$ if $\sqrt{\frac{l}{w}} < l$.*

Combining Theorem 3.1 with Lemma 3.3, 3.4, and 3.5, we have the bounds for $|S_v|$ in the following theorem:

Theorem 3.2 *If $\sqrt{\frac{l}{w}} < l$ then*

1. $|S_{v_k}| > \frac{1}{20\sqrt{2}}\sqrt{lw} + 1$ for $k \geq 0$ and $k \neq i$,

2. $|S_{v_i}| < 2\sqrt{2}\sqrt{lw} + 1$,

3. $|S_{v_{k,a}}| > \frac{1}{120\sqrt{2}}\sqrt{lw} + 1$ for $k > 0$ and $0 < a < a_k$,

4. $|S_{v^*}| > \frac{1}{120\sqrt{2}}\sqrt{lw} + 1$;

otherwise, $|S_{v_i}| < 2\sqrt{2} + 1$.

Assume P contains at least one grid point in it. That is, $|S_v| \geq 1$ for any v. If $\sqrt{\frac{l}{w}} \geq l$, then

$$|S_v| > 1 = \frac{2\sqrt{2}+1}{2\sqrt{2}+1} > \frac{1}{2\sqrt{2}+1}|S_{v_i}|.$$

If $\sqrt{\frac{l}{w}} < l$, for $v \neq v_i$

$$|S_v| > \frac{\sqrt{lw}}{120\sqrt{2}} = \frac{2\sqrt{2}\sqrt{lw}}{480} > \frac{2\sqrt{2}\sqrt{lw}+1}{960} = \frac{1}{960}|S_{v_i}|.$$

v_i is then a quasi-optimal direction. Let A denote the area of P. Then $A = O(\sqrt{lw})$. We have the following theorem:

Theorem 3.3 v_i *is a quasi-optimal direction. That is, there exits a constant c such that $c|S_{v_i}| \leq |S_v|$ for all possible v. Moreover, the size of S_{v_i} is $O(\sqrt{A})$ where A is the area of P.*

4 Algorithm and Time Complexity

Following discussion in previous sections, we the following algorithm for a quasi-optimal direction:

Algorithm :

> **Step 1:** Determine a diametral pair of P and then compute the diameter l and the width w of P. Let α be the slope of line through the diametral pair.
>
> **Step 2:** Compute the closest convergent of α with denominator less than or equal to $min\{\sqrt{\frac{l}{w}}, l\}$. Slope of set of parallel lines is then equal to this convergent.

Step 1 can be done in $O(n)$ time [10]. By property 1 of Lemma 2.2, we know there are at most $O(logmin\{\sqrt{\frac{l}{w}}, l\})$ convergents with denominators no greater than $min\{\sqrt{\frac{l}{w}}, l\}$. Step 2 can then be done in time $O(logmin\{\sqrt{\frac{l}{w}}, l\})$. Total time of the algorithm is $O(n + logmin\{\sqrt{\frac{l}{w}}, l\})$ If P is contained in a $U \times U$ lattice, then $min\{\sqrt{\frac{l}{w}}, l\} \leq U$. Total time for the algorithm will then be $O(n + logU)$.

As an example, let $\alpha = \frac{123}{638}$. Its convergents are $\frac{0}{1}, \frac{1}{5}, \frac{5}{26}, \frac{11}{57}, \frac{16}{83}, \frac{123}{638}$. If $l = 640$ and $w = 10$, then the closest convergent to α with denominator less than $8(= \sqrt{\frac{l}{w}})$ is $\frac{1}{5}$. Quasi-optimal direction of the lines is then $(5, 1)$.

5 Conclusion

In this paper, we have tried to find a direction such that the number of parallel lines of equal spacing with this direction is minimal to cover grid points in a convex polygon. If directions are represented by vectors of two relative prime integers, we propose a classification of possible directions by continued fraction expansions. Various bounds of size of the parallel lines of different types are derived. We have also shown one particular direction is a quasi-optimal and this direction can be found efficiently by a search as effective as binary search.

Problem considered in this paper is interesting in itself. It remains open whether there exists an efficient way to determine the direction of real minimal. Since set of lines found is equally spaced, some of them may be redundant, that is, covering no grid points. We then pose the question of whether a set of minimal non-redundant parallel lines can be found efficiently to cover grid points in a convex polygon. An extension to higher dimensions is also a challenging topic.

References

[1] M. R. Garey and D. S. Johnson, *Computers and Intractability: A Guide to the Theory of NP-Completeness*, Fremman, San Francisco, California.

[2] D. H. Greene and F. F. Yao, Finite-resolution computational geometry, *Proc. 27th Symposium on Foundations of Computer Science*, pp. 143-152, 1986.

[3] R. Hassin and N. Megiddo, Approximation algorithms for hitting objects with straight lines, *Discrete Applied Mathematics*, Vol. 30 No. 1, pp. 29-24, 1991.

[4] D. E. Knuth, *The Art of Computer Programming*, Addison-Wesley, second edition, 1981.

[5] H. S. Lee and R. C. Chang, Weber problem on a grid, Techn. Rep., Dept. of Comp. Science and Info. Engineering, NCTU, 1991.

[6] W. J. LeVeque, *Fundamentals of Number Theory*, Addison-Wesley, 1977.

[7] L. Lovasz, *An Algorithmic Theory of Numbers, Graphs and Complexity*, Society for Industrial and Applied Mathematics, Pennsylvania, 1986.

[8] S. Mehta, M. Mukherjee and G. Nagy, Constrained integer approximation to 2-D line intersections, *Second Canadian Conf. on Computational Geometry*, Ottawa, Canada, pp. 302-305, August 1990.

[9] M. H. Overmars, Computational geometry on a grid: an overview, in *Theoretical Foundations for Computer Graphics and CAD*, Springer-Verlag, Berlin, pp. 167-184, 1988.

[10] F. P. Preparata and M. I. Shamos, *Computational Geometry: An Introduction*, Springer-Verlag, New York, 1985.

Appendix

Proo of Lemma 3.1

As shown in Figure 7, let q_1, q_2, q_3, q_4 be vertices of rectangle P'. Let A denote the area of P. Length and width of P' are l and w. We have $A \geq \frac{1}{2}lw$. It is known $|\overline{p_2 p_3}|S(P, v^\perp) \geq A$.

Therefore $|\overline{p_2p_3}|S(P,v^\perp) \geq \frac{1}{2}lw$. That is $w \leq 2S(P,v^\perp)\frac{\overline{p_2p_3}}{l}$. Since $|\overline{p_2p_3}| \leq l$, we have $w \leq 2S(P,v^\perp)$. Since sides of P' with length w must intersect with rectangle $p_1p_2p_3p_4$, we have $S(P',v^\perp) < S(P,v^\perp) + 2w$. Therefore $S(P',v^\perp) < 5S(P,v^\perp)$. It is trivial that $S(P,v^\perp) \leq S(P',v^\perp)$. Hence $S(P,v^\perp) < S(P',v^\perp) < 5S(P,v^\perp)$. □

Proof of Lemma 3.2

Without loss of generality, assume $0 \leq \theta_v \leq \frac{\pi}{2}$. It is trivial that $f(v) \geq L(v)$. Only $f(v) \leq 2L(v)$ needs to be proved. Let $K(\theta_v) = \frac{1}{2}w\sin\theta_v + w\cos\theta_v$. $K(\theta_v)$ will achieve its minimum, $\frac{1}{2}w$, when $\theta_v = \frac{\pi}{2}$. Since $\frac{1}{2}l\sin\theta_v + w\cos\theta_v \geq K(\theta_v) = \frac{1}{2}w\sin\theta_v + w\cos\theta_v$, we have $\frac{1}{2}w \leq \frac{1}{2}l\sin\theta_v + w\cos\theta_v$. Therefore, $\frac{1}{2}l\sin\theta_v + \frac{1}{2}w \leq l\sin\theta_v + w\cos\theta_v$. Hence, we have $f(v) \leq 2L(v)$. □

Proof of Lemma 3.3

Upper bound for $f(v_i)$ is first shown. Consider following two cases:

1. v_i is last convergent of α:
   That is, $\frac{g_i}{h_i} = \alpha$. Then $f(v_i) = w\|v_i\| < \sqrt{2}wh_i \leq \sqrt{2}\min\{\sqrt{\frac{l}{w}}, l\}w$. If $\sqrt{\frac{l}{w}} \leq l$ then $f(v_i) < \sqrt{2}\sqrt{lw}$; otherwise, $f(v_i) < \sqrt{2}lw < \sqrt{2}$.

2. v_i is not last convergent of α:
   Let $\theta_{v_i,v_{i+1}}$ denote angle between v_i and v_{i+1}. From Lemma 2.3, $\theta_{v_i} \leq \theta_{v_i,v_{i+1}}$. Also, $\sin\theta_{v_i,v_{i+1}} = \frac{\|v_i \times v_{i+1}\|}{\|v_i\|\|v_{i+1}\|}$ if these vectors are treated as three dimensional vectors, letting $\times$ denote the cross product. Since $\|v_i \times v_{i+1}\| = 1$, $\sin\theta_{v_i,v_{i+1}} = \frac{1}{\|v_i\|\|v_{i+1}\|}$. Then,

$$
\begin{aligned}
f(v_i) &= (l\sin\theta_{v_i} + w)\|v_i\| \\
&\leq (\frac{l}{\|v_i\|\|v_{i+1}\|} + w)\|v_i\| \\
&= \frac{l}{\|v_{i+1}\|} + w\|v_i\| \\
&< \frac{l}{h_{i+1}} + w\sqrt{2}h_i \\
&< \frac{l}{\min\{\sqrt{\frac{l}{w}}, l\}} + w\sqrt{2}\min\{\sqrt{\frac{l}{w}}, l\}.
\end{aligned}
$$

If $\sqrt{\frac{l}{w}} < l$ then $f(v_i) < \sqrt{lw} + \sqrt{2}\sqrt{lw} < 2\sqrt{2}\sqrt{lw}$; otherwise, $f(v_i) < 1 + \sqrt{2}lw < 2\sqrt{2}$.

If $\sqrt{\frac{l}{w}} < l$, lower bound of $f(v_k)$ for $k \neq i$ and $k \geq 0$ is now shown. Consider following two cases:

1. v_k is last convergent of α:
   Then $f(v_k) = (l\sin\theta_{v_k} + w)\|v_k\| = w\|v_k\| > wh_k$. $k < i$ is impossible. That is,

$k > i$. $h_k > min\{\sqrt{\frac{l}{w}}, l\} = \sqrt{\frac{l}{w}}$. Substituting this for previous result, we have $f(v_k) > w\sqrt{\frac{l}{w}} = \sqrt{lw}$.

2. v_k is not last convergent of α:

v_{k+1} then exists. Let $v_{k+1}^* = v_{k+1} + v_k$. From Lemma 2.3, it is apparent $\theta_{v_k} > \theta_{v_{k+1}^*, v_k}$ (Figure 8). Also, $\|v_k \times v_{k+1}^*\| = 1$ and $\|v_{k+1}^*\| < 2\|v_{k+1}\|$. Hence

$$sin\theta_{v_{k+1}^*, v_k} = \frac{\|v_k \times v_{k+1}^*\|}{\|v_k\|\|v_{k+1}^*\|} = \frac{1}{\|v_k\|\|v_{k+1}^*\|} > \frac{1}{2\|v_k\|\|v_{k+1}\|}.$$

Therefore,

$$\begin{aligned} f(v_k) &= (lsin\theta_{v_k} + w)\|v_k\| \\ &> \frac{l}{2\|v_{k+1}\|} + w\|v_k\| \\ &> \frac{l}{2\sqrt{2}h_{k+1}} + wh_k. \end{aligned}$$

If $k < i$ then $h_{k+1} \leq h_i \leq \sqrt{\frac{l}{w}}$. That is, $f(v_k) > \frac{l}{2\sqrt{2}\sqrt{\frac{l}{w}}} = \frac{1}{2\sqrt{2}}\sqrt{lw}$. If $k > i$ then $h_k > \sqrt{\frac{l}{w}}$. That is, $f(v_k) > wh_k > \sqrt{lw}$. $\square$

Proof of Lemma 3.4

Assume $\sqrt{\frac{l}{w}} < l$. Let $\theta_{v_{k,a}, v_{k,a+1}}$ denote angle between vectors $v_{k,a}$ and $v_{k,a+1}$. Since $a < a_k$, by Lemma 2.3, it is known $\theta_{v_{k,a}} > \theta_{v_{k,a}, v_{k,a+1}}$. As the proof in Lemma 3.3, vectors are treated as vectors in 3-space. Then,

$$\begin{aligned} f(v_{k,a}) &= (lsin\theta_{v_{k,a}} + w)\|v_{k,a}\| \\ &> (lsin\theta_{v_{k,a}, v_{k,a+1}} + w)\|v_{k,a}\| \\ &= \frac{l\|(av_{k-1} + v_{k-2}) \times ((a+1)v_{k-1} + v_{k-2})\|}{\|(a+1)v_{k-1} + v_{k-2}\|} + \|av_{k-1} + v_{k-2}\|w \\ &= \frac{l}{\|(a+1)v_{k-1} + v_{k-2}\|} + \|av_{k-1} + v_{k-2}\|w \\ &> \frac{l}{\|v_k\|} + \|v_{k-1}\|w \end{aligned} \qquad (2)$$

If $k \leq i$, $\|v_k\| < \sqrt{2}h_k \leq \sqrt{2}min\{\sqrt{\frac{l}{w}}, l\} = \sqrt{2}\sqrt{\frac{l}{w}}$.

If $k > i+1$, $\|v_{k-1}\| > h_{k-1} > min\{\sqrt{\frac{l}{w}}, l\} = \sqrt{\frac{l}{w}}$. Hence,

$$\text{if } k \leq i, \quad (2) > \frac{l}{\|v_k\|} > \frac{1}{\sqrt{2}}\sqrt{lw};$$

$$\text{if } k > i+1, \quad (2) > \|v_{k-1}\|w > \sqrt{lw}.$$

That is, for $k \neq i+1$ and $1 \leq a \leq a_k - 1$,

$$f(v_{k,a}) > \frac{1}{\sqrt{2}}\sqrt{lw}. \qquad (3)$$

Lower bound of $f(v_{i+1,a})$ for $1 \leq a \leq a_{i+1}-1$ is now to be shown. Since $\|v_i \times v_{i+1,a}\| = 1$ for $1 \leq a \leq a_{i+1}$, $\{v_i, v_{i+1,a}\}$ is a basis for the lattice Z^2 (See [7]). Therefore,

$$L(v_i)L(v_{i+1,a}) \geq \text{ the area of } P' = lw.$$

Since $f(v_i) > L(v_i)$ and $f(v_{i+1,a}) > L(v_{i+1,a})$, we have

$$f(v_i)f(v_{i+1,a}) > lw. \tag{4}$$

Similar to the proof of Lemma 3.3, we have

$$
\begin{aligned}
f(v_{i-1}) &< (l\sin\theta_{v_{i-1},v_i} + w)\|v_{i-1}\| \\
&< \frac{l}{h_i} + \sqrt{2}wh_{i-1} \\
f(v_i) &< (l\sin\theta_{v_i,v_{i+1}} + w)\|v_i\| \\
&< \frac{l}{h_{i+1}} + \sqrt{2}wh_i.
\end{aligned}
\tag{5}
$$

As for $f(v_{i+1})$, if v_{i+1} is the last convergent, then

$$f(v_{i+1}) = w\|v_{i+1}\| < \sqrt{2}wh_{i+1};$$

otherwise,

$$
\begin{aligned}
f(v_{i+1}) &< (l\sin\theta_{v_{i+1},v_{i+2}} + w)\|v_{i+1}\| \\
&< \frac{l}{h_{i+2}} + \sqrt{2}wh_{i+1}
\end{aligned}
$$

Hence, if v_{i+1} is the last convergent then $f(v_i)f(v_{i+1}) < \sqrt{2}lw + 2w^2 h_i h_{i+1}$; otherwise,

$$
\begin{aligned}
f(v_i)f(v_{i+1}) &< \frac{l^2}{h_{i+1}h_{i+2}} + \sqrt{2}lw + \sqrt{2}lw\frac{h_i}{h_{i+2}} + 2w^2 h_i h_{i+1} \\
&< 4lw + 2w^2 h_i h_{i+1}
\end{aligned}
$$

In other words,

$$f(v_i)f(v_{i+1}) < 4lw + 2w^2 h_i h_{i+1} \tag{6}$$

always holds. However,

$$
\begin{aligned}
f(v_i)f(v_{i-1}) &< \frac{l^2}{h_i h_{i+1}} + \sqrt{2}\frac{h_{i-1}}{h_{i+1}}lw + \sqrt{2}lw + 2w^2 h_i h_{i-1} \\
&< \frac{l^2}{h_i h_{i+1}} + 5lw
\end{aligned}
\tag{7}
$$

If $h_i h_{i+1} > \frac{l}{w}$ then $(7) < 6lw$; otherwise $(6) \leq 6lw$. In other words, either

$$f(v_i)f(v_{i-1}) < 6lw \tag{8}$$

or

$$f(v_i)f(v_{i+1}) < 6lw. \tag{9}$$

By (4), (8) and (9), we have either $f(v_i)f(v_{i+1,a}) > lw > \frac{1}{6}f(v_i)f(v_{i-1})$ or $f(v_i)f(v_{i+1,a}) > lw > \frac{1}{6}f(v_i)f(v_{i+1})$. That is, $f(v_{i+1,a}) > \frac{1}{6}f(v_{i-1})$ or $f(v_{i+1,a}) > \frac{1}{6}f(v_{i+1})$. We already show $f(v_{i-1}), f(v_{i+1}) > \frac{1}{2\sqrt{2}}\sqrt{lw}$. Therefore,

$$f(v_{i+1,a}) > \frac{1}{12\sqrt{2}}\sqrt{lw} \tag{10}$$

By (3) and (10), we can conclude that $f(v_{k,a}) > \frac{1}{12\sqrt{2}}\sqrt{lw}$ for $k \geq 1$ and $0 < a < a_k$. □

Proof of Lemma 3.5

Assume $\sqrt{\frac{l}{w}} < l$. First assume v_i is not last convergent of α. Put all $\frac{g_k}{h_k}$ and $\frac{g_{k,a}}{h_{k,a}}$ for $0 \leq k \leq i+1$ and $1 \leq a \leq a_k - 1$ on the real line. Note that by Lemma 2.3, if i is even then, $0 = \frac{g_0}{h_0} < \frac{g_2}{h_2} < \cdots < \frac{g_{i-2}}{h_{i-2}} < \frac{g_i}{h_i} < \alpha \leq \frac{g_{i+1}}{h_{i+1}} < \frac{g_{i-1}}{h_{i-1}} < \cdots < \frac{g_3}{h_3} < \frac{g_1}{h_1} < \frac{g_{-1}}{h_{-1}} = \infty$; otherwise, $0 = \frac{g_0}{h_0} < \frac{g_2}{h_2} < \cdots < \frac{g_{i-1}}{h_{i-1}} < \frac{g_{i+1}}{h_{i+1}} \leq \alpha < \frac{g_i}{h_i} < \frac{g_{i-2}}{h_{i-2}} < \cdots < \frac{g_3}{h_3} < \frac{g_1}{h_1} < \frac{g_{-1}}{h_{-1}} = \infty$. It is also known that $\frac{g_{k,a}}{h_{k,a}}$ is lying in the interval bounded by $\frac{g_{k-2}}{h_{k-2}}$ and $\frac{g_k}{h_k}$ for $1 \leq a \leq a_k - 1$. $\frac{t}{s}$ then either falls in an open interval bounded by $\frac{g_i}{h_i}$ and $\frac{g_{i+1}}{h_{i+1}}$ or falls in an open interval bounded by $\frac{g_{k-2}}{h_{k-2}}$ and $\frac{g_k}{h_k}$ for $1 \leq k \leq i+1$.

Consider the case where $\frac{t}{s}$ falls in an open interval bounded by $\frac{g_i}{h_i}$ and $\frac{g_{i+1}}{h_{i+1}}$. Since $|g_i h_{i+1} - h_i g_{i+1}| = 1$, $\frac{g_i}{h_i}$ and $\frac{g_{i+1}}{h_{i+1}}$ are adjacent in $F_{h_{i+1}}$. s must then be greater than h_{i+1}. Hence $f(v^*) = (l\sin\theta_{v^*} + w)\|v^*\| > w\sqrt{s^2 + t^2} > wh_{i+1} > \sqrt{lw}$.

If $\frac{t}{s}$ falls in an open interval bounded by $\frac{g_{k-2}}{h_{k-2}}$ and $\frac{g_k}{h_k}$ for $1 \leq k \leq i+1$, $\frac{t}{s}$ must fall in an open interval bounded by b_1 and b_2, where $\{b_1, b_2\} = \{\frac{g_{k,a}}{h_{k,a}}, \frac{g_{k,a+1}}{h_{k,a+1}}\}$ for $0 \leq a \leq a_k - 1$ if $k > 1$ or $1 \leq a \leq a_1 - 1$ otherwise. Note that $\frac{g_{k,0}}{h_{k,0}} = \frac{g_{k-2}}{h_{k-2}}$ and $\frac{g_{k,a_k}}{h_{k,a_k}} = \frac{g_k}{h_k}$. If $v_{k,a}$ is treated as a 3-space vector, then $\|v_{k,a} \times v_{k,a+1}\| = 1$ for $0 \leq a \leq a_k - 1$. That is $|g_{k,a} h_{k,a+1} - h_{k,a} g_{k,a+1}| = 1$. In other words, $\frac{g_{k,a}}{h_{k,a}}$ and $\frac{g_{k,a+1}}{h_{k,a+1}}$ are adjacent in $F_{h_{k,a+1}}$ when $k \geq 2$ or $k = 1$ but $a \geq 1$. Hence, we have $\|v^*\| > \|v_{k,a+1}\| > \|v_{k,a}\|$. Also, angle between v^* and $(1, \alpha)$, θ_{v^*}, is greater than $\theta_{v_{k,a}}$ or $\theta_{v_{k,a+1}}$. Therefore, either $f(v^*) > f(v_{k,a})$ or $f(v^*) > f(v_{k,a+1})$. Since $f(v_{k,a}), f(v_{k,a+1}) > \frac{1}{12\sqrt{2}}\sqrt{lw}$ for $0 \leq a \leq a_k - 1$, $f(v^*) > \frac{1}{12\sqrt{2}}\sqrt{lw}$.

Finally, if v_i is last convergent of α, that is, $\frac{g_i}{h_i} = \alpha$, then convergents next to $\frac{g_i}{h_i}$ will be $\frac{g_{i-2}}{h_{i-2}}$ and $\frac{g_{i-1}}{h_{i-1}}$. That is, if i is even, $\frac{g_{i-2}}{h_{i-2}} < \frac{g_i}{h_i} < \frac{g_{i-1}}{h_{i-1}}$; while the inequality signs are reversed when i is odd. To handle this exception, a pseudo v_{i+1} can be defined as follows: let a_{i+1} be the smallest integer such that $a_{i+1}h_i + h_{i-1} > \sqrt{\frac{l}{w}}$. If we can show $f(v_{i+1,a}) > \frac{1}{12\sqrt{2}}\sqrt{lw}$ for $1 \leq a \leq a_{i+1} - 1$, then above discussions are also applicable to this degenerate case. Since v_i is last convergent of α, we have

$$f(v_i) = w\|v_i\| < \sqrt{2}wh_i \tag{11}$$

By (5) and (11), we have $f(v_{i-1})f(v_i) < \sqrt{2}lw + 2w^2 h_i h_{i-1} < 4lw$. As (4) in the proof of Lemma 3.4, it can be shown $f(v_i)f(v_{i+1,a}) > lw$ for $1 \leq a \leq a_{i+1} - 1$. Hence, $f(v_i)f(v_{i+1,a}) > lw > \frac{1}{4}f(v_{i-1})f(v_i)$. Therefore, $f(v_{i+1,a}) > \frac{1}{4}f(v_{i-1}) > \frac{1}{8\sqrt{2}}\sqrt{lw} > \frac{1}{12\sqrt{2}}\sqrt{lw}$. □

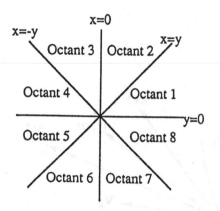

Figure 1. Covering grid points in a convex
polygon with a set of
parallel lines of equal spacing.

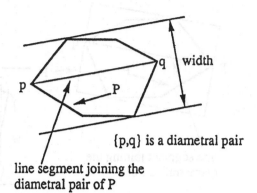

{p,q} is a diametral pair

line segment joining the
diametral pair of P

Figure 2. An illustration of the diameter,
width and diametral pair of a convex
polygon.

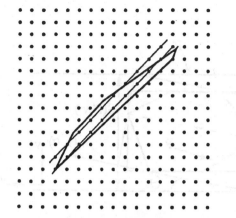

Figure 3. Octants of plane.

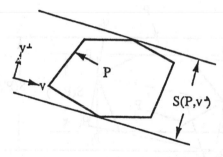

Figure 4. The span of P in direction $v^\perp$.

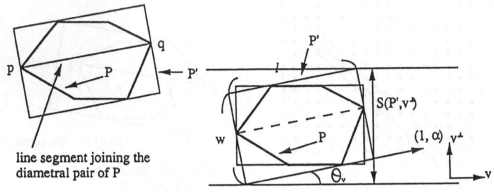

{p,q} is a diametral pair of P

Figure 5. The smallest rectangle P' containing P with two sides parallel with the line joining the diametral pair of P.

Figure 6. The length of $S(P',v^\perp)$.

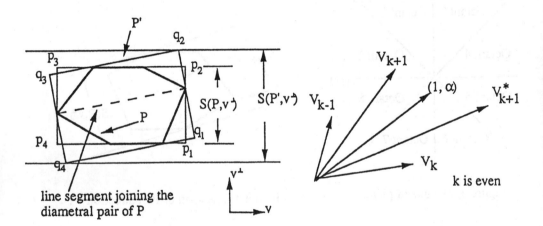

Figure 7 The relationship between $S(P,v^\perp)$ and $S(P',v^\perp)$.

Figure 8. The relations of V_{k-1}, V_k, V_{k+1}, V_{k+1}^* and $(1, \alpha)$.

On the Complexity of some Hamiltonian and Eulerian Problems in Edge-Colored Complete Graphs (Extended Abstract)

A. Benkouar[*]   Y. G. Manoussakis[†]   V. Th. Paschos[††]

R. Saad[†]

1 Introduction

We study the existence of *alternating Eulerian cycles* in edge-colored graphs and the existence of *alternating Hamiltonian cycles* and paths in edge-colored complete graphs.

In what follows, we let K_n^c denote an edge-colored complete graph of order n, with vertex set $V(K_n^c)$ and edge set $E(K_n^c)$. The set of used colors is denoted by X. If A and B are subsets of $V(K_n^c)$, then AB denote the set of edges between A and B. An AB-edge is an edge between A and B, i.e. it has one extremity in A and the other in B. Whenever the edges between A and B are monochromatic, then their color is denoted by $\chi(AB)$. If $A = \{x\}$ and $B = \{y\}$ then for simplicity we write xy (resp. $\chi(xy)$) instead of AB (resp. $\chi(AB)$. If x denotes a vertex of K_n^c and χ_i is a color of X, then we define the χ_i-degree of x to be the number of vertices y such that $\chi(xy) = \chi_i$. The χ_i-degree of x is denoted $\chi_i(x)$.

A *path P is alternating* if it has length at least *two* and any two adjacent edges of P have different colors. Similary, we define alternating cycles and altenating Hamiltonian (Eulerian) cycles and paths. An *alternating factor* F is a collection of pairwise vertex-disjoint alternating cycles $C_1, C_2, ..., C_m$, $m \geq 1$, covering the vertices of the graph. All cycles and paths considered in this paper are elementary, i.e. they go through a vertex exactly once, unless otherwise specified.

The notion of alternating paths was originally raised by Erdos and Bollobas in [3]. Results in almost the same vein are proved in [5].

In Section 2, we present constructive results regarding the existence of alternating Eulerian cycles and paths in edge-colored graphs.

In Section 3, we deal with Hamiltonian problems on 2-edge-colored complete

[*]Université de Paris-XII, Av. du Général de Gaule 94000 Créteil cedex, France
[†]LRI, Université de Paris-Sud, Centre d'Orsay, 91405 Orsay, France.
[‡]CERMSEM, Université de Paris I-Sorbonne, 12 pl. du Panthéon 75005 Paris, France.

graphs K_n^c. Namely, by using known results on matchings, we obtain $O(n^{2.5})$ algorithms for finding alternating factors with a minimum number of alternating cycles in K_n^c. As an immediate consequence, we obtain $O(n^{2.5})$ algorithms for finding alternating Hamiltonian cycles and paths. As a byproduct of this latter result we obtain an $O(n^{2.5})$ algorithm for finding Hamiltonian cycles in *bipartite tournaments* (another $O(n^{2.5})$ is given in [14]).

In Section 4, we give some NP-complete results for Hamiltonian problems on k-edge-colored complete graphs, $k \geq 3$, and we propose related conjectures. The following result is used in Section 3.

Theorem 1. *[M. BANKFALVI and Z. BANKFALVI ([1])]. Let K_{2p}^c be a two edge-colored complete graph with vertex-set $V(K_{2p}^c) = \{x_1, x_2, ..., x_{2p}\}$. Assume that $r(x_1) \leq r(x_2) \leq ... \leq r(x_{2p})$. The graph K_{2p}^c contains an alternating factor with a minimum number m of alternating cycles, if and only if there are m numbers k_i, $2 \leq k_i \leq p - 2$, such that for each i, $i = 1, 2, ..., m$, we have:*

$$r(x_1) + r(x_2) + ... r(x_{k_i}) + b(x_{2p}) + b(x_{2p-1}) + b(x_{2p-2}) + ... + b(x_{2p-k_i+1}) = k_i^2.$$

The results here are given without detailed proofs which can be found in [2].

2 Eulerian alternating cycles and paths.

In this section we study the existence of alternating Eulerian cycles in edge-colored graphs. In fact, we give a constructive proof of the following theorem (an non constructive proof of the same result is given by [12]. See also [8], page VI-1, Theorem VI-1. Up to now, no algorithm was known for this problem. We notice that in this section a cycle (respectively path) is not necessarily elementary, i.e. it goes through an edge once but it can go through a vertex many times.

In view of Theorem 2, we establish in [2] an $O(n \log n)$ procedure that finds a perfect matching in a specified family of complete k-partite graphs and we call it *matching procedure*.

Theorem 2. *Let G^c be an edge colored graph. There exists an alternating Eulerian cycle in G^c if and only if it is connected, and for each vertex x and for each color i, the total degree of x is even and $\chi_i(x) \leq \sum_{j \neq i} \chi_j(x)$.*

In the proof of theorem 2 in [2], for every vertex v and every edge e incident to v, we will associate an edge denoted $M_v(e)$ incident to v such that $\chi(e) \neq \chi(M_v(e)$. This guarantees that each time we visit e through v a vertex we can leave it through $M_u(e)$. In order to determine this association, for each vertex v we define a new graph G_v such that the vertices of G_v are the edges adjacent to v. Furthermore, two vertices are connected in G_v if their corresponding edges in G^c have different colors. It is clear that associating e to $M_v(e)$ is the same as finding a *perfect matching* in G_v. We remark that G_v verify Tutte's condition and that G_v is a complete k-partite graph.

Algorithm 1.

Input: An edge colored graph G^c satisfying the hypotheses of Theorem 2.
Output: An alternating Eulerian cycle.

1. for each v do apply *"matching procedure"* to G_v.
   (We suppose that $y_0 y_1$ is an edge of G^c.)
   $i \leftarrow 0;$

2. $P \leftarrow y_0 y_1$
   while there exists $M_{y_m}(y_{m-1} y_m) \notin E(P)$ do
   $P \leftarrow P \cup \{M_{y_m}(y_{m-1} y_m)\};$
   $m \leftarrow m + 1;$
   mark that y_m belongs to P
   od
   $C_i \leftarrow P;$
   $G^c \leftarrow G^c \setminus E(C_i)$ (we delete the edges but not their extremities)

3. if $E(G^c)$ is not empty, then find an edge wz in $E(G^c)$;
   $i \leftarrow i + 1;$
   $y_0 \leftarrow w;$
   $y_1 \leftarrow z;$
   go to step 2.

4. We stack a cycle and we start walking around it until a vertex that is intersection point with another cycle is found. We stack the new cycle and we start walking now around it by preserving the alternance of colors on the point we have changed the cycle we are walking (we notice here that this preservation is always possible).
   We continue this procedure until a cycle is entirely walked out in which case is unstacked.
   We continue in this way until the stack becomes empty.
   The above walk determines an Eulerian cycle.

3 Alternating Hamiltonian cycles and paths in 2-edge-colored complete graphs.

In this section, we suppose that the edges of K_n^c are colored *red* and *blue*. It turns out to be convenient for technical reasons to divide the vertices of an alternating cycle C of K_n^c into two classes $X = \{x_1, x_2, \ldots, x_s\}$ and $Y = \{y_1, y_2, \ldots, y_s\}$ such that the edge $x_i y_i$ is *red* and the edge $y_i x_{i+1}$ is *blue*, for each $i = 1, 2, \ldots, s$ (i is considered *modulo s*).

Notation: if C_1 and C_2 denote two pairwise vertex-disjoint alternating cycles of K_n^c (if any) with classes X_1, Y_1 and X_2, Y_2 respectively, then we say that C_1 *dominates* C_2, if either all $X_1 C_2$ edges are *red* and all $Y_1 C_2$ edges are *blue* or all $X_1 C_2$ edges are *blue* and all $Y_1 C_2$ edges are *red*.

In a first part of this section we prove two lemmas (Lemmas 1 and 2) and we establish Procedure 2 for complete graphs whose vertices are covered by two pairwise vertex-disjoint alternating cycles. These results are usefull for Algorithm 2 given later. The basic ideas used in the proofs of Lemmas 1 and 2 are originally exposed in [1]. The following lemma is almost trivial.

Lemma 1. *Let K_n^c be a 2 edge-colored complete graph. Assume that there existe two pairwise vertex-disjoint alternating cycles $C_1 = \langle x_1 y_1 \ldots x_t y_t x_1 \rangle$ and $C_2 = x_1' y_1' \ldots x_s' y_s' x_1'$ in K_n^c covering all its vertices. Furthermore, assume that there are two red edges $x_i x_j'$ and $y_i y_j'$. Then K_n^c contains an alternating Hamiltonian cycle.*

Lemma 2. *Let K_n^c, C_1 and C_2 be as they are defined in Lemma 1. Furthermore, assume that there are at least two $X_1 X_2$ (or $X_1 Y_2$ or $Y_1 X_2$ or $Y_1 Y_2$) edges with different colors. Then K_n^c contains an alternating Hamiltonian cycle.*

By using Lemmas 1 and 2 we establish in [2] Procedure 1 which finds an alternating Hamiltonian cycle (if any) in the particular case where K_n^c has an alternating factor consisting of two alternating cycles. This procedure will be usefull for Algorithm 2 given later.

Procedure 1.
Input. A two edge-colored graph K_n^c and two alternating cycles C_1 and C_2 covering the vertices of K_n^c.
Output. Either an alternating Hamiltonian cycle of K_n^c or else an answer that either C_1 dominates C_2 or else C_2 dominates C_1.

1. We look if there are two edges e_1 and e_2 between X_1 and X_2 (or between Y_1 and Y_2, or X_1 and Y_2, or Y_1 and X_2) with different colors. This can be done by a simple examination of the edges between C_1 and C_2. Now, if the edges e_1 and e_2 exist, then we find an alternating Hamiltonian cycle by Lemma 2. Otherwise go to next step.

2. Since no alternating Hamiltonian cycle was found in Step 1, all $X_1 X_2$ (resp. $Y_1 Y_2$, $X_1 Y_2$, $Y_1 X_2$) edges are monochromatic. Assume w.l.ofg. that all $X_1 X_2$ edges are *red*. Now by using Lemma 1, we may conclude that either any $Y_1 Y_2$ edge is *blue*, $X_1 Y_2$ edge is *red* and $Y_1 X_2$ edge is *blue*, that is, C_1 dominates C_2 or else any $Y_1 Y_2$ edge is blue, $X_1 Y_2$ edge is *blue* and $Y_1 X_2$ edge is *red*, that is, C_2 dominates C_1.

Procedure 2.
Input. A two edge-colored graph K_n^c and two alternating cycles C_1, C_2 such that C_1 dominates C_2.
Output. Either an alternating Hamiltonian cycle or else conclude that all $X_1 X_1$ and $X_1 C_2$ edges are monochromatic.

- Assume w.l.ofg. that all $X_1 C_2$ edges are *red*. We look now if there exists a blue edge $x_i x_j$, where $i \neq j$ and x_i, $x_j \in X_1$. If it is the case, then $y_{i-1}x_{i-1}\ldots x_j x_i y_i x_{i+1}\ldots y_{j-1}y'_h x'_h y'_{h-1}\ldots y'_{h+1}x'_{h+1y_{i-1}}$, is an alternating cycle of K_n^c, where y'_h is an appropriate vertex of Y_2. Similarly, If there exists a *red* edge $y_i y_j$, where $i \neq j$ and $y_i, y_j \in Y_1$, then by using the same arguments, we may find, once more, an alternating Hamiltonian cycle of K_n^c.

In the remaining part of this section we shall deal with 2-edge-colored complete graphs which contain an alternating factor consisting of m alternating cycles $C_1, C_2, \ldots, C_m$, $m \geq 2$. It seems convenient to use the following notation: For each i, $i = 1, 2, \ldots, m$, we put $|V(C_i)| = 2c_i$. Furthermore, we let X_i and Y_i denote the classes of a cycle C_i, where we define $X_i = \{x_{i1}, x_{i2}, \ldots, x_{ic_i}\}$ and $Y_i = \{y_{i1}, y_{i2}, \ldots, y_{ic_i}\}$.
The following lemma will be used in Algorithm 2 given later.

Lemma 3. *Let K_n^c be a 2-edge-colored complete graph containing an alternating factor F consisting of cycles $C_1, C_2, \ldots, C_m$, $m \geq 2$. Assume that C_i dominates C_j for each i and j, $1 \leq i < j \leq m$. Assume that all $X_1 C_2$ edges are red. There exists an alternating Hamiltonian path with begin in Y_1 and terminus in C_m such that both its first and last edges are blue.*

Algorithm 2 below finds an alternating factor with a minimum number of alternating cycles in 2 edge-colored complete graphs in $O(n^{2.5})$ steps. Our result was essentially inspired by Theorem 1.

Algorithm 2.
Input. A complete graph K_n^c on n vertices. The edges of K_n^c are colored *red* and *blue.*
Output. Either an alternating factor F_μ of K_n^c with a minimum number of alternating cycles $R_1, \ldots, R_\mu$, $\mu \geq 1$, or else an answer that K_n^c has no alternating factor at all.
We suppose that K_n^c has an even number of vertices, since otherwise it has no alternating factor. Furthermore, we initialize μ to zero.

1. Find a *blue* maximum matching M_b and a *red* one M_r in K_n^c. If either $|M_b| < n/2$ or $|M_r| < n/2$ then stop; K_n^c has no alternating factor. Otherwise, form an alternating factor F by considering the union of M_b and M_r. Then go to next step.

2. Let $C_1, C_2, \ldots, C_m$, $m \geq 1$, be the alternating cycles of F (in what follows we shall shortly write $F \leftarrow C_1, C_2, \ldots, C_m$). If $m = 1$, then we have finished by setting $\mu = 1$ and $R_1 = C_1$. Assume $m \geq 2$. Apply Procedure 1 on the subgraph of K_n^c induced by $V(C_i) \cup V(C_j)$, for each i and j, $i, j = 1, 2, \ldots, m$ and $i < j$. If Procedure 1 produces an alternating cycle, say C', with vertex set $V(C_i) \cup V(C_j)$, then put $C_i \leftarrow C'$, $C_h \leftarrow C_{h+1}$ for all h, $j \leq h \leq m-1$, $m \leftarrow m-1$, $F \leftarrow C_1, C_2, \ldots, C_m$ and then we go to the beginning of this step.
When this step terminates, if $m = 1$, we put $m = 1$ and $R_1 = C_1$

and then stop the algorithm. If $m = 2$, we put $m = 2$, $R_1 = C_1$ and $R_2 = C_2$, and then we go to Step 4. If $m > 3$, then we go to next step.

3. Define from K_n^c a new graph T as follows: Replace any cycle C_i by a new vertex c_i and then add an arc $c_i c_j$, $i \neq j$, $i, j = 1, 2, ..., m$, in T, iff C_i dominates C_j in K_n^c. Clearly T is a tournament. Now, we find a Hamiltonian cycle in each nontrivial component T_i, i.e. $|V(T_i)| \geq 3$, of K_n' by using the algorithm of [13]. By using these cycles we define the corresponding $C_1, C_2, ..., C_m'$ alternating cycles by considering appropriate colored edges of K_n^c.
   If $m' = 1$, then we terminate the algorithm by setting $\mu = 1$ and $R_1 = C_1$. On the other hand, if $m' \geq 2$, then we set $m \leftarrow m'$, $F \leftarrow C_1, C_2, ..., C_m$ and we go to next step.

4. By the structure of K_n^c, C_i dominates C_j, for all i, j, $1 \leq i < j \leq m$. In particular, C_1 dominates any other cycle C_i, $i = 2, 3, ..., m$. Assume w.l.ofg. that all $X_1 C_2$ edges are *red*.
   If $m = 2$, then we apply Procedure 3 on K_n^c and then we stop the algorithm.
   If $m \geq 3$, then we determine the largest integer h, $3 \leq h \leq m$, such that the $X_1 C_h$ edges are *blue*. By using Lemma 3, we find an alternating path with vertex set $V(C_1) \cup V(C_2) \cup ... \cup V(C_h)$ such that both its first and last edges are *blue* and its begin is in Y_1 and its terminus is in C_h. By using this path and the fact that all $Y_1 C_h$ edges are *red*, we construct easily an alternating cycle C with vertex set $V(C_1) \cup V(C_2) \cup ... \cup V(C_h)$. Then we go to next step.

5. If for some i, j and l, $1 \leq i, j \leq h$ and $h + 1 \leq l \leq m$, the edges $X_i C_l$ are *red* and the edges $X_j C_l$ are *blue*, then by applying the arguments of Lemma 2 on C and C_l, we find a new alternating cycle C, where $C \leftarrow C \cup C_l$. Repeat this step between C and all cycles $C_{h+1}, ..., C_{l-1}, C_{l+1}, C - l + 2, ..., C_m$. Then we go to next step.

6. Rename the current cycles so that C_i dominates C_j for all i and j, $1 \leq i < j \leq m$. Clearly the edges $X_1 C_i$ are *red*, for all $i = 1, 2, ..., m$. Look if there is a *blue* edge e inside $X_1 X_1$ (or a *red* edge inside $Y_1 Y_1$).
   If e does not exist, then set $\mu \leftarrow \mu + 1$, $R_\mu = C_1$. Next set $V(K_n^c) \leftarrow V(C_2) \cup ... \cup V(C_m)$ $m \leftarrow m - 1$, rename appropriatelly $C_2, ... C_m$ and go back to Step 4.
   If e exists, say in $X_1 X_1$, set $e = x_{1i} x_{1j}$. By using the arguments of Lemma 3, find an alternating path with begin y_{i-1}, terminus in C_m such that its first and last edges are both *red*. Then by using the segment $y_{i-1} x_{i-1} ... x_j x_i y_i y_{i+1} ... y_{j-1}$ of C_1, define an alternating Hamiltonian cycle of K_n^c.

All the proofs on the complexities and the completenesses of the above procedures and algorithms are found in [2].

From Algorithm 2 we obtain the following corollaries.

Corollary 1. *There exists an $O(n^{2.5})$ algorithm for finding Hamiltonian cycles in a 2-edge-colored complete graph K_n^c.*

Corollary 2. *([14]). There exists an $O(n^{2.5})$ algorithm for finding Hamiltonian cycles in bipartite tournaments.*

In the following theorem, we characterize 2-edge-colored complete graphs admitting alternating Hamiltonian paths.

Theorem 3. *Any 2-edge-colored complete graph K_n^c has a Hamiltonian path if and only if the graph K_n^c has:*
(i) an alternating factor or
(ii) an "almost alternating factor", that is, a spanning subgraph which differs from a factor by the color of exactly one edge e or finally
(iii) an odd number of vertices and, furthermore, K_n^c has a red matching M_r and a blue one M_b, each one having cardinality $(n-1)/2$.

Relying on Lemmas 1, 2, 3 and Theorem 1 we deduce an $O(n^{2.5})$ algorithm for finding alternating Hamiltonian paths. The techniques used for this algorithm are pretty much similar to the ones of algorithm 2 and thus it is omitted.

We shall conclude this section with the following problem.

Problem.

Let x and y be two specified vertices in a 2-edge-colored complete graph. Is there a polynomial algorithm for finding an alternating Hamiltonian path (if any) between x and y?

4   Some NP-completeness results

In this section we consider the problem of finding Hamiltonian configurations with specified edge-colorings in k-edge-colored complete graphs, $k \geq 3$. We prove that some of these problems are NP-complete.

Notation. Let p be an integer and $X = \{\chi_1 \chi_2 ... \chi_k\}$ be the set of used colors. An $(\chi_1 \chi_2 ... \chi_k)$ cycle is a cycle of length pk such that the sequence of colors $\langle \chi_1 \chi_2 ... \chi_k \rangle$ is repeated p times.

Problem P.

Instance: Let K_n^c be a 3-edge colored complete graph.
Question: Does there exist a Hamiltonian (123) cycle in K_n^c?

Theorem 4. *The problem P is NP-complete.*

Problem P'.

Instance: K_n^c a 3-edge-colored complete graph, a set $S \subset V(K_n)$ of six vertices of K_n^c.
Question: Does there exist an $(\chi_1 \chi_2 \chi_3)$ cycle in K_n^c containing the vertices of S?

Theorem 5. *P' is NP-complete.*

The technique proposed in the above theorems can be extended to obtain the same results even when $k \geq 4$. However because of the difficulty of the proofs we propose in [2] some other more "direct" and straightforward reductions.

Theorem 6. *The following problem is NP-complete.*
Instance: Positive integeres p and k, $k \geq 4$, a k-edge-colored complete graph K_n^c such that $n = kp$.
Question: Does K_n^c contain a Hamiltonian $\chi_1\chi_2...\chi_k$ cycle C?

By using arguments, similar to these of the proof of Theorem 6, we may prove the following result on Hamiltonian paths.

Theorem 7. *The following problem is NP-complete.*
Instance. Positive intgeres p and k, $k \geq 4$, a k-edge colored complete graph K_n^c such that $n = kp + 1$.
Question. Does K_n^c contain a Hamiltonian $\chi_1\chi_2...\chi_k$ path P (respectively, a Hamiltonian $\chi_1\chi_2...\chi_k$ path P' with specified extremities ?

An extension of th. 6 could be the following.

Theorem 8. *The following problem is NP-complete.*
Instance. A complete graph K_n^c, a set of distinct colors $X = \{\chi_1, \chi_2, ..., \chi_k\}$, $k \geq 4$, a function $f : E(K_n) \to X$, a positive integer p.
Question. Does K_n^c contain a cycle of length pk such that the sequence of colors $\langle \chi_1\chi_2...\chi_k \rangle$ is repeated p times in C?

In fact, as it can be seen in the sequel, even if the *ordering prerequisite* is relaxed the problem remains NP-hard provided that a *frequency* on the occurence of the colors is maintained.
Let us consider the following problem:
Problem PF.
Instance: Positive integers p and k, $k \geq 3$, a k-edge-colored complete graph K_n^c such that $n = kp$.
Question: Does K_n^c contain an alternating Hamiltonian cycle C such that each color appears at least p times on C?

Theorem 9. *PF is NP-hard.*

We shall conclude this paper with the following open problem.
Problem. *Instance: An edge-colored complete graph K_n^c.*
Question: Does K_n^c admit an alternating Hamiltonian cycle?

References

[1] M. BANKFALVI and Z. BANKFALVI, *Alternating Hamiltonian Circuit in Two-Coloured Complete Graphs*, Theory of Graphs (Proc. Colloq. Tihany 1968), Academic Press, New York, pp. 11-18.

[2] A. BENKOUAR, Y. G. MANOUSSAKIS, V. Th. PASCHOS and R. SAAD, *On the Complexity of some Hamiltonian Problems in Edge-Colored Complete Graphs*, Rapport de Recherche, LRI, Université de Paris-Sud, Centre d'Orsay, 1990.

[3] B. BOLLOBAS and P. ERDOS, *Alternating Hamiltonian Cycles*, Israel Journal of Mathematics, Vol. 23, 1976.

[4] A. BONDY and U.S.R. MURTY, *Graph Theory with Applications*, MCMILLAN PRESS LTD, 1976.

[5] C. C. CHEN and D. E. DAYKIN, *Graphs with Hamiltonian Cycles having Adjacent Lines Different Colors*, J. of Combinatorial Theory, (B) 21, pp. 135-139, 1976.

[6] J. EDMONDS and R. M. KARP, *Theoritical Improvements in Algorithmic Efficiency for Network Flow Problems*, J. ACM Vol. 19, No 2, 1972, pp. 248-264.

[7] S. EVEN and O. KARIV, *An $O(n^{2.5})$ Algorithm for Maximum Matching in General Graphs*, in Proceedings of the 16th Annual Sumposium on Foundations of Computer Science (Berkeley, 1975), pp 100-112.

[8] H. FLEISHNER, *Eulerian Graphs and Related Topics*, Part 1, Volume 1, Series book of the Annals of Discrete Mathematics, North-Holland, 1990.

[9] S. FORTUNE, J. HOPCROFT and J. WYLLIE, *The Directed Subgraph Homeomorphism Problem*, Theor.Comput.Science 10 1980, pp 111-121.

[10] M. GAREY and D. JOHNSON, *Computers and Intractability - A Guide to the Theory of NP-Completeness*, Freeman, New York, 1979.

[11] P. HELL, Y. MANOUSSAKIS and Z. TUZA, *On the Complexity of of Some Packing Problems in Edge-Colored Complete Graphs*, manuscript.

[12] A. KOTZIG, *Moves Without Forbidden Transitions in a Graph*, Mat. Fyz. Časopis 18 1968, No 1, pp 76-80.

[13] Y. MANOUSSAKIS, *A Linear Algorithm for Finding Hamiltonian Cycles in Tournaments*, to appear in Discrete Mathematics.

[14] Y. MANOUSSAKIS and Z. TUZA, *Polynomial Algorithms for Finding Cycles and Paths in Bipartite Tournaments*, SIAM Journal on Discrete Mathematics, Vol. 3 No 2, 1990, pp 537-543.

[15] N. ROBERTSON and P.D. SEYMOUR, *Graph Minors*, to appear in J. Combinatorial Theory series B.

Dynamic Programming on Intervals

Takao Asano *

Abstract

We consider problems on intervals which can be solved by dynamic programming. Specifically, we give an efficient implementation of dynamic programming on intervals. As an application, an optimal sequential partition of a graph $G = (V, E)$ can be obtained in $O(m \log n)$ time, where $n = |V|$ and $m = |E|$. We also present an $O(n \log n)$ time algorithm for finding a minimum weight dominating set of an interval graph $G = (V, E)$, and an $O(m \log n)$ time algorithm for finding a maximum weight clique of a circular-arc graph $G = (V, E)$, provided their intersection models of n intervals (arcs) are given.

1 Introduction

Dynamic programming is one of the most popular techniques for designing efficient algorithms and has been applied to many problems. For example, an optimal sequential partition of a graph and a maximum weight clique of a circular-arc graph can be obtained by dynamic programming on intervals [4, 6]. In this paper we propose an efficient implementation of dynamic programming on intervals.

For an interval z, we denote by $x(z)$ and $y(z)$ the left and right endpoints of z, respectively. We assume $x(z)$ and $y(z)$ are integers from 1 to p. Then the problem we consider is: for a set Z of intervals, a set $SZ = \{z_1, z_2, \ldots, z_p\}$ with $y(z_i) = i$ and a real-valued cost function $c(z)$ for $z \in Z$, compute

$$E[i] = \min\{D[j] + C(j, i)\}, \tag{1}$$

for each $1 \le i \le p$. Here we assume that the minimum is taken over all the integers $j < i$ contained in $z_i \in SZ$. We also assume that $D[i]$ can be computed from $E[i]$ in

---

*Department of Mechanical Engineering, Sophia University, Chiyoda-ku, Tokyo 102, Japan. This work was supported in part by Grant in Aid for Scientific Research of the Ministry of Education, Science and Culture of Japan under Grant-in-Aid for Co-operative Research (A) 02302047 (1990,1991).

constant time and that $C(j, i)$ is a constant multiple of the total cost of intervals $z \in Z$ that contain both i and j, i.e.,

$$C(j, i) = \alpha \sum_{j, i \in z \in Z} c(z).$$

Furthermore, we assume $E[1] = 0$. This problem can easily be solved by the following Algorithm DP based on dynamic programming:

For $i := 1$ to p, compute $E[i]$ by (1) and $D[i]$.

It is clear that Algorithm DP correctly computes all $E[i]$ (and $D[i]$) since all $D[j]$ $(j < i)$ are already computed when computing $E[i]$, and if we let $q = |Z|$, it requires $O(q + p^2)$ time. In this paper we shall present an efficient implementation of Algorithm DP and thus obtain an $O(q \log p)$ time algorithm for computing all $E[i]$.

As an application, an optimal sequential partition of a graph $G = (V, E)$ can be obtained in $O(m \log n)$ time, where $n = |V|$ and $m = |E|$. Similarly, we can obtain efficient algorithms on interval graphs and circular-arc graphs. For example, a minimum-weight dominating set of a weighted interval graph (circular-arc graph) can be obtained in $O(n \log n)$ ($O(n^2 \log n)$) time and a maximum-weight clique of a weighted circular-arc graph can be obtained in $O(m \log n)$ time, provided that their intersection models are given. These except for the clique algorithm improve the complexity of existing algorithms [3, 5, 6]. For the maximum-weight clique problem on circular-arc graphs, Shih and Hsu [8] has already proposed a faster algorithm with $O(n \log n +_* m \log\log n)$ time. However, we believe our algorithm proposed here is easier to implement and more practical.

2 Efficient Implementation

In this section, we present an efficient implementation of Algorithm DP. By this implementation, all $E[i]$ in (1) can be computed in $O(q \log p)$ time and $O(p)$ space.

For the efficient implementation, we consider the following problem. We are given a set of numbers $X = \{1, 2, \dots, p\}$ ($p \geq 2$) and a set DZ of intervals each of which has its both endpoints in X. Also each interval $z \in DZ$ has a cost $cost(z)$. For a number x, $csum(x, DZ)$ is defined as $csum(x, DZ) = \sum_{x \in z \in DZ} cost(z)$ if $DZ \neq \emptyset$ and $csum(x, DZ) = \infty$ if $DZ = \emptyset$. We consider the following three operations.

$$findmin(z, DZ) : \text{ return } \min_{x \in z}\{csum(x, DZ)\},$$

$$insert(z, DZ) : \text{ replace } DZ \text{ with } DZ \cup \{z\},$$

$$delete(z, DZ) : \text{ replace } DZ \text{ with } DZ - \{z\}.$$

A crucial observation is:

Observation 1. Each of the operations $findmin$, $insert$ and $delete$ can be done in $O(\log p)$ time with a data structure of $O(p)$ space constructed by an $O(p)$ time preprocessing.

Although this observation can be easily shown by using the segment tree in computational geometry [2],[7], we will prove briefly later in Section 2.2 for completeness. Now we assume Observation 1 and show that each $E[i]$ and $D[i]$ can be computed in $O(\log p)$ time.

2.1 Implemetation of Algorithm DP

In this subsection we give an implementation of Algorithm DP. It is based on a kind of (line) sweep. In the procedure, M is supposed to be a sufficiently large positive constant.

Procedure IMPLDP;
begin {comment computing $E[i]$ and $D[i]$;}
1   $X := \{1, 2, \ldots, p\}$; $DZ := \emptyset$;
2   $z_0 := [1, p]$; $cost(z_0) := M$; $insert(z_0, DZ)$;
3   for $i := 1$ to p do begin
4   if $i = 1$ then $E[i] := 0$ else $E[i] := findmin(z_i, DZ)$;
5   compute $D[i]$ from $E[i]$;
6      $y_i := [i, i]$; $cost(y_i) := D[i] - M$; $insert(y_i, DZ)$;
7   if i is the right endpoint of an interval in Z then
8   for each $z \in Z$ with $y(z) = i$ do $delete(z, DZ)$;
9   if i is the left endpoint of an interval in Z then
10   for each $z \in Z$ with $x(z) = i$ do
11   begin $cost(z) := \alpha c(z)$; $insert(z, DZ)$ end
   end
end;

The following observations are useful to show the correctness of Procedure IMPLDP.

(A) If we choose M such that $M > 2\sum_{z \in Z} |\alpha c(z)|$, then just before the i-th iteration of Line 3, $csum(x, DZ) > M/2$ for all points $x \in [1, p]$ except integer points $x < i$.

(B) Just before the i-th iteration, $csum(x, DZ) = D[x] + C(x, i)$ for an integer point $x < i$.

We will prove Observations (A) and (B) by induction on i.

Initially (just after Line 2), $csum(x, DZ) = M > M/2$ for all $x \in [1, p]$ and thus, just before the first iteration ($i = 1$), (A) and (B) hold.

We assume that (A) and (B) hold just before the i-th iteration and consider the i-th iteration. Note that $csum(x, DZ) < M/2$ holds only if virtual interval y_x has been inserted into DZ in Line 6. Thus, during the i-th iteration, only $csum(i, DZ)$ can newly become less than or equal to $M/2$ (this will happen when $D[i]$ is less than or equal to $M/2$). Thus, (A) holds immediately after i-th iteration (i.e., just before the $i + 1$-th iteration). Next we consider (B). Define $Z_R(x)$ to be the set of intervals in Z having x as the right endpoint, i.e.,

$$Z_R(x) = \{z \in Z \mid y(z) = x\}.$$

Similarly, define $Z_L(x)$ and $Z(x,y)$ for $x < y$ as follows.

$$Z_L(x) = \{z \in Z \mid x(z) = x\},$$

$$Z(x,y) = \{z \in Z \mid x, y \in z\}.$$

Note that $C(x,y) = \alpha \sum_{z \in Z(x,y)} c(z)$. During the i-th iteration, all intervals in $Z_R(i)$ are deleted from DZ and all intervals in $Z_L(i)$ are inserted to DZ. Furthermore, $D[x]$ is not changed for an integer $x < i$. Thus, $Z(x, i+1) = Z(x,i) - Z_R(i)$ and just before the $i + 1$-th iteration, we have

$$csum(x, DZ) = D[x] + C(x, i+1),$$

for an integer $x < i$. Similarly, for a sufficiently small $\epsilon > 0$,

$$Z(i, i+1) = (Z(i - \epsilon, i) - Z_R(i)) \cup Z_L(i).$$

Thus,

$$csum(i, DZ) = D[i] + C(i, i+1)$$

after i-th iteration, since y_i is inserted. This implies that (B) holds just before the $i + 1$-th iteration. Thus, we complete our proof of (A) and (B).

These observations imply that each $E[i]$ (and $D[i]$) is correctly computed by $E[i] := findmin(z_i, DZ)$ (or $E[i] := 0$) in Line 4 during the i-th iteration when $i = y(z_i)$ and $findmin(z_i, DZ) < M/2$, since $csum(x, DZ) = D[x] + C(x, i)$ for an integer point $x < i$ and

$$findmin(z_i, DZ) = \min_{x \in z_i}\{csum(x, DZ)\}$$

$$= \min_{x < i, x \in z_i, x:integer}\{D[x] + C(x, i)\},$$

by Observations (A) and (B) (note that $csum(i, DZ) > M/2$ just before the i-th iteration). If $findmin(z_i, DZ) > M/2$, then the correct value of $E[i]$ is ∞ and in the procedure $E[i]$ (and $D[i]$) is set a value $> M/2$. Thus we can obtain the following theorem by Observation 1 since the number of operations $findmin$, $insert$ and $delete$ is $O(q)$.

Theorem 1. All $E[i]$ can be correctly computed in $O(q \log p)$ time.

2.2 Data Structure Supporting $findmin$, $insert$ and $delete$

In this subsection, we shall describe a data structure supporting the operations $findmin$, $insert$ and $delete$ defined before. Since $delete(z, DZ)$ can be replaced by $insert(z', DZ)$ using $z = z'$ and $cost(z') = -cost(z)$, we consider only $findmin$ and $insert$.

We consider a balanced binary search tree $T(X)$ for a set $X = \{1, 2, \ldots, p\}$. That is, $T(X)$ satisfies the following (i)-(iii).

(i) Each node of $T(X)$ has either no children or two children, its left child $\ell(v)$ and its right child $r(v)$ (a node with no children is a *leaf* and a node with children is an *inner node*).

(ii) Each node v of $T(X)$ contains a key $key(v) \in X$ and keys are arranged in symmetric order: each $i \in X$ is stored in exactly one leaf of $T(X)$ as a key, and each inner node v has $key(v)$ satisfying

$$\max_{u \in L(v)} \{key(u)\} = key(v) < \min_{u \in R(v)} \{key(u)\},$$

where $L(v)$ denotes the set of descendants of the left child $\ell(v)$ of v and $R(v)$ denotes the set of descendants of the right child $r(v)$ of v.

(iii) $depth(v) = O(\log p)$ for each node v of $T(X)$.

Note that this balanced binary search tree $T(X)$ can be constructed in $O(p)$ time and $O(p)$ space. For simplicity, we write x_i to denote the leaf containing i as a key. We consider two values $cmin(v)$ and $cin(v)$ for each node v of $T(X)$. We maintain $cmin(v)$ and $cin(v)$ so that they satisfy

$$cmin(v) = \min\{cmin(\ell(v)) + cin(\ell(v)), \ cmin(r(v)) + cin(r(v))\} \qquad (2)$$

for each inner node v of $T(X)$. Let $P(u,v)$ be the simple path of $T(X)$ from node u to node v. We also use $P(u,v)$ to denote the set of nodes on the path $P(u,v)$. Initially (when $DZ = \emptyset$), $cmin(v) = 0$ and $cin(v) = 0$ for each node v of $T(X)$. Equation (2) together with $cmin(x_i) = 0$ for each leaf x_i will imply $cmin(v) = \min\{\sum_{u \in P(v,x_i) - \{v\}} cin(u)\}$ for each inner node v of $T(X)$, where the minimum is taken over all the leaves x_i that are descendants of v.

We first consider $insert(z, DZ)$. Let $z = [i,j]$ $(i \leq j)$. Then $insert(z, DZ)$ can be done as follows. Let r_{ij} be the nearest common ancestor of x_i and x_j. Consider the paths $P(r_{ij}, x_i)$ and $P(r_{ij}, x_j)$ from r_{ij} to x_i and x_j. Let v_i be the deepest node on $P(r_{ij}, x_i)$ such that v_i is the right child of a node on $P(r_{ij}, x_i)$ if such v_i exists, otherwise we set $v_i := r_{ij}$. Similarly, let v_j be the deepest node on $P(r_{ij}, x_j)$ such that v_j is the left child of a node on $P(r_{ij}, x_j)$ if such v_j exists, otherwise we set $v_j := r_{ij}$. Define $R(r_{ij}, v_i)$ and $L(r_{ij}, v_j)$ as follows. If $v_i = v_j = r_{ij}$ then $R(r_{ij}, v_i) = L(r_{ij}, v_j) = \{r_{ij}\}$. If $v_i = r_{ij} \neq v_j$ then $R(r_{ij}, v_i) = \{\ell(r_{ij})\}$. If $v_i \neq r_{ij} = v_j$ then $L(r_{ij}, v_j) = \{r(r_{ij})\}$. If $v_i \neq r_{ij}$ then

$$R(r_{ij}, v_i) = \{v = r(p(v)) \mid p(v) \in P(\ell(r_{ij}), p(v_i)), \ v \text{ is not in } P(\ell(r_{ij}), p(v_i))\}$$

that is, $v \in R(r_{ij}, v_i)$ if and only if v is the right child of a node on the path $P(\ell(r_{ij}), p(v_i))$ but v itself is not on the path $P(\ell(r_{ij}), p(v_i))$. Thus, $v_i \in R(r_{ij}, v_i)$ if $v_i \neq r_{ij}$. If $v_j \neq r_{ij}$ then

$$L(r_{ij}, v_j) = \{v = \ell(p(v)) \mid p(v) \in P(r(r_{ij}), p(v_j)), \ v \text{ is not in } P(r(r_{ij}), p(v_j))\}$$

that is, $v \in L(r_{ij}, v_j)$ if and only if v is the left child of a node on the path $P(r(r_{ij}), p(v_j))$ but v itself is not on the path $P(r(r_{ij}), p(v_j))$. Thus, $v_j \in L(r_{ij}, v_j)$ if $v_j \neq r_{ij}$. Note that, for each $k \in X$, $k \in z = [i,j]$ if and only if there is exactly one node $v \in R(r_{ij}, v_i) \cup L(r_{ij}, v_j)$ such that x_k is a descendant of v. We first set $cin(v) := cin(v) + cost(z)$ for each $v \in R(r_{ij}, v_i) \cup L(r_{ij}, v_j)$. Next, we have to modify $cmin(u)$ for all $u \in P(r_{ij}, p(v_i)) \cup P(r_{ij}, p(v_j)) \cup P(r, r_{ij})$, since $cin(v)$ for all $v \in R(r_{ij}, v_i) \cup L(r_{ij}, v_j)$ are now changed ($P(r, r_{ij})$ is the path from the root r to r_{ij}). This can be done by traversing nodes along the paths $P(r_{ij}, p(v_i))$, $P(r_{ij}, p(v_j))$ and $P(r, r_{ij})$ in decreasing order of their *depth* and looking their children. Thus, $insert(z, DZ)$ can be done in $O(\log p)$ time, since the $depth(v)$ of each node v of $T(X)$ is $O(\log p)$.

Next we consider $findmin(z, DZ)$. Let $z = [i, j]$ ($i \leq j$). Let r_{ij}, $P(r_{ij}, x_i)$, $P(r_{ij}, x_j)$, v_i, v_j, $R(r_{ij}, v_i)$, $L(r_{ij}, v_j)$ and $P(r, r_{ij})$ be the same as above. If we let

$$cs(v) := cmin(v) + \sum_{u \in P(r,v)} cin(u)$$

for each $v \in R(r_{ij}, v_i) \cup L(r_{ij}, v_j)$, then $\min_{v \in R(r_{ij}, v_i) \cup L(r_{ij}, v_j)}\{cs(v)\}$ is the desired value $findmin(z, DZ)$. This value can be computed in $O(\log p)$ time by traversing nodes along the paths $P(r_{ij}, p(v_i))$, $P(r_{ij}, p(v_j))$ and $P(r, r_{ij})$ in decreasing order of their *depth* and looking their children. Thus, $findmin(z, DZ)$ can also be done in $O(\log p)$ time. Note that $T(X)$ with $cmin(v)$ and $cin(v)$ for each node v of $T(X)$ can be represented in $O(p)$ space and is almost the same as the segment tree in computational geometry [2], [7].

3 Applications and Related Problems

In this section, we present some applications of an efficient implementation of Algorithm DP. The first application is to optimal sequential partitions of graphs proposed by Kernighan [6].

3.1 Optimal Sequential Partitions of Graphs

We are given a graph $G = (V, E)$, a cost function $c : E \rightarrow R^+$, a weight function $w : V \rightarrow R^+$ and a positive number K. Here we assume the vertices of the graph are labeled from 1 to n. For a subset $B = \{b_1, b_2, \ldots, b_s\}$ of V such that $b_1 = 1$ and $b_i < b_j$ for $i < j$, we obtain a family of graphs $G(B) = \{G_1, G_2, \ldots, G_s\}$, where each G_i is the subgraph of G induced by the set $\{b_i, b_i + 1, \ldots, b_{i+1} - 1\}$ (for convenience, we assume $b_{s+1} - 1 = n$). We will call $P = G(B)$ a sequential partition (or partition, for short) of the graph $G = (V, E)$ defined by B. The cost of partition $P = G(B)$, denoted by $cost(P)$, is defined to be the total cost of the edges of G joining different graphs in $G(B)$. A partition $G(B)$ is called *admissible* if all the graphs in $G(B)$ have the weights $< K$ (the weight of $G_i \in G(B)$ is equal to $w(b_i) + w(b_i + 1) + \cdots + w(b_{i+1} - 1)$). An admissible partition P of G is called *optimal* if $cost(P) \leq cost(P')$ for all admissible partitions P' of G.

We would like to find an optimal partition of G. Kernighan [6] first proposed an algorithm for finding an optimal partition and later Kaji and Ohuchi [5] corrected the complexity analysis of his algorithm. They showed that the complexity of the Kernighan's algorithm is $O(n^2)$.

Here we will show that the Kernighan's algorithm can be implemented to run in $O(m \log n)$ time, where $m = |E|$. We first review the Kernighan's algorithm. For a subset B_i of $\{1, 2, \ldots, i\}$ containing 1 and i, the corresponding partition $G(B_i)$ of G will be called a *partial admissible partition* if all graphs G_j in $G(B_i)$ except the graph containing the vertex i have the weights $\leq K$. Let $T(i)$ be the minimum cost of a partial admissible partition $G(B_i)$ of G for a subset $B_i \subset \{1, 2, \ldots, i\}$ containing 1 and i. Then Kernighan observed the following.

$$T(i) = \min\{T(j) - A(j-1, i) + A(i-1, i)\}, \tag{3}$$

where minimum is taken over all j's such that $w(j) + w(j+1) + \cdots + w(i-1) \leq K$ and $A(x, y)$ $(x < y)$ is the total cost of the edges which contain both x and y (we consider an edge $e = (i, j)$ to be an interval $[i, j]$ on the real line).

Thus, if we let $E[i] := T(i) - A(i-1, i)$, $D[j] := T(j)$ and $C(j-1, i) := -A(j-1, i)$ by setting $\alpha := -1$, then we can rewrite (3) as follows.

$$E[i] = \min\{D[j] + C(j-1, i)\}.$$

This is almost the same as (1) and can be solved by a similar method described before. Here we reduce (3) to the same form as in (1). We consider the following intervals and their costs.

$$Z(E) = \{z(e) = [2x+1, 2y] \mid e = [x, y] \in E\},$$

$$Z(V) = \{z(i) = [2x(i), 2i] \mid i \in V, x(i) = \min\{j \mid w(j) + w(j+1) + \cdots + w(i-1) \leq K\}\},$$

and

$$c(z(e)) = c(e) \text{ for } z(e) \in Z(E).$$

Let $Z = Z(E)$ and $SZ = Z(V) \cup \{z = [2i-1, 2i-1] \mid i = 1, 2, \ldots, n\}$. We use the same notation as in Section 3 ($q = m$ and $p = 2n$). Note that if we set $\alpha := -1$ then $C(2j, 2i) = -A(j-1, i)$. Thus if we let $E[2i] := T(i) - A(i-1, i)$ and $D[2j] := T[j]$, then we have the problem of computing $E[2i]$ and $D[2i]$ defined by (1) ($E[2i-1]$ and $D[2i-1]$ will become ∞). Note that all $A(i-1, i) = -C(2i, 2i)$ can be computed in $O(m)$ time in advance and we store them in a table. Thus we can compute $D[2i]$ from $E[2i]$ in costant time. Since there are $m + n$ intervals and $2n$ endpoints, all $E[2i]$ and $D[2i]$ can be computed in $O(m \log n)$ time by the method in Section 2.

3.2 Maximum Weight Clique of a Circular-Arc Graph

There are several variations of the problem described in Section 1. Here we consider one variation which appears in Hsu's algorithm for finding a maximum weight clique of a circular-arc graph [4] and dominates the complexity of the algorithm. We are given a sequence $Z = (z_1, z_2, \ldots, z_q)$ of intervals and a subsequence $SZ = (z_{\sigma(1)}, z_{\sigma(2)}, \ldots, z_{\sigma(r)})$ of Z and a real-valued cost function $c(z)$ for $z \in Z - SZ$, and compute

$$E[i] = \max\{D[j] + C(j, i)\}, \tag{4}$$

for each $1 \leq i \leq r$. Here we assume that the maximum is taken over all the integers $j < i$ such that $\ell(z_{\sigma(j)})$ is contained in $z_{\sigma(i)} \in SZ$. We also assume that $D[i]$ can be computed from $E[i]$ in constant time and that $C(j, i)$ is a constant multiple of the total cost of intervals $z_k \in Z - SZ$ such that z_k appears after $z_{\sigma(j)}$ and before $z_{\sigma(i)}$ in the sequence Z (i.e., $\sigma(j) < k < \sigma(i)$) and contains both $\ell(z_{\sigma(i)})$ and $\ell(z_{\sigma(j)})$. Furthermore, we assume $E[1] = 0$.

This problem can also be solved by a similar method as one described before. Thus, a maximum weight clique of a circular-arc graph $G = (V, E)$ can be obtained in $O(m \log n)$ time and $O(n)$ space provided that G is given by its intersection model of arcs of a circle, where $m = |E|$ and $n = |V|$. See [1] for details. It should be noted that Shin and Hsu [8] has already proposed a faster algorithm with $O(n \log n + m \log\log n)$ time based on another elegant observation. However, we believe that our algorithm is easier to implement and practically runs faster.

3.3 Minimum Weight Dominating Set

A dominating set of a graph $G = (V, E)$ is a subset U of V such that, for every vertex $v \in V$, $v \in U$ or there is an edge $e = (u, v) \in E$ connecting a vertex $u \in U$ and v. A minimum weight dominating set is a dominating set of minimum weight, where each vertex has a positive weight. To find a minimum dominating set of an interval graph, we first consider a shortest path problem on the interval graph. Here the length of a path is the sum of the weights of vertices on the path. We assume an interval graph $G = (V, E)$ is given by its intersection model of intervals $Z = \{z_1, z_2, \ldots, z_n\}$ on the line. We also assume $x(z_i) \leq x(z_j)$ for $i < j$. We now want to compute the distances of shortest paths from z_1 to other vertices z_i. Let $D[i]$ be the distance of a shortest path from z_1 to z_i. Then $D[1] = w(z_1)$ and, for $i = 2, 3, \ldots, n$,

$$D[i] = \min_{z_j \cap z_i \neq \emptyset} \{D[j] + w(z_i)\}.$$

Thus, if we let $E[i] := D[i] - w(z_i)$ then we have

$$E[i] = \min_{z_j \cap z_i \neq \emptyset} \{D[j]\}.$$

This can be solved by a similar method described before. In fact, a simpler method can be applied and all $E[i]$ can be computed in $O(n \log n)$ time.

To find a minimum weight dominating set of an interval graph $G = (V, E)$, we consider new intervals obtained from Z. Specifically, we consider an interval z_i' with $w(z_i') = w(z_i)$ corresponding to z_i as follows: $x(z_i') = x(z_i)$ and $y(z_i')$ is the leftmost right endpoint $y(z)$ of the intervals $z \in Z$ such that $y(z_i) < x(z)$. Let z_0 be a virtual interval with weight 0 such that $x(z_0) = x(z_1)$ and $y_0 = \min\{y(z) \mid z \in Z\}$. Then the distance of a shortest path from z_0 to z_n' is the weight of a minimum dominating set of the graph $G = (V, E)$. Thus, a minimum weight dominating set of an interval graph can be obtained in $O(n \log n)$ time. Since a minimum weight dominating set of a circular-arc graph $G = (V, E)$ can be obtained by solving n times the minimum weight dominating set problems of interval graphs, it can be obtained in $O(n^2 \log n)$ time. This improves the previous complexity [3].

References

[1] T. Asano, *An faster algorithm for finding a maximum weight clique of a circular-arc graph*, Technical Report of Institut für Operations Research, Universität Bonn, 90624-OR, 1990.

[2] J.L. Bentley, *Decomposable searching problems*, Information Processing Letters, 8 (1979), pp.244-251.

[3] A.A. Bertossi and S. Moretti, *Parallel algorithms on circular-arc graphs* Information Processing Letters, 33 (1989/1990), pp.275-281.

[4] W.-L. Hsu, *Maximum weight clique algorithms for circular-arc graphs and circle graphs*, SIAM Journal on Computing, 14 (1985), pp. 224-231.

[5] T. Kaji and A. Ohuchi, *Optimal sequential partitions of graphs by branch and bound*, Technical Report 90-AL-10, Information Processing Society of Japan, 1990.

[6] B.W. Kernighan, *Optimal sequential partitions of graphs*, Journal of ACM, 18 (1971), pp.34-40.

[7] F.P. Preparata and M.I. Shamos, *Computational Geometry: An Intorduction*, Springer-Verlag, New York, 1985.

[8] W.-K. Shih and W.-L. Hsu, *An $O(n \log n + m \log\log n)$ maximum weight clique algorithm for circular-arc graphs*, Information Processing Letters, 31 (1989), pp.129-134.

COMBINATORIAL OPTIMIZATION THROUGH ORDER STATISTICS

Extended Abstract

Wojciech Szpankowski[*]

Department of Computer Science

Purdue University

West Lafayette, IN 47907

Abstract

A mathematical model studied in this paper can be formulated as follows: find the optimal value of $Z_{\max} = \max_{\alpha \in B_n} \{\Xi_{i \in S_n(\alpha)} w_i(\alpha)\}$ (respectively $Z_{\min}$), where Ξ is an operator, n is an integer, B_n is the set of all feasible solutions, $S_n(\alpha)$ is the set of all objects belonging to the α-th feasible solution, and $w_i(\alpha)$ is the weight assigned to the i-th object. Our interest lies in finding an asymptotic solution to this optimization problem in a probabilistic framework. Using some novel results from order statistics, we investigate in a uniform manner a large class of combinatorial optimization problems such as: the assignment problem, the traveling salesman problem, the minimum spanning tree, the minimum weighted k-clique, the bottleneck assignment and traveling salesman problems, location problems on graphs, and so forth. For example, we provide some sufficient conditions that assure asymptotic optimality of a greedy algorithm.

1. INTRODUCTION

We consider in this paper a class of generalized optimization problems in a probabilistic framework. A general mathematical model is as follows: for some integer n define $Z_{max} = \max_{\alpha \in B_n} \{\Xi_{i \in S_n(\alpha)} w_i(\alpha)\}$ (Z_{min} respectively), where Ξ is an operator (e.g., $\Xi = \Sigma$ or $\Xi = \min$, etc.), B_n is the set of all feasible solutions, $S_n(\alpha)$ is the set of all objects belonging to the α-th feasible solution, and $w_i(\alpha)$ is the weight assigned to the i-th object. For example, in the traveling salesman problem [BOR62, KAR76, LLK85, WEI80] the operator Ξ becomes a sum Σ operator, B_n represents the set of all Hamiltonian paths, $S_n(\alpha)$ is the set of edges that fall into the α-th Hamiltonian path, and $w_i(\alpha)$ is the length of the i-th edge; for the bottleneck traveling salesman problem [GaG78, WEI80] the operator Ξ becomes "min" operator. Some other examples include the assignment problem [BOR62, FHR87, WAL79, WEI80, LLK85], the minimum spanning tree [BOL85, KNU73], the minimum weighted k-clique problem [LUK81, BOL85] the bottleneck and capacity assignment problems [WEI80], geometric location problems [PAP81], and some others not directly related to optimization such as the height and depth of digital trees [KNU73, SZP88a, SZP91], the maximum queue length, hashing with lazy deletion, pattern matching, edit distance [AHU74, UKK90] and so forth. In our probabilistic framework, we assume that the weights $w_i(\alpha)$ are random variables drawn from a common distribution function $F(\cdot)$. Our interest lies in finding an asymptotic solution of Z_{max} (Z_{min}) and the best k-th solution $Z_{(k)}$ in some probability sense for a large class of distribution functions $F(\cdot)$, and apply these findings to design efficient heuristic algorithms that achieve asymptotically the optimal performance.

Our analysis *does not* assume any *a priori* information regarding the distribution of inputs. We, however, identify two classes of distributions that lead to precise asymptotic expansions for Z_{max} and Z_{min}. We present several new results that are grouped into two categories, namely *general results* (Section 3.2) and *specific solutions* (Section 3.3). All of these results are obtained in a systematic way using a variety of tools from *order statistics* (Section 3.1). We have novel results in all three facets mentioned above. In particular, Lemma 1a of Section 3.1 establishes a simple characteristic of a probabilistic behavior of the k-th best solution $Z_{(k)}$ of our optimization problem. Such a generalization of the problem is necessary in

---

[*]This research was supported by AFOSR grant 90-0107, in part by the NSF grant CCR-8900305, and in part by grant R01 LM05118 from the National Library of Medicine.

many fields of science, most notably in molecular biology (DNA and protein foldings), pattern recognition, and so forth. Moreover, Lemma 3 presents a systematic approach to analyze the limiting distribution of the optimal solutions Z_{max} and Z_{min}. These probabilistic tools are next used to design practical heuristic algorithms for some optimization problems. This is simply achieved by comparing (in some probability sense) the performance of the optimal solution with the performance of a specific solution, and therefore controlling the performance quality of the heuristic. In particular, we present sufficient conditions under which a greedy algorithm achieves asymptotically the same performance as the optimal one (cf. Theorem 4). Our next result concentrates on a large class of bottleneck optimization problems and demonstrates a constructive probabilistic approach to design semi-optimal algorithms for arbitrary distribution functions (cf. Theorem 5). Finally, the list of our general results is concluded by a finding concerning the additive objective function (i.e., $\Xi = \Sigma$) that is virtually without probabilistic assumptions however it restricts the size of the input (cf. Theorem 6). We have also several specific results and suggest several approximate algorithms. In particular, we discuss the linear assignment problem (cf. Problem 1), maximal properties of digital trees (cf. Problem 2), the optimal weighted k-clique problem (cf. Problem 3), the optimal location problem (cf. Problem 4). In all categories we have obtained new results, and more importantly all of these problems – previously treated by disparate methods – are analyzed in this paper in a uniform manner.

2. PROBLEM STATEMENT

Let n be an integer (e.g., number of vertices in a graph, number of keys in a digital tree, etc.), and S a set of objects (e.g., set of vertices, keys, etc). We shall investigate the optimal values Z_{max} and Z_{min} defined as follows

$$Z_{max} = \max_{\alpha \in B_n} \left\{ \sum_{i \in S_n(\alpha)} w_i(\alpha) \right\} \qquad Z_{min} = \min_{\alpha \in B_n} \left\{ \sum_{i \in S_n(\alpha)} w_i(\alpha) \right\}, \qquad (2.1)$$

where B_n is a set of all feasible solutions, $S_n(\alpha)$ is a countable set of objects from S belonging to the α-th feasible solution, and $w_i(\alpha)$ is the weight assigned to the i-th object in the α-th feasible solution (in addition, by w_{ij} we denote a weight assigned to a pair of objects (i,j) in S). Throughout this paper, we adopt the following assumptions:

(A) The cardinality $|B_n|$ of B_n is fixed and equal to m. The cardinality $|S_n(\alpha)|$ of the set $S_n(\alpha)$ does *not* depend on $\alpha \in B_n$ and for all α it is equal to N, i.e., $|S_n(\alpha)| = N$.

(B) For all $\alpha \in B_n$ and $i \in S_n(\alpha)$ the weights $w_i(\alpha)$ (i.e., the weights w_{ij}) are identically and independently distributed (i.i.d) random variables with common distribution function $F(\cdot)$, and the mean value μ and the variance σ^2.

The assumption (B) defines a *probabilistic model* of our problem (2.1). We shall explore the asymptotic behaviors of Z_{max} and Z_{min} as n becomes large (*in probability* and/or *almost surely* sense).

There are many combinatorial problems that fall into our formulation (2.1). For example, the linear assignment problem [BOL85, WAL79, FHR87], the traveling salesman problem [LLK85], the minimum spanning tree [AHU74, BOL85], the minimum weighted k-clique problem [BOL85, LUE81], and so forth. In particular, in the linear assignment problem $|B_n| = n!$, $|S_n(\alpha)| = n$ and the weights $w_i(\alpha)$ are elements of a matrix; in the traveling salesman problem B_n is the set of all Hamiltonian paths and $S_n(\alpha)$ is a set of n edges in the α-th Hamiltonian path, that is, $N = |S_n(\alpha)| = n$; in the k-clique problem B_n is defined as a set of all k-cliques in a graph and $S_n(\alpha)$ the set of edges belonging to the α-th k-clique, and so forth.

So far, we have restricted our attention to problems which can be represented as (2.1). In practice some other objective functions are important. For example, in a class of bottleneck and capacity problems [GaG78, HoS86 SZP90] the operator 'Σ' in (2.1) is replaced by 'max' and 'min', respectively. Therefore, we extend (2.1) to the following

$$Z_{max} = \max_{\alpha \in B_n} \left\{ \Xi_{i \in S_n(\alpha)} w_i(\alpha) \right\} \qquad Z_{min} = \min_{\alpha \in B_n} \left\{ \Xi_{i \in S_n(\alpha)} w_i(\alpha) \right\} \qquad (2.2)$$

where Ξ is an operator applied to a set $\{w_i(\alpha), i \in S_n(\alpha)\}$, e.g., in (2.1) $\Xi = \Sigma$. For example, in the bottleneck and capacity assignment problems, and bottleneck and capacity traveling salesman problems [GaG78, HoS86] the operator Ξ becomes either "max" or "min" respectively.

3. MAIN RESULTS

To formulate our problem in terms of *order statistics*, we define a random variable X_α as $X_\alpha = \Xi_{i \in S_n(\alpha)} w_i(\alpha)$ where α is a feasible solution and $\alpha \in \{1, 2, ..., m = |B_n|\}$. Then $Z_{max} = \max_{1 \le \alpha \le m} \{X_\alpha\}$ and $Z_{min} = \min_{1 \le \alpha \le m} \{X_\alpha\}$. We denote the distribution of X_α by $F_N(x) = P\{X_\alpha < x\}$. In some cases there exists explicit relationship between the distribution function $F_N(\cdot)$ and the original distribution $F(\cdot)$ of weights. For example, for $\Xi = \Sigma$ we have $F_N(x) = F^{*N}(x)$, where $*$ is the convolution operator, while for if $\Xi = \min$ one obtains $F_N(x) = F^N(x)$. In some other instances, we can apply either the *Central Limit Theorem* (CLT) or the *Extreme Statistic Distributions* to obtain the limiting distribution of $F_N(x)$.

3.1 Main Tools

It should be clear that a successful solution of our problem depends on establishing asymptotics for some order statistics; in particular, for maximum and minimum of dependent random variables. So, let us consider the following abstract problem: *given n random variables $X_1, X_2, ..., X_n$ with marginal distribution functions $G_1(x), ..., G_n(x)$ respectively, evaluate for large n the behavior of $Z_{max} = \max_{1 \le k \le n} \{X_k\}$ and $Z_{min} = \min_{1 \le k \le n} \{X_k\}$ and the k-th best solution $Z_{(k)}$, that is, $Z_{(k)}$ represents the kth order statistic of the sequence $\{\vec{X}\}_{\alpha \in B_n}$. It turns out that the solution to such a problem largely depends on the behavior of the distribution function at infinity. We shall consider three types of distribution functions described in the next definitions.

Definition 1. (i) A general distribution function $G(\cdot)$ is called Type I distribution.

(ii) A distribution function satisfying the following two conditions:

$$G(x) < 1 \quad \text{for} \quad x \to \infty \tag{3.1}$$

$$\lim_{x \to \infty} \frac{1 - G(xc)}{1 - G(x)} = 0 \quad \text{for all} \quad c > 1 \tag{3.2}$$

is called Type II distribution (or exponential-tail distribution). If conditions (3.1) are replaced by $G(x) > 0$ for $x \to -\infty$ and (3.2) by $\lim_{x \to -\infty} G(xc)/G(x) = 0$ for $c < 1$ then we have Type II' distributions.

(iii) Assuming (3.1) holds and

$$\lim_{x \to \infty} \frac{1 - G(x + c)}{1 - G(x)} = 0 \quad \text{for all} \quad c > 0 \tag{3.3}$$

we obtain Type III distributions (or superexponential-tail distribution). Finally, we require condition $\lim_{x \to -\infty} G(x + c)/G(x) = 0$ for $c < 0$ for Type III' distributions. ∎

Definition 2. A sequence of distribution functions $G_1(x), ..., G_n(x)$ is *uniformly* Type II and III (resp. II' and III') if conditions (3.1) and (3.3) hold uniformly in n, that is,

$$\lim_{x \to \infty} \sup_n \frac{1 - G_n(xc)}{1 - G_n(x)} = 0 \quad \text{for all} \quad c > 1 \tag{3.4}$$

implies uniformly Type II distributions. ∎

Next lemma contains a gallery of results on the highest (lowest) order statistics (i.e., maximum and minimum) that are useful in establishing asymptotic results for our optimization problems. Most of these results are known (but not widely known, and never systematically applied to optimization problems), and we only slightly generalize some of them to include non-identical distributions. Define a_n and b_n as the smallest and the largest solutions of the following equations

$$\sum_{k=1}^{n} [1 - G_k(a_n)] = 1 \quad , \quad \sum_{k=1}^{n} G_k(b_n) = 1 \tag{3.5}$$

respectively. Then, we can prove the following lemma (cf. [ALD89, GAL87, LaR77]).

Lemma 1. (i) *Let $G_1(\cdot), ..., G_n(\cdot)$ belong to Type I distributions. Then the following bounds for the r-th moments EZ_{max}^r and EZ_{min}^r of Z_{max} and Z_{min} hold*

$$EZ_{max}^r \le \check{a}_n + \sum_{k=1}^{n} \int_{\check{a}_n}^{\infty} [1 - G_k(x^{1/r})] dx \quad ; \quad EZ_{min}^r \ge \check{b}_n - \sum_{k=1}^{n} \int_{-\infty}^{\check{b}_n} G_k(x^{1/r}) dx \tag{3.6}$$

where $\bar{a}_n$ and $\bar{b}_n$ are the smallest and the largest solutions of the following

$$\sum_{k=1}^{n} [1 - G_k(\bar{a}_n^{1/r})] = 1; \qquad \sum_{k=1}^{n} G_k(\bar{b}_n^{1/r}) = 1 \qquad (3.7)$$

respectively. Note that for $r = 1$, $\bar{a}_n$ and $\bar{b}_n$ coincide with a_n and b_n defined in (3.5) respectively.

(ii) *For uniformly Type II and II' distributions, the following holds*

$$\lim_{n\to\infty} Z_{max}/a_n \leq 1 \qquad \lim_{n\to\infty} Z_{min}/b_n \geq 1 \qquad in \ probability \ (pr.) \qquad (3.8)$$

respectively. If, in addition, the following inequality (the so called mixing condition)

$$P\{X_1 < x_1, X_2 < x_2, ..., X_n < x_n\} \leq \delta \cdot G_1(x_1) \cdot G_2(x_2) \ ... \ G_n(x_n) \qquad (3.9)$$

(or $P\{X_1 > x_1, ..., X_n >_n\} \leq \delta \cdot [1 - G_1(x_1)]...[1 - G_n(x_n)]$ for minimum) holds for some $\delta = O(1)$, then respectively

$$\lim_{n\to\infty} Z_{max}/a_n = \lim_{n\to\infty} Z_{min}/b_n = 1 \qquad in \ probability . \qquad (3.10)$$

Finally, if for some r we have $\int_{-\infty}^{\infty} x^r \cdot g_k(x) \, dx < \infty$, where $g_k(x)$ is the density function of X_k, then the convergence in probability can be replaced by convergence in mean, namely

$$\lim_{n\to\infty} E|Z_{max}/a_n - 1|^r = 0 \quad , \qquad \lim_{n\to\infty} E|Z_{min}/b_n - 1|^r = 0 . \qquad (3.11)$$

(iii) *For uniformly Type III and III' distributions stronger asymptotics can be established, namely all results from (ii) hold with $Z_{max}/a_n \to 1$ replaced by $Z_{max} - a_n \to 0$.*

Proof. Details are omitted and can be found in [GAL87, SZP90, SZP91]. ∎

Higher order statistics $Z_{(r)}$ can be analyzed in a similar manner, however, one needs information regarding joint distributions. Define for the r-th order statistic $R^{(r)}(x_1, ..., x_r) = \Pr\{X_1 > x_1, ..., X_r > x_r\}$. We write $R^{(r)}(x)$ if $x_1 = ... = x_r = x$. Lemma 1 can be easily extended to include rth order statistics (for brevity of presentation we only extend part (ii) of Lemma 1).

Lemma 1a. *Let $\{X_i\}_{i=1}^{n}$ be exchangeable random variables,[†] and let $R^{(r)}(x)$ satisfies (3.1) and (3.2). Define $a_n^{(r)}$ as the smallest solution of*

$$\binom{n}{r-1} R^{(n-r+1)}(a_n^{(r)}) = 1 . \qquad (3.12)$$

Then, $Z_{(r)} \leq a_n^{(r)}$ (pr.). If, in addition, X_i are i.i.d., then $Z_{(r)} \sim a_n^{(r)}$ (pr.) for $n \to \infty$.

Proof. Apply Boole's inequality to $\Pr\{Z_{(r)} > x\} = \Pr\{\bigcup_{j_1,...,j_{n-r+1}} \bigcap_{i=1}^{n-r+1} (X_{j_i} > x)\}$ for all distinct $j_1, ..., j_{n-r+1} \in \{1, ..., n\}$. ∎

In many applications the lower bound is rather hard to prove, and the mixing condition (3.9) are neither satisfied nor easy to verify. Then, the following second moment method is useful (below a version due to Chung and Erdös (cf. [ALD89]) is presented).

Lemma 2. The Second Moment Method. *The following holds*

$$\Pr\{Z_{max} > r\} = \Pr\{\bigcup_{i=1}^{n}(X_i > r)\} \geq \frac{(\sum_i \Pr\{X_i > r\})^2}{\sum_i \Pr\{X_i > r\} + \sum_{(i \neq j)} \Pr\{X_i > r \ \& \ X_j > r\}} . \qquad (3.13)$$

If the RHS of the above tends to one for $r = (1 - \epsilon)a_n$ with a_n given in (3.5), then we have $Z_{max} \geq a_n$. ∎

Finally, when X_i are strongly dependent, then the following idea can be applied. We call it the *Sieve Method* since the *inclusion-exclusion* rule is used [BOL85]. Let X_i be identically distributed random variables for which the joint distributions depend only on the number of variables (e.g., exchangeable

---

[†]A sequence $\{X_i\}_{i=1}^{n}$ is exchangeable if for any k-tuple $\{j_1, .., j_k\}$ of the index set $\{1, ..., n\}$ the following holds: $\Pr\{X_{j_1} < x_{j_1}, ..., X_{j_k} < x_{j_k}\} = \Pr\{X_1 < x_1, ..., X_k < x_k\}$, that is, joint distribution depends only on the *number* of variables.

random variables). Then, from the inclusion-exclusion rule applied to the event $\{Z_{max} > k\} = \{\bigcup_{i=1}^n \{X_i > k\}$ one obtains

$$\Pr\{Z_{max} > k\} = \frac{1}{n}\sum_{r=2}^n (-1)^r \binom{n}{r} r\Pr\{X_1 > k, \ldots, X_r > k\} \, . \tag{3.14}$$

Define $Z_n(z)$ as the probability generating function of Z_{max}, and let $F_r(z) = \sum_{k=0}^\infty z^k \Pr\{X_1 > k, \ldots, X_r > k\}$. Then, after some algebra one proves the following lemma.

Lemma 3. Sieve Method. *For $|z| < 1$ one shows*

$$Z_n(z) = 1 - \frac{1-z}{n}\sum_{r=2}^n (-1)^r \binom{n}{r} r F_r(z) \, . \tag{3.15}$$

For large n (and some class of functions $F_r(z)$ that do not grow too fast at infinity; for details see [SZP88b])

$$Z_n(z) \sim \frac{1}{2\pi i} \int_{-1/2-\infty}^{-1/2+\infty} \Gamma(z) F_r(1-z) n^{-z} dz \tag{3.16}$$

where $\Gamma(z)$ is the gamma function.

Proof. Equation (3.15) is a direct consequence of (3.14). Indeed, it suffices to multiply both sides of (3.14) by z^k and sum up. Then, (3.16) follows directly from the Mellin-like asymptotics derived in [SZP88b]. It worth mentioning that the RHS of (3.16) can be usually easily evaluated by appealing to the Cauchy's residue theorem. ∎

3.2 General Results

Some general results can be formulated for two important classes of operators Ξ, namely, the so called *nondecreasing* operators defined by the following property

$$w_i(\alpha) \leq w_i'(\alpha) \quad \Rightarrow \quad \Xi_{i \in S_n(\alpha)} w_i(\alpha) \leq \Xi_{i \in S_n(\alpha)} w_i'(\alpha) \, , \tag{3.17}$$

and *ranking-dependent* operators defined by the identity $f(Z_{max}) = \max_{\alpha \in B_n}\{\Xi_{i \in S_n(\alpha)} f(w_i(\alpha))\}$ for every increasing function $f(\cdot)$. For example, the operator Σ, "max" are nondecreasing operators, but "max" and "min" are ranking-dependent operators.

We first discuss some general bounds on Z_{max}, and we start with some upper bounds. We recall that $F_N(\cdot)$ denotes the distribution function of X_α and we define $R_N(x) = 1 - F_N(x)$. Then Lemma 1(i) immediately implies $Z_{max} \leq R_N^{-1}(1/m)$ and $Z_{min} \geq F_N^{-1}(1/m)$ (pr.).

To establish another bound on Z_{max} we assume that Ξ is nondecreasing. Since $w_k \leq \max_{i \in S}\{w_i\}$ for every $k \in S$ where S is a set of all objects, we immediately obtain

$$Z_{max} = \max_{\alpha \in B_n}\{\Xi_{i \in S_n(\alpha)} w_i(\alpha)\} \leq \Xi_{i \in S_n(\alpha)} \max_{\alpha \in B_n}\{w_i(\alpha)\} \, . \tag{3.18}$$

To simplify the above we note that the sequence $\{w_i(\alpha)\}_{\alpha \in B_n}$ contains many identical weights which, if deleted, do not change the value of $\max_{\alpha \in B_n}\{w_i(\alpha)\}$. Let us define the set $O(i)$ by the following identity $\max_{\alpha \in B_n}\{w_i(\alpha)\} = \max_{j \in O(i)}\{w_{ij}\}$ where $O(i)$ enumerates all *distinct* weights $w_i(\alpha)$ over all feasible solutions $\alpha \in B_n$, and we denote these distinct weights by w_{ij} ("direct neighbors" of the i-th object). Naturally, by Lemma 1 one shows that $\max_{j \in O(i)} w_{ij} \sim R^{-1}(1/K)$ (pr.), provided $|O(i)| = K \to \infty$ with $n \to \infty$. This can be translated into Z_{max} for some operators. In particular, for $\Xi = \Sigma$ it is easy to see that $Z_{max} \leq NR^{-1}(1/K)$ and $Z_{min} \geq NF^{-1}(1/K)$ (pr.) provided $K, N \to \infty$ with $n \to \infty$.

Lower bounds are usually harder to establish, and one has to appeal either to the mixing condition or the second moment method. This means that the joint distribution of X_α must be estimated. However, for some optimization problems a very simple approach exists that is based on the following observation: *for any solution $\alpha \in B_n$ the following holds $Z(\alpha) \leq Z_{max}$.* The challenge is how to choose α. We present two methods. One is based on some known results for unweighted random structures (e.g., graphs), and the other method assumes a greedy solution α_{grd}. In the first method (see [WEI80, LUE81] for applications in graphs) we consider a random structure which mimics our structure under consideration, except that objects do not have weights. It is assumeed that the following is known:

- *A feasible set B_n is nonempty (a.s.) or (pr.) if objects of a random (unweighted) structure are selected with at least probability p_n.*

Then, it is easy to prove the following $Z_{max} \geq \Xi_{i \in S_n} R^{-1}(p_n)$ and $Z_{min} \leq \Xi_{i \in S_n} F^{-1}(p_n)$ (pr.) or (a.s.). We must note however, that a successful application of the above depends on *a priori* knowledge about the critical probability p_n.

At last, we turn our attention to the *greedy algorithm* approach. We note that for any solution $\alpha \in B_n$ we can write $\alpha = x_1 x_2 ... x_N$ where $x_i \in S$. A greedy algorithm is a sequential procedure that selects in the k-th step an object x_k which locally optimizes the objective function. To be more formal, we extend the set of feasible solution B_n to a set $\tilde{B}_n$ such that $\beta = x_1 x_2 ... x_k \subset \alpha$ implies $\beta \in \tilde{B}_n$ provided $\alpha \in \tilde{B}_n$. Note that if $\beta \in \tilde{B}_n$ and $x_k \in S$ then $\beta x_k \in \tilde{B}_n$ if βx_k is a partial feasible solution.

Greedy Algorithm:

1. For $k = 1, 2, ..., N$ select an element $x_k \in S$ such that $\beta \in \tilde{B}_n$ implies $\beta x_k \in \tilde{B}_n$ and

$$\Xi_{i \in S_n(\beta x_k)} w_i(\beta x_k) = \max_{y \in S} \left\{ \Xi_{i \in S_n(\beta y)} w_i(\beta y) \right\} .$$

2. Stop if no such x_k exists. ∎

The application of the above greedy procedure depends on the random structure under consideration and the type of operator Ξ. To avoid heavy notations, we focus on graphs, and assume $\Xi = \Sigma$. Let $\beta_i = x_1 x_2 ... x_i \in \tilde{B}_n$ be a partial solution found up to the i-th step in a greedy procedure. Then, the $i + 1$-st object x_{i+1} is selected from the (presumely nonempty) neighborhood $A(i)$ of the i-th object such that $\beta_i x_{i+1}$ is a feasible subsolution, and $A(i) \subset O(i)$. The greedy algorithm for $\Xi = \Sigma$ might look as follows $Z(\beta_{i+1}) = Z(\beta_i) + \max_{j \in A(i)} \sum_{k=1}^{M_i} w_{kj}$, where in the i-th step the algorithm selects M_i objects (see Problem 3 in Section 3.3). Then, Lemma 1, after some algebra, implies the following asymptotics

$$Z_{grd} \sim \sum_{i \in S'_n} R_{M_i}(|A(i)|^{-1}) + (N - N')\mu \tag{3.19}$$

provided $|A(i)| \to \infty$, where $F_{M_i}(x) = 1 - R_{M_i}(\cdot)$ denotes the convolution of M_i distribution functions of the weights, $\mu = E w_{ij}$, and $N' = |S'_n| \leq |S_n(\alpha_{grd})| = N$. In the above S'_n is the set of objects found by the greedy algorithm that constitutes a partial solution to the problem, *and* that assures the extension of this partial solution to a feasible solution.

Our results find many applications in the design of effective heuristic (approximate) algorithms for some problems. The effectiveness of a heuristic can be – in the spirit of Karp's idea [KAR76] – measured by the relative error e_n defined as follows $e_n = |(Z_{opt} - Z(\alpha))/Z_{opt}|$. Then, a solution α as near-optimal (a good heuristic) if $e_n \to 0$ as $n \to \infty$ almost surely, or in probability or in mean. We formulate our conclusions in the form of the following theorem.

Theorem 4. (i) *Let α be a feasible solution $\alpha \in B_n$ such that*

$$\lim_{n \to \infty} R_N^{-1}(1/m)/Z(\alpha) = 1 \qquad (pr.) \qquad \lim_{n \to \infty} F_N^{-1}(1/m)/Z(\alpha) = 1 . \tag{3.20}$$

Then α is near-optimal, provided $m \to \infty$ as $n \to \infty$ and $F(\cdot)$ is type II and III (II' or III') distribution function. In addition, for $\Xi = \Sigma$ this remains true if (3.20) is replaced by

$$\lim_{n \to \infty} N R^{-1}(1/K)/Z(\alpha) = \lim_{n \to \infty} N F^{-1}(1/K)/Z(\alpha) = 1 \qquad (a.s.) , \tag{3.21}$$

and for $\Xi = \min (\max)$

$$\lim_{n \to \infty} F^{-1}(N^{-1/K})/Z(\alpha) = \lim_{n \to \infty} R^{-1}(N^{-1/K})/Z(\alpha) \qquad (pr.) \tag{3.22}$$

respectively, provided $K, N \to \infty$ as $n \to \infty$ and $F(\cdot)$ is type II (II') or III (III') distribution functions.

(ii) **When is greedy near optimal?** *Let $\Xi = \Sigma$. Then, a greedy algorithm α_{grd} is near optimal (i.e., $Z_{grd} \sim Z_{max}$) if*

$$\lim_{n \to \infty} \frac{R_N^{-1}(1/m)}{\sum_{i \in S_n} R^{-1}(|A(i)|^{-1}) + (N - N')\mu} = 1 , \tag{3.23}$$

and the above conclusion also holds when the numerator is replaced by $N \cdot R_N^{-1}(1/K)$.

Proof. It is a consequence of our previous discussions. ∎

We can strengthen our results in the case of two important operators, namely, for $\Xi = \max$ (min) and $\Xi = \Sigma$. We first discuss an optimal solution for a class of bottleneck problems (e.g., $\Xi = \max$). To avoid heavy notation, the following theorem presents a sample of two general results for bottleneck problems in which the weights are distributed according to any strictly continuous distribution function $F(\cdot)$ (cf. [SZP90]).

Theorem 5. *Let $\Xi = \max(\min)$, and K, $N \to \infty$ with $n \to \infty$. In addition, we assume that the distribution function $F(\cdot)$ of weights is a strictly continuous, but otherwise arbitrary, function.*

(i) **Bottleneck (Capacity) Assignment Problem** [HoS86, KAR87, FHR87]. *If $F^{-1}(\cdot)$ denotes the inverse function, then asymptotically the optimal solution of the bottleneck (capacity) assignment problem becomes*

$$Z_{min} \sim F^{-1}\left(\frac{\log n}{n}\right) \quad (pr.) \qquad Z_{max} \sim F^{-1}\left(1 - \frac{\log n}{n}\right) \tag{3.24}$$

respectively. The same holds for the Bottleneck (Capacity) Traveling Salesman Problem.

(ii) **Bottleneck k-Clique Problem** [HoS86]. *Asymptotically the following holds*

$$Z_{min} \sim F^{-1}(n^{-2/(k-1)}) \quad (pr.) . \tag{3.25}$$

Proof. All the above asymptotic expressions can be proved in a similar manner using our tools established in Section 3.1. For brevity we only sketch the proof for the Bottleneck Assignment Problem (BAP). The lower bound on Z_{min} comes from the following obvious inequality $Z_{min} \geq \max_{1 \leq j \leq n} \min_{1 \leq i \leq n} a_{ij}$. Since $\Xi = \max$ is a ranking-dependent operator, we can prove our results for a selected distribution. We choose an exponential distribution, and then by Lemma 1 we can see that a_n is a solution of $ne^{-na_n} = 1$ (cf. (3.5)), so the lower bound follows. For the upper bound we apply an algorithmic strategy similar to the greedy approach. Let us sort all elements a_{ij} of a matrix A in a nondecreasing order. Define m^* such that the first m^* elements of A form *almost surely* a feasible solution. From the *Coupon Collector Problem* [ALD89] we know that $m^* = n \log n + cn$ for some c [ALD89]. Then, naturally, $Z_{min} \leq a_{(m^*)}$ where $a_{(m^*)}$ is the m^*-th order statistics of n^2 elements from the matrix A. Again we are free to select distribution of a_{ij}. For this purpose we choose the uniform distribution. But then , $a_{(m^*)} = (n \log n + cn)/n^2 \sim \log n/n$ (pr.) [GAL87], and this completes the proof of Theorem 5. ∎

Finally, we present one general result regarding $\Xi = \Sigma$ operator. This theorem is virtually without probabilistic assumptions, but restricts the size of the problem.

Theorem 6. *Let $\Xi = \Sigma$, and let m be polynomially related to N, that is, for some $d > 0$ we have $m = O(N^d)$, and $N \to \infty$ with $n \to \infty$. If the average weight $\mu = Ew \neq 0$, then for large n the following holds*

$$Z_{min} = N\mu(1 + o(1)) = Z_{max} \quad (pr.) \tag{3.26}$$

In fact, the lower (upper) bound for Z_{max} (Z_{min}) holds almost surely.

Proof. We consider only Z_{max}. The lower bound is trivial since $EZ_{max} \geq NE\{w_i(\alpha)\} = N\mu$. For the upper bound, we represent our problem (2.1) as $Z_{max} = N\mu + \sigma\sqrt{N}Y$ where $Y = (\sum_{i \in S_n(\alpha)} w_i(\alpha) - N\mu)/\sigma\sqrt{N}$. From the *Central Limit Theorem* (CLT) one knows that $Y \Rightarrow N(0,1)$, where $N(0,1)$ is the standard normal distribution. Lemma 1 implies that $Z_{max} \leq a_n$ where a_n solves $m(1 - F_N(a_n)) = 1$. It is easy to see that under our assumption a_n is asymptotically equivalent to a solution of $m(1 - \Phi(\bar{a}_n)) = 1$ where $\Phi(\cdot)$ is the standard normal distribution function. This follows from Feller's representation of $F_N(x)$ as (cf. [FEL70])

$$F_N(x) = \Phi(x) + \frac{e^{x^2/2}}{\sqrt{2\pi}} \sum_{k=3}^{r} N^{1/2 \cdot k + 1} P_k(x) + o(N^{1/2 \cdot k + 1}) \tag{3.27}$$

where $P_k(x)$ is a polynomial of order k that does not depend on N. To complete the proof we note that the maximum of m standard normal random variables behaves asymptotically like $\sqrt{2 \log m}$ (pr.) [GAL87], hence $Z_{max} \leq N\mu + \sigma\sqrt{2N \cdot d \cdot \log N}$ for $m = O(N^d)$. ∎

3.3 Specific Problems and Their Solutions

PROBLEM 1. *Linear Assignment Problem*

In this case $m = n!$ and $N = n$. We assume that the distribution function of weights (elements of a matrix) $\{a_{ij}\}_{i,j=1}^n$ is either of type II or of type III. Then, $Z_{max} \leq R_n^{-1}(1/n!)\,(pr.)$ and $Z_{max} \leq nR^{-1}(1/n)$ (a.s.) respectively. A lower bound follows from the greedy approach (cf. [BOR62]). We have $|A(k)| = n-k$, $N = N'$, and finally

$$\sum_{k=1}^{n} R^{-1}\,(1/k) \leq Z_{max} \leq R_n^{-1}(1/n!) \qquad (pr.) . \tag{3.28}$$

In particular, for the gamma distributions $gamma(\beta, \lambda)$ and the normal distributions $N(\mu, \sigma)$ the following almost sure convergence can be easily derived from the above

$$Z_{max} = \frac{n}{\lambda}\log n + \frac{n\beta}{\lambda}\log\log n + O(n) \qquad (a.s.) \tag{3.29a}$$

$$Z_{max} = n\sigma\sqrt{2\log n} + n\mu + o(n) \qquad (a.s.) \tag{3.29b}$$

These asymptotics *cannot* be, in general, extended to distribution functions of type I (cf. [WAL79]).

PROBLEM 2. *Characteristics of Digital Trees*

Let $X_1, ..., X_n$ represent n strings of symbols from a V-ary alphabet. The i-th symbol occurs independently with probability p_i. The strings $\{X_i\}_{i=1}^n$ – which may represent keys – are used to build a digital tree called *trie* [AHU74]. Let D_n and H_n denote the depth and the height of such a trie. The depth is defined as the length of a path from the root of the tree to a randomly selected external node, while the height is simple the maximum of all n depths. An alternative definitions of the above is more useful for estimating the asymptotics of D_n and H_n. Define an alignment C_{ij} between X_i and X_j as the length of the longest common prefix of these two strings. Of course, $\Pr\{C_{ij} > k\} = P^{k+1}$, where $P = \sum_{i=1}^V p_i^2$ is the probability of a match in any given position. Then the depth and the height can be defined as $D_n = \max_{1 \leq \ell \leq n} C_{i\ell}$ and $H_n = \max_{1 \leq i < \ell \leq n} C_{i\ell}$.

Let us first discuss the height H_n. Lemma 1 suggests that $H_n \leq a_n$ (pr.) where $a_n \sim -2\log n/\log P$. To prove the lower bound we need to appeal to Lemma 2. For this we have to evaluate the joint distributions of the alignments. Fortunately, the following is easy to prove. For simplicity of presentation we restrict hereafter to a binary case with $p_1 = p$ and $p_2 = q = 1 - p$. Then, [SZP91] $\Pr\{C_{12} > k, ..., C_{1r} > k\} = (p^r + q^r)^{k+1}$, and after some algebra one shows that $H_n \sim -2\log n/\log P$.

If one tries to match the upper bound for the depth, then surprisingly enough it turns out to be non-doable. We prove that $D_n \sim \log n/h$ (pr.) where $h = -p\log p - q\log q$ is the entropy of the alphabet. In fact, we use Lemma 3 to show much stronger result, namely a limiting distribution of the depth. From the above and Lemma 3 we easily prove that the probability generating function $D_n(z)$ of the depth D_n becomes

$$D_n(z) = 1 - \frac{1-z}{n}\sum_{r=2}^{n}(-1)^r \binom{n}{r}\frac{r(p^r + q^r)}{1 - z(p^r + q^r)} . \tag{3.30}$$

To establish the limiting distribution of D_n we appeal to (3.16) from Lemma 3. But, first we can show how (3.30) can be used to evaluate the average depth ED_n and the variance $var D_n$. One obtains

$$ED_n = \frac{1}{n}\sum_{r=2}^{n}(-1)^r \binom{n}{r}\frac{r(p^r + q^r)}{1 - (p^r + q^r)} . \tag{3.31}$$

But [SZP88b]

$$ED_n \sim \frac{1}{2\pi i}\int_{-1/2-\infty}^{-1/2+\infty} \frac{\Gamma(z)n^{-z}(p^{1-z} + q^{1-z})}{1 - p^{1-z} - q^{1-z}}dz ,$$

and by residues calculus one immediately proves that $ED_n = \log n/h + O(1)$. In a similar manner, we can show that $var D_n = \beta\log n + O(1)$ (asymmetric case $p \neq q$) where $\beta = (h_2 - h^2)/h^3$ and $h_2 = p\log^2 p + q\log^2 q$ [SZP88a]. For the limiting distribution we apply directly the Mellin-like formula (3.16), and simple calculus of residues shows that for the asymmetric case $(D_n - ED_n)/\sqrt{var D_n} \Rightarrow \mathcal{N}(0,1)$ where $\mathcal{N}(0,1)$ is standard normal distribution. Finally, we mention that the above results can be extended to the cases: (i) when symbols in any string have Markov dependency (the so called *Markovian model*) (cf. [JsS91]); (ii) when the strings $X_1, ..., X_n$ are dependent, e.g., as in suffix tree the X_i string is the i-th suffix of the first string (cf. [SZ91]).

PROBLEM 3. *The Optimal Weighted k-Clique*

This is an interesting problem since $N = C_k^2$, and for bounded k the size N of a feasible solution is also bounded. Therefore, most of our methods do not give precise estimates of the asymptotic behavior, and a greedy algorithm does not perform asymptotically as good as the optimal solution. Following Lueker [LUE81] and Bollobas [BOL85], we first note that a random (unweighted) graph possesses almost surely a k-clique if the probability of an edge p is not smaller than $n^{-2/(k-1)} + \epsilon$, where $\epsilon > 0$. So by our previous discussion

$$(a.s.) \quad C_k^2 \cdot R^{-1}(n^{-2/(k-1)}) \leq Z_{max} \leq R_{C_k^2}^{-1}(1/C_n^k) \quad (pr.)$$

We recall that $R_{C_k^2}(\cdot)$ represents the C_k^2 convolutions of the weight distributions. For example, the following can be derived from the above:

- for the gamma distribution *gamma* (β, λ) we have $Z_{max} \sim k \log n$ (pr.)

- for the normal distribution $N(\mu, \sigma)$ one proves $Z_{max} \sim k\sigma\sqrt{(k-1)\log n}$ (pr.)

- for the uniform distribution $U(0, 1)$ we show $[(k!C_k^2)^{1/C_k^2} - 1/(C_k^2 + 1)] \cdot n^{-\frac{2}{k-1}} \leq EZ_{min} \leq C_k^2 \cdot n^{-\frac{2}{k-1}}$

Now we investigate the greedy approach and suggest a practical solution for this problem. One can invent at least three different greedy approaches, and below we present one of them.
Greedy 2.

begin
select an edge with the maximum weight
do for $i = 2$ **to** k
select i **edges of the maximum** *total* **weight and add them to subgraph formed in the first** $(i-1)$**st steps**
end

For the gamma distribution one finds $Z_{grd} \sim k \log n \sim Z_{max}$, so the greedy asymptotically performs as good as the optimal solution. This is not, however, any longer true for the normal distribution. The greedy algorithm gives in this case $Z_{grd} \sim 2 + \sum_{i=2}^{k-1} \sqrt{2i} \cdot \sqrt{\log n}$, which agrees with Lueker's result [LUE81]. Finally for the uniform distribution, we find the *lightest* k-clique. After some algebra we come to the following solution $EZ_{grd} \sim 1/n^2 + \sum_{l=1}^{k-1}(l!/n)^{1/l}(1 - 1/(l+1))$.

PROBLEM 4. *Location Problem on Graphs*

We *do* restrict our investigation to a location problem *on graphs*. The objective function is as follows $Z_{max} = \max_{\alpha \in B_n}\{\sum_{i \in \mathcal{M}-\alpha} \max_{j \in \alpha}\{w_{ij}\}\}$ where $\mathcal{M}$ represents the set of all vertices, and a feasible solution $\alpha = (c_1, c_2, ..., c_L)$ consists of L selected vertices that maximizes the distance between them and all other vertices in the graph. Naturally, $|B_n| = \binom{n}{L} \sim n^L/L!$ for bounded L. Let us define $W_i(\alpha) = \max_{j \in \alpha}\{w_{ij}\}$, and we note that $W_i(\alpha)$ is i.i.d., random sequence. The distribution function $F_W(x)$ of $W_i(\alpha)$ is equal to $F^L(x)$, since α has cardinality L. If the weights are exponentially distributed, then $EW = H_L$ [GAL87] where H_L is the L-th harmonic number [KNU73]. Next, we note that $m = |B_n| = n^L/L!$, so m is polynomially related to n, and one may consider applying Theorem 6, since $N = |\mathcal{M} - \alpha| = n - L \to \infty$ as $n \to \infty$. In summary, by Theorem 6 we obtain

$$Z_{max} \sim Z_{min} \sim (n-L)EW + \sigma_W\sqrt{2nL\log n}$$

providing $EW \neq 0$. In particular, $Z_{max} \sim Z_{min} \sim (n-L)H_L/\lambda$ for the exponential distribution of weights.

REFERENCES

AHU74] Aho, A., Hopcroft, J. and Ullman, J., *The design and analysis of computer algorithms*, Addison-Wesley, Reading (1974).

[ALD89] Aldous, D., *Probability approximations via the Poisson clumping heristics*, Springer Verlag, New York 1989.

[BOL85] Bollobas, B., *Random Graphs* Academic Press, London (1985).

[BOR62] Borovkov, A., A probabilistic formulation of two economic problems, *Soviet Mathematics,* 3, pp. 419–427 (1962).

[FEL71] Feller, W., *An introduction to probability theory and its applications,* Vol. II., John Wiley & Sons, New York (1971).

[FHR87] Frenk, J., van Houweninge, M. and Rinnooy Kan, A., Order statistics and the linear assignment problem, *Computing,* 39, pp. 165–174 (1987).

[GaG78] Garfinkel R., Gilbert K., The bottleneck traveling salesman problem: Algorithms and probabilistic analysis, *Journal of the ACM,* 25, 435-448 (1978).

[GAL87] Galambos, J., *The asymptotic theory of extreme order statistics,* R.E. Krieger Publication Comp., Malabar 1987.

[HoS86] D. Hochbaum and D. Shmoys, A Unified Approach to Approximate Algorithms for Bottleneck Problems, *J. of the ACM,* 33, 533-550, 1986.

[JaS91] Jacquet, P., and Szpankowski, W., Analysis of digital tries with Markovian dependency, *IEEE Trans. on Information Theory,* 37, 5, 1991.

[KAR76] Karp, R., The probabilistic analysis of some combinatorial search algorithms, in: *Algorithms and Complexity,* (ed. J.F. Traub), Academic Press, New York (1976).

[KNU73] Knuth, D., *The art of computer programming. Sorting and searching,* Addison-Wesley, Reading (1973).

[LaR77] Lai, T. and Robbins, H., A class of dependent random variables and their maxima, *Z. Wahrschein-hich,,* 42, pp. 89–111 (1978).

[LLK85] Lawler, E., Lenstra, J.K., Rinnooy Kan, A.H., and Shmoys, D., *The traveling salesman problem,* John Wiley & Sons, Chichester (1985).

[LUE81] Lueker, G., Optimization problems on graphs with independent random edge weights, *SIAM J. Computing,* 10, pp. 338–351 (1981).

[PAP81] Papadimitriou, C., Worst-case and probabilistic analysis of a geometric location problem, *SIAM J. Computing,* 10, pp. 542–557 (1981).

[SZP88a] Szpankowski, W., Some results on asymmetric V-ary tries, *J. Algorithms,* 8, 224 - 244 (1988).

[SZP88b] Szpankowski, W., The evaluation of an alternating sum with applications to the analysis of some data structures, *Information Processing Letters,* 28, 13-19, (1988).

[SZP90] Szpankowski, W., Another unified approach to some bottleneck and capacity optimization problems, Purdue University, CSD TR-1022, 1990.

[SZP91] Szpankowski, W., Height of a digital trees and related problems, *Algorithmica,* 6, pp.256-277 (1991).

[UKK90] Ukkonen, A linear-time algorithm for finding approximate shortest common superstring, *Algorithmica,* 5, 313-323 (1990).

[WAL79] Walkup, D., On the expected value of a random assignment problem, *SIAM J. Computing,* 8, pp. 440–442 (1979).

[WEI80] Weide, B., Random graphs and graph optimization problems, *SIAM J. Computing,* 9, pp. 552–557 (1980).

Combinatorics and Algorithms of Geometric Arrangements

Leonidas J. Guibas
Standford University

ABSTRACT

In this talk we survey the most important combinatorial and algorithmic results on arrangements. Arrangements are the subdivisions of Euclidean space defined by collections of algebraic manifolds. They have turned out to be a crucial concept in Computational Geometry with numerous unforeseen applications to a wide variety of fundamental geometric problems. We will discuss some key areas, such as zone and lower envelope theorems, many cell problems, and point location. Many of the fundamental algorithmic techniques in geometry can be illustrated on problems in arrangements. Examples include randomized algorithms and epsilon-nets, parametric search, partition trees, and duality.

An Analysis of Randomized Shear Sort on the Mesh Computer

(Extended Abstract)

Susumu Hasegawa, Hiroshi Imai and Koji Hakata

Department of Information Science, University of Tokyo
Hongo, Tokyo 113, Japan

Abstract

A simple and practical randomized sorting algorithm on $n \times n$ mesh computer is proposed by Iwama, Miyano and Kambayashi [3] without giving any rigorous analysis on its randomized complexity. This algorithm can be regarded as a randomized version of the shear sort [7], which requires $O(\log n)$ iterations of row and column sorting operations and in total $O(n \log n)$ parallel time. We prove that the randomization improves the time complexity from $O(n \log n)$ to $O(n \log \log n)$. Although our analysis is insufficient about the constant factor, experimental results of the algorithm are fairly good even compared with $O(n)$ algorithm using only row and column sorting [5]. Moreover since the policy of this algorithm is very simple, it can easily be extended to the three-dimensional case.

1. Introduction

Sorting is the most fundamental and useful problem in computer science, and there have been developed many efficient sequential and parallel algorithms for sorting. In this paper, we are interested in parallel sorting algorithms which are simple enough and useful in practice. Parallel algorithms depend highly upon parallel computation models. PRAM is the most powerful model, and is widely used as a standard model to discuss the inherent parallel complexity. In 1980's, optimal parallel sorting algorithms for N numbers are given which run in $O(\log N)$ time using $O(N)$ processors on PRAM (e.g., [1]). However, these algorithms are fairly complicated, and have large constant factors in the time complexity function. Also, PRAM might be too powerful to realize as it is.

This paper aims at developing simple sorting algorithms on a much simpler parallel computation model, the two-dimensional mesh connected computer. This model is often slower than other machines such as the hypercube, but is ideally suited to VLSI implementation. Considering VLSI implementation, the simplicity of the algorithm is quite important.

For the two-dimensional mesh computer, there have been proposed many sorting algorithms. Thompson and Kung [8] give the first $O(n)$ parallel time algorithm on the $n \times n$ mesh machine. Later, several $O(n)$ time algorithms are proposed, among which Schnorr and Shamir [7] present an optimal algorithm with $3n + o(n)$ parallel steps. However, the former algorithm is based on the divide-and-conquer principle, and is rather complicated. The latter algorithm is non-recursive, but its low order term is relatively large for moderate values of n.

Schnorr and Shamir [7] also present two simple algorithms which use only row and column operations. Roughly, these algorithms alternately sort the rows and columns. Since sorting rows (columns) can be performed in a completely synchronized manner and there are only sorting operation as well as its slight variants for each row and column, the algorithms are very simple. Among the two algorithms, one is called shear sort, and requires $\lceil \log n \rceil$ steps of alternately sorting rows and columns. The other is called revsort, and requires $\lceil \log \log n \rceil$ steps plus 3 iterations of a step of the shear sort. The revsort additionally uses shift operations for rows and columns. In 1986 Marberg and Gafni [5] present an $O(n)$ parallel time sorting algorithm, named rotatesort, which also uses only row and column operations. Iwama, Miyano and Kambayashi [2] describe a randomized sorting algorithm, which we call algorithm RS here since it may be viewed as a randomized version of the shear sort. In that paper, they consider another much complicated algorithm with its complexity analysis, and do not give any analysis on this algorithm.

In this paper we analyze the randomized algorithm RS in detail, and show that by randomization the $O(n \log n)$ parallel time complexity of the shear sort is improved to $O(n \log \log n)$. Our analysis is insufficient concerning constant factors and analytical time complexity is worse than revsort. However, experimental results are fairly good and even better than rotatesort when n is of moderate size. We also extend the algorithm to the three-dimensional case.

The paper proceeds as follows. In section 2.1 we analyze the complexity of algorithm RS and prove that randomization improves the computational time from $O(n \log n)$ to $O(n \log \log n)$. In section 2.2 experimental results are compared with our analysis and other existing algorithms. In section 3 we extend RS to $n \times n \times k$ three-dimensional case and prove that the extended algorithm is also an $O(n \log \log n)$ algorithm.

2. Algorithm RS
2.1. Analysis of RS

We now describe the sorting algorithms on the mesh machine. Initially, n^2 numbers are given, one at each processor of the $n \times n$ mesh computer. We consider sorting these numbers in the row-major snake-like order as shown in Figure 2.1(e). Rows (columns) can be sorted simultaneously in $O(n)$ parallel time. In the sequel, we suppose that sorting rows (columns) can be performed in a unit time, and will be concerned with the number of times row and column operations are executed. The shear sort [7], to be called algorithm S, can be described as follows:

Algorithm S: [7]
1. Shear the rows (sort odd-numbered rows to the right and sort even-numbered rows to the left);
   Check if sorting is done; if yes, halt;
2. Sort each column from the top to the bottom;
   Return to 1;

$$
\begin{array}{cccc}
3 & 16 & 12 & 4 \\
1 & 9 & 5 & 14 \\
11 & 8 & 2 & 7 \\
10 & 6 & 13 & 15 \\
\end{array}
\text{(column sort)} \Longrightarrow
\begin{array}{cccc}
1 & 6 & 2 & 4 \\
3 & 8 & 5 & 7 \\
10 & 9 & 12 & 14 \\
11 & 16 & 13 & 15 \\
\end{array}
\text{shear} \Longrightarrow
$$

(a)   (b)

$$
\begin{array}{cccc}
1 & 2 & 4 & 6 \\
8 & 7 & 5 & 3 \\
9 & 10 & 12 & 14 \\
16 & 15 & 13 & 11 \\
\end{array}
\text{column sort} \Longrightarrow
\begin{array}{cccc}
1 & 2 & 4 & 3 \\
8 & 7 & 5 & 6 \\
9 & 10 & 12 & 11 \\
16 & 15 & 13 & 14 \\
\end{array}
\text{shear} \Longrightarrow
\begin{array}{cccc}
1 & 2 & 3 & 4 \\
8 & 7 & 6 & 5 \\
9 & 10 & 11 & 12 \\
16 & 15 & 14 & 13 \\
\end{array}
$$

(c)   (d)   (e)

Figure 2.1 How algorithm S proceeds.

An example is given in Figure 2.1. Within $\lceil \log n \rceil$ iterations, algorithm S completes sorting [7], where throughout the paper log and ln denote the logarithm of base 2 and e, respectively.

Another sorting algorithm for $n \times n$ mesh computer is proposed by Iwama, Miyano and Kambayashi [3]. The algorithm is a randomized algorithm, but no detailed analysis of the randomized time complexity is given. This algorithm, we call it algorithm RS in this paper, is as follows.

Algorithm RS: [3]

1. Randomize each row;
2. Sort each column from the top to the bottom;
3. Shear rows;
   Check if sorting is done; if yes, halt;
4. Sort each column from the top to the bottom;
   Return to 1;

Figure 2.2 shows how algorithm RS proceeds. Randomizing each row in step 1 means that numbers of each row are shuffled randomly. We can get the same effect if each processor produces a random number, and emulate row sorting of these numbers.

RS can be regarded as a randomized version of algorithm S. We will prove that the randomization improves the time complexity from $O(n \log n)$ to $O(n \log \log n)$ in this section.

We would like to analyze sorting of n^2 general numbers (real, integer, etc.), though, the following lemma reduces the problem to that for n^2 binary numbers.

Lemma 2.1 [2] If any n^2 binary numbers (0 or 1) can be sorted on the mesh machine using row sorting in the row-major snake-like order in $t(n)$ time with probability $1 - p(n)$, for the general number sorting problem, every number goes to its correct row in $t(n)$ time with probability $1 - np(n)$. □

Based on the above lemma we will only investigate the problem of sorting n^2 binary numbers on the $n \times n$ mesh machine. We call a row *dirty* if there are both 0 and 1 in the row. A row which is not dirty is called *clean*. If the total number of dirty rows becomes at most one, sorting is over with one more row sorting.

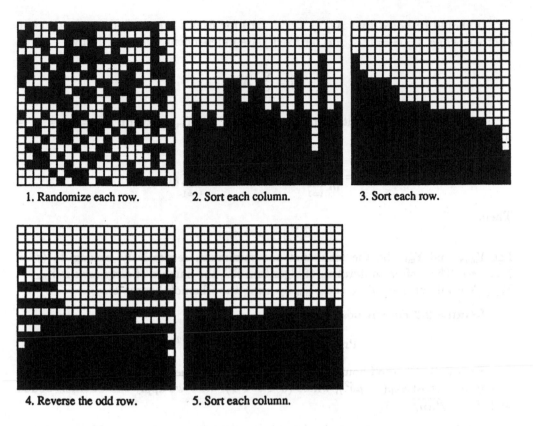

1. Randomize each row.   2. Sort each column.   3. Sort each row.

4. Reverse the odd row.   5. Sort each column.

Figure 2.2 How algorithm RS proceeds. (0 is represented by a black box and
1 is represented by a white box.)

By the algorithms considered here, once a row becomes clean, the row will never
become dirty again. Also, regarding all the rows in the beginning as dirty, all dirty
rows are consecutive at any time due to column sorting. Lower and upper rows become
clean first, and then middle rows become clean. Hence, at each iteration, it suffices to
evaluate how the number of dirty rows will be reduced. We extract a submesh consisting
of dirty rows at each stage, and consider the sorting problem for the submesh. Let m
be the number of dirty rows at some stage, i.e., the number of rows in the submesh.
Initially, m is n.

Consider the $m \times n$ submesh computer. Suppose that rows are re-indexed from
1 to m. In the beginning of the stage, let n_i be the number of 1's in row i, and let
$p_i = n_i/n$. By randomizing rows, the binary numbers are distributed at each row and
events on different rows are independent.

The probabilistic behavior of the algorithm can be modeled as follows. Let X_{ij} be
a random variable, taking 0 or 1 as its value, at the (i,j)-element of the mesh computer.

It should be noted that X_{ij} $(j = 1, \ldots, n)$ for fixed i are not independent to one another, while X_{ij} $(i = 1, \ldots, m)$ for fixed j are independent. Defining Y_j by

$$Y_j = \sum_{i=1}^{m} X_{ij},$$

Y_j is the sum of independent Bernoulli trials $X_{1j}, \ldots, X_{mj}$ with $\Pr(X_{ij} = 1) = p_i = n_i/n$. Define μ and p by

$$p = \frac{1}{m} \sum_{i=1}^{m} p_i, \qquad \mu = mp = \sum_{i=1}^{m} p_i.$$

Then,

$$E[Y_j] = \mu = mp.$$

Let $Y_{\max}$ and $Y_{\min}$ be the maximum and minimum, respectively, among Y_j $(j = 1, \ldots, n)$. Then, after an iteration of algorithm RS, the number of dirty rows is $Y_{\max} - Y_{\min}$. We will estimate $Y_{\max}$ and $Y_{\min}$ by using the Chernoff bound [6].

Lemma 2.2 For any positive constant c,

$$\Pr(Y_j < \mu - c\sqrt{\mu \ln n}) < n^{-c^2/2}.$$

Proof: By Chernoff bound [6], the number of successes in Bernoulli trials is below $\mu - \delta\mu$ is at most $\exp(-\frac{1}{2}\mu\delta^2)$, where $(0 < \delta < 1)$. Now we apply the Chernoff bound with $\delta = c\sqrt{\ln n/\mu}$. $\square$

Let $\bar{\mu} = m - \mu = m(1 - p)$. Applying Lemma 2.2 for a random variable $\bar{Y}_j = 1 - Y_j$ with expectation $\bar{\mu}$, we have the following.

Lemma 2.3 For any positive constant c,

$$\Pr(Y_j > \mu + c\sqrt{\bar{\mu} \ln n}) < n^{-c^2/2}. \quad \square$$

Lemma 2.4 An iteration of randomizing rows and sorting columns in algorithm RS reduces the number m of dirty rows to $\sqrt{(2 + \epsilon)m \ln n}$ with probability at least $1 - 2n^{-1-\epsilon}$.

Proof: Setting $c = \sqrt{4 + 2\epsilon}$ in Lemmas 2.2 and 2.3 and considering that here the union of $2n$ events are taken below, we obtain the following for any positive constant ϵ:

$$\Pr\Big((Y_{\min} < \mu - \sqrt{(4 + 2\epsilon)\mu \ln n}) \cup (Y_{\max} > \mu + \sqrt{(4 + 2\epsilon)\bar{\mu} \ln n})\Big)$$

$$= \Pr\Big(\bigcup_{j=1}^{n} [(Y_j < \mu - \sqrt{(4 + 2\epsilon)\mu \ln n}) \cup (Y_j > \mu + \sqrt{(4 + 2\epsilon)\bar{\mu} \ln n})]\Big) \leq 2n^{-1-\epsilon}.$$

This implies that, by steps 1 and 2 of RS, m is reduced to

$$\sqrt{(4 + 2\epsilon)m \ln n}(\sqrt{p} + \sqrt{1 - p}) \leq \sqrt{(8 + 4\epsilon)m \ln n}$$

with probability at least $1 - 2n^{-1-\epsilon}$, where the last inequality follows from Lemma 2.5 below. Since steps 3 and 4 reduce the number of dirty rows by halves [7], the lemma is proved. □

Lemma 2.5 For $0 < a \le b$ and $0 \le \alpha \le \beta \le 1$,

$$\max\{\sqrt{ax} + \sqrt{b(1-x)} \mid \alpha \le x \le \beta\} = \begin{cases} \sqrt{a\alpha} + \sqrt{b(1-\alpha)} & a/(a+b) < \alpha \\ \sqrt{a+b} & \alpha \le a/(a+b) \le \beta \\ \sqrt{a\beta} + \sqrt{b(1-\beta)} & a/(a+b) > \beta. \end{cases} \quad □$$

Theorem 2.1 In algorithm RS, $2\lceil \log\log n \rceil + 5$ iterations complete sorting n^2 general numbers with probability at least $1 - 2\lceil \log\log n \rceil n^{-3.5}$

Proof: By the above lemma with $\epsilon = 3.5$ each iteration succeeds to reduce dirty rows from m to $\sqrt{5.5m \ln n}$ with probability at least $1 - 2n^{-4.5}$. Therefore with probability at least $1 - 2\lceil \log\log n \rceil n^{-4.5}$ each of the first $\lceil \log\log n \rceil$ iterations succeeds, and the number of dirty rows becomes

$$(5.5 \ln n)^{1/2 + 1/2^2 + \cdots + 2^{\lceil \log\log n \rceil}} n^{1/2^{\lceil \log\log n \rceil}} \le 16 \log n.$$

Since steps 3 and 4 reduce the number of dirty rows by halves, additional 4 iterations complete the sorting of binary numbers, and thus by Lemma 2.1, general number sorting completes within another step with probability at least $1 - 2n^{-\epsilon}$. □

The experimental result in section 2.2 will reveal that our probabilistic analysis is insufficient. However, more detailed analysis may improve the time complexity. For example, if $Y_{med} = (Y_{max} + Y_{min})/2$ holds after step 2 of the algorithm RS, where Y_{max}, Y_{med} and Y_{min} are the maximum, median and minimum respectively of the number of 0's in each column, the number m of dirty rows reduces to $m/4$, less than the bound $m/2$ for shear sort. In fact Y_{med} is near $(Y_{max} + Y_{min})/2$ with high probability. To improve the time complexity by using such analysis is a future work.

2.2. Computational experiments

We performed computational experiments of algorithm RS with several mesh sizes. Sorting n^2 general numbers are executed with $n = 2^i$ ($i = 4, \cdots, 10$) 100 times for each n and the numbers of row or column sortings are counted. The result is shown in Table 2.1.

In Figure 2.3 these results are compared with following three cases.
1. REVSORT ($4\lceil \log\log n \rceil + 6$ row or column operations)
2. ROTATE SORT (14 row or column operations)
3. RS the result of section ($8\lceil \log\log n \rceil + 20$ row or column operations with probability at least 0.99996)

The analytical probabilistic bound derived in section 2.1 is worst among the bounds shown in Figure 2.3. However, the experimental result for RS is much better, and is better than the bound for revsort. It is even better than rotatesort for $n \le 32$.

Table 2.1. The maximum m_{max}, average m_{av} and minimum m_{min} numbers of column or row sorting required to complete the n^2 general numbers for sorting 100 different test sets by RS.

n	32	64	128	256	512	1024
m_{max}	15	15	15	15	15	15
m_{av}	12.36	14.52	15.00	15.00	15.00	15.00
m_{min}	11	11	15	15	15	15

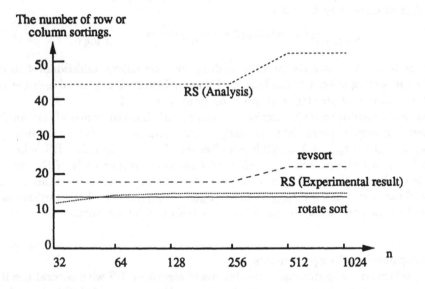

Figure 2.3 Comparison of the number of row or column sortings of RS and several algorithms.

3. Extension of RS to layered mesh computer

Since the policy of RS is very simple, we can easily extend RS to the three-dimensional case. We consider the k-layered mesh machine, each layer constituting the $n \times n$ two-dimensional mesh computer. The processors of each layer are indexed from $(1,1)$ to (n,n). We call the k processors of same position (i,j) in each layer (i,j)-tower. Two processors of consecutive layers can communicate in $O(1)$ time, if these processors are in the same tower. We can then assume that each tower can be sorted in $O(k)$ time. Our goal is, when each processor has a general number, sorting them such that every number of a layer is larger than the one in the lower layer and each layer is sorted row-major snake-like order.

The three-dimensional algorithm we propose is as follows:

Algorithm 3DRS:

Phase A

Repeat $12 + \lceil 3.5 \log \log k \rceil$ times from step 1 to step 4.

1. Randomize each layer. This can be done by randomizing each row and then randomizing each column;
2. Sort each tower from the top to the bottom;
3. In each layer sort each column from the top to the bottom and sort each row from the left to the right; then for the i-th layer rotate the layer by $\frac{\pi}{2} \times (i \bmod 4)$;
4. Sort each tower from the top to the bottom; Go to 1;

Phase B

5. Sort each layer using rotatesort from the largest to the smallest in odd-numbered layers, and from the smallest to the largest in even-numbered layers; then check to see if sorting is over. If so, the algorithm terminates;
6. Sort each tower from the top to the bottom; then go to step 5;

Similar to the two-dimensional case, we can only consider binary data. That is, the following lemma holds.

Lemma 3.1 If any $n \times n \times k$ binary numbers can be sorted on the k-layered mesh machine in t time with probability $1 - p$, then the general numbers go to the correct layers in t time with probability at least $1 - kp$. $\square$

Now we turn to the analysis of 3DRS for binary data. We call a layer is *dirty* if both 0 and 1 are in the layer. Similar to RS the following two lemma can be obtained.

Lemma 3.2 Step 1 and step 2 of 3DRS reduce the number m of dirty layers to $\sqrt{(8 + 4\epsilon)m \ln n}$ with probability at least $1 - 2n^{-\epsilon}$. $\square$

Lemma 3.3 Step 3 and step 4 of 3DRS reduce the number m of dirty layers to $3\lceil m/4 \rceil + 1$ deterministically.

Proof: Consider the $n \times n \times m$ subsystem consisting of m dirty layers. Let $h(i,j)$ be the number of 0's in (i,j)-tower, and $\alpha = h(\lceil n/2 \rceil, \lceil n/2 \rceil)$. Further, let $\bar{i} = n - i$. After step 2, owing to the sorting we have $h(1,1) = 0$ and $h(n,n) = m$, and

$$h(i,j) \leq h(i, j+1), \ h(i+1,j) \quad (\forall i,j)$$

Therfore

$$h(i,j) \leq \alpha \quad (\forall (i,j) \in S)$$

$$h(i,j) \geq \alpha \quad (\forall (i,j) \in L)$$

where $S = \{(i,j) \mid i,j \leq \lceil n/2 \rceil\}$ and $L = \{(i,j) \mid i,j \geq \lceil n/2 \rceil\}$. By step 3, 0's in $h(i,j)$ is distributed to $(i,j),(\bar{i},j),(i,\bar{j})$ and $(\bar{i},\bar{j})$. Therefore after step 3 about 1/4 of 0's in towers of $(i,j),(\bar{i},j)$,$(i,\bar{j})$ and $(\bar{i},\bar{j})$ go to (i,j)-tower and the number of 0's become $h'(i,j)$. Since one of the four tower is in S and another is in L, the following relations hold.

$$\left\lfloor \frac{\alpha}{4} \right\rfloor \leq h'(i,j) \leq \left\lceil \frac{\alpha}{4} \right\rceil + 3 \left\lceil \frac{m}{4} \right\rceil$$

Therefore the number of dirty layers becomes at most $3\lceil \frac{m}{4} \rceil + 1$. $\square$

From Lemma 3.3 we can estimate the number of iterations to reduce the number m of dirty layers to a constant number:

Lemma 3.4 After $\lceil \log_{4/3} m \rceil$ iterations of steps 3 and 4 the number m of dirty layers is reduced to 8 deterministically.

Proof: By Lemma 3.3 one iteration of steps 3 and 4 reduce the number m of dirty layers to

$$3\left\lceil \frac{m}{4} \right\rceil + 1 \leq \frac{3}{4}m + 1.75.$$

Therefore $\lceil \log_{4/3} m \rceil$ iterations reduce m to

$$m\left(\frac{3}{4}\right)^{\lceil \log_{4/3} m \rceil} + 1.75\left(1 + \frac{3}{4} + \cdots + \left(\frac{3}{4}\right)^{\lceil \log_{4/3} m \rceil}\right) \leq 8. \quad \square$$

Now from Lemmas 3.2 and 3.4 we can obtain the theorem.

Theorem 3.1 With probability at least $1 - 2\lceil \log\log k \rceil kn^{-2}$, 4 iterations of step 5 and 6 in phase B complete sorting the general number. Therefore algorithm 3DRS sorts n^2 general numbers in $O((n + k)\log\log k)$ time with this probability.

Proof: By Lemma 3.2 with $\epsilon = 2$ the first $\lceil \log\log k \rceil$ iterations reduce the number of dirty layers to at most $32\ln k$ with probability at least $1 - 2\lceil \log\log k \rceil n^{-2}$. Next, from Lemma 3.4 at the end of phase A, by extra $\lceil \log_{4/3}(32\ln k) \rceil \leq \lceil 2.5\log\log k + 11 \rceil$ iterations, the number of dirty layers become at most 8 with the same probability. Then the phase B, which can be regarded as an m-column n^2-row two-dimensional shear sort, completes the sorting with additional 4 iterations or, $4(k + O(n))$ parallel time with probability at least $1 - 2\lceil \log\log k \rceil kn^{-2}$. $\quad \square$

4. Conclusion

We have analyzed randomized sorting algorithms on the mesh computer and prove that the time complexity is $O(n\log\log n)$. Although the constant factor is fairy large, experimental results are much better. Our analysis indicate that even the simplest shear sort or its randomized version has better average complexity, and for n of practical size, this algorithm is comparable to the other existing methods. Further, using the simple policy of algorithm RS, we have extended this algorithm to the layered mesh computer and proved that the extended algorithm also has a time complexity $O(n\log\log n)$.

Acknowledgment

The authors would like to thank Professor Yasuura of Kyoto University for his useful comments. This work was partially supported by the Grant-in-Aid of the Ministry of Education, Science and Culture of Japan.

References

[1] R. Cole: Parallel Merge Sort. *SIAM Journal on Computing*, Vol.17, No.4 (1988), pp.770–785.

[2] S. Hasegawa, H. Imai and K. Hakata: Randomized Sorting Algorithms on the Mesh Computer Using Only Row and Column Sorting. *Technical Report SIGAL 21-1*, IPSJ, May 1991.

[3] K. Iwama, E. Miyano and Y. Kambayashi: A Parallel Sorting Algorithm on the Mesh-Bus Machine. *Technical Report SIGAL 18-2*, IPSJ, November 1990.

[4] D. Knuth: *The Art of Computer Programming: Vol.3: Sorting and Searching.* Addison-Wesley, 1973.

[5] J. M. Marberg and E. Gafni: Sorting in Constant Number of Row and Column Phases on a Mesh. *Proceedings of the 24th Annual Allerton Conference on Communication, Control and Computing*, 1986, pp.603-611.

[6] P. Raghavan: Lecture Notes on Randomized Algorithms. *IBM Research Report RC 15340*, IBM Research Division, 1990.

[7] C. P. Schnorr and A. Shamir: An Optimal Sorting Algorithm for Mesh Connected Computers. *Proceedings of the 18th Annual ACM Symposium on Theory of Computing*, 1986, pp.255-263.

[8] C. Thompson and H. Kung: Sorting on a Mesh-Connected Parallel Computer. *Communications of the ACM*, Vol.20 (1977), pp.263-271.

Efficient Parallel Divide-and-Conquer
for a Class of Interconnection Topologies. [1]

I-Chen Wu

School of Computer Science, Carnegie Mellon University
Pittsburgh, PA 15213, U.S.A.

Abstract: In this paper, we propose an efficient scheduling algorithm for expanding any divide-and-conquer (D&C) computation tree on k-dimensional mesh, hypercube, and perfect shuffle networks with p processors. Assume that it takes t_n time steps to expand one node of the tree and t_c time steps to transmit one datum or convey one node. For any D&C computation tree with N nodes, height h, and degree d (maximal number of children of any node), our algorithm requires at most $(N/p + h)t_n + \varphi d h t_c$ time steps, where φ is $O(\log_2 p)$ on a hypercube or perfect shuffle network and is $O(\sqrt[k]{p})$ on a $n_{k-1} \times \cdots \times n_0$ mesh network, where $n_{k-1} = \cdots = n_0 = \sqrt[k]{p}$. This algorithm is general in the sense that it does not know the values of N, h, and d, and the shape of the computation tree as well, *a priori*. Most importantly, we can easily obtain a linear speedup by nearly a factor of p, especially when $N \gg ph(1 + \varphi d t_c/t_n)$.

1. Introduction

Divide and conquer (D&C) is a common computation paradigm in which the solution to a problem is obtained by solving subproblems recursively. Examples of D&C computations include various sorting methods such as quick sort [8], computational geometry procedures such as convex hull calculation [14], AI search heuristics such as constraint satisfaction techniques [6], adaptive data classification procedures such as generation and maintenance of quadtrees [16], and numerical methods such as multigrid algorithms [12] for solving partial differential equations.

A D&C computation can be viewed as a process of expanding and shrinking a tree. In this paper, we only consider expanding a tree for simplicity of discussion. Each node in the tree corresponds to a problem instance, and expanded children of the node correspond to its subproblems. If a node expands no children, it is a leaf in the tree. At any time, nodes on a wavefront that cuts across paths from the root to leaves can be active in performing the node expansion operation.

At first glance, one might think that it should be straightforward to perform D&C in parallel, because nodes on the wavefront can all be processed independently. However, if one wants to achieve good load balancing between the processors, then parallelizing D&C becomes nontrivial. In fact, doing efficient D&C on any real parallel machines has been a major challenge to many researchers [4, 5, 9, 18] for many years.

The difficulties are due to the fact that many D&C computations are highly dynamic in the sense that these computations are data-dependent. During computation, a problem instance can be expanded into any number of subproblems depending on the data that have been computed so far. In fact, the trees of D&C computations can be expected to be sparse and totally irregular. As a result, load balancing must be adaptive to the tree structure and must be done dynamically at run

[1] This research was supported in part by the Defense Advanced Research Projects Agency, Information Science and Technology Office, under the title *Research on Parallel Computing* issued by DARPA/CMO under Contract MDA972-90-C-0035, ARPA Order No. 7330, in part by the National Science Foundation and the Defense Advanced Research Projects Agency under Cooperative Agreement NCR-8919038 with the Corporation for National Research Initiatives, and in part by the Office of Naval Research under Contract N00014-90-J-1939.

time. This implies that computation loads need to be moved around between processors during computation. This paper presents efficient scheduling algorithms on several different architectures which can achieve good load balancing while minimizing the communication overhead of moving computations around.

1.1. Definitions and Notation

A D&C algorithm described above needs a *node expansion procedure* to *generate* subproblems recursively. Given a problem instance as the input, we can view the computation as a process of expanding a tree by applying the procedure.

A *tree of a computation* is defined as the tree whose nodes have all been expanded by the node expansion procedure. When a node is *expanded*, zero or more children may be *generated*. If a node does not generate any children, the node is a *leaf*; if a node generates one or more (up to d) children, the node is an *internal node*. Each newly generated node will in turn be expanded by some processor in the future. In the tree, the number of nodes, the *height*, and the *degree*, defined as the maximal number of children of any node, are denoted by N, h, and d respectively. A node is said to be at *tree level i* if it is the i-th node on the path from the root to the node. (For example, the root is at level 1.)

For the parallel system that will carry out the computation, we have the following assumptions.

- The parallel system has p processors. By enumerating these processors, we can let P_i denote the $(i + 1)$-st processor in the system, where $0 \le i < p$. (We call i the ID of P_i.) For convenience, $P_{i:j}$ denotes the processors, P_i, P_{i+1}, $\cdots$, P_{j-1}, and P_j.

- It takes t_n *time steps* for a processor to expand a node. Note that for simplicity the time for storing/scheduling a node locally is assumed to be negligible (or is included in the t_n time steps).

- It takes t_c time steps for a sending/receiving processor to send/receive one datum or convey one node over a network connection. We also assume that handling incoming messages *preempts* the node expansion procedure.

  We distinguish between t_n and t_c because t_n can vary a lot depending on the node granularity and t_c can vary a lot depending on the parallel architecture. Also, for simplicity, it takes $O(1)$ time steps for other primitive operations, e.g., context switching and setting a specific time in the future to do some operation.

In this paper, we will consider three different kinds of networks:

1. *k-Dimensional (k-D) Mesh Network:* In this network, there is a k-D array of $n_{k-1} \times n_{k-2} \times \cdots \times n_0(= p)$ processors, where n_i is the size in dimension i. The processor at the coordinate $(i_{k-1}, i_{k-2}, \cdots, i_0)$ is $P_{i_{k-1}n_{k-2}\cdots n_1 n_0 + \cdots + i_1 n_0 + i_0}$, where $0 \le i_j < n_j$ for each j. In each dimension j, the processor is connected to processors at the coordinate $(i_{k-1}, \cdots, i_j \pm 1, \cdots i_0)$, if they exist. For simplicity, assume $n_{k-1} = \cdots = n_0 = \sqrt[k]{p}$. An example is the iWarp system [3] with an 8×8 mesh network. If in each dimension the last element is connected back to the first one the network is called a *k-D torus network*.

2. *Hypercube Network:* Let $p = 2^q$. This network can be viewed as a q-D mesh network with size two in each dimension. Examples are the hypercube systems described in [7, 17].

3. *Perfect Shuffle Network:* Let $p = 2^q$ and $i_{q-1}i_{q-2}\cdots i_0$ be the binary representation of i. Every processor P_i is connected to $P_{i_{q-2}\cdots i_0 i_{q-1}}$, $P_{i_{q-2}\cdots i_0 \overline{i}_{q-1}}$, $P_{i_0 i_{q-1}\cdots i_1}$, and $P_{\overline{i}_0 i_{q-1}\cdots i_1}$, where $\overline{i}_b$ is the complement of i_b.

In a parallel system, a *scheduling algorithm* for a D&C computation schedules frontier nodes on processors for expansion. A *frontier node* is a node which has been generated but has not yet been expanded. The scheduling algorithm proposed in this paper is general in the sense that it does not know the values of N, h and d, and the shape of the computation tree as well, *a priori*.

For the algorithm, the total time for a D&C computation is the *latency* (the total number of time steps) to complete the whole computation on a parallel system. Assume that one processor can do one *work unit* in one time step. The *aggregate overhead* of a communication operation or processor idling refers to the total number of work units which would have been done if the operation had not been done or the processors had not been idle. For example, if one processor executes one operation and all the other processors are idle in some time step, the aggregate overhead for processor idling is $p - 1$ within this time step.

1.2. Previous Work

There have been several approaches in performing parallel D&C. A simple approach (e.g., in [2]) is to expand all the nodes above a fixed level on one processor and then distribute nodes at this level to other processors. Load balancing would be done poorly in this approach when the tree is irregular. Another approach [10, 15, 18] is to distribute generated nodes to balance the load. For this scheme, the communication overhead can be very high. For example, the execution times for the algorithms in [10, 15] are all $O(N(t_n + t_c)/p)$ with high probability.

Recently, some researchers have made efforts to reduce communication overhead. A popular approach [4, 5, 9, 19, 20] is based on the "donate-highest-subtree" strategy, in which an idle processor will be given frontier nodes near the root. Since a subtree rooted near the top usually has many descendants and these descendants may all be expanded locally, this strategy tends to reduce the amount of interprocessor communication. Without using global information, the scheme in [5] with round-robin scheduling requires at most $(N/p + h)t_n + \phi t_c$, where $\phi = O(p^2 h)$, according to our estimation. In contrast, by using global information, Wu and Kung [19] designed an algorithm whose computation time is at most $(N/p + h)t_n + C_{wk} t_c$, where $C_{wk} = O(pdh)$. The computation time for Wu and Kung's algorithm is still high because their algorithm uses one processor to balance the load (for simplicity of their discussion about communication complexity). Obviously, the processor may become a "hot spot". This paper will present an efficient D&C algorithm which uses the *concentration route* technique [11] to reduce the communication overhead on various networks.

1.3. Main Result

Theorem 1 *A scheduling algorithm can be devised such that the total time for a computation tree is*

$$T \leq (N/p + h)t_n + \varphi dh t_c$$

where N, h, and d are the number of nodes, the height, and the degree of the tree respectively; $\varphi = O(\lg p)$ on a hypercube or perfect shuffle network[2] and $\varphi = O(\sqrt[k]{p})$ on a k-D mesh network.

---

[2] $\lg b$ denotes $\log_2 b$

For the upper bound of this theorem, the left item, $(N/p + h)t_n$, represents the time mainly spent for computation, while the right item, φdht_c, represents the time mainly spent for communication overhead. We will examine how close these items are to optimum.

In the left item, the two values Nt_n/p and ht_n are the inherent lower bounds of the computation tree because: (1) p processors require at least Nt_n/p time steps to expand N nodes; (2) it takes at least ht_n time steps to expand h nodes on the farthest path from the root to a leaf. Nt_n/p dominates the item, as N usually is an exponential function of h and therefore is much larger than ph.

For the communication overhead, Wu and Kung [19] have proved that the communication cost is $\Omega(C_{wk})$ if the total computation time ignoring communication overhead is near the optimum Nt_n/p. The *communication cost* is defined as the number of cross nodes, where a *cross node* is a node that is generated by one processor but expanded by another processor. Let the right item of the upper bound in Theorem 1 be rewritten as $\varphi C_{wk}t_c/p$. In this new formula, the value $\Theta(C_{wk}t_c)$ is the optimal aggregate overhead for transmitting cross nodes; the value φ is dependent on the network. It appears that φ is related to a network's diameter, the maximum distance between any pair of processors. This diameter causes longer response time to move a node to an idle processor.

By Theorem 1, we can easily obtain a linear speedup by nearly a factor of p, especially when $N \gg ph(1 + \varphi dt_c/t_n)$. This is not like other theoretical results [10, 15, 20]: a linear speedup by a factor of cp can be obtained, where $c(< 1)$ may be a very small constant. Hence, our algorithm is practical for implementation. We will use this algorithm as a basis to develop a programming system on the 26-processor Nectar system [1] developed at Carnegie Mellon University.

In the rest of the paper, Wu and Kung's algorithm [19] on which our algorithm is based will be reviewed in Section 2. To avoid the "hot spot" problem, our algorithm will use a more sophisticated and efficient control mechanism based on the *concentration route* (see [11]), in Section 3. Finally, this algorithm is proved to satisfy Theorem 1 in Section 4.

2. Wu-Kung's Algorithm

In their algorithm, each processor maintains a *local stack* of frontier nodes for depth-first traversal. For each processor, nodes generated by the processor are pushed onto its local stack. Every processor schedules a node from the top of its local stack (if not empty) for expansion, so nodes at deeper levels of the tree will be expanded by the processor earlier.

In addition to local stacks, the algorithm also uses a data structure, called a *global pool* (abbr. *GP*), to keep track of nodes at a particular tree level which has not been scheduled by any processor for expansion. Note that the GP is maintained by some processor. This level, identified by a variable gl, has the following property: nodes at higher levels have all been scheduled by processors. Every processor will try to take a node from the GP to work on whenever it becomes idle (i.e. its local stack is empty).

Initially, the GP contains only the root and the value of gl is zero. When all the nodes at level gl in the GP have been scheduled by the processors, the GP becomes empty. Then, all the processors are requested to send in frontier nodes at level $gl + 1$ from their local stacks after t_n time steps when all the nodes at level $gl + 1$ have been generated. The GP is refilled with this set of new nodes, and gl is increased by one. This process repeats until all the nodes have been expanded.

For the algorithm, we will do a depth-first traversal on each processor by using its local stack while globally we use the donate-highest-subtree strategy by using the GP. Consequently, the processor will exhaust all possible work locally before asking for a new node (from the GP) which has high probability of having the most descendants. This is the basic reason why Wu and Kung prove that the communication cost can be as low as *pdh*, in spite of the fact that the computation time can be exponential in *h*. Another advantage of this local depth-first strategy is that it uses the minimum amount of memory.

3. Our Algorithm

Our algorithm based on Wu-Kung's algorithm distributes the GP over all the processors to avoid the hot spot problem. Each processor has a local GP, where the frontier nodes of the GP generated on the processor are stored. In order to refill the GP with frontier nodes at the next level, each processor simply moves these nodes from its local stack to its local GP. When a processor runs out of its local stack, the processor will send out a request for a new node from the GP. Such a node request operation can be viewed as a load balancing problem: move a node of the GP from a (high load) processor containing some nodes of the GP to a (low load) processor being idling.

To perform the load balancing operation, the whole system will synchronize once per *period* (whose value will be determined later). The synchronization for load balancing is "passive" in the sense that the operation will be invoked only when the high load status or the low load status on a processor has been changed. The high load status and the low load status are determined as follows.

1. If no node is being expanded and the local stack is empty, the processor is said to have the *low load* status.

2. If the local GP is not empty, the processor is said to have the *high load* status.

Note that a processor can have both low load status and high load status at the same time. If the status of the processor during some period has been changed, the processor will invoke an operation for load balancing at the beginning of the next period.

The main technique used for load balancing is called *token concentration*. Based on the technique, we concentrate nodes from high load processors to a contiguous block of processors starting at P_0. Then we deconcentrate these nodes to low load processors. In Section 3.1, we will describe how to embed on each network a binary tree which will be used by the technique of token concentration, described in Section 3.2. The details of load balancing are described in Section 3.3.

3.1. Embedded Binary Trees on Networks

For load balancing, we want to embed a binary tree of *processor nodes* (abbr. *p-nodes*) on each network. (Note that several p-nodes can be embedded on the same processor.) For each network, the embedded binary tree, called *an Inorder Leaves binary tree* (abbr. *IL-tree*), must satisfy the following properties:

P1 Each p-node above the lowest level has one or two children; each processor at the lowest level has no child. (If one p-node has only one child, we let it be the left child.)

P2 Each edge of the tree requires at most one physical connection of the network.

P3 At each level, a processor cannot be present twice.

P4 All the processors must be present at the lowest level (once) in the order of processor sequence $(P_0, P_1, ..., P_{p-1})$ from left to right.

Property P1 implies that all the leaves are at the lowest tree level. Property P2 ensures that two p-nodes connected by an edge will need at most one hop to communicate with each other. Property P3 ensures that all the processors at the same level can perform in parallel. Property P4 implies inorder leaves.

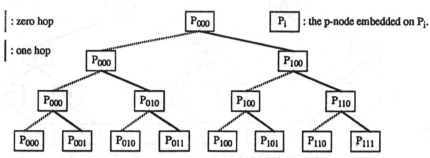

Figure 1: Embedding an IL-tree on a hypercube network for the case $p = 8$.

For the hypercube network, we use the following method to find an IL-tree. Let the root (at level one) be on P_0 and the tree height be $q + 1$. (Recall that $q = \lg p$.) By recurrence, we define that for a p-node on P_i at level $l(\leq q)$ its left child is on $P_{i_{q-1}\cdots i_{q-l+1}0i_{q-l-1}\cdots i_0}$ and its right child is on $P_{i_{q-1}\cdots i_{q-l+1}1i_{q-l-1}\cdots i_0}$. An example for an eight-processor hypercube is illustrated in Figure 1.

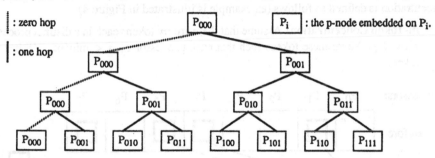

Figure 2: Embedding an IL-tree on a perfect shuffle network for the case $p = 8$.

For the perfect shuffle network, we use the following method to find an IL-tree. Let the root be P_0 and the tree height be $q + 1$. By recurrence, we define that for a p-node on P_i at level $l(\leq q)$ its left child is on $P_{i_{q-2}\cdots i_0 0}$ and its right child is on $P_{i_{q-2}\cdots i_0 1}$. An example for an eight-processor hypercube is illustrated in Figure 2.

For the k-D mesh network, we use the following method to find an IL-tree. Let the root be P_0 and the tree height be $k(n_0 - 1) + 1 = O(n_0)$. (Recall that $n_0 = \sqrt[k]{p}$.) By recurrence, we define that for an internal p-node on P_i at level l, $(m - 1)(n_0 - 1) < l \leq m(n_0 - 1)$, it has the left child on P_i, but it has the right child on $P_{i+n_0^{k-m}}$ only if P_i is at the coordinate $(i_{k-1}, \cdots, i_{k-m}, \cdots, i_0)$ where $i_{k-m} = l - (m - 1)(n_0 - 1) - 1$. An example for a 3×3 mesh network is illustrated in Figure 3.

Note that each of the above trees satisfies properties P1-P4. This can be easily verified by the help of Figures 1, 2, and 3. In addition, for convenience, the height of the IL-tree for each network is denoted by h_{IL}, which is $q + 1$ for the hypercube or perfect shuffle network and is $O(n_0)$ for a k-D mesh network.

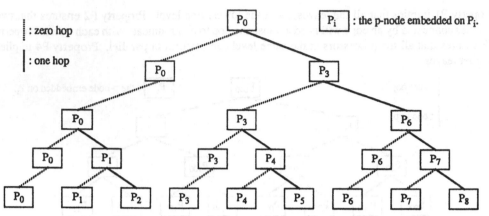

Figure 3: Embedding an IL-tree on a 3×3 network.

3.2. Token Concentration

Token concentration is an important technique used for load balancing (see Leighton's algorithm in Chapter 4 of [13]). Before the operation of token concentration, each processor may create one token. (Note that one processor has no more than one token at any time.) The problem of token concentration is defined as follows (an example is illustrated in Figure 4).

> **m-Token Concentration:** Assume that there are m' tokens each in a distinct processor P_i, $i < p$. Move these tokens such that each processor P_i, $i < \min(m, m')$, has one token.

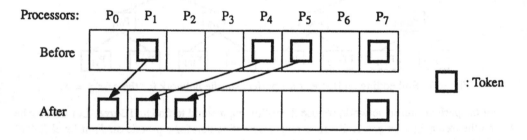

Figure 4: 3-token concentration.

Token concentration consists of two stages: (1) select a destination processor for each token; (2) send the token to the processor. The simplest method for the first stage is that for each token on P_i its destination processor is P_j where j is the number of tokens on processors $P_{0:i-1}$, as illustrated in Figure 4. This method requires a wave-up procedure COUNT and a wave-down procedure LABEL. The second stage uses a route procedure CONCENTRATE. The details of these procedures are described as follows.

COUNT

For each p-node v in the IL-tree, procedure COUNT counts the number S_v of tokens on all the leaves in the subtree rooted at v. For a leaf p-node v, $S_v = 0$ if v has no token; $S_v = 1$ if v has one token. For each internal p-node v, $S_v = S_{lc} + S_{rc}$, where p-node lc (rc) is the left (right) child of v. We can simply let $S_{rc} = 0$ if the right child does not exist. These operations can be done level by level from leaves up to the root in the IL-tree.

The total latency for procedure COUNT is $T_{count} = O(h_{IL}t_c)$ because all processors at the same level can work in parallel by property P3. Hence, the aggregate overhead of the procedure is at most pT_{count}. To reduce the upper bound of the overhead, we let the operation for the procedure be "passive" as described before. Let $t_p(= O(t_c))$ be the maximal propagation time for a processor to receive messages from two children and T_{start} be the starting time of the procedure. Initially, the leaf p-node v will report S_v to his parents only if v has created (or deleted) a token since the previous COUNT for the same type of tokens. (Note we will use two different types of tokens in the next section.) For each internal p-node v at level l, if it receives some values from its children by $T_{start} + (q - l)t_p$, it (except for the root) will report the new S_v to its parent at $T_{start} + (q - l)t_p$. Otherwise, v will do nothing. If for all processors their tokens have not been changed, no operation will be invoked by the procedure. If the number of tokens on a processor has been changed, it will cause at most h_{IL} processors to update their values Ss. Thus, the aggregate overhead of the procedure becomes at most δT_{count}, where δ is the number of processors changing their tokens.

LABEL(m)

After finishing procedure COUNT, we use procedure LABEL to label $m(> 0)$ tokens, each on P_i with $L_i(< m)$, where L_i is the number of tokens in $P_{0:i-1}$. (P_{L_i} is the processor that the token will be sent to.) For each p-node v, let D_v be the number of tokens on $P_{0:j-1}$, by assuming that P_j is the leftmost leaf in the subtree rooted at v. For the root r, let $D_r = 0$. For each p-node v, let $D_{lc} = D_v$ and let $D_{rc} = D_v + S_{lc}$, where lc (rc) is the left (right) child, if any. Obviously, for a leaf v on P_i with a token, $L_i = D_v$. To derive these values D_v, we can do these operations level by level from root down to the leaves in the IL-tree.

The latency for procedure LABEL is $T_{label} = O(h_{IL}t_c)$ because all processors at the same level can work in parallel. Then, the aggregate overhead of the procedure is at most pT_{label}. Again, to reduce the aggregate overhead, we let the parent of each p-node v not send D_v to v if $S_v = 0$ or $D_v \geq m$. Since only the ancestors of the first m tokens need to send D_v down, the aggregate overhead becomes at most mT_{label}.

CONCENTRATE

Procedure CONCENTRATE moves the $m(> 0)$ labeled tokens to a contiguous block of processors $P_{0:m-1}$. More precisely, each token associated with a label $L_i(< m)$ by procedure LABEL will be moved to processor P_{L_i}.

Nassimi and Sahni [11] have designed the collision-free concentration route in $O((\lg p)t_c)$ on hypercube and perfect shuffle networks. For a hypercube network, their method is very simple: at the i-th stage starting at the $(i \cdot 2t_c)$-th time step in this procedure, each processor with a token compares the i-th bits of the current processor's ID and the token's label L_i; if both are not the same, the token will be sent to the other processor in dimension i, otherwise no move is required. They proved that this kind of routing is collision-free (see [11]). So, the latency is $T_{conc} = O((\lg p)t_c)$ and the aggregate overhead for the procedure is at most mT_{conc}. This collision-free concentrate route can also be implemented on a perfect shuffle in the same manner. As for a k-D mesh (or torus) network, we will use the same idea: at the i-th stage, each token is moved to the right position in the i-th dimension. For a k-D torus network, tokens can be sent in the same direction without collision. For a k-D mesh network, since tokens may need to be moved in the opposite directions, collisions may happen. To solve the problem, we divide a stage into two substages and we move tokens in one direction at each substage so that it becomes collision-free. Since each substage takes at most $O(n_0 t_c)$ time steps, for the whole operation the latency is $T_{conc} = O(n_0 t_c)$ and the aggregate overhead is at most $m T_{conc}$. Interestingly, for these three networks, the latencies T_{conc} are all $O(h_{IL} t_c)$ and the aggregate overheads are at most $m T_{conc}$.

We will also need one more procedure, DECONCENTRATE, which just reverses the concentration route. For this procedure, each processor in $P_{0:m-1}$ must have one token associated with the ID of a distinct destination processor. This procedure will move these tokens to the corresponding destination processors. Since this procedure reverses the CONCENTRATE procedure, it is collision free and the time complexity is the same as that for procedure CONCENTRATE.

3.3. Load Balancing

For each period, we will do load balancing between high load processors and low load processors. At the beginning of one period, a high load processor, called a *busy processor* in the period, creates a *B-token* associated with a node selected from the GP unless there has been a B-token. Note that the processor which creates a B-token during the previous period may still has the B-token if the number of high load processors are more than that of low load processors. A low load processor, say P_i, called an *idle processor* in the period, creates a *I-token* with the value i unless there has been an I-token. (Note that we need two different token concentration operations for B-tokens and for I-tokens.) Let p_B (p_I) be the number of busy (idle) processors and $p_{min} = \min(p_B, p_I)$.

The basic idea of load balancing is that during each period we will concentrate B-tokens and I-tokens and then deconcentrate B-tokens by using the values of I-tokens as the IDs of the destination processors. Thus, some p_{min} idle processors will each obtain a node (or a B-token) for expansion.

Before presenting the details of load balancing, we introduce one more relevant procedure, called REFILL, to refill the GP when the GP becomes empty. Initially, the procedure broadcasts a message from the root to each processor. This message causes each processor to increase gl by one. After t_n time steps (the time taken to finish expanding a node), all the nodes above the new level gl must have been expanded, that is, all the nodes at level gl must have been generated. At that time, each processor can refill its local GP with nodes at level gl from its local stack. Since each processor does depth-first search locally, there are at most d frontier nodes at level $l(\leq gl)$. So, each local GP can be refilled with at most d frontier nodes. Then, restart a new period and reset all the tokens.

For a D&C computation tree, initially, we let the GP have the root (on some local GP), gl be 1, and all the local stacks be empty. Then, we create B-tokens and I-tokens to begin the first load balancing period. During each period of time, the load balancing operation uses the following steps.

Step 1: For B-tokens, call procedure COUNT. Then the root in the IL-tree knows p_B.

Step 2: For I-tokens, call procedure COUNT. Then the root knows p_I.

For all processors, if their tokens have been the same as those during the previous period, nothing has been done in these two steps and then all the following time steps can be skipped.

Step 3: If $p_I = p$ and $p_B = 0$ at this time, finish the whole computation. If $p_B = 0$, call procedure REFILL and then restart a new period. (Go back to step 1.)

Step 4: For B-tokens, call procedure LABEL(p_{min}).

Step 5: For I-tokens, call procedure LABEL(p_{min}).

Step 6: For B-tokens, call procedure CONCENTRATE.

Step 7: For I-tokens, call procedure CONCENTRATE.

Step 8: For each of processors $P_{0:p_{min}-1}$, use the value of I-token as the ID of the destination processor and call procedure DECONCENTRATE to deconcentrate B-tokens. Then, each of the first p_{min} idle processors obtains one node (in a B-token) and then removes both I-token and B-token.

Steps 1, 4, and 6 concentrate B-tokens to processors $P_{0:p_{min}-1}$; steps 2, 5, and 7 concentrate I-tokens to processors $P_{0:p_{min}-1}$. Then step 8 deconcentrates B-tokens by I-tokens so that each idle processor obtains one B-token, a node from the GP. As for step 3, we check if the computation is complete or the GP is empty. If the GP is empty, the GP will be refilled with nodes at $gl + 1$ by procedure REFILL.

Note that if all the tokens on processors have been the same as those during the previous period, steps 3-8 can be skipped (see step 2). This is because $p_I = 0$, i.e., $p_{min} = 0$. The reason why $p_I = 0$ is explained as follows. Assume that (1) for all processors their tokens have been the same as those during the previous period (but not for the period before the previous one) and (2) the $p_I \neq 0$ for the previous and current periods. If $p_B = 0$ in the previous period, the computation have been finished or the procedure REFILL has been called by step 3 during the previous period. If the procedure REFILL has been called, I-tokens have been reset. If $p_B \neq 0$ during the previous period, some busy processor must have sent a node to one idle processor. Therefore, the token with the node has been moved out during the previous period. These contradict the first assumption. So, if all the tokens on processors have been the same as those during the previous period, $p_I = 0$.

For the following discussion, we call this period a *refill period* if $p_B = 0$ (see step 3), a *heavy load period* if $p_I \leq p_B(\neq 0)$, or a *light load period* if $p_I > p_B(\neq 0)$. In addition, the time between two consecutive REFILL procedures is called *one round*.

One property is that one round has at most d light load periods. For each light load period, each busy processor must send one node from its local GP to a distinct idle processor; therefore the number of nodes in the local GP will decrease by one. Since each local GP will be refilled with at most d nodes at the beginning of one round, the GP will become empty again (i.e. the round is over) after at most d light load periods.

Now, we can define the time $\mathcal{P}$ for one period as the maximal latency to finish the above operations in steps 1-8 except for procedure REFILL. Since the latency for each step except for procedure REFILL is $O(h_{IL}t_c)$, we know that $\mathcal{P} = O(h_{IL}t_c)$.

4. Proof of Theorem 1

A D&C computation tree of height h requires h rounds. To obtain the total time, we will derive the aggregate overhead for load balancing or processor idling during each round. Note that if a processor is not expanding a node, the processor must get idle or do the load balancing operation. Since each round has only one refill period (including procedure REFILL) and at most d light load periods, the aggregate overhead for these periods is at most $p(t_n + (d + 1)\mathcal{P})$.

Now we want to derive the aggregate overhead for processor idling and load balancing during all the heavy load periods in one round. The processor becomes idle when the local stack is empty and a requested node has not come up yet. When the local stack becomes empty, the processor will wait at most $\mathcal{P}$ time before creating a new I-token in the next load balancing period. In addition, during each heavy load period, each idle processor must be able to receive a new node. Thus, it takes at most $2\mathcal{P}$ time to receive a requested node within heavy load periods. Since we can refill the GP with at most dp nodes at the beginning of one round, at most dp nodes will be requested. Therefore, during all the heavy load periods of one round, the aggregate overhead of processor idling is at most $2dp\mathcal{P}$.

Now, we will consider the aggregate overhead for load balancing. For the ith heavy load period in the round, let p_B^i, p_I^i, and p_{min}^i denote p_B, p_I, and p_{min} in this period respectively. In addition, let δ_B^i (δ_I^i) denote the number of B-tokens (I-tokens) which have been changed. Then, the aggregate overhead for steps 1 and 2 is at most $(\delta_B^i + \delta_I^i)T_{count}$. The aggregate overhead for step 3 is zero if both p_B and p_I are zeros, and is $O(1)$ otherwise. The aggregate overhead for steps 4-8 is at most $p_{min}^i(2T_{label} + 3T_{conc})$. Since $p_{min}^i = p_I^i$ for a heavy load period, the aggregate overhead of load balancing in the period is $(\delta_B^i + \delta_I^i + p_I^i)T_{heavy}$, where $T_{heavy} = O(h_{IL}t_c)$. In total, the aggregate overhead for all the heavy load periods in one round is $T_p \leq \sum_i(\delta_I^i + \delta_B^i + p_I^i)T_{heavy}$. Since the GP has at most dp nodes at the beginning of one round, $\sum_i(\delta_B^i + \delta_I^i) \leq 4dp$. Since during each heavy load period each idle processor will receive a node, $\sum_i p_I^i \leq dp$. Thus, $T_p \leq 5dpT_{heavy}$.

Thus, for each round, the aggregate overhead for load balancing and processor idling is, in total, $pt_n + \varphi dpt_c$, where $\varphi = O(h_{IL})$. Since there are h rounds for the computation, the entire aggregate overhead is at most $pht_n + \varphi pdht_c$. Thus, the total latency for the computation is at most $(Nt_n + (pht_n + \varphi pdht_c))/p = ((N/p + h)t_n + \varphi dht_c)$. □

Acknowledgement

I would like to thank H.T. Kung for his encouragement and discussion for this work. I would also like to thank Greg C. Plaxton for a helpful discussion about the concentration route.

References

[1] E. A. Arnould, F. J. Bitz, E. C. Cooper, H. T. Kung, R. D. Sansom, and P. A. Steenkiste. The design of Nectar: a network backplane for heterogeneous multicomputers. In *Third Intern. Conf. on Architectural Support for Programming Languages and Operating Systems (ASPLOS III)*, Boston, Massachusetts, April 1989.

[2] H.E. Bal. *The shared data-object model as a paradigm for programming distributed systems*. PhD thesis, Vrije Universiteit, Amsterdam, Netherlands, 1989.

[3] S. Borkar, R. Cohn, G. Cox, S. Gleason, T. Gross, H. T. Kung, M. Lam, B. Moore, C. Peterson, J. Pieper, L. Rankin, P. S. Tseng, J. Sutton, J. Urbanski, and J. Webb. iWarp: An integrated solution to high-speed parallel computing. In *Proceedings of Supercomputing '88*, pages 330–339, Orlando, Florida, November 1988. IEEE Computer Society and ACM SIGARCH.

[4] C. Ferguson and R.E. Korf. Distributed tree search and its application to alpha-beta prunning. In *Proceedings of the 7th National Conference on Artificial Intelligence (AAAI 1988)*, pages 128–132, Saint Paul, August 1988.

[5] R. Finkel and U. Manber. DIB – a Distributed Implementation of Backtracking. *ACM Transactions on Programming Languages and Systems*, 9(2):235–256, April 1987.

[6] R.M. Haralick and G.L. Elliott. Increasing tree search efficiency for constraint satisfaction problems. *Artificial intelligence*, 14(3):263–313, October 1980.

[7] J. P. Hayes, T. Mudge, Q. F. Stout, S. Colley, and J. Palmer. A microprocessor-based hypercube supercomputer. *IEEE MICRO*, 6(5):6–17, October 1986.

[8] C.A.R. Hoare. Quicksort. *Computer Journal*, 5:10–15, 1962.

[9] V. Kumar and V. N. Rao. Parallel depth-first search, part I: implementation. *International Journal of Parallel Programming*, 16(6):479–499, 1987.

[10] T. Leighton, M. Newman, A. G. Ranade, and E. Schwabe. Dynamic tree embeddings in butterflies and hypercubes. In *Proceedings of the ACM Symposium on Parallel Algorithms and Architectures*, pages 224–234, June 1989.

[11] D. Nassimi and S. Sahni. Parallel permutation and sorting algorithms and a new generalized connection network. *Journal of the ACM*, 29(3):642–667, July 1982.

[12] J.M. Ortega and R.G. Voigt. Solution of partial differential equations on vector and parallel computers. *SIAM Review*, 27(2):149–240, 1985.

[13] C.G. Plaxton. *Efficient Computation on Sparse Interconnection Networks*. PhD thesis, Dept of CS, Stanford University, September 1989.

[14] F.P. Preparata and M.I. Shamos. *Computational Geometry: an Introduction*. Springer-Verlag, New York, 1985.

[15] A. G. Ranade. Optimal speedup for backtrack search on a butterfly network. In *Proceedings of the ACM Symposium on Parallel Algorithms and Architectures*, pages 40–48, July 1991.

[16] H. Samet. *Applications of Spatial Data Structures: Computer Graphics, Image processing, and GIS*. Addison-Wesley, Reading, MA., 1990.

[17] C. L. Seitz. The cosmic cube. *Communications ACM*, 28(1):22–33, January 1985.

[18] W. Shu and L. V. Kale. A dynamic scheduling strategy for Chare-Kernel system. In *Proceedings of Supercomputing '89*, pages 389–398, New York, NY, November 1989.

[19] I.-C. Wu and H.T. Kung. Communication complexity for parallel divide-and-conquer. To appear in *1991 Symposium on Foundations of Computer Science*, October 1991.

[20] Y. Zhang. *Parallel Algorithms for Combinatorial search problems*. PhD thesis, U.C. Berkeley, November 1989.

OPTIMAL SPECIFIED ROOT EMBEDDING OF FULL BINARY TREES IN FAULTY HYPERCUBES

- Extended Abstract -

M.Y. Chan, F.Y.L. Chin and C.K. Poon
Department of Computer Science
University of Hong Kong
Hong Kong
(mychan@csd.hku.hk, chin@csd.hku.hk, ckpoon@csd.hku.hk)

ABSTRACT

We study the problem of running full binary tree based algorithms on a hypercube with faulty nodes. The key to this problem is to devise an algorithm for embedding a full binary tree in the faulty hypercube. Supposing that the root of the tree must be mapped to a specified hypercube node, we show how to embed an $(n-1)$-tree (a full binary tree with $2^{n-1}-1$ nodes) into an n-cube (a hypercube with 2^n nodes) having up to $n-2$ faults. Our embedding has unit dilation and load, and the result is optimal in the sense that the algorithm is time-optimal, the $(n-1)$-tree is the largest full binary tree that can be embedded in an n-cube, and $n-2$ faults is the maximum number of faults that can be tolerated when the root is fixed. Furthermore, we also show that any algorithm for this problem cannot be totally recursive in nature.

1. Introduction

The hypercube parallel multiprocessor architecture has been the topic of much recent research. It has been shown to be a very versatile architecture [LS] capable of efficiently simulating networks such as rings [CL2], grids [C, CC1, CC2, HJ] and trees [BCLR, BI, CL1, LNRS, MS, WCM, W]. At the same time, the hypercube has been shown to be very robust [A, BS, BCS, HLN, LSGH].

This paper attempts to further demonstrate the versatility and robustness of the hypercube by showing how a full binary tree can be embedded into a faulty hypercube. We consider a strong fault model in which a faulty node can neither compute nor communicate with its neighbors. Thus the embedding should avoid all the faulty nodes and faulty links as well as links which are nonfaulty but adjacent to a faulty node.

2. Definitions and Notations

A *hypercube* of n dimensions (called an n-cube) is an undirected graph of 2^n nodes each having a unique n-bit label. Two nodes are connected by a link if and only if their labels differ in exactly one bit position. We shall refer to nodes by their labels. Furthermore, a node having a 0 (or 1) as the kth bit of its label is said to have a 0 (or 1) in dimension k, with the first bit or dimension taken to mean the leftmost bit. Two nodes differing in the ith dimension only are said to be neighbors of each other on dimension i.

To specify subcubes of the n-cube, we use strings of length n consisting of 0's, 1's and *'s only. A string of length n with exactly m *'s describes an m-cube within an n-cube. For example,

*01** denotes the 3-cube comprised of the eight nodes 00100, 00101, 00110, 00111, 10100, 10101, 10110 and 10111 of a 5-cube.

A *full binary tree* with n levels (or 2^n-1 nodes) is called an *n-tree*. The levels are numbered from 0 to $n-1$, with the root at level 0. Nodes at level i are denoted by strings of length i consisting of L's and R's only. In particular, the empty string ε specifies the root and if T is a string denoting a tree node, LT and RT specify respectively the left and right son of that node.

An *embedding* of a binary tree into a hypercube is a mapping from nodes of the tree to nodes of the hypercube. It is said to have *unit dilation and load* if and only if at most one tree node is mapped to each hypercube node and adjacent nodes in the tree are mapped to adjacent nodes in the hypercube. From now on, *embedding* refers to one with unit dilation and load.

To specify the (unit dilation and load) embedding, we define $H[T]$ as the hypercube node to which tree node T is mapped. As our embedding methods are recursive, it will be convenient to have some notations for specifying the subproblems. Thus we define $C[T]$ as the subcube in which we want to embed the subtree rooted at T. Furthermore, *T-embedding* refers to the embedding of the subtree rooted at T within $C[T]$, with T mapped to $H[T]$.

3. Specified Root Tree Embedding

Theorem 3.1: For all $n \geq 2$ and $0 \leq f \leq n-2$, there exists an $O(2^n)$ time-optimal algorithm to embed an $(n-1)$-tree into an n-cube containing f faulty nodes/links with the root of the tree mapped to a specified nonfaulty node S in the hypercube.

Note that $n-2$ faults are the maximum that can be tolerated when the root is specified. For example, if there were $n-1$ faulty nodes neighboring the specified node S, embedding would be impossible. Theorem 3.2 shows that faults can be ignored if they are far away from the root.

Theorem 3.2: The embedding of any m-tree with root mapped to hypercube node S can never be affected by faulty node F if the Hamming distance between S and F is $\geq m$.

Proof: A level i tree node can only be mapped to a hypercube node at Hamming distance $\leq i$ from S. Thus, the whole m-tree can only be mapped to hypercube nodes at Hamming distance $\leq m-1$ from S. Hence, it is impossible to use F in any embedding. □

To simplify the proof for Theorem 3.1, we assume, without loss of generality, that (i) exactly $n-2$ faults with Hamming distance $\leq n-2$ from S are in the n-cube, (ii) all of the faults are node faults, as link faults can be handled by treating either of the two nodes connected by the link as faulty, and (iii) $S = 0^n$. We prove the theorem by actually constructing an embedding, i.e., performing the ε-embedding with $H[\varepsilon] = 0^n$ and $C[\varepsilon] = *^n$. The construction is based on two techniques: *cube splitting* and *node borrowing*. It can be shown that the embedding algorithm takes time no more than the size of the $(n-1)$-tree, i.e., $O(2^n)$, which is obviously time-optimal.

3.1. Cube Splitting and Node Borrowing

The idea of cube splitting is to map L and R to two nonfaulty neighbors of $H[\varepsilon]$ and then embed the subtrees rooted at L and R into two disjoint subcubes. With $H[\varepsilon] = 0^n$ and $C[\varepsilon] = *^n$, we say that $C[\varepsilon]$ is *split* on dimension i if we set $C[L] = *^{i-1}1*^{n-i}$, $C[R] = *^{i-1}0*^{n-i}$ and $H[L] = 0^{i-1}10^{n-i}$. (In other words, $H[L]$ is the neighbor of $H[\varepsilon]$ on dimension i, $C[L]$ is the subcube not containing $H[\varepsilon]$, and $C[R]$ is the other subcube. See Figure 3.1. Note that $H[L]$ lies in $C[L]$ while $H[R] = 0^{j-1}10^{n-j}$ for some $j \neq i$ must be in $C[R]$.)

By recursively splitting the subcubes on suitable dimensions, we can derive an embedding in most circumstances. This type of embedding, in which the left and right subtrees of every internal node of the $(n-1)$-tree are mapped to disjoint subcubes, is called a *recursive embedding* in [WCM].

We can show that there is a certain fault pattern in which recursive embedding is impossible. Fortunately, we find that only some leaves of the subtree rooted at R cannot be mapped within $C[R]$. We overcome this problem by *borrowing* some nodes from $C[L]$, i.e., mapping these leaves to nonfaulty nodes in $C[L]$. While performing the L-embedding, the borrowed nodes are treated as faults. This avoids mapping the subtree of L to these nodes. As only a few nodes in $C[L]$ are borrowed, the L-embedding is not much affected and hence the whole embedding can be done.

3.2. Constructing the Embedding

We construct the ε-embedding recursively. The base cases, where $n \leq 5$, are considered in [CCP]. For $n \geq 6$, the first step is to find a suitable dimension to split $C[\varepsilon]$. Suppose we split on some dimension i such that $C[L]$ contains p faults and $C[R]$ contains q faults (see Figure 3.1). Then the subtree in $C[L]$ has to avoid p faults while that in $C[R]$ has to avoid q faults and the assigned node $H[\varepsilon]$. Ideally, we want to split $C[\varepsilon]$ so that $p \leq n-3$ and $q \leq n-4$. Then we can perform the L- and R-embeddings recursively. However, this is not always achievable. Thus our strategy is to ensure $p \leq n-3$ first. More precisely, we split $C[\varepsilon]$ on dimension i provided

(A1) $H[L]$ is nonfaulty,

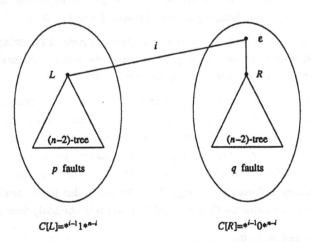

$$C[L]=*^{i-1}1*^{n-i} \qquad\qquad C[R]=*^{i-1}0*^{n-i}$$

Fig. 3.1. Splitting $C[\varepsilon] = *^n$ on dimension i.

(A2) $C[L]$ has at most $n-3$ faults and

(A3) $C[L]$ has as many faults as possible while satisfying conditions (A1) and (A2).

Theorem 3.3 shows that a dimension i which satisfies conditions (A1) and (A2) can always be found. Among all the different possibilities for i, we choose the one which *maximizes* p. Hence, the splitting can always be done.

To illustrate the idea of conditions (A1)-(A3), let us consider the following examples.

Example 3.1: $n = 6$ and there are 4 faults:
$$\beta_1 = 100000$$
$$\beta_2 = 010010$$
$$\beta_3 = 111000$$
$$\beta_4 = 101100$$

By condition (A1), we cannot split on dimension 1 because of β_1. Also, all the remaining dimensions satisfy condition (A2). Among them, we can choose dimensions 2 or 3 because both *1**** and **1*** contains 2 faults. Hence we have $p = 2$ in this case. □

Example 3.2: $n = 6$ and the 4 faults are:
$$\beta_1 = 100011$$
$$\beta_2 = 010011$$
$$\beta_3 = 001011$$
$$\beta_4 = 000011$$

This time, no neighbor of $H[\varepsilon]$ is faulty, thus condition (A1) is always satisfied. However, we cannot split on dimensions 5 and 6 for violation of condition (A2). Among the remaining dimensions, we can choose dimensions 1, 2 or 3 because each of the subcubes 1*****, *1**** and **1*** contains 1 fault. Hence for this example, we have $p = 1$. □

Example 3.3: $n = 6$ and the 4 faults are:
$$\beta_1 = 010000$$
$$\beta_2 = 001000$$
$$\beta_3 = 000100$$
$$\beta_4 = 011000$$

In this case, we cannot split on dimensions 2, 3 and 4 because of condition (A1). We can, however, split on dimensions 1, 5 or 6. Thus $p = 0$ in this case. □

Based on p, we have two cases in general: (1) $0 \leq p \leq 1$ and (2) $2 \leq p \leq n-3$. For case (2) where $2 \leq p \leq n-3$, we have $q = n-2-p \leq n-4$. Hence both the L- and R-embeddings can be done recursively. For case (1) where $0 \leq p \leq 1$, the L-embedding is still easy but the R-embedding is more complicated. There are effectively $q+1 = n-1-p > n-3$ faults in $C[R]$ (the extra fault is from $H[\varepsilon]$) and we cannot perform the R-embedding as simply as in case (2). The details for this case are described in following section.

3.3. Case (1) where $n \geq 6, 0 \leq p \leq 1$.

In this case, we have to choose the dimension on which $C[\varepsilon]$ is split more carefully. As will be shown later, that dimension has to satisfy more contraints besides conditions (A1)-(A3). After that, we split $C[R]$ again so as to separate the $n-1-p > n-3$ faults/assigned nodes. In particular, we ensure that $C[LR]$ contains as many faults as possible while not exceeding its *fault-tolerance capacity* (dimensionality of subcube minus 2). Then the LR-embedding can be done recursively. For the RR-embedding, we split $C[RR]$ again if it still contains too many faults. In general, we keep on splitting $C[R^i]$ until the R^i-embedding can be done recursively.

Before describing the embedding strategy, we shall first arrange the dimensions so that the faults are of a standard pattern (Section 3.3.1). The exact algorithm is then given in Section 3.3.2.

3.3.1. Arranging Dimensions

We choose dimension 1 so that $1*^{n-1}$ contains p faults and 10^{n-1} is nonfaulty. This allows us to split $C[\varepsilon]$ on dimension 1. Consider the remaining $n-2-p$ faults in $0*^{n-1}$. Let there be $n-m$ dimensions k such that $2 \le k \le n$ and $0*^{k-2}1*^{n-k}$ contains all the $n-2-p$ faults. If $n-m > 0$, then let these $n-m > 0$ dimensions be the rightmost $n-m$ dimensions (Figure 3.2a and Example 3.2) so that we can easily choose nonfaulty nodes from $0*^{m-1}0^{n-m}$ during the embedding. In the case where $n = m$, we would instead find a dimension k such that $0*^{k-2}0*^{n-k}$ contains all $n-2-p$ faults and swap it to the rightmost dimension (see Figure 3.2b and Example 3.3). This allows us to find nonfaulty nodes from $0*^{n-2}1$ for embedding. Note that this dimension k may not exist if we have not chosen dimension 1 carefully. However, it can be proved [CCP] that such a dimension k can always be found by properly choosing dimension 1.

The next step is to arrange dimensions 2 through m such that for all i, $2 \le i \le m$, $0^{i-1}1*^{n-i}$ contains as many faults as that in $0^{i-1}*^{k-i}1*^{n-k}$ for all k where $i \le k \le m$. The arrangement allocates more faults to larger subcubes and guarantees that $0^{i-1}1*^{n-i}$ contains no more than $n-i-2$ faults. Hence we can recursively embed an $(n-i-1)$-tree within this $(n-i)$-cube.

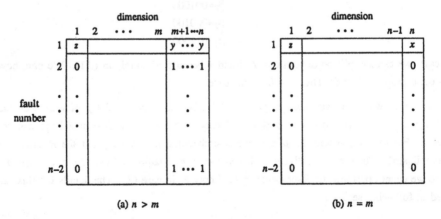

(a) $n > m$ (b) $n = m$

Fig. 3.2. The bit pattern of faults after arranging dimension 1 and the last $\max(n-m,1)$ dimensions. The faults are listed row by row. z is 0 or 1 depending on p is 0 or 1. The value of x and each of the y's can be 0 or 1 subject to other conditions.

Finally, we may need to swap dimensions $\lceil m/2 \rceil$ and $\lceil m/2 \rceil+1$ to guarantee that a particular node used in the embedding is nonfaulty.

More precisely, the dimensions are arranged such that

(B1) subcube $1*^{n-1}$ contains p faults and 10^{n-1} is nonfaulty;

(B2) subcube $\begin{cases} 0*^{m-1}1^{n-m} & \text{if } n > m \\ 0*^{m-2}0 & \text{if } n = m \end{cases}$ contains $n-2-p$ faults;

(B3) for all i, $2 \le i \le m$, subcube $0^{i-1}1*^{n-i}$ contains at least as many faults as in $0^{i-1}*^{k-i}1*^{n-k}$ for all k such that $i \le k \le m$;

(B4) hypercube node $\begin{cases} 0^{\lceil m/2 \rceil}1_{\lfloor m/2 \rfloor}0^{n-m-1}1 & \text{if } n > m \\ 0^{\lceil m/2 \rceil}1_{\lfloor m/2 \rfloor-1}0 & \text{if } n = m \end{cases}$ is nonfaulty.

It is proved in [CCP] that the dimensions can always be arranged to satisfy the above conditions.

3.3.2. Description of the Embedding

We first split $C[R^{i-1}]$ on dimension i for $i = 1, ..., \lfloor m/2 \rfloor$ so that $C[LR^{i-1}]$ equals the $(n-i)$-cube, $0^{i-1}1*^{n-i}$, and $C[RR^{i-1}]$ equals $0^{i-1}0*^{n-i}$. That means the subtree rooted at LR^{i-1} is to be embedded within $0^{i-1}1*^{n-i}$. When splitting $C[R^{i-1}]$, some of the faults within the subcube fall into $C[LR^{i-1}]$. With the arrangement of dimensions in Section 3.3.1, $C[LR^{i-1}]$ gets as many faults as possible but no more than $n-i-2$ (the fault-tolerance capacity of $C[LR^{i-1}]$). Moreover, it does not contain any assigned nodes except $H[LR^{i-1}]$ itself. Hence, the LR^i-embedding can be done recursively.

On the other hand, $C[RR^{i-1}] = 0^{i-1}0*^{n-i}$ will get the remaining faults together with the assigned nodes: $H[\varepsilon]$, $H[R]$, ..., $H[R^{i-1}]$. Note that these are the only assigned nodes we need to take care of while mapping the tree node RR^{i-1} in $C[RR^{i-1}]$. It is because the subtrees rooted at L, LR, ..., LR^{i-1} will not be embedded within $C[RR^{i-1}]$ and the descendants of RR^{i-1} have not been assigned yet. Moreover, it will be proved that these assigned nodes will not fall into $C[LR^{i-1}]$. Consider dimensions $i+1$ to n. We must map RR^{i-1} to a neighbor of $H[R^{i-1}]$ on one of these dimensions. Among them, we choose dimension $m+1-i$ so that the Hamming distances between $H[RR^{i-1}]$ and most of the faults are greater than those between $H[R^{i-1}]$ and the corresponding faults. This way, the roots of lower subtrees will be further and further away from most of the faults.

By the time $i = \lfloor m/2 \rfloor$, there will be approximately $m/2$ assigned nodes and at most $n-2-m/2$ faults within $C[R^{\lfloor m/2 \rfloor}]$. By splitting on dimension n and choosing a suitable $H[RR^{\lfloor m/2 \rfloor}]$ (depending on whether m is even or odd), it can be shown that $C[LR^{\lfloor m/2 \rfloor}]$ contains all the remaining faults while $C[RR^{\lfloor m/2 \rfloor}]$ contains all the assigned nodes. However, many of the faults will be very far from $H[LR^{\lfloor m/2 \rfloor}]$. In fact, when m is even, most of the faults will be too far to affect the embedding. When m is odd, there are also several faults which can be ignored. Anyway, the number of *effective* faults/assigned nodes within each subcube does not exceed their fault-tolerance capacities. Hence the $LR^{\lfloor m/2 \rfloor}$-, and $RR^{\lfloor m/2 \rfloor}$-embeddings can also be done recursively.

Refer to Figure 3.3 for a pictorial view of the general embedding (ε-embedding).

Step (1) For $i = 1, \ldots, \lfloor m/2 \rfloor$, let

$$H[LR^{i-1}] = 0^{i-1}10^{m-2i}01^{i-1}0^{n-m} \qquad\qquad C[LR^{i-1}] = 0^{i-1}1*^{n-i}$$
$$H[RR^{i-1}] = 0^{i-1}00^{m-2i}11^{i-1}0^{n-m} \qquad\qquad C[RR^{i-1}] = 0^{i-1}0*^{n-i}$$

Step (2) For $i = 1, \ldots, \lfloor m/2 \rfloor$, recursively perform the LR^{i-1}-embedding, i.e., embed the $(n-1-i)$-tree rooted at LR^{i-1} into the $(n-i)$-cube $C[LR^{i-1}]$.

Step (3) Suppose $H[R^{\lfloor m/2 \rfloor}] = 0^{\lceil m/2 \rceil}1_1{}^{\lfloor m/2 \rfloor}0^{n-m} = 0^{\lfloor m/2 \rfloor}a_{\lfloor m/2 \rfloor + 1}\ldots a_n$.

Then, we let

$$H[LR^{\lfloor m/2 \rfloor}] = 0^{\lfloor m/2 \rfloor}a_{\lfloor m/2 \rfloor + 1}\ldots a_{n-1}\bar{a}_n,$$
$$C[LR^{\lfloor m/2 \rfloor}] = 0^{\lfloor m/2 \rfloor}*^{n-\lfloor m/2 \rfloor - 1}\bar{a}_n,$$

$$H[RR^{\lfloor m/2 \rfloor}] = \begin{cases} 0^{m/2}a_{m/2+1}\ldots a_{n-2}\bar{a}_{n-1}a_n & \text{for } m \text{ even} \\ 0^{\lfloor m/2 \rfloor}\bar{a}_{\lfloor m/2 \rfloor + 1}a_{\lfloor m/2 \rfloor + 2}\ldots a_n & \text{for } m \text{ odd} \end{cases} \quad \text{and}$$

$$C[RR^{\lfloor m/2 \rfloor}] = 0^{\lfloor m/2 \rfloor}*^{n-\lfloor m/2 \rfloor - 1}a_n.$$

Step (4) Recursively perform the $RR^{\lfloor m/2 \rfloor}$-embedding, i.e., embed the $(n-\lfloor m/2 \rfloor - 2)$-tree rooted at $RR^{\lfloor m/2 \rfloor}$ in the $(n-\lfloor m/2 \rfloor - 1)$-cube, $C[RR^{\lfloor m/2 \rfloor}]$.

Step (5) Recursively perform the $LR^{\lfloor m/2 \rfloor}$-embedding, i.e., embed the $(n-\lfloor m/2 \rfloor - 2)$-tree rooted at $LR^{\lfloor m/2 \rfloor}$ in the $(n-\lfloor m/2 \rfloor - 1)$-cube, $C[LR^{\lfloor m/2 \rfloor}]$ (node borrowing may be required).

There are a number of observations we can make to argue the validity of the above strategy (the proofs of these observations can be found in [CCP]):

(1) $m \geq 3$

(2) For $i = 1, \ldots, \lfloor m/2 \rfloor + 1$, $H[LR^{i-1}]$ and $H[RR^{i-1}]$ are adjacent to $H[R^{i-1}]$;

(3) For $i = 1, \ldots, \lfloor m/2 \rfloor + 1$, $H[LR^{i-1}]$ and $H[RR^{i-1}]$ are nonfaulty;

(4) For $i = 1, \ldots, \lfloor m/2 \rfloor + 1$, $H[LR^{i-1}]$ and $H[RR^{i-1}]$ are distinct;

(5) It is always possible to perform the $RR^{\lfloor m/2 \rfloor}$-embedding;

(6) It is always possible to perform the LR^{i-1}-embedding for $i = 1, \ldots, \lfloor m/2 \rfloor$;

(7) It is always possible to perform the $LR^{\lfloor m/2 \rfloor}$-embedding with node borrowing.

The above observations allow us to conclude that the embedding of an $(n-1)$-tree into an n-cube containing $n-2$ faults, with the root of the tree mapped to a specified nonfaulty hypercube node S, can be done when $n \geq 6, p \leq 1$.

As far as the time complexity is concerned, since there are at most $n-2$ faults each with a n-bit label, the arrangement of dimensions takes at most $O(n^2)$ steps. Let $T(n)$ be the number of steps required to embed an $(n-1)$-tree into an n-cube with $(n-2)$ faults, it can shown easily that $T(n) = O(2^n)$ with the following recurrence,

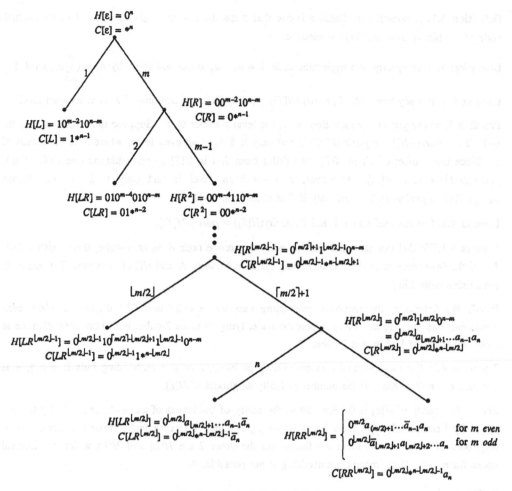

$H[\varepsilon] = 0^n$
$C[\varepsilon] = *^n$

$H[L] = 10^{m-2}10^{n-m}$
$C[L] = 1*^{n-1}$

$H[R] = 00^{m-2}10^{n-m}$
$C[R] = 0*^{n-1}$

$H[LR] = 010^{m-4}010^{n-m}$
$C[LR] = 01*^{n-2}$

$H[R^2] = 00^{m-4}110^{n-m}$
$C[R^2] = 00*^{n-2}$

$H[R^{\lfloor m/2 \rfloor - 1}] = 0^{\lceil m/2 \rceil + 1}1^{\lfloor m/2 \rfloor - 1}0^{n-m}$
$C[R^{\lfloor m/2 \rfloor - 1}] = 0^{\lfloor m/2 \rfloor - 1}*^{n - \lfloor m/2 \rfloor + 1}$

$H[LR^{\lfloor m/2 \rfloor - 1}] = 0^{\lfloor m/2 \rfloor - 1}10^{\lceil m/2 \rceil - \lfloor m/2 \rfloor + 1}1^{\lfloor m/2 \rfloor - 1}0^{n-m}$
$C[LR^{\lfloor m/2 \rfloor - 1}] = 0^{\lfloor m/2 \rfloor - 1}1*^{n - \lfloor m/2 \rfloor}$

$H[R^{\lfloor m/2 \rfloor}] = 0^{\lceil m/2 \rceil}1^{\lfloor m/2 \rfloor}0^{n-m}$
$= 0^{\lfloor m/2 \rfloor}a_{\lfloor m/2 \rfloor + 1}...a_{n-1}a_n$
$C[R^{\lfloor m/2 \rfloor}] = 0^{\lfloor m/2 \rfloor}*^{n - \lfloor m/2 \rfloor}$

$H[LR^{\lfloor m/2 \rfloor}] = 0^{\lfloor m/2 \rfloor}a_{\lfloor m/2 \rfloor + 1}...a_{n-1}\bar{a}_n$
$C[LR^{\lfloor m/2 \rfloor}] = 0^{\lfloor m/2 \rfloor}*^{n - \lfloor m/2 \rfloor - 1}\bar{a}_n$

$H[RR^{\lfloor m/2 \rfloor}] = \begin{cases} 0^{m/2}a_{(m/2)+1}...\bar{a}_{n-1}a_n & \text{for } m \text{ even} \\ 0^{\lfloor m/2 \rfloor}\bar{a}_{\lfloor m/2 \rfloor + 1}a_{\lfloor m/2 \rfloor + 2}...a_n & \text{for } m \text{ odd} \end{cases}$

$C[RR^{\lfloor m/2 \rfloor}] = 0^{\lfloor m/2 \rfloor}*^{n - \lfloor m/2 \rfloor - 1}a_n$

Fig. 3.3. Description of the Embedding

$$T(n) \leq \begin{cases} \sum_{k=\lfloor n/2 \rfloor}^{n-1} T(k) + T(\lfloor n/2 \rfloor) + c_1 n^2 & n > 5 \\ c_2 & n \leq 5 \end{cases}$$

4. Recursive Embeddings

In this section, we study an interesting relationship between our embedding strategies and the *recursive embedding* mentioned in [WCM]. It is easy to see that our specified root embedding is basically a recursive embedding. The only violation is the occasional use of node borrowing. We shall show in Theorem 4.2 that node borrowing is necessary for the specified root embedding to tolerate n–2 faults.

Definition 4.1: A *recursive embedding* is one that maps the left and right subtrees of every internal node of the binary tree into disjoint subcubes.

Definition 4.2: The parity of a hypercube node $A = a_1...a_n$ is defined as $parity(A) = (\sum_{i=1}^{n} a_i) \bmod 2$.

Lemma 4.1: For any tree node T, $parity\,(H[T]) = parity\,(H[\varepsilon])$ if and only if T is at an even level.

Proof: It is easily proved by induction on n, the level number of T. Suppose the lemma is true for $n-1$. Then $parity(H[S]) = parity(H[\varepsilon])$ if and only if S is at an even level, where S is the parent of T. Since the number of 1's in $H[T]$ must differ from that in $H[S]$ by one (dilation one embedding), $parity(H[T]) \neq parity(H[S])$. Moreover, T is at even level if and only if S is not. Hence $parity(H[T]) = parity(H[\varepsilon])$ if and only if T is at an even level. □

Lemma 4.2: For any leaf nodes S and T, $parity(H[S]) = parity(H[T])$.

Lemma 4.3:[WCM] For all $n \geq 2$ and for every hypercube node A in an n-cube, there exists a leaf, T_A, of the $(n-1)$-tree, such that the Hamming distance between A and $H[T_A]$ is at most 2 where H is a recursive embedding.

Proof: By definition, the recursive embedding cuts the n-cube into 2^{n-2} disjoint 2-cubes, each containing one leaf. Hence every hypercube node, lying in some 2-cube, is at Hamming distance at most two from the leaf in that 2-cube. □

Theorem 4.2: For the specified root embedding problem, recursive embedding fails if $n \geq 5$, n is odd and $x = n-2$ where x is the number of faulty neighbors of $H[\varepsilon]$.

Proof: The parity of $H[\varepsilon]$ is 0. For odd n, the parity of the leaves of an $(n-1)$-tree is 1 (by Lemma 4.1). By Lemma 4.3, $H[\varepsilon]$ must be at Hamming distance 1 from a leaf. However, among the n neighbors of $H[\varepsilon]$, $n-2$ of them are faulty and the other 2 are $H[L]$ and $H[R]$ which are internal nodes for $n \geq 4$. Thus, recursive embedding is not possible. □

5. Conclusions

In this paper, we present an optimal embedding algorithm for finding a unit dilation and load embedding of an $(n-1)$-tree into an n-cube that contain some faulty processors and/or links. For the specified root embedding problem, we show that up to $n-2$ faults can be tolerated. This is optimal in the sense that $n-2$ faults are the maximum number of faults that can be tolerated when the root is specified. Moreover, the $(n-1)$-tree is the largest full binary tree that can be embedded into an n-cube even when there are no faults.

For the problem where the root can be mapped to any nonfaulty hypercube node (*variable root embedding problem*), we show in [CCP] that up to $2n-3-\lceil \log n \rceil$ faults can be tolerated. It is not surprising to see that more faults can be tolerated in this problem because there are fewer restrictions on the embedding. Also, our method is classified as a recursive embedding in [WCM]. It was shown in [WCM] that no recursive embedding can tolerate more that $2n-3$ faults in the worst case. Hence our result achieves their bound asymptotically. Recently, [WCM] has derived a non-recursive embedding method that can tolerate $\Omega(n^2/\log n)$ faults.

References

[A]   F. Annexstein. Fault Tolerance of Hypercube Derivative Networks. *Proc. 1st Annual ACM Symp. on Parallel Algorithms and Architectures*, 1986, pp 179-188.

[BS]   B. Becker and H. U. Simon. How Robust is the n-Cube? *Proc. 27th Annual IEEE Symp. on Foundations of Computer Science*, 1986, pp 283-291.

[BCLR]   S. Bhatt, F. Chung, T. Leighton and A. Rosenberg. Optimal Simulations of Tree Machines. *Proc. of 27th Annual Symposium on Foundations of Computer Science*, 1986, pp 274-282.

[BI]   S. N. Bhatt, I. C. F. Ipsen. How to Embed Trees in Hypercubes. *Research Report* YALEU/DCS/RR-443.

[BCS]   J. Bruck, R. Cypher and D. Soroker. Running Algorithms Efficiently on Faulty Hypercubes. *Proc. 2nd Annual ACM Symposium on Parallel Algorithms and Architectures*, 1990, pp 37-44.

[C]   M. Y. Chan. Embedding of Grids into Optimal Hypercubes. *SIAM Journal on Computing, SICOMP 20-5*, October 1991, to appear.

[CC1]   M. Y. Chan and F. Chin. On Embedding Rectangular Grids in Hypercubes. *IEEE Transactions on Computers*, 37, 1988, pp 1285-1288.

[CC2]   M. Y. Chan and F. Chin. Parallelized Simulation of Grids by Hypercubes. *International Computer Symposium 1990*, Taiwan, December 1990, pp 535-544.

[CCP]   M. Y. Chan, F. Chin and C. K. Poon. Optimal Simulation of Full Binary Trees on Faulty Hypercubes. *Technical Report TR-91-06*, July 1991.

[CL1]   M. Y. Chan and S. J. Lee. Fault-Tolerant Embeddings of Complete Binary Trees in Hypercubes, *IEEE Transaction on Parallel and Distributed Systems*, to appear.

[CL2]   M. Y. Chan and S. J. Lee. On the Existence of Hamiltonian Circuits in Faulty Hypercubes, *SIAM Journal on Discrete Mathematics*, to appear.

[HLN]   J. Hastad, T. Leighton, and M. Newman. Fast Computation Using Faulty Hypercubes. *Proc. 21st Annual ACM symposium on Theory of Computing*, 1989, pp 251-284.

[HJ]   C. T. Ho and S. L. Johnsson. Embedding Meshes in Boolean Cubes by Graph Decomposition. *Journal of Parallel and Distributed Computing*, April 1990, pp 325-339.

[LNRS]   T. Leighton, M. Newman, A. G. Ranade and E. Schwabe. Dynamic Tree Embeddings in Butterflies and Hypercubes. *Proc. of 2nd Annual ACM Symposium on Parallel Algorithms and Architectures*, 1989, pp 224-234.

[LS]   M. Livingston and Q. Stout. Embeddings in Hypercubes. *Mathematical Computer Modelling* Vol. 11, pp 222-227.

[LSGH]   M. Livingston, Q. Stout, N. Graham, and F. Harary. Subcube Fault-Tolerance in Hypercubes. *Technical Report CRL-TR-12-87, U. of Michigan Computing Research Laboratory*, September 1987.

[MS]   B. Monien and I. H. Sudborough. Simulating Binary Trees on Hypercubes. *Proc. of the 3rd Aegean Workshop on Computing*, 1988, pp 170-180.

[WCM]   A. Wang, R. Cypher and E. Mayr. Embedding Complete Binary Trees in Faulty Hypercubes. *Technical Report RJ 7821 (72203)*, November 1990.

[W]   A. Y. Wu. Embedding of Tree Networks into Hypercubes. *Journal of Parallel and Distributed Computing 2*, 1985, pp 238-249.

A Tight Lower Bound for the Worst Case of Bottom-Up-Heapsort [1]

(Extended Abstract)

by

Rudolf Fleischer [2]

ABSTRACT   Bottom-Up-Heapsort is a variant of Heapsort. Its worst case complexity for the number of comparisons is known to be bounded from above by $\frac{3}{2}n\log n + O(n)$, where n is the number of elements to be sorted. There is also an example of a heap which needs $\frac{5}{4}n\log n - O(n\log\log n)$ comparisons. We show in this paper that the upper bound is asymptotical tight, i.e. we prove for large n the existence of heaps which need at least $c_n \cdot (\frac{3}{2}n\log n - O(n\log\log n))$ comparisons where $c_n = 1 - \frac{1}{\log^2 n}$ converges to 1. This result also proves the old conjecture that the best case for classical Heapsort needs only asymptotical $n\log n + O(n\log\log n)$ comparisons.

1. Introduction

Bottom-Up-Heapsort is a variant of the classical Heapsort algorithm and was presented in 1989 by Ingo Wegener ([W90]). The input to the algorithms is an array $a[1..n]$ of n elements from an ordered set S which are to be sorted. We will measure the complexity of the algorithms in terms of number of comparisons; firstly, because comparisons are usually the most expensive operations, and secondly, because each comparison is preceded by only a (small) constant number of other calculations.

First the elements will be arranged in form of a heap (Heap Creation Phase) with the biggest element at the root. This means that the array is considered as a binary tree where node i has children $2i$ and $2i+1$, and that a parent node contains a bigger element than its children (see the full paper [F91] or [W90],[FSU] for details). This requires $O(n)$ time ([Wi64]). Then follows the Selection Phase which consists of n Rearrangement Steps. In each Rearrangement Step the root element changes place with the last active element in the array and becomes inactive; then the heap is rearranged with respect to the remaining active elements. So the size of the heap decreases by one. Since the root always contains the biggest heap element, the array will be filled step by step from the end with elements in decreasing order.

The classical rearrangement procedure works as follows ([M84]). At the beginning, the root contains a former leaf element (the last active array element is always a leaf). This element is repeatedly swapped with the bigger one of its children until it is bigger

---

[1] This work was supported by the ESPRIT II program of the EC under contract No. 3075 (project ALCOM)

[2] Department of Computer Science, University of Saarland, D-6600 Saarbrücken, Germany

than both of its children or it is a leaf. At each level two comparisons are made. Hence the total complexity of the Selection Phase might be as big as $2n \log n$.

In Bottom-Up-Heapsort, the rearrangement procedure is changed in the following way. We first compute the **special path** ([W90]) which is the path on which the leaf element would sink in the classical rearrangement procedure. This is the unique path with the property that any node on it (except the root) is bigger than its sibling, and costs only one comparison per level. Then we let our leaf element climb the special path up to its destination node at the additional cost of one comparison per level.

This algorithm tries to make use of the intuitive idea that leaf elements are likely to sink back down almost to the bottom of the heap, so one can expect climbing up to be cheaper than sinking down. In fact, Wegener ([W90]) showed an upper bound of $\frac{3}{2}n \log n + O(n)$ for Bottom-Up-Heapsort. He also conjectured a tighter upper bound of $n \log n + o(n \log n)$, but this was disproved by [FSU] who constructed a heap with an asymptotic lower bound of $\frac{5}{4}n \log n - O(n \log \log n)$. In this paper we give a construction of a heap that improves the lower bound to asymptotical $\frac{3}{2}n \log n - O(n \log \log n)$ which matches the upper bound. This bound also implies an asymptotic upper bound of $n \log n + O(n \log \log n)$ for the best case of the classical Heapsort algorithm, as has been suspected for many years. This conjecture has recently been proven independently by [SS] using very similar methods.

The construction is an improvement on our previous work ([FSU]). In that paper we constructed a heap where the first $\frac{n}{2}$ rearrangement steps of Bottom-Up-Heapsort needed nearly $\frac{3}{4}n \log n$ comparisons. After that, the active heap was the initial heap without its leaf-level which then contained the $\frac{n}{2}$ biggest elements in sorted order. Unfortunately, we could say nothing nontrivial about the remaining rearrangement steps; hence we were limited to a lower bound of $\frac{5}{4}n \log n$.

In this paper we will use the same general ideas but many details are quite different. Also the proof techniques have changed completely. The advantage is that we can now iterate the above procedure, i.e. we can prove that not only the leaf-level but many levels of the heap are expensive. This gives the asymptotic optimal lower bound.

The paper is organized as follows. In Section 2 we give our definitions and prove some simple properties of heaps. In Section 3 we explain the heap construction which is followed by the complexity analysis in Section 4. We conclude with some remarks in Section 5.

2. Basic Properties

In this section we will give our definitions and prove some basic lemmas about heaps. W.l.o.g. we only construct heaps of size $n = 2^m - 1$, i.e. the heap is a complete binary tree of height m. After the first 2^{m-1} steps of Bottom-Up-Heapsort the leaves of the initial tree are deleted (and filled with the 2^{m-1} biggest elements) and the remaining active heap is a complete binary tree of height $m - 1$. We call this deletion of the leaf-level a **stage** of the algorithm. Bottom-Up-Heapsort consists of $m - 1$ stages.

To achieve high complexity in a stage, many leaf elements must climb up the tree to high destination nodes. The upper bound proof ([W90],[FSU]) shows that only about one half of the leaf elements can have this property at all. We will construct a heap, such that for many consecutive stages nearly half of the leaf elements are sent to high destinations. Thus its complexity will be very close to the upper bound. Our construction is based on the two types of heaps defined below, both of which have the above property for the first stage. After the first stage the resulting heap will be of the other type (at least if it is still big enough).

First we need some notations for some particular parts of a heap (see Fig. 2.1). Let k be a fixed constant to be defined later and H_m be a complete heap of height m with $k + \log k + 1 \leq m \leq 2k - 4$.

- The root is called r.
- The first k levels of H_m are a complete binary tree called B (which will always contain big elements).
- A_1 is a binary tree rooted at $leftson(r)$ of height $m - k$ ($\leq k - 4$) with the $k - 2$ rightmost nodes of the lowest level missing, i.e. A_1 contains only $a := 2^{m-k} - k + 1 (\geq k+1)$ nodes. A_2 is defined as being symmetrical. (Remark: For proof of correctness, any other sets A_1 and A_2 in the two halves of B would work as well; but this definition yields optimal results in the complexity analysis).
- The leaves of B are from left to right $v_1, v_2, \ldots, v_{2l}$ with $l := 2^{k-2}$.
- The left subtree of v_i is called D_i, the right subtree is called E_i. D_i and E_i both have height $m - k$ and size $2^{m-k} - 1 \geq 2k - 1$.

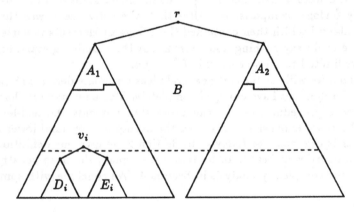

Figure 2.1

We distinguish between **Type-I** and **Type-II Heaps** according to the following conditions (of course, there are also heaps which are of neither type but they do not occur in our construction). Where no ambiguity is possible in the context we also write w for the element stored in a node w and F for the set of elements stored in a heap F.

Def. 2.1 Let F be some heap.

(1) F^{left} is the leftmost leaf of F.

(2) Let G be a heap of smaller size than F. F is a **predecessor** of G if Bottom-Up-Heapsort started with input heap F will end up with heap G after some rearrangement steps.

(3) If F and G are complete heaps of the same size, we say G is **below** F $(G \prec F)$ iff there exists a heap H' as shown in Fig. 2.2 which is a predecessor of heap H (remark : $G < F \Rightarrow G \prec F$ but not vice versa !).

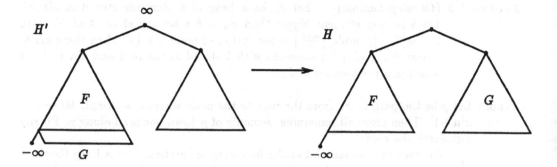

Figure 2.2

(4) If (I.1)–(I.6) are satisfied then F is a

Type-I Heap

   (I.1) $D_i \prec E_{i-1}$, $2 \le i \le 2l$

   (I.2) $E_2 < A_2$

   (I.3) $\exists e \in D_1 : e < E_1^{\text{left}}$

   (I.4) $\exists f \in D_1 \backslash \{e\} : f < E_2^{\text{left}}$

   (I.5) $father^{(2)}(v_1) := father(father(v_1)) < E_{2l}$

   (I.6) $v_i < E_{i+1}^{\text{left}}$, $2 \le i \le 2l-2$

(5) If (II.1)–(II.3) are satisfied then F is a

Type-II Heap

   (II.1) $E_i \prec D_i$, $1 \le i \le 2l$

   (II.2) $E_i < B$, $1 \le i \le 2l$

   (II.3) $v_2 < A_2$

First we show that these definitions make sense, i.e. we show

Lemma 2.2 For all m, $k + \log k + 1 \le m \le 2k - 4$, exists a type-I (type-II) heap of height m.

Proof : Place the smallest elements in D_i, $1 \leq i \leq 2l$, and then fill the heap from left to right with elements in increasing order, i.e. always choose the first node in symorder with no empty children to fill next. Then (I.1)–(I.6) are satisfied. Type-II heaps are similar (smallest elements in E_i instead). □

Now we make the following crucial observation.

Lemma 2.3 (*Onestep-Lemma*)   Let F_n be a heap of n elements stored in $a[1..n]$. Let b be any element bigger than F_n and c be any element of F_n with $c \leq a[\lfloor \frac{n+1}{2} \rfloor]$ (node $\lfloor \frac{n+1}{2} \rfloor$ is the father of node $n+1$). Then there exists a heap F_{n+1} of $n+1$ elements with b stored at the root and $a[n+1] = c$ which is a predecessor of F_n.

Proof : Let p be the path in F_n from the root to the node where c is stored. Move c to $a[n+1]$. Then move all remaining elements of p down one level along p. Finally put b into the root.

The tree thus obtained has the following properties: It is a heap (because it was a heap before) and p is the upper part of the special path (any element of p was replaced by its father, thus becoming the bigger sibling). Hence the first rearrangement step of Bottom-Up-Heapsort will transform F_{n+1} into F_n. □

From this two more useful lemmas follow immediately.

Lemma 2.4 (*Filling-Lemma*)   Let F be a complete heap and F_1 be some subset of F with $F_1 < F - F_1$. Then there exists a heap which is a predecessor of F and whose lowest level consists only of the $|F_1|$ leftmost positions filled with the elements of F_1 (see Fig. 2.3).

Proof : An element of F_1 is called wrong if it is not at the final lowest level or if it has an F_1-father. Apply the *Onestep-Lemma* $|F_1|$ times to the smallest wrong F_1-element. □

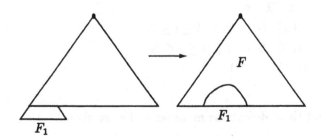

Figure 2.3

Lemma 2.5 (*Below-Lemma*)   Let F and G be complete heaps of the same size. Let $F = F_1 \cup F_2$ where F_2 are some leaves of the left subtree of F, and let $G = G_1 \cup G_2$ where G_2 are some leaves of G (see Fig. 2.4). If $F_1 > G_1$, $F_2 > G_2$ and $|G_2| \geq 2|F_2|$ then $G \prec F$.

Proof: The leaves of G can all be placed as new leaves below the left subtree of F. Hence we can repeatedly apply the *Onestep-Lemma* to heap H of Fig. 2.2, first moving all leaves of G in an appropriate order below F.   □

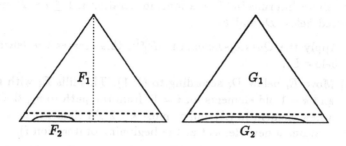

Figure 2.4

3. Heap Construction

In this section we show how to construct a heap of n elements which forces Bottom-Up-Heapsort to make many comparisons.

Theorem 3.1 *For any $m \geq 50$ we can construct a heap H_m of height m which forces Bottom-Up-Heapsort to make $\frac{n}{2} \cdot (3 \log n - 8 \log \log n - 2) \cdot (1 - \frac{1}{\log^2 n})$ comparisons, where $n = 2^m - 1$ is the size of the heap.*

Proof: Choose $k := m - \lfloor 4 \log m \rfloor$ and let $H_{m - \lfloor 2 \log m \rfloor}$ be a type-II heap of height $m - \lfloor 2 \log m \rfloor$ (which exists for $m \geq 50$ by Lemma 2.2). Then apply alternately Theorems 3.2 and 3.3 until a heap H_m of height m is constructed. These theorems can be applied as long as the height of the current heap is at most $2k - 5$, which is always the case if $m \geq 50$. The complexity of Bottom-Up-Heapsort started with H_m will be analyzed in Section 4.   □

We remark that there is a tradeoff (within some limits) between the factor $c_n := 1 - \frac{1}{\log^2 n}$ of the whole term and the factor 8 of the $\log \log n$ term by choosing other values of k together with another number of stages.

Theorem 3.2 *Let H_m be an arbitrary type-II heap of height m, $k + \log k + 1 \le m \le 2k - 5$. Then we can construct a type-I heap H_{m+1} of height $m + 1$ which is a predecessor of H_m.*

Proof : The elements of H_m are called old elements whereas the new added elements are called **new elements**. New elements are bigger than any old element and they are added step by step in increasing order.

Algorithm 1

The algorithm runs in $2l$ iterations; in iteration i, $1 \le i \le 2l$, new leaves are created below D_i and E_i.

(i) Apply the *Onestep-Lemma* to D_i^{left}; this creates the leftmost new leaf below D_i.

(ii) Move E_i below D_i according to (II.1). This fills E_i with new elements and $k - 1$ old elements for $i = 1$ (from the path to v_1; the root received a new element in (i)) and at most $k - 2$ old elements for $i \ge 2$ (the root contains a new element at the beginning of iteration i).

(iii) If $i = 1$ then apply the *Filling-Lemma* to E_1 and its $k - 1$ smallest elements.
If $i \ge 2$ then apply the *Filling-Lemma* to E_i and its $k - 2$ smallest elements; then apply the *Onestep-Lemma* to the leftmost new leaf (which contains one of these elements). Now E_i and its leftmost new leaf contain only new elements.

(iv) Hence the elements of A_2 (if $i \le l$) or A_1 (if $i \ge l + 1$) can be placed as new leaves below E_i (apply the *Onestep-Lemma*).

Algorithm 1 has constructed a heap H_{m+1} of height $m + 1$. It remains to show that it is a type-I heap. An upper index II denotes the old elements of H_m.

(I.1) We have to show $D_i \prec E_{i-1}$, $2 \le i \le 2l$.
D_i only contains the old elements v_i^{II}, D_i^{II} and E_i^{II}; v_i^{II} is the biggest of them.
E_1 contains only the $k - 1$ old elements from the path to $father(v_1^{II}) = father(v_2^{II})$ and A_2^{II}; (II.3) implies $D_2 < E_1$.
If $i \ge 3$ then E_{i-1} contains at most $k - 2$ old elements from the path to $father(v_{i-1}^{II})$; they are stored in the left half of E_{i-1} and they are bigger than E_i^{II} by (II.2); since $|E_i^{II}| \ge 2(k - 2)$ the *Below-Lemma* can be applied.

(I.2) A_2 was filled with new elements in iterations $i \ge l + 1$ whereas E_2 can only contain new elements of iterations $i \le l$. Hence $E_2 < A_2$.

(I.3) E_1^{left} is an old element from the path to $father(v_1^{II})$ which is bigger than D_1.

(I.4) E_2^{left} is a new element and hence bigger than D_1.

(I.5) E_{2l} only contains new elements which were added in iterations $i \geq l + 1$ whereas $father^{(2)}(v_1^{II})$ can only contain a new element of iterations $i \leq l$ because it is below A_1.

(I.6) After step (iii) of the algorithm, E_i^{left} contains a new element which was added during iteration i. Hence it is bigger than any element to the left of the path to E_i^{left}. □

The other construction is similar but some details are more complicated.

Theorem 3.3 *Let H_m be an arbitrary type-I heap of height m, $k + \log k + 1 \leq m \leq 2k - 5$. Then we can construct a type-II heap H_{m+1} of height $m + 1$ which is a predecessor of H_m.*

Proof : We first give the algorithm. As in Algorithm 1 we have old and new elements.

Algorithm 2

The algorithm runs in $2l + 1$ iterations; in iteration i, $1 \leq i \leq 2l + 1$, new leaves are created below E_{i-1} and D_i (if they exist).

(a) $i = 1$:

Analogous to the *Filling-Lemma* we can move $D_1 \cup \{v_1\}$ into the new leaves below D_1 by filling the path to v_1 and the upper part of D_1 with 2^{m-k} new elements.

(b) $i = 2$:

(i) Apply the *Onestep-Lemma* to the element $e \in D_1$ of (I.3) which now is in one of the new leaves below D_1.

(ii) Move D_2 below E_1 according to (I.1). This fills D_2 with new elements and one old element, v_2.

(iii) Apply the *Filling-Lemma* to D_2 and its $k - 1$ smallest elements. Now D_2 contains only new elements; the $k - 1$ new leaves contain v_2 and new elements.

(iv) Hence all elements of A_2 can be placed as new leaves below D_2 (use the *Onestep-Lemma*).

(c) $3 \leq i \leq l$:

(i) Apply the *Onestep-Lemma* to the smallest element of D_{i-2}. This is possible by (I.4) for $i = 3$, by (I.6) and (b)(iii) for $i = 4$ and by (I.6) and (c)(iii) for $i \geq 5$.

(ii) Move D_i below E_{i-1} according to (I.1). This fills D_i with new elements and at most $k - 1$ old elements (from the path to v_i; the root received a new element in (i)).

(iii) Apply the *Filling-Lemma* to D_i and its $k - 1$ smallest elements. Now D_i contains only new elements; some of the $k - 1$ new leaves contain old elements, one of them is v_i.

(iv) Hence all elements of A_2 can be placed as new leaves below D_i (use the *Onestep-Lemma*).

(d) $l + 1 \leq i \leq 2l$:

Same as (c) but in (iv) use A_1 instead of A_2.

(e) $i = 2l + 1$:

The new leaves of D_1 now contain some elements of $D_1 \cup \{v_1, father(v_1), father^{(2)}(v_1)\}$. By (I.5) we can apply the *Onestep-Lemma* to move all these leaf-elements below E_{2l}.

Algorithm 2 has constructed a heap H_{m+1} of height $m + 1$. It remains to show that it is a type-II heap. An upper index I denotes old elements of H_m.

(II.1) We have to show $E_i \prec D_i$ for all i.

> $\underline{i = 1}$: D_1 contains only $k - 3$ old elements (which were on the path to $father^{(3)}(v_1^I)$). E_1 only contains old elements; these are smaller than the old elements of D_1 because they were below $father(v_1^I)$. Hence $E_1 < D_1$.
>
> $\underline{i = 2}$: D_2 contains only the old elements A_2^I. E_2 only contains the old elements E_2^I, D_3^I and a $f^I \in D_1^I$ (I.4). Since $E_2^I < A_2^I$ by (I.2) we conclude $E_2 < D_2$.
>
> $\underline{3 \leq i \leq 2l - 1}$: D_i contains at most $k - 1$ old elements from the path to v_i^I. E_i only contains the old elements E_i^I, D_{i+1}^I and v_{i-1}^I. Since $E_i^I < v_i^I$ we conclude $E_i < D_i$.
>
> $\underline{i = 2l}$: D_{2l} contains only one old element, v_{2l}^I. E_{2l} only contains the old elements E_{2l}^I and some elements smaller than $father^{(2)}(v_1^I)$. Since $E_{2l}^I < v_{2l}^I$ we conclude $E_{2l} < D_{2l}$.

(II.2) B is completely filled with new elements whereas each E_i only contains old elements. Hence $E_i < B$.

(II.3) A_2 was filled with new elements in iterations $i \geq l + 1$ whereas v_2 can only contain a new element of iterations $i \leq l$. Hence $v_2 < A_2$.   □

4. Complexity Analysis

First we will prove that the heaps constructed in Theorems 3.2 and 3.3 have an expensive first stage.

Lemma 4.1 Let H_{m+1} be the type-I heap constructed in Theorem 3.2. If $m \geq k + 2\log k$ then the first stage of H_{m+1} needs at least $2^{m-1} \cdot (2m + k - 2)$ comparisons.

Proof : We have to search 2^m special paths at the cost of at least $m - 1$ each. For each special path the leaf elements will climb up to their destination nodes. In each of the 2^{k-1} iterations, $a = 2^{m-k} - k + 1$ leaf elements climb up to A_1 or A_2 which costs at least k; the other leaf elements need at least one comparison each. Hence

$$T_I \geq 2^{k-1} \cdot [a \cdot k + 2^{m-k} + (k - 1)] + 2^m \cdot (m - 1)$$

$$= 2^{m-1} \cdot (k + 2m - 2) + 2^{k-1} \cdot (2^{m-k} + (k - 1) - (k - 1) \cdot k)$$

$$\geq 2^{m-1} \cdot (2m + k - 2) \qquad \text{for } m \geq k + 2\log k. \qquad \square$$

Lemma 4.2 Let H_{m+1} be the type-II heap constructed in Theorem 3.3. If $k \geq 4$ and $m \geq k + 2\log k + 1$ then the first stage of H_{m+1} needs at least $2^{m-1} \cdot (2m + k - 2)$ comparisons.

Proof : We have to search 2^m special paths at the cost of at least $m - 1$ each. Then the leaf elements will climb up the special path to their destination nodes. In iterations i, $2 \leq i \leq 2l$, $a = 2^{m-k} - k + 1$ leaf elements climb up to A_1 or A_2 which costs at least k; all other leaf elements need at least one comparison each. Hence

$$T_{II} \geq (2^{k-1} - 1) \cdot [a_2 \cdot k + (k - 1)] + (2^{k-1} + 1) \cdot 2^{m-k} + 2^m \cdot (m - 1)$$

$$\geq 2^{m-1} \cdot (k + 2m - 2) + 2^{k-1} \cdot [(k - 1) - (k - 1) \cdot k]$$

$$-(2^{m-k} - k + 1) \cdot k - (k - 1) + 2^{m-1}$$

$$\geq 2^{m-1} \cdot (2m + k - 2) + 2^{m-1} - 2^{k-1} \cdot k^2 - 2^{m-k} \cdot k$$

$$\geq 2^{m-1} \cdot (2m + k - 2) \qquad \text{for } k \geq 4 \text{ and } m \geq k + 2\log k + 1. \qquad \square$$

Proof of Theorem 3.1 :

We apply Lemma 4.1 to heaps of height $m - \lfloor 2\log m \rfloor$, $m - \lfloor 2\log m \rfloor + 2, \ldots$ and Lemma 4.2 to heaps of height $m - \lfloor 2\log m \rfloor + 1$, $m - \lfloor 2\log m \rfloor + 3, \ldots$ until the construction stops with a heap of height m (with $k = m - \lfloor 4\log m \rfloor$ and $m \geq 50$ all constraints about m and k are satisfied). Hence we have total complexity

$$
\begin{aligned}
T_m &\geq \left(2(m - \lfloor 2\log m\rfloor) + k - 2\right) \cdot \left(2^{m-2} + 2^{m-3} + \cdots + 2^{m-\lfloor 2\log m\rfloor - 1}\right) \\
&\geq 2^{m-1} \cdot (3m - 8\log m - 2) \cdot \left(1 - \tfrac{1}{m^2}\right) \\
&= \tfrac{n}{2} \cdot (3\log n - 8\log\log n - 2) \cdot \left(1 - \tfrac{1}{\log^2 n}\right) .
\end{aligned}
$$

$\square$

5. Conclusions

For any given m, we showed how to construct a heap of height m which forces Bottom-Up-Heapsort to make nearly $\frac{3}{2}n\log n$ comparisons. This matches the upper bound asymptotical up to low-order terms ([W90]). Furthermore, this problem is closely related to the old problem of finding the best case for the classical Heapsort algorithm; the immediate consequence is that Heapsort needs only asymptotical $n\log n + O(n\log n)$ comparisons for our heap. Another open problem about both variants of Heapsort is their average running time. We refer to [W90] for details and [SS] for a good bound on the average running time.

Acknowledgements

We would like to thank Ingo Wegener for suggesting the problem and his interest in our result. We would also like to thank C. Uhrig and B.P. Sinha for their previous joint work. And we are grateful to S. Meiser for providing some tools to produce nice TeX-pictures.

References

[C87a] S. Carlsson
        "A variant of HEAPSORT with almost optimal number of comparisons"
        Information Processing Letters **24** (1987), 247–250

[C87b] S. Carlsson
        "Average-case results on HEAPSORT"
        BIT **27** (1987), 2–17

[F91]   R. Fleischer
        "A tight lower bound for the worst case of bottom-up-heapsort"
        Technical Report MPI-I-91-104, Max-Planck-Institut für Informatik,
        W-6600 Saarbrücken, Germany, April 1991

[FSU]   R. Fleischer, B. Sinha, C. Uhrig
        "A lower bound for the worst case of bottom-up-heapsort"
        Technical Report No. A23/90, University Saarbrücken, December 1990

[MDR]   C.J.H. McDiarmid, B.A. Reed
"Building heaps fast"
Journal of Algorithms **10** (1989), 352–365

[M84]   K. Mehlhorn
"Data Structures and Algorithms, Vol. 1, Sorting and Searching"
Springer Verlag, Berlin, 1984

[SS]   R. Schaffer, R. Sedgewick
"The analysis of heapsort"
Technical Report CS-TR-330-91, Princeton University, January 1991

[W90]   I. Wegener
"BOTTOM-UP-HEAPSORT, a new variant of HEAPSORT beating on
average QUICKSORT (if n is not very small)"
MFCS'90, Lecture Notes in Computer Science, 516–522, 1990

[W91]   I. Wegener
"The worst case complexity of McDiarmid and Reed's variant of
BOTTOM-UP-HEAPSORT is less than $n \log n + 1.1n$"
Proc. STACS 1991, 137–147

[Wi64]   J.W.J. Williams
"Algorithm 232"
Communications of the ACM **7** (1964), 347–348

Historical Searching and Sorting

Alistair Moffat*   Ola Petersson†

Abstract: A 'move to the front' dictionary data structure that supports $O(\log t)$ time access to objects last accessed t operations ago is described. This 'Historical Search Tree' is then used in two adaptive sorting algorithms. The first algorithm, 'Historical Insertion Sort', exploits the temporal locality present in a nearly sorted list rather than the more normally exploited spatial locality. The second of the new algorithms, 'Regional Insertion Sort', exploits both temporal and spatial locality. Regional Insertion Sort also gives rise to a new measure of presortedness *Reg* that is superior to all known measures of presortedness, in that any sequence regarded as nearly sorted by any other measure will also be regarded as nearly sorted by the measure *Reg*.

1 Introduction

A sorting algorithm is *adaptive* if sequences that are initially 'close' to sorted are processed faster than those that are not, where the 'distance' from totally sorted is quantified by some *measure of presortedness* [7].

Here we describe two new adaptive sorting algorithms, both implemented using a data structure we call a 'Historical Search Tree'. The Historical Search Tree has the property of providing fast re-access to recently accessed or inserted items using a tree-based analogy of a move-to-front list. In particular, an item last accessed t operations ago can be re-accessed in $O(\log t)$ time.

Both of the new sorting algorithms make use of the insertion sort paradigm of one by one inserting the items of the input sequence into a sorted list. This is the paradigm used in Straight Insertion Sort [6]; in A-Sort [9]; and in Local Insertion Sort [7]. The novel idea in the new algorithms is of adding a time dimension to better capture the intuitive notion of presortedness. Measures of presortedness exploited by previous insertion sort algorithms measure distances only in space, and judge a sequence to be nearly sorted if most of the items appear in the sorted list 'not too far' away (in space) from the previous item in the unsorted sequence.

The first of the new algorithms, 'Historical Insertion Sort', judges a sequence to be nearly sorted if most of the items are inserted adjacent to an item that was itself inserted into the sorted list 'not too far' ago in *time*. This leads directly to a new measure of presortedness based upon the accumulation of the 'historical distance' crossed by each insertion; we will show that this new measure is more useful for quantifying presortedness than both the number of inversions and the number of blocks [1] in the input sequence.

The second new algorithm, 'Regional Insertion Sort', combines the space based search of Local Insertion Sort and the time based search of Historical Insertion Sort, and is efficient on sequences where most of the items are inserted 'not too far' away in space from some item that was itself inserted 'not too long' ago in time. We are able to show that Regional Insertion Sort requires asymptotically fewer comparisons than a wide range of other adaptive sorting

---

*Dept. Comp. Sc., The University of Melbourne, Victoria 3052, Australia. alistair@cs.mu.OZ.AU.

†Dept. Comp. Sc., Lund University, Box 118, S-221 00 Lund, Sweden. ola@dna.lth.SE.

algorithms, and is something of a 'swiss army knife' for adaptive sorting. The corresponding measure of presortedness is also surprisingly versatile, and is shown to be superior to all other known measures of presortedness.

2 Historical Searching

The new sorting algorithms are based upon a 'Historical Search Tree', a dictionary data structure that supports logarithmic-time key-based access to all items stored, but is particularly efficient when recently inserted or accessed items are re-accessed.

Define an *object* to be either an actual item that is stored in the dictionary or the 'gap' between two stored items, and define a *reference* to an object to be any operation that either accesses the object or creates the object. That is, a successful search references the item found; an unsuccessful search references a gap; an insertion destroys a gap and references the newly inserted item and two new gaps; and a deletion creates one newly referenced gap. Suppose that the i'th operation in a sequence of operations accesses object x_i, and let t_i be the number of distinct objects that have been referenced since x_i was last referenced.

The data structure we describe allows for all queries, both successful and unsuccessful, in[1] $O(\log t_i)$ time in the worst case. The structure also allows insertions and deletions in $O(\log t_i)$ comparisons in the worst case and $O(\log t_i + \log \log n)$ amortised time, where n is the number of items in the dictionary. Since $t_i \leq 2n + 1$, all of these bounds are $O(\log n)$.

The data structure is similar to the implicit structure described by Frederickson [2, 3]; but allows for faster insertion and provides bounds that are valid in an amortised sense rather than an average case sense. A similar structure has recently been described by Martel [8]; his structure does not support fast insertion, requiring $\Theta(\log n)$ time. The splay trees of Sleator and Tarjan [11] achieve similar time bounds for access operations in an amortised sense (the 'Working Set Theorem').

The data structure for Historical Searching consists of a forest T of K finger search trees [9]. For $k < K$ tree T_k contains one node for each of the (undeleted) $N(k) - 1$ most recently accessed items, where $N(1) = 2$, and, for $k \geq 2$, $N(k) = 2^{2^{k-1}} = (N(k-1))^2$. The K'th tree T_K contains one node for every item in the dictionary and is required to contain at least $N(K-1)/2$ nodes. If n_k is the actual number of nodes in tree T_k we have

$$
\begin{array}{rcccl}
n_{k-1} & \leq & n_k & < N(k) & 1 \leq k < K \\
N(k-1)/2 & \leq & n = n_k & < N(k) & k = K
\end{array}
$$

and that $2 + \lfloor \log \log n \rfloor \leq K \leq 2 + \lfloor \log \log 2n \rfloor$, where n is the number of items currently in the dictionary. An item last accessed t_i operations ago and still in the dictionary occurs within all trees T_k for $2 + \lfloor \log \log t_i \rfloor \leq k \leq K$, and might occur in earlier trees as well. For example, tree T_1 always contains a node for the most recently accessed item, which in turn is always represented in every one of the K trees. If $n_K = N(K-1)/2$ we will say that T_K is at the point of *underflow*, and if T_K has $N(K) - 1$ nodes it is at the point of *overflow*.

Let C_i represent the collection of nodes collectively representing dictionary item x_i. The nodes C_i are threaded together by a doubly linked circular list in order of occurrence in the trees of the forest. If x_i is represented by a node in tree T_k and not in tree T_{k-1} we say that this is the 'top' node for x_i, and it is specially marked. Hence, when the node in T_k representing x_i has been determined, the nodes representing x_i in trees T_{k-1} and T_{k+1} (if they exist) can be determined in $O(1)$ time by following a single thread. More generally, by following the threads in order of decreasing k until they 'wrap around' one step past the top node, the node in T_K

---

[1]It is convenient to let $\log x$ denote the function $\log_2(\max\{2, x\})$ when used within the big-Oh notation.

for item x_i can be found in $O(k)$ time once the node for x_i is found in T_k. Note that, although an item may correspond to as many as $\log\log n$ tree nodes, the total number of nodes is given by $\sum_{k=1}^{K} n_k = O(n)$.

Within each tree T_k insertions and deletions immediately adjacent to a fingered node can be carried out in $O(1)$ amortised time [9] not counting the cost of maintaining the threads. Insertions and deletions of a node not known to be immediately adjacent to a fingered node require $O(\log n_k) = O(2^k)$ time in the worst case.

One final set of pointers is required. These pointers are required only on the nodes of T_K, and implement a doubly linked 'move-to-front' linear list P. An array L of K fingers point into this list, with L_k pointing to the node (in T_K) representing the item x that is the 'oldest' item in T_k. Using this auxiliary structure and the circular threads the oldest node in T_k can be accessed in $O(1)$ time.

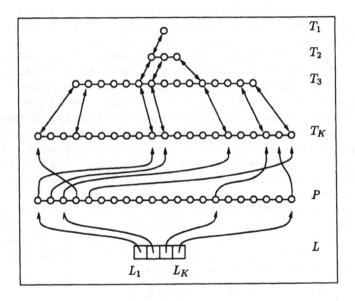

Figure 1: Structure for Historical Searching

Figure 1 shows this combination of forest T, move-to-front list P, and fingers L. For convenience each of the trees T_k is drawn as a list.

2.1 Successful Search

To search for an item the trees of T are searched in order $T_1, T_2, \ldots$ until the item is either found or can be known to not be in the dictionary. Suppose that item x_i is in the dictionary and is first encountered in tree f. The cost of searching trees $T_1, \ldots, T_f$ is bounded by

$$\sum_{k=1}^{f} O(2^k) = O(2^f) = O(\log t_i)$$

since $f \le 2 + \lfloor \log\log t_i \rfloor$.

Once a node representing x_i has been found the forest must be updated to reflect the fact that x_i is now the most recently accessed item. To do this, nodes for x_i must be inserted into all trees $T_1, \ldots, T_{f-1}$, and threaded into the list C_i. We call this operation the *promotion* of x_i. This will also require $O(\log t_i)$ time. Next, the 'oldest' item in each of trees $T_1, \ldots, T_{f-1}$ should be *demoted*, so that n_k is unchanged for each of the K trees. This is done in each tree T_k by deleting the node for the oldest item, found using the finger L_k. Each of these fingers then moves one position closer to the front of P to point at what was the next oldest node in T_k and is now the oldest node, leaving the node that was deleted from tree T_k as the 'newest' node in tree T_{k+1}. The actual deletion of each oldest node requires $O(2^k)$ time in the worst case. Finally, to mark x_i as the 'newest' item in the dictionary, it should be unlinked from its current position in P and 'moved-to-front' to become the newest node in tree T_1 and thus the newest item in every tree.

This sequence of changes is sufficient to ensure that at all times the t'th most recently accessed item appears in all trees T_k with $2 + \lfloor \log \log t \rfloor \leq k \leq K$, as required.

2.2 Unsuccessful Search

To handle unsuccessful search efficiently, the searching procedure must become a little more complex. Rather than just search each of the trees T_k in turn, the searching process must pause between trees to determine whether it is necessary to continue.

Suppose that tree T_f has just been searched, and that the immediate predecessor in tree T_f of the key value x_i for which we are searching is x_p and the immediate successor is x_s. By following the circular threads for these two nodes the corresponding nodes in tree T_K for x_p and x_s can be located in $O(f)$ time, which is $O(2^f)$. Moreover, since tree T_K contains a node for every item in the dictionary, an additional $O(1)$ time suffices to determine whether either (or both) of x_p or x_s is immediately adjacent to the gap containing x_i. If it is, and x_i is found to be less than the T_K successor of x_p or greater than the T_k predecessor or x_s, the search can be terminated with the knowledge that x_i is not in the dictionary. If x_i is not immediately adjacent to x_p or x_s the search continues into tree T_{f+1}.

Use of this modified strategy will not affect the asymptotic cost of a successful search, and means that an unsuccessful search will terminate after at most f trees have been examined, where f is the index of the first tree that contains either of the two items immediately adjacent to and defining the gap containing x_i. Suppose, without loss of generality, that it is x_p that is found in tree T_f to be the immediate predecessor in tree T_K of x_i. To be sure that subsequent operations will be fast, the object searched for — in this case the gap — is promoted into T_1 by promoting x_p as if that item had been the target of a successful search. Hence the worst case cost of each unsuccessful search will be $O(\log t_i)$.

We have now proved a worst case equivalent of the Working Set Theorem [11]:

Theorem 1 *In a Historical Search Tree each access operation, whether successful or unsuccessful, requires $O(\log t)$ time in the worst case, where t is the number of distinct objects referenced since the last reference to the object accessed.*

2.3 Insertion

To insert an item into the dictionary we first perform a search for the gap that item x_i splits. Suppose again that item x_p is first found in tree T_f and is established to be the immediate predecessor in the dictionary of item x_i. Rather than promoting x_p as would happen for an unsuccessful search, the new item x_i is inserted into the gap and is then itself promoted, effectively promoting both of the new gaps that are created. The promotion is then followed by demotions in each of $T_1, \ldots, T_{f-1}$, leaving n_k for $1 \leq k < f$ unchanged. However item x_i must also be inserted into trees $T_f, \ldots, T_K$ and a complete set of threads established for the K

nodes in C_i. We do this by following the threads on item x_p, inserting a new node immediately adjacent to x_p in each of trees $T_f, \ldots, T_K$. One item must also be demoted from each of these trees.

As T_K is required to contain one node for every item in the dictionary, deletion of the oldest node in T_K is not permitted, and if tree T_K was at the point of overflow before the insertion then the forest must grow by one tree. This is done by making an exact duplicate and linking each of the nodes in the new tree into the corresponding circular thread. Then K is incremented, the new tree labelled T_K, and finally the oldest item in the (still) over-full tree T_{K-1} is demoted.

The cost of the dictionary insertion is calculated as follows. First, there is an unsuccessful search. This requires $O(\log t_i)$ time in the worst case, including the cost of the first $f-1$ node insertions and demotions. Then there are another $O(K-f+1) = O(\log \log n)$ node insertions and demotions. Each of the node insertions is immediately to the right (or left) of a fingered node, and each of the deletions is of a fingered node. Hence each of these operations can be carried out without any comparisons on items, and in $O(1)$ amortised time. When the forest expands by one tree another $O(n)$ time must be expended for the duplication and re-threading of the old T_K. However after this operation at least $\Theta(n^2)$ insertions must be performed before tree T_K again reaches the point of overflow, at which time $\Theta(n^2)$ time would be required for the cloning operation. Similarly, since we only require that $n_K \geq N(K-1)/2$ and immediately after the cloning we will have $n_K = n = N(K-1)$, at least $\Theta(n)$ deletions are required before tree T_K can reach the point of underflow. The amortised cost of the cloning of tree T_K is thus $O(1)$ per dictionary insertion.

Summing these various contributions, an insertion into the dictionary requires at most $O(\log t_i)$ comparisons in the worst case and $O(\log t_i + \log \log n)$ amortised time.

2.4   Deletion

To delete an item we must first pay for a successful search, requiring $O(\log t_i)$ time. Then, exactly as it would be in a successful search, item x_i is promoted, again requiring $O(\log t_i)$ time. By following the threads and using tree T_K it is easy to find x_s, the inorder successor (or predecessor if x_i is the greatest item) in the dictionary of x_i, and rather than delete x_i directly we first replace each of the (now) K nodes representing x_i by a corresponding node representing x_s. Finally, the original threaded list of nodes representing x_s is deleted. These last two steps require $O(K) = O(\log \log n)$ amortised time and no comparisons; the reason we first replace x_i by x_s is to ensure that x_s appears in T_1 and thus that the enlarged gap now adjacent to x_s created and referenced by the deletion of x_i can be rapidly accessed by a subsequent unsuccessful search.

Deleting nodes may result in some trees having less than $N(k)$ nodes, and so although all subsequent operations will still be $O(\log t_i)$, it might be that t_i becomes asymptotically greater than n. To protect against this possibility we enforce the lower bound on the number of nodes in the last tree T_K. When tree T_K underflows we remove tree T_{K-1} entirely and replace it by tree T_K, decrementing K and reducing the size of the forest by one. This does not violate any of our constraints, and ensures that we will always have $\sum \log n_k = O(\log n)$. The removal of T_{K-1} will cost $\Theta(N(K-1)) = \Theta(n)$ time, but this can again be amortised over all item deletions, and it will take another $\Theta(n)$ item deletions before $\Theta(\sqrt{n})$ time is required for the next tree removal. The amortised time required for each item deletion is thus $O(\log \log n)$. When this is added to the $O(\log t_i)$ cost of the initial search, we have that item deletion requires $O(\log t_i)$ comparisons in the worst case, and $O(\log t_i + \log \log n)$ amortised time. We have now proved Theorem 2:

Theorem 2 *Starting with an empty historical search tree, a sequence of n insertion and/or deletion operations and m access operations can be accomplished in*

$$O(\sum_{i=1}^{n+m} \log t_i)$$

comparisons and

$$O(\sum_{i=1}^{n+m} \log t_i + n \log \log n)$$

time in the worst case.

3 Historical Insertion Sort

The first of the new adaptive sorting algorithms is now obvious. We create an empty historical search tree, and one by one insert the items of the input sequence X into the structure. At the completion of the insertion phase the final tree T_K can then be traversed to recover the sequence in sorted order.

To analyse this algorithm we need a few definitions. Let $d_{i,j}$ be the number of items (the distance in space) between x_i and x_j at the time x_i is inserted. That is,

$$d_{i,j} = \|\{k \mid 1 \le k < i \text{ and } \min\{x_i, x_j\} < x_k < \max\{x_i, x_j\}\}\| + 1.$$

Then $t_i = \min\{j \mid 1 \le j < i \text{ and } d_{i,i-j} = 1\}$. If we further define

$$Hist(X) = \prod_{i=2}^{n} t_i$$

then 'Historical Insertion Sort' requires $O(n + \log Hist(X))$ comparisons and $O(n \log \log n + \log Hist(X))$ time to sort any n-sequence X.

4 Regional Insertion Sort

The Historical Insertion Sort described above searches only in time, and so even if the insertion point of item x_i is close in space to a recent item the insertion might still require $\Theta(\log n)$ comparisons. Conversely, Mannila's Local Insertion Sort searches only in space, and the insertion of an item might be expensive even if it is eventually inserted immediately adjacent to an item that was itself inserted only a short time ago. To make this difference in abilities even more apparent, we include an analogy made in [10]. If 'remembering' and 'watching' correspond to searching in time and in space respectively, then Mehlhorn's A-Sort has good eyes but no memory at all; Local Insertion Sort adds one unit of memory but still relies mainly upon its eyes; and Historical Insertion Sort has a good memory but is almost blind. A further insertion sort was described by Katajainen, Levcopoulos, and Petersson [5]; their Multiple Finger Insertion Sort has a limited amount of memory, but suffers the affliction of not being permitted to look around until it has finished remembering.

Taking the analogy to its conclusion, we desire an algorithm that alternately remembers and then looks and, if that is not enough, remembers further back and then looks further away, searching in two-dimensional space and time until the insertion point for each item is found. The algorithm will be most efficient when most of the insertions are somewhere close (in space) to an item that was inserted not too long ago (in time). This is the basis for our 'Regional Insertion Sort'.

The algorithm is based upon a modified Historical Search Tree. Since the entire algorithm consists of a sequence of n insertions, we only need to describe the searching and insertion strategies.

Suppose that x_i is the next item to be inserted. The searching strategy already described for Historical Insertion Sort pauses after each of the trees T_k and performs a 'proximity check' in T_K to see whether or not the gap containing x_i has been located. This check consists of examining the T_K successor of the T_k predecessor of x_i, and the T_K predecessor of the T_k successor of x_i. These two checks require $O(k)$ time each using the threads.

In fact we can afford to spend more time in this checking phase, since $O(2^k)$ time has just been spent searching in tree T_k. Suppose we allow the proximity check in T_K to also spend $\Theta(2^k)$ time checking to see if the insertion point for x_i has been located yet. Since T_K is a finger search tree, it is possible in $\Theta(2^k)$ time to search a distance (in space) of 2^{2^k} items. The searching phase stops whenever x_i is found in T_K to lie within 2^{2^k} items of either x_p, the T_k predecessor of x_i, or x_s, the T_k successor of x_i. This improved searching algorithm is shown in Figure 2.

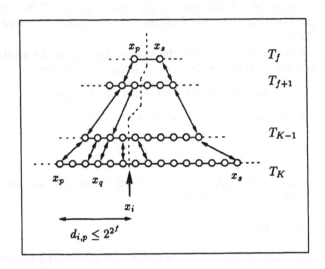

Figure 2: Regional Searching

Before discussing the actual insertion of x_i let us consider the computational requirements of this revised searching phase. Suppose that the search stops after tree f, and that $t = 2^{2^f}$. The fact that the search has stopped implies that

$$\min_{i-t \leq p < i} \{d_{i,p}\} \leq t$$

and the cost of the search into and including T_f is proportional to $\log t + \log d_{i,p}$. In [10] we show that this latter quantity is $O(\log r_i)$, where

$$r_i = \min_{1 \leq t < i} \{t + d_{i,i-t}\}.$$

Consider for a moment this quantity r_i. The t component corresponds to a search in time. When the minimising value of t is small a recently inserted item x_{i-t} is being used as the 'anchor' point of the search. When $d_{i,i-t}$ is small the insertion point is close to this t'th most recently inserted item. The quantity r_i is thus the minimum two-dimensional distance in the time-space plane from any previous item to the insertion point of x_i. Moreover, $r_i \leq 1 + d_{i,i-1}$ and $r_i \leq t_i + 1$, and the searching cost cannot exceed the $O(\log d_{i,i-1})$ cost of Local Insertion Sort and the $O(\log t_i)$ cost of Historical Insertion Sort.

The insertion of x_i into trees $T_1, \ldots, T_f$ has already been taken care of during the searching process. The insertion of x_i into T_K will take $O(1)$ amortised time, since the insertion point has already been located. The remainder of the insertion process, for Historical Searching relied on the knowledge that the node for x_i should be inserted into trees $T_{f+1}, \ldots, T_{K-1}$ adjacent to the node for x_p. In Regional Searching this assumption is no longer valid. However the following lemma shows that all intervening nodes can be removed from trees $T_{f+1}, \ldots, T_{K-1}$, and so after a preliminary pruning stage the threads C_p can again be used as a guide for the insertion of x_i.

Lemma 3 *Let $q < p < i$ and $x_p < x_q < x_i$. Then for $j > i$, x_q will never be required as an anchor point during the insertion of x_j.*

Proof (Sketch) If $d_{i,q} < i - q$ then for $j > i$ the quantity r_j will not be uniquely minimised at $t = j - q$, since there always exists an i', $i \leq i' < j$ such that $t = j - i'$ results in a value for r_j that is at least as small. $\diamond$

We now describe the insertion algorithm in detail. First, the insertion point of x_i in T_K is determined by the search strategy described above. This search also establishes the anchor point, x_p say. We then use the threads to move into tree T_{K-1} to the node representing x_p, and begin a linear search looking for the T_{K-1} insertion point of x_i. Each node that is passed over during this linear search is deleted from T_{K-1} and, using the threads, from all other trees that it is a member of except T_K. This will take 1 comparison and $O(\log \log n)$ amortised time for each deleted item. Once all of the intervening items have been deleted a node for x_i can be inserted in each of trees $T_f, \ldots, T_{K-1}$ and threaded into the list C_i using $O(\log \log n)$ time and no further comparisons, as before. If tree T_f contains the top node for x_p then x_p can also be deleted, once x_i has been inserted using x_p as a guide. Since each item in T_{K-1} can only be deleted once, the cost of the item deletions and the comparisons can be amortised over all insertions, and in total contributes $O(n)$ comparisons and $O(n \log \log n)$ time.

All items are retained in T_K, and so, when the forest is required to grow by one level we clone T_{K-1} rather than T_K.

This revised structure leads directly to Theorem 4:

Theorem 4 *Let*

$$Reg(X) = \prod_{i=2}^{n} (r_i - 1).$$

Then Regional Insertion Sort requires $O(n + \log Reg(X))$ comparisons and $O(n \log \log n + \log Reg(X))$ time.

This bound is in fact a little more pessimistic than it could be. When x_i is inserted immediately adjacent to x_{i-1} Lemma 3 says that x_{i-1} can be removed, since we will have $p = i - 1$ and $d_{i,i-1} = 1$. Because of these immediate removals the forest T will in the end contain at most one item for each 'block' of the input sequence, where a block is a subsequence of consecutive items in the unsorted sequence that are also consecutive in the sorted sequence. More formally, we define [1]

$$Block(X) = \|\{1 \leq i < n \mid x_i \text{ gets a new right neighbour during the sorting}\}\| + 1.$$

Then $n_K \leq Block(X) \leq n$ at all times during Regional Insertion Sort, and the bound on the running time can be improved to $O(n \log \log Block(X) + \log Reg(X))$. On a sorted sequence, where $Block(X) = 1$, the running time of Regional Insertion Sort will be $O(n)$.

5 Measures of Presortedness

We now compare *Hist* and *Reg* with other measures of presortedness, using the framework established in [10].

Theorem 5 *Suppose that X is some n-sequence. Then*

$$Hist(X) = O\left(n^{Block(X)}\right).$$

Proof For the first item of each block we must have $t_i \leq n$. The second and subsequent items within any block will have $t_i = 1$. The result follows. ⋄

Theorem 6 *Suppose that X is some n-sequence. Then*

$$Hist(X) = O\left((1 + \frac{Inv(X)}{n})^n\right).$$

Proof (Sketch) We consider the number of inversions induced *on* x_i and the number of inversions induced *by* x_i, $inv_l(x_i)$ and $inv_r(x_i)$ respectively:

$$inv_l(x_i) = \|\{j \mid 1 \leq j < i \text{ and } x_j > x_i\}\|$$
$$inv_r(x_i) = \|\{j \mid i < j \leq n \text{ and } x_j < x_i\}\|$$

It can be shown that $t_i \leq 1 + inv_l(x_i) - inv_l(x_{i-t_i}) + inv_r(x_{i-t_i}) - inv_r(x_i)$, and thus that $\sum t_i \leq n + 2 \cdot Inv(X)$. The result then follows. ⋄

These two theorems are important in that they show that the measure *Hist* is as least as good as *Inv* and *Block* for quantifying presortedness. Mannila [7] introduced the concept of an algorithm that *optimally* adapts to some measure of presortedness. The bounds of Theorems 5 and 6 show that any *Hist*-optimal algorithm, requiring $O(n + \log Hist(X))$ time, will also be optimal with respect to the measures *Block* and *Inv*, since algorithms optimal with respect to these two measures should take time $O(n + Block(X) \log n)$ and $O(n \log(Inv(X)/n))$ respectively [1, 4].

To show that *Hist* is strictly superior it remains to give sequences that *Hist* regards as nearly sorted but are regarded poorly by *Inv* and *Block*. The sequence

$$X = (1, n/2, 2, n/2 + 2, \ldots, n/2, n)$$

meets this requirement for both measures. It is straightforward to give other sequences that distinguish *Hist* from all known measures of presortedness.

Similar superiority results for the measure *Reg* [10, Thm. 13] show that any algorithm that is optimally adaptive with respect to *Reg*, requiring $O(n + \log Reg(X))$ time, is optimally adaptive with respect to *all known measures of presortedness*.

6 Remarks

The obvious open problem is this: are there sorting algorithms that are optimally adaptive with respect to *Hist* or *Reg*, requiring $O(n + \log Hist(X))$ or $O(n + \log Reg(X))$ time? The Historical and Regional Insertion Sorts we have described here attain these bounds for *comparisons*, and so are sufficient to show that the measure *Reg* is superior to all other known measures of presortedness. However in terms of running time the algorithms are *not* optimally adaptive, and this line of research cannot be closed until optimal algorithms have been described. A data structure supporting historical searching that allows insertions and deletions in $O(\log t_i)$ time would also be interesting in its own right.

Acknowledgements

The authors would like to thank Arne Andersson, who suggested the inorder threading of the items in the historical search tree after reading an early description of the data structure. This work was in part supported by the Australian Research Council.

References

[1] S. Carlsson, C. Levcopoulos, and O. Petersson. Sublinear merging and Natural Mergesort. In *Proc. SIGAL International Symposium on Algorithms*, pages 251–260. LNCS 450, Springer-Verlag, 1990.

[2] G.N. Frederickson. Self-organizing heuristics for implicit data structures. *SIAM Journal on Computing*, 13:277–291, 1984.

[3] G.N. Frederickson. Implicit data structures for weighted elements. *Information and Computation*, 66:61–82, 1985.

[4] L.J. Guibas, E.M. McCreight, M.F. Plass, and J.R. Roberts. A new representation of linear lists. In *Proc. 9th Annual ACM Symposium on Theory of Computing*, pages 49–60, 1977.

[5] J. Katajainen, C. Levcopoulos, and O. Petersson. Local Insertion Sort revisited. In *Proc. International Symposium on Optimal Algorithms*, pages 239–253. LNCS 401, Springer-Verlag, 1989.

[6] D.E. Knuth. *The Art of Computer Programming, Vol. 3: Sorting and Searching*. Addison-Wesley, Reading, Mass., 1973.

[7] H. Mannila. Measures of presortedness and optimal sorting algorithms. *IEEE Transactions on Computers*, C-34(4):318–325, 1985.

[8] C. Martel. Self-adjusting multi-way search trees. *Information Processing Letters*, 38(3):135–142, 1991.

[9] K. Mehlhorn. *Data Structures and Algorithms, Vol 1: Sorting and Searching*. Springer-Verlag, Berlin/Heidelberg, F.R.Germany, 1984.

[10] O. Petersson and A.M. Moffat. A framework for adaptive sorting. Manuscript, February 1991.

[11] D.D. Sleator and R.E. Tarjan. Self-adjusting binary search trees. *Journal of the ACM*, 32(3):652–686, 1985.

Comparison-efficient and Write-optimal Searching and Sorting

Arne Andersson
Lund University
Lund, Sweden

Tony W. Lai
NTT Communication Science Labs
Kyoto, Japan

Abstract

We consider the problem of updating a binary search tree in $O(\log n)$ amortized time while using as few comparisons as possible. We show that a tree of height $\lceil \log(n+1) + 1/\sqrt{\log(n+1)} \rceil$ can be maintained in $O(\log n)$ amortized time such that the difference between the longest and shortest paths from the root to an external node is at most 2.

We also study the problem of sorting and searching in the *slow write* model of computation, where we have a constant size cache of fast memory and a large amount of memory with a much slower writing time than reading time. In such a model, it is important to sort using only $\Theta(n)$ writes into the slower memory. We say that such algorithms are *write optimal*, and we introduce a $O(n \log n)$ time, write-optimal sorting algorithm that requires only $n \log n + O(n)$ comparisons in the worst case. No previous sorting algorithm that performs $n \log n + o(n \log n)$ comparisons in the worst case had previously been shown to be write optimal.

The above results are based on a class of trees called k-stratum trees, which can be viewed as a generalization of stratified search trees.

1 Introduction

In this abstract, we introduce the *slow write* model of computation. In this model, we have a constant size cache of fast memory and a large amount of memory with a much slower writing time than reading time. We justify the model by noting that certain types of memory such as flash memories and EEPROMs have much slower writing times than reading times.

We consider the problem of efficiently maintaining a comparison-based dictionary, while minimizing the number of comparisons and writes per update. In particular, we want updates and searches to require only $\log n + O(1)$ comparisons in the worst case; we say that a dictionary with this property is *comparison-efficient*. In addition, we want updates to require only a constant number of writes; a dictionary with this property is *write-optimal*.

In the area of comparison-efficient bounds, we show that a binary search tree of height at most $\lceil \log(n+1) + \epsilon(n) \rceil$ can be maintained in $O(\log n)$ amortized time, such that $\lim_{n\to\infty} \epsilon(n) = 0$. No such upper bound on the number of comparisons has been previously demonstrated. The exact value of $\epsilon(n)$ we obtain is $1/\sqrt{\log(n+1)}$.

In the area of write-optimal bounds, we show that a binary search tree of height at most $\lceil \log(n+1) + \epsilon \rceil$ can be maintained in $O(\log n)$ amortized time, such that the amortized number of writes is $O(1/\epsilon^3)$. Thus, by inserting n elements into such a structure, we obtain an $O(n \log n)$ time sorting algorithm that uses only $n \log n + O(n)$ comparisons and $\Theta(n)$ writes. Although there are both write-optimal sorting algorithms [5, 6, 7] and sorting algorithms with an optimal number of comparisons (with respect to the leading term) [3, 8], no previous solution has been shown to combine both properties.

The presented results are based on a new data structure called a k-stratum tree. They can be viewed as a generalization of stratified search trees [7].

Before proceeding further, we define some more terms. We define the length of a path in a tree to be the number of internal nodes on the path. We define the *height*, height(T), of a tree T to be the length of the longest path from the root to an external node. We define the *number of complete levels*, short(T), of a tree T to be the length of the shortest path from the root to an external node. The *incompleteness* of T is given by height(T) − short(T). A tree T is k-incomplete if height(T) − short$(T) \leq k$. Thus, a complete tree is 0-incomplete, and a tree of minimal internal path length is 1-incomplete. The *weight*, $|T|$, of a tree T is the number of external nodes of T. A tree T is *perfectly balanced* if T is an external node, or T's subtrees are perfectly balanced and their weights differ by at most 1.

2 Comparison efficiency

To maintain an extremely well-balanced tree, we use the following basic ideas.

- Although the amortized cost of maintaining a binary search tree of minimal internal path length (1-incomplete tree) can be shown to be $\Omega(n)$ in general, it is possible to obtain a polylogarithmic amortized cost when the size of the tree is not close to a power of 2.

- By keeping two layers of 1-incomplete trees on top of each other, we achieve a 2-incomplete tree. By careful maintenance algorithms consisting of partial rebuilding within both layers, 3-way splitting and merging of subtrees, and occasional rebuilding of the entire tree, we guarantee that the sizes of the trees in both layers are favorable all the time. We also make the frequency of splitting, merging, and global rebuilding low enough to achieve a total amortized cost of $O(\log n)$ per update.

Note that we also achieve an incompleteness of 2; this incompleteness can be shown to be the smallest maintainable in $O(\log n)$ time in general, so we obtain matching upper and lower bounds on the incompleteness of a binary search tree.

Since we use 1-incomplete trees in two layers or strata, our tree is called a *2-stratum tree*. We give a formal definition of a 2-stratum tree below.

Definition 2.1 *Given two positive integers H_1 and H_2, a 2-stratum tree consists of a topmost tree whose external nodes are replaced by leaf subtrees, such that the following properties hold.*

1. Let T_1 be the topmost tree. Then, $\text{short}(T_1) \geq H_1 - 1$ and $\text{height}(T_1) \leq H_1$.

2. Let T_2 be a leaf subtree. Then, $\text{short}(T_2) \geq H_2 - 1$ and $\text{height}(T_2) \leq H_2$.

3. $H_1 + H_2 \leq \left\lceil \log(n+1) + \frac{1}{\sqrt{\log(n+1)}} \right\rceil$, where n is the number of elements in the tree.

We say that the topmost tree is the apex of the tree and is in stratum 1, and the leaf subtrees are in stratum 2.

The maintenance algorithms ensure a low height by utilizing the values of H_1 and H_2, which are changed only during global rebuildings of the tree. During most updates, H_1 and H_2 remain unchanged. When the tree is globally rebuilt, we make sure that $H_1 + H_2 \leq \left\lceil \log(n+1) + \frac{\epsilon}{2} \right\rceil$. Since the height of the tree is at most $H_1 + H_2$, it cannot increase between global rebuildings, so the only way the tree can become too tall is when a large number of deletions causes the number of stored elements to become too small. If this occurs, we make a new global rebuilding. This ensures that the height of the tree satisfies Definition 2.1 all the time.

To determine which stratum a node belongs in, we store one bit in each node.

The rest of the maintenance algorithms for a 2-stratum tree are similar to the algorithms for B-trees [2] in that updates are performed in the lowest layer, and leaf subtrees are split or merged when they become too large or too small. However, special care has to be taken due to the fact that both the topmost tree and the leaf subtrees are 1-incomplete. In order to keep a tree 1-incomplete at a low cost, we must ensure that its size is not close to a power of two. For this reason, the sizes of involved trees have to be carefully controlled when splitting and merging leaf subtrees. This is achieved by using three-way splitting and merging. The constants in the algorithms may seem complicated, but this is due to the above restrictions.

2.1 Construction

Recall that we want to maintain a tree of height $\lceil \log(n+1) + \epsilon(n) \rceil$, where $\epsilon(n) = \frac{1}{\sqrt{\log(n+1)}}$. Let $\delta(n) = \frac{4}{5}(1 - 2^{-\epsilon(n)/2})$. For brevity, in the following we refer to $\epsilon(n)$ as ϵ and $\delta(n)$ and δ.

We first present the global rebuilding algorithm that allows us to obtain a well-shaped 2-stratum tree for each value of n. Second, we present the algorithms for insertion and deletion.

In the following, for convenience, we assume that n is sufficiently large to make our formulas work.

To construct a tree of n elements (with weight $N = n + 1$), we choose the weight N_1 of the topmost tree and construct a tree such that the minimum and maximum weights of the leaf subtrees are $\lfloor N/N_1 \rfloor$ and $\lceil N/N_1 \rceil$, respectively. From N_1, we also determine the values of H_1 and H_2. There are three cases.

1. $2^{\lfloor \log N \rfloor} \leq N < \frac{(1+\delta)(2-\delta)}{2} 2^{\lfloor \log N \rfloor}$. We choose $N_1 = \left\lceil (1+\delta)2^{\lfloor \log N \rfloor - 2\lceil \log \log N \rceil - 1} \right\rceil$, which implies that $H_1 = \lfloor \log N \rfloor - 2\lceil \log \log N \rceil$ and $H_2 = 2\lceil \log \log N \rceil + 1$.

2. $\frac{(1+\delta)(2-\delta)}{2} 2^{\lfloor \log N \rfloor} \le N \le (2 - \frac{5}{2}\delta) 2^{\lfloor \log N \rfloor}$. In this case, we choose $N_1 = \lfloor (2 - \delta) 2^{\lfloor \log N \rfloor - 2 \lceil \log \log N \rceil - 1} \rfloor$, which implies that $H_1 = \lfloor \log N \rfloor - 2 \lceil \log \log N \rceil$ and $H_2 = 2 \lceil \log \log N \rceil + 1$.

3. $(2 - \frac{5}{2}\delta) \cdot 2^{\lfloor \log N \rfloor} < N < 2 \cdot 2^{\lfloor \log N \rfloor}$. We choose $N_1 = \lceil (1+\delta) 2^{\lfloor \log N \rfloor - 2 \lceil \log \log N \rceil} \rceil$, which implies that $H_1 = \lfloor \log N \rfloor - 2 \lceil \log \log N \rceil + 1$ and $H_2 = 2 \lceil \log \log N \rceil + 1$.

Lemma 2.1 *Immediately after a global rebuilding, $H_1 + H_2 \le \lceil \log(n+1) + \epsilon/2 \rceil$.*

Proof: The proof follows in a straightforward manner from the construction algorithm described above. □

2.2 Insertion

Before proceeding further, we define some more notation. For a node p in stratum i, we define the weight $|p|$ to be the weight of the subtree in stratum i rooted at p. Similarly, we define the height $\text{height}(p)$ to be the height of the subtree in stratum i rooted at p, and $\text{short}(p)$ to be the number of complete levels of the subtree in stratum i rooted at the node p.

Initially, we perform an insertion in a subtree in the bottommost stratum, that is, stratum 2. To insert a node x into stratum i, first insert x into an appropriate subtree T_i in stratum i. If $\text{height}(T_i) \le H_i$, then exit. Otherwise, determine the lowest ancestor p of the inserted node such that $\text{height}(p) > \left\lceil \left(1 + \frac{\log(\frac{1}{1-\delta/4})}{H_i}\right) \log |p| \right\rceil$. We consider two cases.

Case 1: Such a node p exists. Rebuild the subtree inside stratum i rooted at p to perfect balance.

Case 2: No such node p exists. The subtree T_i is too large. If $i = 1$, then globally rebuild the entire tree. Otherwise, redistribute nodes or split leaf subtrees as follows. Let T_i' denote T_i's closest leaf subtree. (As a measure of distance between leaf subtrees, we use the length of the path between their roots.) Two subcases occur.

   Case 2.1: $|T_i| + |T_i'| < \lceil 3.5 \cdot 2^{H_i} \rceil$. Redistribute nodes between T_i and T_i' such that they have the same weight (within ± 1) and terminate.

   Case 2.2: $|T_i| + |T_i'| \ge \lceil 3.5 \cdot 2^{H_i} \rceil$. Split T_i and T_i' into three leaf subtrees U_i, U_i', and U_i'' of equal weight (within ± 1). Splitting two leaf subtrees into three implies that some node x' is inserted into the apex, that is, into stratum 1. Insert x' into stratum 1 using this algorithm.

2.3 Deletion

We use a deletion algorithm similar to the insertion algorithm described above. Note that we may always transform a deletion of an internal node x into a deletion of a leaf by replacing x by its inorder successor y and deleting y. Hence, to perform a deletion, we first delete a node from a subtree from stratum 2.

To delete a node x from stratum i, first delete x from an appropriate subtree T_i in stratum i. If short$(T_i) \geq H_i - 1$, then exit. Otherwise, determine the lowest ancestor p of the deleted node such that short$(p) < \left\lfloor (1 - \frac{\log(\frac{1}{1-\delta/4})}{H_i}) \log |p| \right\rfloor$. Two cases occur.

Case 1: Such a node p exists. Rebuild the subtree inside stratum i rooted at p to perfect balance.

Case 2: No such node p exists. The subtree T_i is too small. If $i = 1$, then globally rebuild the tree. Otherwise, redistribute nodes or merge leaf subtrees as follows. Let T_i' and T_i'' denote T_i's two closest leaf subtrees. Two subcases occur.

Case 2.1: $|T_i| + |T_i'| + |T_i''| > \left\lceil \frac{11}{3} \cdot 2^{H_i} \right\rceil$. Redistribute nodes between T_i, T_i', and T_i'' such that they have the same weight (within ± 1) and terminate.

Case 2.2: $|T_i| + |T_i'| + |T_i''| \leq \left\lceil \frac{11}{3} \cdot 2^{H_i} \right\rceil$. Merge T_i, T_i', and T_i'' into two leaf subtrees U_i and U_i' of equal weight (within ± 1). Merging three leaf subtrees into two implies that some node x' is deleted from the apex, that is, from stratum 1. Delete x' from stratum 1 using this algorithm.

We also globally rebuild the entire tree whenever $H_1 + H_2 > \left\lceil \log(n+1) + \frac{1}{\sqrt{\log(n+1)}} \right\rceil$.

2.4 Analysis

From the description of the algorithms above it can be shown that the above algorithms correctly maintain a 2-stratum tree. We have the following lemma.

Lemma 2.2 *After each update, the tree satisfies Definition 2.1.*

It remains to show that the amortized cost of maintaining a 2-stratum tree is logarithmic.

From the description of the maintenance algorithms, the following technical lemmas can be shown. The cumbersome proof of Lemma 2.3 is omitted; details of the proofs can be found in [4].

Lemma 2.3 *The following is true for a 2-stratum tree.*

(a) *When a partial rebuilding is made at a node p in a leaf subtree, at least $\left\lceil \frac{c|p|}{48H_2} \right\rceil$ updates have been made below p since the last time p was involved in a partial or global rebuilding, split, merge, or redistribution.*

(b) *When an update in a leaf subtree v causes a split, merge, or redistribution, at least $\left\lceil \frac{c \cdot 2^{H_2}}{35} \right\rceil$ updates have been made since the last time v was involved in a split, merge, or redistribution.*

(c) *When a partial rebuilding is made at a node p in the topmost tree, at least $\frac{c^2}{840} H_1 |p|$ updates have been made below p since the last time p was involved in a partial or global rebuilding.*

(d) When an update causes a global rebuilding, at least $\left\lceil \frac{\epsilon^2 \cdot 2^{H_1 + H_2}}{1225} \right\rceil$ updates have been made since the last global rebuilding.

Lemma 2.4 *The restructuring operations used in a 2-stratum tree have the following costs, including the depth first search performed to decide where to make the restructuring.*

Partial rebuilding at a node p, located in a leaf subtree: $O(|p|)$.

Split or merge of leaf subtrees: $O(2^{H_2})$.

Partial rebuilding at a node u, located in the topmost tree: $O(|u|)$.

Global rebuilding of the tree: $O(2^{H_1 + H_2})$.

Proof: Immediate from the fact that a rebuilding, split, or merge takes linear time in the sizes of the subtrees involved, including the time of a depth first search. □

Theorem 2.5 *A 2-incomplete binary search tree of height $\left\lceil \log(n+1) + \frac{1}{\sqrt{\log(n+1)}} \right\rceil$ can be maintained with an amortized cost of $O(\log n)$ per update.*

Proof: The theorem can be proved by a straightforward amortized analysis. From Lemma 2.3 follows that the number of updates between rebuilding, split, or merge operations is large enough to cover the costs of these operations, given by Lemma 2.4. □

3 Write optimality

We generalize the 2-stratum trees above by allowing k strata, for $k \geq 3$. This allows us to reduce the amortized amount of restructuring, since expensive updates high in the tree are paid by many updates lower in the tree. For convenience, we ensure that, for each stratum i, the subtrees in stratum i have the same height and are of minimum height, but we do not impose any restriction on their incomplete levels. Note that the resulting k-stratum trees can be viewed as a generalization of stratified search trees [7].

Observe that if all subtrees in stratum i have height H and size at least $2^{H-\epsilon}$, then they contribute at most ϵ to the height (above $\log n$). Thus, to obtain a tree of nearly optimal height, we make the bottom stratum contribute at most $\epsilon/2$ to the height, the second bottommost stratum contribute at most $\epsilon/4$ to the height, and so forth. However, we can ensure only that apex has minimum height, so it contributes 1 to the height. Thus, the total height is less than $\log n + 1 + \epsilon$, or less than or equal to $\lceil \log(n+1) + \epsilon \rceil$.

3.1 Construction

We define $\log^{(0)} n = n$, and, for $i \geq 1$, we define $\log^{(i)} n = \log \log^{(i-1)} n$. We define $\log^* n$ to be the smallest integer such that $\log^{(\log^* n)} n \leq 1$.

Without loss of generality, assume that $\epsilon < 1$. Let $k(N)$ be a varying parameter, where $N = n + 1$ is the weight of the tree; we refer to $k(N)$ as k for brevity. Assume that $3 \leq k \leq \log^* N - \log^*(41/\epsilon) + 1$. It is possible to construct a k-stratum tree if $\log^* N < k + 3$, but this is not necessary for our purposes.

We construct the tree as follows. Let $H_3 = 2 \lceil \log \log N \rceil + 1$, and, for $3 < i \leq k$, let $H_i = 2 \lceil \log H_{i-1} \rceil + 1$. Let $N_k = N$, and, for $1 \leq i \leq k - 1$, let N_i be the weight of the tree if the subtrees in strata $i + 1, \ldots, k$ are removed. Let $\epsilon_i = \epsilon/2^{k+1-i}$ and $\delta_i = \frac{4}{5}(1 - 2^{-\epsilon_i})$, for $2 \leq i \leq k$. For $2 \leq i \leq k - 1$, we choose $N_i = \lceil N_{i+1}/W_{i+1} \rceil$, where $W_{i+1} = \left\lceil (2 - 2\delta_{i+1})2^{H_{i+1}-1} \right\rceil + 1$; that is, we choose the weights of subtrees in stratum i to be either $W_i - 1$ or W_i. We also choose $N_1 = \lceil N_2/W_2 \rceil$, where $H_2 = \left\lfloor \frac{1}{2} \log N_2 \right\rfloor$ and $W_2 = \left\lceil (2 - 2\delta_2)2^{H_2-1} \right\rceil + 1$. We construct the tree such that the weight of the apex is N_1; for $i = 2, \ldots, k$, there are N_{i-1} subtrees in stratum i, and each subtree has weight either $\lfloor N_i/N_{i-1} \rfloor$ or $\lceil N_i/N_{i-1} \rceil$; and, for $i = 1, \ldots, k$, each subtree in stratum i is perfectly balanced and has height H_i.

Lemma 3.1 *After the construction of a k-stratum tree, for $2 \leq i \leq k$, the weight of any subtree in stratum i is either $W_i - 1$ or W_i, where $W_i = \left\lceil (2 - 2\delta_i)2^{H_i-1} \right\rceil + 1$.*

3.2 Insertion

We use an insertion algorithm similar to that of the 2-stratum tree insertion algorithm. However, we handle apex updates differently, and we use a more general multiway splitting scheme instead of a 3-way splitting scheme.

To insert a node into the apex, we perfectly rebalance the apex. To insert a node x into stratum i in a tree T, for $i > 1$, we apply the algorithm below. Note that insertions are performed in a bottom-up manner, so we first update stratum k.

1. Insert x into a subtree T_i in stratum i.

2. If height$(T_i) \leq H_i$, then exit.

3. Otherwise, find the lowest ancestor p of node x in T_i such that height$(p) > \left\lceil (1 + \frac{\gamma_i}{H_i}) \log |p| \right\rceil$, where $\gamma_i = \log(\frac{1}{1-\delta_i/8})$.

4. If such a node exists, rebalance the maximal subtree of T_i rooted at p.

5. Otherwise, no such node exists.

   (a) Locate the subtrees $U_{i1}, \ldots, U_{i,m-1}$ in stratum i closest to T_i, where $m = \left\lceil \frac{4}{3\delta_i} \right\rceil$.

   (b) If possible, redistribute nodes in T_i, $U_{i1}, \ldots, U_{i,m-1}$ to obtain T_i', $U_{i1}', \ldots, U_{i,m-1}'$, such that the following conditions hold.

      i. $|T_i'| = \left\lceil (1 - \frac{\delta_i}{4})2^{H_i} \right\rceil$.

      ii. For all j, if $|U_{ij}| \geq \left\lceil (1 - \frac{\delta_i}{4})2^{H_i} \right\rceil$, then $|U_{ij}'| = |U_{ij}|$.

      iii. For all j, if $|U_{ij}| < \left\lceil (1 - \frac{\delta_i}{4})2^{H_i} \right\rceil$, then $|U_{ij}| \leq |U_{ij}'| \leq \left\lceil (1 - \frac{\delta_i}{4})2^{H_i} \right\rceil$.

      iv. T_i', $U_{i1}', \ldots, U_{i,m-1}'$ are all perfectly balanced.

(c) Otherwise, split $T_i, U_{i1}, \ldots, U_{i,m-1}$ into $m+1$ subtrees $V_{i1}, \ldots, V_{i,m+1}$ of equal weight (within ± 1); some node x_{i-1} must be inserted into stratum $i-1$. Insert x_{i-1} using this algorithm.

3.3 Deletion

As in the case of the insertion algorithm, the deletion algorithm for k-stratum trees is similar to the deletion algorithm for 2-stratum trees. However, we do not use partial rebuilding, and we use general multiway merging.

To delete a node from the apex, we perfectly rebalance the apex. To delete a node x from stratum i in a tree T, for $i > 1$, we apply the algorithm below. Note that deletions are performed in a bottom-up manner, so we first update stratum k.

1. Delete x from a subtree T_i in stratum i.

2. If $|T_i| \geq \left\lceil (1 - \frac{5}{4}\delta_i)2^{H_i} \right\rceil$, then exit.

3. Otherwise, perform the following steps.

   (a) Locate the subtrees $U_{i1}, \ldots, U_{i,m-1}$ in stratum i that are closest to T_i, where $m = \left\lceil \frac{4}{3\delta_i} \right\rceil$.

   (b) If possible, redistribute nodes in $T_i, U_{i1}, \ldots, U_{i,m-1}$ to obtain $T_i', U_{i1}', \ldots, U_{i,m-1}'$, such that the following conditions hold.

      i. $|T_i'| = \left\lfloor (1 - \delta_i)2^{H_i} \right\rfloor$.

      ii. For all j, if $|U_{ij}| \leq \left\lfloor (1 - \delta_i)2^{H_i} \right\rfloor$, then $|U_{ij}'| = |U_{ij}|$.

      iii. For all j, if $|U_{ij}| > \left\lfloor (1 - \delta_i)2^{H_i} \right\rfloor$, then $|U_{ij}| \geq |U_{ij}'| \geq \left\lfloor (1 - \delta_i)2^{H_i} \right\rfloor$.

      iv. $T_i', U_{i1}', \ldots, U_{i,m-1}'$ are all perfectly balanced.

   (c) Otherwise, merge $T_i, U_{i1}, \ldots, U_{i,m-1}$ into $m - 1$ subtrees $V_{i1}, \ldots, V_{i,m-1}$ of equal weight (within ± 1); some node x_{i-1} must be deleted from stratum $i-1$. Delete x_{i-1} using this algorithm.

We also rebuild the tree after $\overline{N}/2$ updates since the last global rebuilding, where $\overline{N}$ is the weight of the tree during the last rebuilding.

3.4 Analysis

The analysis is similar to the one for 2-stratum trees, although the details are more cumbersome. For brevity, we omit all the details.

Lemma 3.2 *The following is true for a k-stratum tree, for $k \geq 3$.*

(a) *When a partial rebuilding is made at a node p in stratum i, for $i \geq 2$, at least $\frac{|p|}{40\sqrt{2H_i}} \cdot \frac{\epsilon}{2^{k+1-i}} \cdot \prod_{j=i+1}^{k} \left(\frac{\ln 2}{20\sqrt{2}} \cdot \frac{\epsilon}{2^{k+1-j}} 2^{H_j} \right)$ updates have been made below p since the last time p was involved in a partial or global rebuilding, split, merge, or redistribution.*

(b) *When an update in subtree v in stratum i causes a split, merge, or redistribution, for $i \geq 2$, at least $\prod_{j=i}^{k}(\frac{\ln 2}{20\sqrt{2}} \cdot \frac{\epsilon}{2^{k+1-j}} 2^{H_j})$ updates have been made since the last time v caused a split, merge, or redistribution, or the last time the tree was globally rebuilt.*

(c) *When the apex is updated, at least $\prod_{j=2}^{k}(\frac{\ln 2}{20\sqrt{2}} \cdot \frac{\epsilon}{2^{k+1-j}} 2^{H_j})$ updates have been made since the apex was last updated or the tree was globally rebuilt.*

Lemma 3.3 *The restructuring operations used in a k-stratum tree have the following costs.*

Partial rebuilding at a node p, located in a subtree in stratum i: $O(|p|)$.

Split, merge, or redistribution of subtrees in stratum i: $O(\lceil \frac{4}{3\delta_i} \rceil 2^{H_i})$.

Updating of the apex: $O(2^{H_1}) = O(2^{H_2})$.

Global rebuilding of the tree: $O(N)$.

Theorem 3.4 *A binary search tree of height at most $\lceil \log(n+1) + \epsilon \rceil$ can be maintained with an amortized cost of $O(\log n)$ per update, for any constant $\epsilon > 0$. Furthermore, the amortized amount of restructuring performed is $O(1/\epsilon^3)$ per update.*

Corollary 3.5 *There exists a sorting algorithm that requires $O(n \log n)$ time, $n \log n + O(n)$ comparisons, and $\Theta(n)$ writes in the worst case.*

Proof: The existence follows immediately from Theorem 3.4. □

4   Comments

We have presented new algorithms for updating a dictionary in $O(\log n)$ amortized time such that all operations require only $\lceil \log(n+1) + \epsilon(n) \rceil$ comparisons in the worst case, where $\epsilon(n) = 1/\sqrt{\log(n+1)}$. Observe that $\lim_{n \to \infty} \epsilon(n) = 0$. This improves upon the best previously known bound of Andersson [1]. He showed that a bound of $\lceil \log(n+1) + \epsilon \rceil$ comparisons can achieved for constant $\epsilon > 0$. However, he obtained efficient worst-case algorithms, which suggests the open question of whether there are matching upper bounds for $O(\log n)$ worst-case time and $O(\log n)$ amortized time update algorithms.

We have also presented new algorithms for updating a dictionary in $O(\log n)$ such that all operations require only $\log n + O(1)$ comparisons (in the worst case) and $O(1)$ writes (amortized). This result implies that sorting can be performed in $O(n \log n)$ time, such that only $n \log n + O(n)$ comparisons and $\Theta(n)$ writes are performed. No analogous result had been shown previously.

It is interesting to compare the power of writes and data movements for sorting. Munro and Raman [5] showed that n elements can be sorted in $O(n \log n)$ expected time with $O(n)$ data movements, a substantially weaker result. Because Munro and

Raman restrict the number of data movements, they also restrict the number of pointers a sorting algorithm can use, since an algorithm with $\Omega(n)$ pointers can sort elements indirectly via pointers, and thus avoid many data movements. In contrast, the slow write model of computation we have proposed does not allow such "loopholes," since pointer assignments require writes. Thus, we can obtain meaningful results while freely exploiting pointers.

References

[1] A. Andersson. *Efficient Search Trees.* Ph. D. Thesis, Lund University, Sweden, 1990.

[2] R. Bayer and E. M. McCreight. Organization and maintenance of large ordered indices. *Acta Informatica*, 1:173–189, 1972.

[3] D. E. Knuth. *The Art of Computer Programming*, volume 3: Sorting and Searching. Addison-Wesley, 1973.

[4] T. W. Lai. *Efficient maintenance of binary search trees.* PhD thesis, University of Waterloo, 1990.

[5] J. I. Munro and V. Raman. Sorting with minimum data movement (preliminary draft). In *Proceedings of the 1st Annual Workshop on Algorithms and Data Structures*, pages 552–562, 1989.

[6] H. J. Olivie. A new class of balanced search trees: Half-balanced binary search trees. *R.A.I.R.O. Informatique Theoretique*, 16:51–71, 1982.

[7] J. van Leeuwen and M. H. Overmars. Stratified balanced search trees. *Acta Informatica*, 18:345–359, 1983.

[8] I. Wegener. The worst case complexity of McDiarmid and Reed's variant of the bottom-up-heap sort in less than $n \log n + 1.1n$. In *Proceedings of the 8th Annual Symposium on Theoretical Aspects of Computer Science*, pages 137–147, 1991.

Nearest Neighbors Revisited

Frances Yao

Xerox Palo Alto Research Center

Abstract

The nearest neighbor graph (NNG), defined for a set of points in Euclidean space, has found many uses in computational geometry and clustering analysis. Yet there seems to be surprisingly little knowledge about some basic properties of this graph. In this talk, we ask some natural questions that are motivated by geometric embedding problems. For example, in the simulation of many-body systems, it is desirable to map a set of particles to a regular data array so as to preserve neighborhood relations, i.e., to minimize the dilation of NNG. We will derive bounds on the dilation by studying the diameter of NNG. Other properties and applications of NNG will also be discussed. (Joint work with Mike Paterson)

Competitiveness and Response Time in On-Line Algorithms

Vladimir Estivill-Castro
Department of Computer Science
York University, North York,
Ontario M3J 1P3, Canada

Murray Sherk
University of Science &
Technology in China,
Hefei, People's Republic of China

Abstract

The study of competitive algorithms has concentrated on competitiveness – comparing on-line algorithms to optimal off-line algorithms on sequences of operations. Published algorithms proven (or suggested) to be competitive invariably have pessimal response time i.e. their worst-case single operation time is as bad as possible. We consider whether or not such algorithms can be adapted to improve the response time without sacrificing competitiveness. We consider lists, off-line static search trees, dynamic search trees, and the k-server problem on a line segment of length L. For lists, pessimal response time is unavoidable. For off-line static search trees our algorithm is 2-competitive and has response time $2 \log n$. For dynamic search trees our algorithm has logarithmic amortized time and is statically optimal (like splay trees), but the response time is $O(\sqrt{n} \log n)$ and $\Omega(\log^2 n)$. For the k-server problem we prove that any algorithm with O(optimal) response time has a competitive ratio of at least $\Omega(L/k)$. This is achieved by a simple on-line algorithm. We also show that even a weak limit on response time for the k-server problem (e.g. response time less than half pessimal) yields an $\Omega(L/k)$ separation between on-line and off-line algorithms. Our results apply to high-performance multi-head disk drives where response time is critical.

1 Introduction

In many applications of data structures, a sequence of operations or requests, rather than just one operation is performed. Therefore, we are interested in the total time spent in executing the sequence. Response time may also be critical and is evaluated by a worst case analysis for the time of each individual operation. However, the sum of the worst-case times of individual operations is a pessimistic estimate of the total time. On the other hand, an average-case analysis may be inaccurate, since the assumptions on the distribution of the operations required to carry out the analysis may be false. Amortized analysis, where the cost over a worst-case sequence of operations is spread over the operations, is suitable for these situations [Meh84,Tar85]. Amortized analysis not only describes the cost of a sequence averaged over the number of operations in the sequence, but it also evaluates the quality of an algorithm, with respect to a class of algorithms, through the notion of competitiveness.

Definition 1.1 *With respect to a class of algorithms C, the* competitive ratio *of algorithm A is the supremum of* $time_A(s)/time_B(s)$ *where s is any sufficiently long sequence of requests, and B is any algorithm in the class C. (When it is clear which class is referred to, we do not specify it.) A is* α*-competitive if its competitive ratio is* α *and A is* competitive *if* α *is a constant.*

In particular, this notion is applied to evaluate *on-line* algorithms where a request must be served before future requests are known. An on-line algorithm is compared to the class of *off-line* algorithms where all requests are known before the first one has to be serviced. In all cases the requests must be serviced in the given order.

The study of on-line competitive algorithms has achieved notable success for some problems. The move-to-front heuristic for sequential lists involves moving elements to the front of the list when they are accessed. Sleator and Tarjan [ST85a] showed that the move-to-front heuristic is competitive. This means that the move-to-front strategy is as fast as any other list maintenance strategy, to within a constant factor, even if the other strategy obtains as input all future requests. Sleator and Tarjan also introduced the *splay tree* [ST83,ST85b], an on-line binary search tree that appears to be competitive in the class of binary search trees. The k-server problem was introduced in 1988 by Manasse, McGeoch and Sleator [MMS88]. This problem models many applications and involves moving k "servers" in a metric space to request points. In 1990, a simple competitive algorithm for k-servers on a line was published by Chrobak, Karloff, Payne and Vishwanathan [CKPV90]. Later the same year Fiat, Rabani, and Ravid showed a competitive solution for all metric spaces [FRR90].

Definition 1.2 *The* response time *of an algorithm is the worst-case single operation time during any sequence of operations starting from a reasonable initial state (e.g. an empty dictionary if insertions are allowed).*

The latter condition simply eliminates cases in which the response time is determined by a bad starting configuration, for example a completely unbalanced tree. Note: "Pessimal" means worst possible, i.e. the opposite of optimal. "O(optimal)" means optimal to within a constant factor.

Research into on-line algorithms has paid little attention to response time, despite the fact that, in practical on-line problems, response time is of great interest. For example, the multi-head disk is an application of the k-server problem on a line where response time is critical [CCJF85, Hof83]. We attempt to improve the response time of on-line algorithms without losing the good competitive ratios. We show that this is inherently impossible for lists, spectacularly successful for static search trees, reasonably successful (though not optimally so) for dynamic binary search trees, and impossible for the k-server problem.

2   Lists and Off-line Static Search Trees

Response time in sequential lists of n elements is always n since any request for an item not in the list forces a check of all items in the list. Hence response time in lists is always both optimal and pessimal. In one sense, move-to-front lists already have optimal response time, but this is only because improving the response time is inherently impossible. We are more interested in classes in which the response time may be improved.

One strategy for the use of search trees is to consider the sequence of requests that will have to be satisfied and choose the static tree that minimizes the total time required for searches. A static tree does not change during the sequence, so there are no update costs. On-line static trees with no knowledge of the request sequence or the distribution of requests are weak – the best we can do is choose a balanced tree. An optimal off-line static tree is a tree that minimizes the time required to fulfill a given sequence. On a sequence of requests R, an optimal off-line static tree is at least as fast as the corresponding balanced static tree and it may be faster by a factor of $\log n$.

It is easy to find a sequence for which the optimal static tree has the worst possible height, which is $n - 1$ for an n-node tree. Hence, though an optimal static tree is, by definition 1-competitive, it has pessimal response time. The following lemma shows that the response time can be improved to $2 \log n$ while increasing the competitive ratio by only a factor of two.

Lemma 2.1 *For any n-node binary search tree T, there is a binary search tree T' which contains the same nodes and has both of the following properties. (1) The height of T' is at most $2\lfloor \log_2 n \rfloor$. (2) For all nodes x in T, the depth of x in T' is at most twice the depth of x in T.*

Proof: Perfectly balance each subtree rooted at depth $\lfloor \log n \rfloor$. □

Consider an optimal static tree T. The lemma shows that it is possible to find another tree T' of logarithmic height which is "nearly" as fast as T on any sequence of any length. By sacrificing a constant factor (two) in the competitive ratio, we can achieve an exponential improvement in response time. Lemma 2.1, though simple, shows that in some classes we can have both competitiveness and $O(\text{optimal})$ response time. Moreover, Lemma 2.1 is applied in the next section where we introduce dynamic tree algorithms.

3 Dynamic Search Trees

Sleator and Tarjan's splay tree [ST83,ST85b] is an on-line binary search tree that appears to be competitive. While the question of splay tree competitiveness is still open, several significant results have supported the conjecture. Sleator and Tarjan proved that the amortized operation time in a splay tree is $O(\log n)$ and that splay trees are statically optimal. (A data structure is *statically optimal* if it is competitive with any static data structure in the same class. Note that AVL trees and red-black trees are *not* statically optimal.) Hence, on any sufficiently long sequence a splay tree is at least as fast (to within a constant factor) as an optimal static tree for the sequence. Our variation of the splay tree, the *deepsplay tree*, also has $O(\log n)$ amortized time and is statically optimal. Moreover, deepsplay tree response time is $O(\sqrt{n} \log n)$, which is a great improvement over the splay tree response time of n. Unfortunately, deepsplay tree response time is also $\Omega(\log^2 n)$, i.e. not $O(\text{optimal})$. We have no evidence that $O(\log n)$ response time is possible while maintaining good performance on sequences.

Figure 1 illustrates the three types of splay steps used in a splay tree. For each type the symmetric variant of the step is not shown, and the tree shown may be only a subtree of the entire tree. The numbers represent subtrees in which no changes occur. A *splay* of node X consists of performing splay zigzig and zigzag steps until X is the root or a child of the root. In the latter case, we do a final zig step to make X the root. A *splay tree* is a binary search tree with the standard FIND, INSERT, and DELETE algorithms except that after performing the standard algorithm we perform a splay of the deepest node accessed in the request.

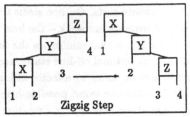
Zigzig Step

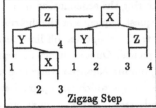
Zigzag Step

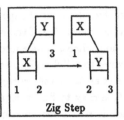
Zig Step

Figure 1: Splay steps in a splay tree

After a FIND of every node in a splay tree in order from largest to smallest, the tree will have height $n - 1$. Hence splay tree response time is pessimal. We now present the *deepsplay tree*, a variation of the splay tree with greatly improved, though not O(optimal), response time.

Definition 3.1 *A deepsplay is a splay of the leftmost deepest node in the tree. A deepsplay tree is a splay tree in which a deepsplay is performed (unrequested) every time the number of requests since the previous deepsplay is at least $\lceil \log N \rceil$, where N is the number of items in the dictionary after the current request. Note that the deepsplays are performed* in addition to *the splays included in every operation in a splay tree.*

To implement a deepsplay tree, we keep counters which record the current number of nodes N, $\lceil \log N \rceil$, and the number of requests since the previous deepsplay. After every requested operation, we update these counters and compare the last two to determine if a deepsplay should be done. The comparison and updates take constant time per operation.

To find the leftmost deepest node for a deepsplay we maintain a height marker at each node recording the height of the subtree rooted at that node. In order to keep the height markers valid, we must update them during both deepsplays and requested operations. Note that in the zigzig and zigzag cases of Figure 1, the height markers at the roots of subtrees 1 to 4 suffice to determine valid height markers for nodes X, Y and Z in constant time. The zig case is similar. Note also that any step invalidates the height markers of only those nodes which were above X at the beginning of the step. Hence we can recalculate height markers as we splay back up the search path to the root, maintaining the property that the only nodes with invalid height markers are proper ancestors of X. This recalculation requires only a constant amount of work per step, and so maintaining the height markers does not increase the time required for a splay by more than a constant factor.

Theorem 3.1 *For all request sequences R of length m, the time to do R in a deepsplay tree with any initial n_0-node configuration is $O(m + \sum_{i=1}^{m} \log n_i + n_0 \log n_0)$, where n_i is the number of nodes in the tree just after the i'th request in R is satisfied.*

Theorem 3.1 can be proven by an easy adaptation of Sleator and Tarjan's proof [ST83] for splay trees. Straight-forward application of Sleator and Tarjan's technique is not sufficient to prove static optimality for deepsplay trees. The difficulty lies in proving that the time required for the (unrequested) deepsplays is at most proportional to the time required for requested operations. To obtain a useful upper bound on the time required for deepsplays, our proof of Theorem 3.2 uses Lemma 2.1. This bound, combined with Sleator and Tarjan's technique yields the proof of deepsplay tree static optimality. Complete proofs of Theorems 3.1 and 3.2 appear in [She90].

Theorem 3.2 *Let n be the number of data items in the dictionary. Let R be any sequence of membership queries. Let t be the time required to perform R in a static tree T. Performing R in any deepsplay tree holding exactly those n data items requires time $O(t + n \log n)$.*

Thus two of the most important results pertaining to competitiveness apply to the deepsplay tree as well as the splay tree, though the competitiveness of both remains an open question. Turning our attention to the worst-case height in a deepsplay tree, we prove that deepsplay trees have dramatically better, though not O(optimal), response time.

Theorem 3.3 *Let R be any sequence of requests. If we satisfy R in an initially empty deepsplay tree, the height of the tree at the end of the sequence is $O(\sqrt{n} \log n)$, where n is the number of nodes in the tree after the sequence.*

Proof: We assume n is large enough; that is, $n/\lceil \log(2n) \rceil > \sqrt{n}$. Let H be the height of the tree after R is satisfied. Assume $H > \sqrt{n}(\lceil \log n \rceil + 1)$. Since the tree was initially empty, there must have been at least n insertions in R, so $|R| \geq n$.

Consider the subsequence R' composed of the last n requests of R (including the deepsplays during those requests). The maximum number of nodes that could be in the tree during the n requests of R' is $2n$, since at most one node can be deleted per request and there are n nodes at the end of R'. Thus, by Theorem 3.1, the time required for R' is $O(2n \log(2n)) = O(n \log n)$.

Since the maximum number of nodes in the tree during R' is at most $2n$, there must have been deepsplays at least every $\lceil \log(2n) \rceil$ requests during R'. Thus, there must have been at least $n/\lceil \log(2n) \rceil > \sqrt{n}$ deepsplays during R'. We determine a lower bound on the time for the last $\sqrt{n}$ deepsplays performed during R. There are at most $\lceil \log(2n) \rceil$ requests between deepsplays, and it is not hard to show that each request (including the deepsplay) can increase the height of the tree by at most two. Hence, the height of the tree at the time of the i'th last deepsplay of R' must have been at least $H - 2i(\lceil \log(2n) \rceil + 1)$, otherwise the height could not have increased to H by the end of R'. Thus, every one of the last $\sqrt{n}$ deepsplays must have involved following a path (to a deepest node) containing at least $H - 2\sqrt{n}(\lceil \log(2n) \rceil + 1)$ nodes. Hence the time required for each of the last $\sqrt{n}$ deepsplays in R' must be $\Omega(H - 2\sqrt{n}(\lceil \log(2n) \rceil + 1)) = \Omega(H - \sqrt{n} \log n)$.

Thus, the time required for R' (which includes those deepsplays) is $\Omega(\sqrt{n}(H - \sqrt{n} \log n)) = \Omega(\sqrt{n}H - n \log n)$. Combining this with our upper bound on the time for R' and solving for H shows that H must be $O(\sqrt{n} \log n)$. $\square$

The next theorem shows that deepsplay trees do not have O(optimal) response time.

Theorem 3.4 *The worst-case height in an n-node deepsplay tree is $\Omega(\log^2 n)$.*

Sketch of proof: Let $\log n$ be even. Identify nodes by their index in an in-order traversal. Let S_0 be the first $\log^2 n$ nodes, S_1 be the next $\log^2 n$ nodes, and so on. For all i, let F_i and L_i be the first and last nodes in S_i. We say S_i is *pure* if all nodes in S_i (except for F_i and L_i) have both F_i and L_i as ancestors. Let S_i be pure and consider the subtree T composed of nodes $F_i + 1$ to $L_i - 1$. We define the *tendril* of S_i recursively as follows. Node $F_i + 1$ is in the tendril if its right subtree is null or a single node. Any other node is in the tendril if its left child is in the tendril and its right subtree is null or a single node. Intuitively, the tendril is a thin lower part of the left spine of T. Note that the tendril may be null. Our proof of Theorem 3.4 uses three facts.

Fact 1: FIND(x) destroys the purity of a pure S_i if and only if $F_i < x < L_i$.

Fact 2: Any S_i can be made pure by a sequence of four FINDs. Combined with Fact 1, this implies that there is a sequence of FIND operations that ends with a deepsplay being performed and results in all but one of the S_i being pure.

Fact 3: The height of the tendril of a pure S_i anywhere in the deepsplay tree can be increased by at least k by a sequence of $2k + 5$ FINDs that leaves S_i pure.

Proof of Fact 3: Let x be the first (lowest value) node *not* in the tendril. Perform FIND $x, x+1, \ldots, x+2k, F_i$. Then if the depth of L_i is even, perform FIND $x + 2k - 1, F_i, L_i$, otherwise perform FIND L_i, L_i, L_i. □

We force the height of some tendril to be $\Omega(\log^2 n)$ using the following adversary strategy.
Algorithm for the Adversary:
```
1)   Make all but one S_i pure while ensuring that a deepsplay
         occurs as the last request (Fact 2).
2)   PURE = all the pure S_i.
3)   while(PURE has more than one S_i)
         begin /* Make one pass of PURE to increase heights of pure S_i */
4)         NEW = null set
5)         while(PURE has more than one S_i)
               begin Remove two of the S_i, say A and B, from PURE
6)                 Use (log n)/2 FIND requests to increase the height of
                   the tendril of A by Ω(log n) as in Fact 3
7)                 Use (log n)/2 FIND requests to increase the height of
                   the tendril of B by Ω(log n) as in Fact 3
8)                 Put A and B in NEW.
                   A deepsplay occurs. By Fact 1 this destroys the purity of
                   at most one pure S_i.
9)                 Remove the impure S_i (if there is one) from PURE or NEW.
           end
10)        PURE = NEW
       end
```

In each pass of PURE (i.e. each iteration of the outer loop), the number of S_i in PURE is reduced by a factor of two. Immediately after Line 2 the size of PURE is $\Omega(n/\log^2 n)$, so there will be $\Omega(\log n)$ passes of PURE. In each pass, every S_i that remains in PURE has its tendril increased in height by $\Omega(\log n)$. Hence the S_i left in PURE at the end of the algorithm has a tendril of height $\Omega(\log^2 n)$. □

4 The k-server Problem on a Line Segment

The k-server problem is the problem of scheduling the motion of k servers so as to serve a sequence of requests, where to serve a request is to move one of the k servers to the request site. The requests must be fulfilled on-line; that is, each request is served before any further requests

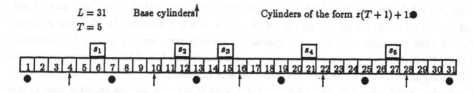

Figure 2: A model for a disk with $L = 31$ cylinders and response time $T = 5$.

are known. We study the k-server problem on a line segment or a linear array of L positions representing the cylinders on a hard disk. The servers represent disk heads. Each server occupies one point or cylinder. Repeatedly a request (a point or a cylinder r) appears. To fulfill r, each server moves along L some non-negative distance, after which the point r must be occupied by one of the servers. The serving algorithm may move the servers once more hoping to speed up later requests. The *cost* incurred in serving the request is the sum of the k distances moved, if this sum is positive; and it is 1 otherwise. A serving algorithm A is said to have *response time T* (denote $RT(A) \leq T$) if for all finite sequences $R = (r_1, \ldots, r_n)$ of requests, when A serves R the cost to serve r_i is never larger than T.

Our first result provides a lower bound on the number of servers required by all on-line algorithms with response time T. Moreover, this lower bound applies to off-line serving algorithms as well, provided that the initial position of the servers is fixed before any request is known. Initially, one may suspect that a server can fulfill requests T units to each side of its current position with cost T or less. Thus, a segment of length L can be covered by approximately $L/2T$ servers with response time T. However, one realizes that, a request to the left of a server s may drag s far enough to the left that s will not serve a request to the right in time T or less. Thus, the servers must cover smaller subintervals, possibly dynamically defined and possibly overlapping so that they can help each other and serve requests within the time bound T.

We show that to guarantee response time T, each server can cover an interval of length at most $T + 1$. We present our result for a disk of L cylinders, but an equivalent result holds for a line segment of length L. The cylinders are numbered from 1 to L and the cost of moving a server from cylinder i to cylinder j is $|i - j|$.

Theorem 4.1 *Let A be a serving algorithm (on-line or off-line with given initial configuration) for a disk with L cylinders and with $RT(A) \leq T$. If $k(T+1) < L$, then there is a sequence $R = (r_1, \ldots, r_n)$ with $r_j \leq k(T+1) + 1$ such that A must use $(k+1)$ servers to service R.*

Proof: Contrary to the claim of the theorem, suppose that A can fulfill all sequences $R = (r_1, \ldots, r_n)$ of requests with $1 \leq r_j \leq k(T+1) + 1$ only moving $q \leq k$ servers and $RT(A) \leq T$.

For a server s, we denote by $p_i(s)$ the position of server s (the cylinder number under head s) just before request r_i is presented to A. Thus, $p_0(s)$ is the initial position of server s. Let $s_1, \ldots, s_q$ be the servers named such that $p_0(s_j) \leq p_0(s_{j+1})$, for all $j = 1, \ldots, q-1$. Note that no server needs to pass over another server; if algorithm A requires server s_j to pass over server s_{j+1}, we replace this move by moving s_j until the position of s_{j+1} and moving s_{j+1} to the desired position. Therefore, $p_i(s_j) \leq p_i(s_{j+1})$, for $j = 1, \ldots, q-1$ and for all $i \geq 0$.

We will propose a sequence R of requests such that, in order to fulfill R with response time T, algorithm A must move its servers into an increasingly uneven spread. The sequence $R = (r_1, \ldots, r_n)$ of requests will be long enough that this would be impossible. In fact, r_i is the cylinder with number of the form $z(T + 1) + 1$ (z a nonnegative integer) that is farthest away from all current positions occupied by the servers of A. We prove the theorem by defining a potential function Φ that measures how unevenly spread the servers are over $k(T + 1) + 1$ cylinders. We show that Φ grows by a constant after each request is served. We obtain a contradiction since, over all configurations of the servers, the value of Φ is bounded.

A cylinder of the form $1 + \lfloor (2j + 1)(T + 1)/2 \rfloor$ with $0 \le j < k$ will be called a base cylinder; see Figure 2. Let $m_i(s)$ be the minimum distance at time i of a server s to a base cylinder:

$$m_i(s) = \min_{0 \le j < k} \left| p_i(s) - 1 - \left\lfloor \frac{(2j + 1)(T + 1)}{2} \right\rfloor \right|.$$

Let Φ_i be the potential before request r_i has appeared and defined by $\Phi_i = \sum_{j=1}^{q} m_i(s_j)$. Let $d_i[r]$ be the distance of cylinder r from the set of q servers before request r_i has appeared; that is,

$$d_i[r] = \min_{1 \le j \le q} |r - p_i(s_j)|.$$

Let μ_i be such that $0 \le \mu_i \le k$ and $\mu_i(T + 1) + 1$ is the cylinder of the form $z(T + 1) + 1$ that is farthest away from all current positions of the servers; that is,

$$d_i[\mu_i(T + 1) + 1] = \max_{0 \le z \le k} d_i[z(T + 1) + 1].$$

We define the sequence R of requests by $r_i = \mu_i(T + 1) + 1$. This implies $d_i[r_i] \ge (T + 1)/2$. Consider algorithm A working on R. Algorithm A must move a server to serve request r_i with no more that T steps. Say A sends server s to serve r_i. This costs $|p_i(s) - r_i|$. Since $RT(A) \le T$ we have $|p_i(s) - r_i| \le T$. Thus, algorithm A is left with $T - |p_i(s) - r_i|$ steps to rearrange servers in preparation for request r_{i+1}.

Suppose $p_i(s) \le r_i$. Thus, $\mu_i > 1$ and, since $d_i[r_i] \ge (T + 1)/2$ we have, $(T + 1)/2 \le d_i[r_i] \le r_i - p_i(s)$. This and $r_i = \mu_i(T + 1) + 1$ imply $p_i(s) \le r_i - (T + 1)/2 = 1 + (2\mu_i - 1)(T + 1)/2$. Since $p_i(s)$ is an integer, we have $p_i(s) \le 1 + \lfloor (2\mu_i - 1)(T + 1)/2 \rfloor$. Moreover, $|p_i(s) - r_i| \le T$ implies that the cylinder $1 + \lfloor (2\mu_i - 1)(T + 1)/2 \rfloor$ is the closest base cylinder to s before request r_i appears. Thus, $m_i(s) = 1 + \lfloor (2\mu_i - 1)(T + 1)/2 \rfloor - p_i(s)$. Once s is at cylinder r_i, no base cylinder is closer to s than base cylinder $1 + \lfloor (2\mu_i + 1)(T + 1)/2 \rfloor$. Therefore, Φ has increased by

$$1 + \left\lfloor \frac{(2\mu_i + 1)(T + 1)}{2} \right\rfloor - r_i - m_i(s) = p_i(s) - r_i + \left\lfloor \frac{(2\mu_i + 1)(T + 1)}{2} \right\rfloor - \left\lfloor \frac{(2\mu_i - 1)(T + 1)}{2} \right\rfloor$$

$$= p_i(s) - r_i + T + 1.$$

We know that algorithm A has at most $T - (r_i - p_i(s))$ steps to reduce this. Thus, A can decrease the increment in Φ to 1 and no less. Therefore, $\Phi_{i+1} \ge \Phi_i + 1$ as required.

The case $p_i(s) > r_i$ is proved similarly. $\qquad\square$

Theorem 4.1 allows us determine the optimal response time for a disk with k servers. We use it with $T = \lceil |L|/k \rceil - 1$ for a lower bound and obtain the following result.

Theorem 4.2 *Consider a disk with L cylinders and k servers, where $k < L$. Then, the optimal response time is $\lceil L/k \rceil - 1$ and this is achieved by the following algorithm. Divide the cylinders into k intervals. For $i = 1$ to $k - 1$ the i-th interval has $\lceil L/k \rceil$ cylinders, namely $(i - 1)\lceil L/k \rceil + 1$, $(i - 1)\lceil L/k \rceil + 2, \ldots i\lceil L/k \rceil$. The last interval has $L - (k - 1)\lceil L/k \rceil$ cylinders. Assign a server to each interval, who fulfills all requests that fall in its interval.*

Let A be a serving algorithm with response time T. Since for each request an adversary is incurring a cost of at least 1 and A is incurring a cost no larger than T, then A has a competitive ratio of T. Hence, the algorithm in Theorem 4.2 achieves optimal response time and has a competitive ratio of $\lceil L/k \rceil - 1$. Moreover, all serving algorithms with O(optimal) response time have a competitive ratio of order L/k as proven by the following theorem.

Theorem 4.3 *Consider a disk with L cylinders and k servers, where L is much larger than k. Let c be a constant with $1 \leq c < k$. If A is a serving algorithm with response time cL/k, (that is, A is a factor of c from optimal response time) then A has a competitive ratio of at least $L/k\left[(k-c)/(k-1/2)\right] - 1/(k-1/2)$; that is, $\Omega(L/k)$.*

Proof: Let $T = cL/k$ be the response time guaranteed by algorithm A. A must keep, at all times, at least one server in the first $T + 1$ cylinders. Let t be the largest integer such that $T + 1 + t + 2t(k-1) \leq L$. Consider an adversary that places its i-th server at cylinder $c_i = T + 1 + t + (i-1)2t$, for $i = 1, \ldots, k$. Then, there is a sequence that shows that A has a competitive ratio of $2t$. Each time the adversary makes a request it chooses the cylinder in $\{c_1, \ldots, c_k\}$ that maximizes A's cost at that time. This implies that the competitive ratio is at least $2t = L/k\left[(k-c)/(k-1/2)\right] - 1/(k-1/2)$ as required. $\square$

This means that the algorithm in Theorem 4.2 not only achieves optimal response time but it also achieves the best possible competitive ratio, to within a constant factor over all algorithms that have O(optimal) response time.

Theorem 4.4 *Any on-line k-server algorithm that limits response time to T, with $L - T \in \Omega(L)$, has competitive ratio $\Omega(L/k)$ compared to off-line algorithms that limit response time to T.*

Proof: Use the adversary strategy in the proof of Theorem 4.3 which places requests only on the last $L - (T+1)$ cylinders. The on-line algorithm must keep a server in the first $T + 1$ cylinders (in case cylinder 1 is requested). The off-line algorithm is under no such restraint since all requests happen to be in the last $L - (T + 1)$ cylinders. $\square$

This last result shows an interesting phenomenon. We know that if response time is not limited, there is an on-line algorithm as fast, to within a constant factor, as any off-line algorithm on any sequence [CKPV90]. By Theorem 4.4, if response time is limited then off-line algorithms are faster than on-line algorithms by a factor of $\Omega(L/k)$ on some sequences.

5 Final remarks

Response time is a critical factor in many applications and it is natural to strive for competitive algorithms that guarantee O(optimal) response time. We have demonstrated that, for some problems, we can achieve optimal or nearly optimal (within a constant factor) response time without sacrificing competitiveness. For other problems, we have shown that we can greatly improve response time without (apparently) sacrificing competitiveness. But, for yet other problems, improving the response time implies sacrificing competitiveness with algorithms that do not limit response time. We demonstrated the effect of response time in the k-server problem on a line. In this problem, ignoring response time allows constant competitive ratios. However, the situation changes completely when response time is considered because the best competitive ratio achievable for algorithms with O(optimal) response time are, unfortunately, large. Moreover, the same large ratios separate on-line algorithms from off-line algorithms when both respect limits on response time (even such weak limits as half pessimal).

Does competitiveness on worst-case sequences give an appropriate measure of the desirability of an algorithm? Consider a k-server on a line situation where the limit on response time is twice optimal i.e. $2L/k$. We can meet this limit with $k/2$ "guard" servers, each of which services requests in a segment of length $1 + 2L/k$. What should be done with the remaining $k/2$ servers? Our results show that there is always at least one bad sequence that forces a competitive ratio of $\Omega(L/k)$; so we may as well use all k servers as guards to reduce response time (and hence competitive ratio) to the optimal L/k. Yet it is apparent that using the extra $k/2$ servers as "rovers" not constrained to any particular place on the line, allows an on-line algorithm to be competitive on many more sequences. We leave open the question of a measure of performance that gives insight into this behaviour.

References

[CCJF85] A.R. Calderbank, E.G. Coffman Jr, and L. Flatto. Sequence problems in two-server systems. *Mathematics of Operations Research*, 10(4):585–598, November 1985.

[CKPV90] M. Chrobak, H. Karloff, T. Payne, and S. Vishwanathan. New results on server problems. In *Proc. of the Symp. on Discrete Algorithms*, pages 291–300. SIAM, 1990.

[FRR90] A. Fiat, Y. Rabani, and Y. Ravid. Competitive k-server algorithms. In *Proceedings of the Symposium on Foundations of Computer Science*, pages 454–463. IEEE, 1990.

[Hof83] M. Hofri. Should the two-headed disk be greedy? – yes, it should. *Information Processing Letters*, 16:83–85, February 1983.

[Meh84] K. Mehlhorn. *Data Structures and Algorithms, Vol 1: Sorting and Searching*. EATCS Monographs on Theoretical Computer Science. Springer-Verlag, 1984.

[MMS88] M. Manasse, L. McGeoch, and D. Sleator. Competitive algorithms for on-line problems. In *Proceedings of the Symposium on Theory of Computing*, pages 322–333, New York, 1988. ACM.

[She90] M. Sherk. *Self-Adjusting k-ary Search Trees and Self-Adjusting Balanced Search Trees*. PhD thesis, U. of Toronto, 1990. Available as Comp. Sci. Technical Report 234/90.

[ST83] D. Sleator and R. Tarjan. Self-adjusting binary trees. In *Proceedings of the Symposium on Theory of Computing*, pages 235–245, New York, 1983. ACM.

[ST85a] D. Sleator and R. Tarjan. Amortized efficiency of list update and paging rules. *Communications of the ACM*, 28(2):202–208, February 1985.

[ST85b] D. Sleator and R. Tarjan. Self-adjusting binary search trees. *Journal of the ACM*, 32:652–686, July 1985.

[Tar85] R. Tarjan. Amortized computational complexity. *SIAM Journal of Alg. Disc. Meth.*, 6(2):306–318, April 1985.

A Linear Time Optimal Via Assignment Algorithm for Three-Dimensional Channel Routing

Jan-Ming Ho[1]
Institute of Information Sciences
Academia Sinica
R. O. C.

Abstract

A three-dimensional channel refers to a 3-D rectangular block with multiple routing layers. Terminals exist only on the top and the bottom layers and they form two well-aligned 2-D rectangular channels. In this paper, we consider a special version in which the three-dimensional channel contains only three layers. The routing algorithm is as follows. First, a channel routing algorithm is applied to both the top and the bottom layer to route the terminals belonging to the same net on the same layer. The second step is to form another channel routing problem as defined below. A net N is said to be an *inter-layer net* if it contains terminals on both the top and the bottom layers. Two via positions in the middle layer are selected for each inter-layer net N. The first via is chosen from the position immediately below one of the terminals belonging to N on the top layer, while the second is chosen from the position immediately above one of the terminals belonging to N on the bottom layer. Notice that it thus defines a channel routing problem containing only two-terminal nets in the three respective layers. The channel routing algorithm can then applied to complete the routing. In this paper, we present a linear time optimal via assignment algorithm for the second step decribed above such that number of incompletely routed nets are minimized.

[1]This work has been supported in part by the National Science Council, R.O.C., under Grant NSC 79-0404-E-001-01.

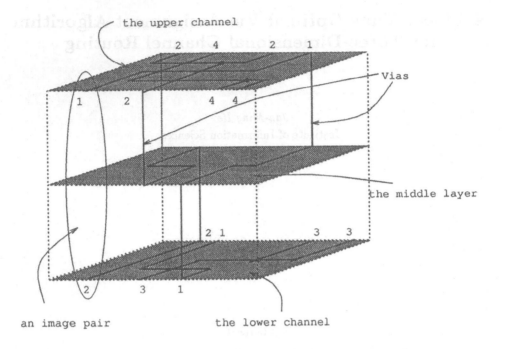

Figure 1: A three-dimensional channel; where nets 1 and 2 are inter-layer nets, while nets 3 and 4 are not.

1 Introduction

The three-dimensional channel routing problem has been studied by C.C. Tong and C.L. Wu [2]. In a two-dimensional channel routing problem, a channel is a rectangular region. Terminals to be connected by wires exist on the upper and the lower boundaries of the channel. Interior of the rectangle is used as the routing region. A three-dimensional channel (see figure 1) is defined as a multilayer rectangular block containing two well-aligned channels on the top and the bottom layers respectively. Terminals does not exist in the middle layers. A net specifies a set of terminals to be connected. A net is said to be an *inter-layer* net if it contains terminals on both the top and the bottom layers. Vias must be used to complete the connection of an inter-layer net. Subset N' of terminals of an inter-layer net N is called the *upper subnet* (*lower subnet*) of N if N' contains the terminals of N on the top (bottom) layer. Both upper and lower subnets are also called *intra-layer* subnets. In [2], channels are defined as having three layers. The routing algorithm first performs channel routing on the top and the bottom layers to complete the connection of the terminals on the same layer. For the inter-layer nets, two vias per net, one for the upper subnet and the other for the lower subnet, are assigned to bring the subnets to the boundaries of the middle layer to form another two-dimensional channel routing problem. The via assigned to a terninal of a upper (lower) subnet refers to the wire segment which is used to bring the terminal to the middle layer. A terminal t' on the upper (lower) layer is said to be the *image* of a terminal t on the lower (upper) layer if t and t' align in the z-direction. Note that a terminal t and its image t' cannot both be assigned the via between them, unless t and t' belong to the same net, to avoid electrically connecting two disjoint nets. A terminal and its image are said to form a *positional conflict* in [2]. The upper (lower) subnet of an inter-layer net N is said to be blocked if no via is assigned to it, i.e., each grid point immediately below (above) its terminals belongs to a via assigned to other nets. An inter-layer net is said to be *blocked* if either its upper subnet or its lower subnet is blocked. The objective of the optimal via assignment problem for three-dimensional channel routing, *OVA problem* for short, is to minimize the number of blocked intra-layer subnets.

In [2], a *constraint graph* is defined as follows. A vertex of the constraint graph is a terminal of an inter-layer net. An edge exists between two vertices if the corresponding terminals either belong to the same upper (or lower) subnet of an inter-layer net or form a positional conflict. That is, an intra-layer subnet forms a clique in the constraint graph. The construction of the constraint graph takes $O(n^2)$ time. Tong and Wu give characterizations of the constraint graph and presented an $O(n^2)$ algorithm for the *OVA problem*. In this paper, we present a linear-time algorithm for the *OVA problem*. In section 2, we give a new formulation of the *OVA problem* based on the notion of a *constraint hyper-graph* and we also describe the linear-time *algorithm OVA*. Note that construction of the initial constraint hyper-graph takes only linear time. Correctness of the *algorithm OVA* is asserted in section 3. We also consider the objective of minimizing the number of incompletely routed nets. In section 4, we present a *modified algorithm OVA* which optimizes the new objective.

Before proceeding, let's consider a special case of the *maximum cardinality bipartite matching problem* which we'll also refer to as the *SBM* problem. A bipartite graph is usually denoted as $G = (X, Y, E)$, where $X \cup Y$ is the set of vertices, $X \cap Y = \emptyset$, and each edge $e = (x, y)$ in E connects a vertex x in X and a vertex y in Y. A matching M of G is a set of edges such that each vertex $v \in X \cup Y$ has degree 1 in M. A maximum cardinality bipartite matching is thus defined as a matching of a bipartite graph having the maximum cardinality. In the *SBM* problem, we consider a special case in which each vertex in X has degree no greater than 2. We'll show in the following that the SBM problem is reducible to the OVA problem.

Lemma 1 *The SBM problem is reducible to the OVA problem.*

Proof: As a preprocessing step to the OVA problem, for each inter-layer net N, we can assign a via which goes across layer 1 and 3 to connect two terminals t and t' if t and t' both belongs N and they align in the z-direction. The two intra-layer subnets of N are then regarded as two independent nets. It can be shown that the optimality of the OVA problem remains the same after the preprocessing step in that the number of blocked intra-layer subnets is not changed. Without loss of generality, we assume that for every pair of terminals t and t' of N, t and t' are not aligned in the z-direction.

For each inter-layer net N, a vertex $y_N \in Y$ is created. For each grid position p in the middle layer, a vertex $x_p \in X$ is created if there is a terminal t right above or below p. For each terminal t in the 3D channel, let y_N denote the vertex in Y correponding to the net N containing t, and x_p be the vertex in X corresponding to the adjacent position p of t in the middle layer. An edge $e = (y_N, x_p)$ is created. Note that it's not difficult to show that an optimal solution to the OVA problem implies an optimal solution of the above instance of the SMB problem, and vice versa. $\diamond$

Note that by this reduction, it immediately follows that the OVA problem is solvable in $O(n^{2.5})$ time [1].

2 The Constraint Hyper-Graph and the OVA Algorithm

The constraint hyper-graph is a dynamic data structure used to describe the *OVA problem* at its current status. An intra-layer subnet is said to be *active* if there has not been any via assigned to it. A *super-node* is associated with each active intra-layer subnet and it can be implemented, say, as a linked list. Note that a super-node corresponds to a clique in the constraint graph defined in [2]. A terminal is said to be *active* if it belongs to an active inter-layer subnet and its adjacent grid position in the middle layer has not been assigned to any via. We associcate a *vertex* with each active terminal. An edge connecting two vertices is called a *P-edge* if the corresponding terminals form an image pair. Note that a P-edge $e = (u, v)$ is the candidate of the via position for one of the active intra-layer subnets containing the terminals corresponding to the vertices u and v. A via can always be assigned to a terminal, if it is a leaf, i.e., the corresponding vertex does not have a P-edge connecting to it. In which case, the terminal is called a *leaf*. A constraint hyper-graph can thus be denoted as a triple $H = (V, N, E)$, where V denotes the set of vertices, N denotes the set of super-nodes, and E is the set of P-edges. Notice that N is a partition of V. The *associated constraint graph* of H is the graph $G_H = (V, E')$, where E' is the union of E and the edges contained in the cliques defined by super-nodes in N. Edges in $E' - E$ are also referred to as the N-edges.

Initialization of the constraint hyper-graph H can be done as the following. We are going to use four arrays *U1, L1, U2,* and *L2* to store information of the terminals belonging to each side of the two channels, where *U1* and *U2* belong to the upper channel and *L1* and *L2* belong to the lower channel. Since a net is specified as a set of terminals in terms of their coordinates in the 3D channel, we can easily determine if a net is an inter-layer net and we can also create the assocaited super-nodes in time proportional to the size of the net. The information of each terminal can be written to one of the four arrays in constant time. *U1* and *L1* are then examined simultaneously to construct the set of P-edges E and the set of vertices V. In this way, the constraint hyper-graph is created in a total of $O(n)$ time, where n is the number of grid point on each side of the upper and lower channels. Tong and Wu's algorithm constructs the *associated constraint graph* G_H explicitly. Since every intra-layer subnet constitutes a clique in G_H, the construction of G_H takes $O(n^2)$ time. Note that our data structure maintains the cliques implicitly in the linked-lists representing the super-nodes to keep the time complexity down to $O(n)$. In this paper, we are going to use the notations super-node and cliques interchangeable since they only differ in the implementation details.

The vertices are given priorities 1, 2 and 3 as the following. A vertex is given priority

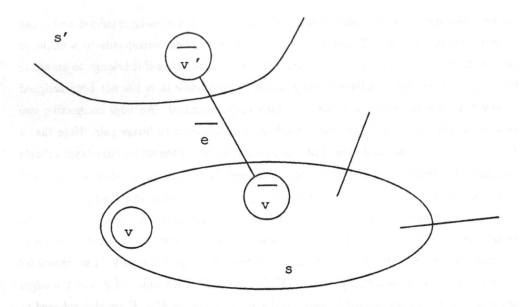

Figure 2: The procedure *CLEAR(v)*.

1 if it is a leaf. If a *P*-edge connects two vertices corresponding to terminals in the same inter-layer net, then the two vertices are also given priority 1. A vertex is given priority 2 if it correponding to the last terminal *t* left in the super-node containing *t*. All the other vertices are given priority 3. Priorities of the vertices can be computed in a total of $O(n)$ time. Note that the priority of each vertex change dynamically during the execution of *algorithm OVA*

Let *s* denote the super-node containing *v*. We first introduce the procedure *CLEAR(v)* as below. The procedure *CLEAR(v)* (see figure 2) first examines every vertex $\bar{v}$ in *s*, $\bar{v} \neq v$. If $\bar{v}$ is a leaf, then it is simply deleted form *s*. Otherwise, let $\bar{v}'$ be the vertex connected to $\bar{v}$ by the *P*-edge $\bar{e}$. Then, $\bar{v}'$ is marked as a leaf with its priority being raised to 1 and the *P*-edge $\bar{e}$ is deleted from *E*. After all the other vertices in *S* are removed, *v* is deleted from *s* and *s* is deleted from *N*.

A variable *INCOMPLETE* is used to store the number of failures in assigning a via to each intra-layer subnet and is initialized to 0. *Algorithm OVA* then iteratively selects from *V* a vertex *v* of the lowest priority, assigns a via to the intra-layer subnet containing *t* and updates the data structures according to the following procedure, where *t* denotes the active terminal corresponding to v. If v is a leaf, *algorithm OVA* assigns the adjacent via to v and calls $CLEAR(v)$. Otherwise, if v is not a leaf and the priority of v is 1, then *algorithm OVA* assigns the via associated with e, the P-edge connecting v and its image v', to v v' and deletes e from E. Procedure $CLEAR(\bullet)$ is then performed on both v and v'.

If the priority of v is 2 or 3, then *algorithm OVA* performs the following before calling $CLEAR(v)$. Let v' be the vertex connected to v by a P-edge e and s' be the super-node containing v'. If s' contains v' as the last vertex (in this case, the priority of v and v' are both 2), then *INCOMPLETE* is incremented by one, v' is removed from s' and e is deleted from E, and s' is removed from N. If s' contains more than two vertices, then v' is deleted from s' and the P-edge e is deleted from E after the following steps are performed. If s' contains two vertices v' and v'', then the priority of v'' is either 1 or 3. If the priority of v'' is 3, then it is decreased to 2 to denote that it now becomes the last vertex in s'. If the priority of v'' is 1, then it remains unchanged. The *algorithm OVA* terminates when there is no vertex in V.

Note that we can maintain V as three lists of vertices, each being distinguished by the priority of the vertices it contains. The operation of selecting the vertex of the lowest priority in V and that of updating the priority of a vertex then both takes $O(1)$ time. We also notice that the priority of a vertex only changes non-increasingly. Obviously, *algorithm OVA* runs in linear time.

Theorem 1 *Algorithm OVA runs in $O(n)$ time.*

In the $O(n^2)$ time algorithm of Tong and Wu's, the bottleneck lies in identifying and breaking loops in the graph obtained by contracting the cliques in G_H. While *algorithm OVA* avoids these operations.

3   Correctness of Algorithm OVA

Let $H = (V, N, E)$ be the constraint hyper-graph describing an instance of the three-dimensional channel routing problem, and $G_H = (V, E')$ be the associated constraint graph. Let $G_1, G_2, \ldots, G_l$ be the connected components of G_H. Tong and Wu [2] has shown that there is at most one failure per connected component in assigning vias to the intra-layer sub-nets. *Algorithm OVA* basically makes use of this characterization. To assert the correctness of *algorithm OVA*, we first point out the sufficient condition for a failure in assigning vias to the intra-layer subnets to occur.

Definition 1 *A connect component G_i is said to be* distinguished *if G_i has the following properties:*
(1) G_i does not have a leaf nor a pair of images belonging to the same net;
(2) G_i is a tree of the cliques corresponding to the intra-layer subnets.

Lemma 2 *There is at least one failure in assigning vias to a distinguished connected component G_i.*

Proof: Let q be the number of cliques in G_i. In other words, G_i contains q intra-layer subnets. Since the cliques form a tree, there are $q - 1$ P-edges. Also note that each P-edge allows at most one via, thus at least one of the intra-layer subnets fails to attaining a via.
◇

Notice that *algorithm OVA* always assigns vias to the intra-layer nets of G_i with at most one failure if G_i is distinguished.

Lemma 3 *Algorithm OVA assigns vias to the intra-layer nets of G_i with at most one failure if G_i is distinguished.*

Proof: We shall prove by induction on the number of vertices contained in G_i. According to the definition of a distinguished component, the minimum size of G_i is two, v and v', each belonging to a distinct super-node s and s', respectively. Obviously only one of the intra-layer subnets corresponding to s and s' will get the via while the other fails. Now, let's assume that for every distinguished component G' containing less than k vertices, there is at most one failure in assigning vias to the intra-layer nets of G' by *algorithm OVA*, where k is the size of G_i and $k > 2$. Since G_i is distinguished, a clique s at the leaf of G_i must be contain only one vertex v. By our priority scheme, v gets the priority 2. Also notice that the priority of a vertex w at an internal clique of G_i gets priority 3. Thus the first vertex selected by *algorithm OVA* has a priority 2. Let's denote this vertex as v and v' as the vertex connected to v by a P-edge e. Denote s' as the super-node containing v'. If s' contains v' as the only vertex, since G_i is a connected component, we immediately induce that there are two vertices in G_i. Which contradicts our previous assumption of $k > 2$. Thus s' contains at

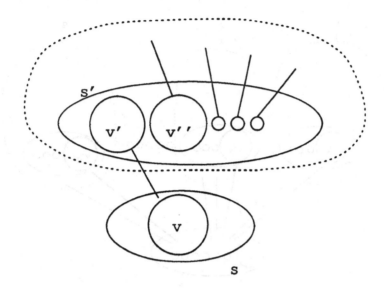

Figure 3: Illustration of the processing of a distinguished G_i.

least two vertices (see figure 3). In processing v, *algorithm OVA* will assign the via associated with e to v and it will delete v' from s'. Obviously, G_i remains distinguished at its new state and its size is decreased by two. Thus, we can apply the principle of induction to conclude that *algorithm OVA* will process G_i with a single failure.

◇

Lemma 4 *If there is a vertex with priority 1 in the connected component G_i, then vias can be assigned to the intra-layer nets of G_i without failure.*

Proof: This can be proved by using induction on the number of vertices in G_i. If v is the only vertex in G_i, then it must be the leaf and we can assign a via to v without conflict. Now, let's assume that the lemma is true for every connected component having less than k vertices, where $k > 1$ is the number of vertices in G_i. Now let's consider the first vertex v in G_i selected by *algorithm OVA*. v is either a leaf or is contained in an image pair both belonging to the same inter-layer net according to the definition of priority 1.

(1) v is a leaf. Then a via can be assigned to v without conflict and we can apply the procedure *CLEAR (v)*. Denote s as the super-node containing v. Note that s contains at

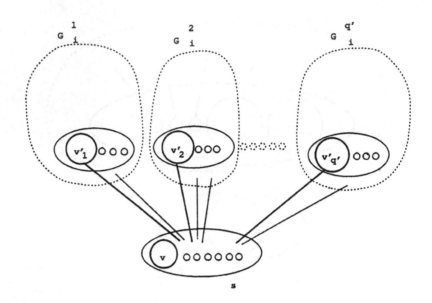

Figure 4: Illustration of the proof of lemma 4-(1).

least two vertices. After applying $CLEAR\ (v)$, G_i is decomposed into smaller components $G_i^1, G_i^2, \ldots, G_i^{q'}$ (see figure 4). Each G_i^j contains at least a leaf v_j' which is the image of a vertex $v_j \neq v$ containing in s. Note that the size of G_i^j is smaller than k. By the hypothesis of induction, vias are also assigned to G_i^j without failures.

(2) v is contained in a P-edge $e = (v, v')$, and both v and v' belong to the same inter-layer net. Denote s and s' as the super-nodes containing v and v' repectively. Note that e can be (a) contained in a cycle of G_i or (b) otherwise (see figure 5). *algorithm OVA* will assign the via associated with e to v, and perform $CLEAR(\bullet)$ on both v and v' and yields several connected components $G_i^1, G_i^2, \cdots, G_i^{q'}$ in both cases (a) and (b). Similar to the discussions in (1), we know that the components each contains at least a leaf and size of each G_i^j is smaller than k. Again, by the hypothesis of induction, vias are successfully assigned to G_i^j.
◇

Lemma 5 *Algorithm OVA assigns vias to G_i optimally.*
Proof: Equivalently, we show that *algorithm OVA* assigns vias to G_i with

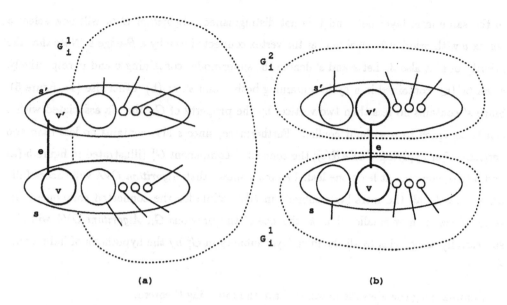

Figure 5: Illustration used in the proofs of lemma 4-(2) and 5.

(1) one failure if G_i is distinguished;

(2) zero failure if otherwise.

Again, we are going to use induction on the number of vertices in G_i. It's obvious that the equivalent statement is satisfied when the size of G_i is one or two. Now, we assume that the equivalent statement is satisfied by every connected component G' of size less than k, the size of G_i. (1) is the result of lemma 2 and 3. If G_i contains a leaf or a P-edge connecting two vertices belonging to the same inter-layer net, then (2) is satisfied as asserted by lemma 4. Thus, we can assume G_i as having neither a leaf nor a P-edge connecting two vertices belonging to the same inter-layer net, and G_i is not distinguished. We are going to show that G_i satisfies (2) under these assumptions. First, we know that the priority of each vertex of G_i must be either 2 or 3. If G_i contains a vertex of priority 2, i.e., the super-node s containing v has only one vertex. Then similar to the arguments we used in the proof of lemma 3, we conclude that *algorithm OVA* will recursively delete such vertices until none is left. Note that G_i is still connected and still holds the three properties of G_i mentioned previously, i.e. it does not have neither a leaf nor a P-edge connecting two vertices belonging to the same inter-layer net, and it is not distinguished. *Algorithm OVA* will now select a vertex v with priority 3. Denote v' as the vertex connected to v by a P-edge e. Note that the priority of v' is also 3. Let s and s' denote the super-nodes containing v and v' respectively. e can be (a) contained in a cycle containing both s and s'; or (b) otherwise (see figure 5). Since s' contains no less than two vertices by the property of G_i, the via associated with e can be assigned to v without conflict. Furthermore, since s also contains no less than two vertices, after applying $CLEAR(v)$, the connected component G_i^1 (illustrated in figure 5-(a) and 5-(b)), contains at least one leaf. Lemma 4 shows that *Algorithm OVA* will successfully assign vias to all the intra-layer subnets in G_i^1. While for the connected component G_i^2, since its size is now smaller than k, the size of the previous G_i, *Algorithm OVA* will also successfully assign vias to all the intra-layer subnets in G_i^2 by the hypothesis of induction.

◇

Summarizing the above lemmas, we have the following theorem.

Theorem 2 *Algorithm OVA optimally assigns vias to the intra-layer subnets of a three-dimensional channel routing problem such that the number of intra-layer subnets failed in getting a via to the middle layer is minimized.*

4   The Modified Algorithm OVA'

In this section, we study the problem of optimally assigning vias to the inter-layer nets such that the number of unrouted nets is minimized. As we've seen in the above lemmas, *algorithm OVA* yields one failure in assigning vias to the intra-layer subnets of a connected component G_i if only if G_i is *distinguished*. There is a potential of blocking both intra-layer subnets s_1 and s_2 of an inter-layer net n if s_1 and s_2 happen to appear in two distinct distinguished components G_i and $G_j, i \neq j$, while *algorithm OVA* fails to assign vias to both s and s'. Whenever one such pair is not selected simultaneously, the number of unrouted nets is decreased by one. It can be easily seen that the problem of minimizing the number of unroutable nets is equivalent to that of finding the *maximum cardinality matching*. We'll first present a formal discussion of the approach, then present the algorithm.

First, we are going to construct a *component graph* $G^c = (V^c, E^c)$, where a vertex v_i in V^c denotes a distinguished component and an edge $e = (v_i, v_j)$ in E^c denotes the set of inter-layer nets with one of its intra-layer subnet in G_i and the other in G_j. Let $M \subseteq E$ be the maximum cardinality matching of G^c. For every edge $e = (v_i, v_j) \in M$, let $s = s_1 \cup s_2$ denote one of the inter-layer nets associated with e, where s_1 in G_i and s_2 in G_j denote the two intra-layer subnets of s. We can modify G_i and G_j by removing s_1 and s_2 respectively which yields several connected components. By removing s_1 and G_i, for each edge $e' = (v, v')$ connecting a vertex v in s_1 to another vertex $v' \in V$, the vertex v' is now designated as a leaf whose priority becomes 1 and the edge e' is removed from E, and we also remove the vertex v from V and s_1 from N. Here $H = (V, N, E)$ is the hyper-graph representing the OVA problem. Each of the connected components, say G_i yielded by removing the super-nodes (the intra-layer subnets) contains at least one leaf and no more failures will be incurred in future assignments of the vias to G_i.

We can now present the *algorithm OVA'* as the following.

1. Perform a depth-first search to identify the connected components of the associated constraint graph G_H;

2. Run *algorithm OVA* to identify those connected components which are distinguished;

3. Construct the *component graph* G^c;

4. Compute the maximum cardinality matching M of G^c;

5. Remove the selected super-nodes from the hyper-graph H according to M;

6. Perform another phase of *algorithm OVA* to assign the vias.

Proof of correctness is not difficult and is omitted.

Theorem 3 *Algorithm OVA minimizes the number of unrouted nets for the three-dimensional channel routing problem.*

Note that every step except step 4 of *algorithm OVA'* takes $O(n)$ time and step 4 can be performed in $O(m^{1/2} n_s)$ time [1], where n_s and m denote the number of edges and the number of vertices in G^c respectively.

Theorem 4 *Algorithm OVA' runs in $O(n + m^{1/2} n_s)$ time, where n_s and m denote the number of edges and the number of vertices in G^c respectively.*

5 Conclusion

In this paper, we've studied the problem of three-dimensional channel routing. Note that the projections of the problem on the top and the bottom layers are each two-dimensional channel routings. The intra-layer subnets can thus be connected simultaneously using traditional (two-dimensional) channel routing algorithms. To electrically connect the intra-layer subnets belonging to the same inter-layer net, we can assign proper vias to one of the terminals per intra-layer subnet to connect it vertically to the middle layer. This defines another (two-dimensional) channel routing problem in the middle layer in which only two-terminal nets are given. Unfortunately, in general, the vias can not be completely assigned to all the intra-layer subnets. We presented a linear time algorithm to minimize the number of failures in doing via assigment. We also presented an almost linear time algorithm to minimize the number of incomplete nets under this routing method.

References

[1] C.H. Papadimitriou and K. Steiglitz. *Combinatorial Optimization: Algorithms and Complexity.* Prentice-Hall, Inc., 1982. p. 247.

[2] C.C. Tong and C.L. Wu. Optimizing positional conflicts in three-dimensional channel routing. manuscript, 1990.

Symmetry of Information and One-Way Functions

Luc Longpré and Sarah Mocas
College of Computer Science
Cullinane Hall, Northeastern University
Boston, MA 02115

Abstract

Symmetry of information (in Kolmogorov complexity) is a concept that comes out of formalizing the idea of how much information about a string y is contained in a string x. The situation is symmetric because it can be shown that the amount of information contained in the string y about the string x is almost exactly the same as that contained in x about y.

In this paper we address symmetry of information in resource bounded environments. While we show that symmetry still holds in space bounded environments, it probably doesn't hold in time bounded environments. We show that if it holds for polynomial time bounds, then one-way functions cannot exist, even using a weak definition for one-way functions, where a function is one-way if any polynomial time computable function fails to compute the inverse on at least a super polynomial fraction of the strings.

1   Introduction

In probability theory, the phenomenon of dependence between random variables is well known. Cast in terms of classical Shannon entropy [Sha48, Sha49], the quantity of information in a random variable Y about another random variable X is $I(X, Y) = H(X) - H(X \mid Y)$, where $H(X)$ is the entropy and $H(X \mid Y)$ is the conditional entropy. An interesting fact is that this quantity is symmetric: $I(X, Y) = I(Y, X)$, so it is called the *mutual information*.

Kolmogorov complexity is one way of formalizing the quantity of information contained in a string. It has the advantage that it depends on the string itself, as opposed to the string as part of a collection of strings and a probability distribution associated with the strings. The Kolmogorov complexity, $K(x)$, of a string x with respect to a Universal Turing Machine M ([Sol64, Kol65, Cha66]) is the size of the smallest program y that outputs x:

$$K(x) = \min\{l \mid l = |y| \text{ and } M(y) = x\}.$$

The *conditional* Kolmogorov complexity, $K(x \mid z)$, of a string x with respect to the Turing Machine M is

$$K(x \mid z) = \min\{l \mid l = |y| \text{ and } M(\langle y, z \rangle) = x\}.$$

The quantity of information contained in y about x can be defined as follows:

$$I(x,y) = K(x) - K(x \mid y).$$

As it was shown in [ZL70] (where the result is attributed to Kolmogorov and Levin independently), this definition is also commutative, if one neglects some small additive quantities, as a result of the following theorem:

Theorem 1 (Symmetry of Information) *To within an additive term of* $O(\log K(xy))$,

$$K(xy) = K(x) + K(y \mid x).$$

Corollary 2

$$|I(x,y) - I(y,x)| \in O(\log(K(xy))).$$

The difference can even be reduced if one considers self-delimiting Kolmogorov complexity, as discovered by Levin [Lev74] and Gács [Gac74] (see also [LV88]).

But there is one problem associated with this notion of dependence between random variables. While two random variables may be dependent, they may look totally independent to a polynomial time bounded observer. Yao [Yao82] addresses this issue by defining *effective entropy* and *effectively independent variables*. He also hints at a connection between these notions and cryptography.

A similar problem arises in the definition of mutual information in strings. We would like to address the issue of the validity of symmetry of information in a resource bounded environment. For this, we need to use resource bounded Kolmogorov complexity.

The following definitions for time and space bounded K-complexity and conditional time and space bounded K-complexity are commonly used. For a specific (universal) Turing machine M, time bound $T(n)$, space bound $S(n)$ and integer m, define

$$KT(x, T(n)) = \min\{l \mid l = |y| \text{ and } M(y) = x, \text{ using at most } T(|x|) \text{ time}\}$$

$$KT(x \mid m, T(n)) = \min\{l \mid l = |y| \text{ and } M(\langle y,m\rangle) = x, \text{ using at most } T(|x|) \text{ time}\}$$

$$KS(x, S(n)) = \min\{l \mid l = |y| \text{ and } M(y) = x, \text{ using at most } S(|x|) \text{ space}\}$$

$$KS(x \mid m, S(n)) = \min\{l \mid l = |y| \text{ and } M(\langle y,m\rangle) = x, \text{ using at most } S(|x|) \text{ space}\}$$

To simplify equations, we need to introduce a collection of functions as parameters of resource bounded K-complexity. For example, we would like to be able to say that

$$KS(x, O(S(n))) \leq KS(y, O(S(n))).$$

This will allow us to talk about symmetry of information in time or space bounded environment without having a very specific bound. Informally, the previous equation should mean that if there is a program that can print y in space $c_1 S(|y|)$, then there is a smaller program that can print x in space $c_2 S(|x|)$, for some c_2 depending on c_1, but not on y. To that effect, we introduce some notation below that will formalize what we mean by that statement. The following definition is in terms of a string x, but could be easily generalized to be in terms of more than one string (it will be used in terms of a pair of strings x and y).

Definition 3 *For a collection of functions C and for a pair of functions f_1 and f_2 that accept as parameters a string and a function from C,*

$$f_1(x, C) \preceq f_2(x, C)$$

if and only if

$$(\forall S_2 \in C)\,(\exists S_1 \in C)\,(\forall x)\ f_1(x, S_1) \le f_2(x, S_2)$$

Definition 4

$$f_1(x, C) \asymp f_2(x, C)$$

if and only if

$$f_1(x, C) \preceq f_2(x, C) \text{ and } f_2(x, C) \preceq f_1(x, C)$$

It is reasonable to ask whether Symmetry of Information is still valid in resource bounded environments. In section 2, we show that Symmetry of Information still holds in a space bounded environment:

Space bounded symmetry theorem. *Let $S(n) \ge n$ and $l(n)$ be a non-decreasing $S(n)$ space computable function. If $|y| = l(|x|)$, then to within an additive term of $O(\log KS(xy \mid |xy|, O(S(n))))$,*

$$KS(xy \mid |xy|, O(S(n))) \asymp KS(x \mid |x|, O(S(n))) + KS(y \mid x, O(S(n)))$$

Whether Symmetry of Information holds for time bounded environments is an even more interesting question. It would seem that here we loose symmetry. A reasonable formulation of Symmetry of Information for time bounded K-complexity that parallels the formulation for space bounds is as follows:

Time bounded symmetry hypothesis. *Let $l(n)$ be a non-decreasing $T(n)$ time computable function. Then to within an additive term of $O(\log KT(xy \mid |xy|, T(n)))$, $\forall x, y$ such that $|y| = l(|x|)$,*

$$KT(xy \mid |xy|, T(n)) \asymp KT(x \mid |x|, T(n)) + KT(y \mid x, T(n))$$

Because we are concerned mainly with the cryptographic significance of this hypothesis we will actually be concerned with a weaker statement, where the polynomials involved can be different on each side of each inequality.

Polynomial time bounded symmetry hypothesis. *Let $l(n)$ be a non-decreasing polynomial time computable function. Then to within an additive term of*

$$O(\log KT(xy \mid |xy|, \text{Poly})),$$

$\forall x, y$ *such that* $|y| = l(|x|)$,

$$KT(xy \mid |xy|, \text{Poly}) \asymp KT(x \mid |x|, \text{Poly}) + KT(y \mid x, \text{Poly})$$

We show in section 3 that if the hypothesis holds, then no one-way functions can exist. We do this by showing that if the hypothesis holds, then one can in polynomial time invert any polynomial time 1-1 function on an arbitrarily large fraction of inputs. This in turn implies that no Public-Key CryptoSystem can exist.

There have been other attempts to connect the existence of one-way functions to randomness. A significant amount of research has been done to connect the existence of one-way functions with the existence of secure pseudo-random number generators. Levin [Lev87] showed that functions that are one-way on their iterates exist if and only if secure generators exist. The concept of a function that is one-way on its iterates is somewhat unnatural. Later, the existence of functions one-way on their iterates have been shown to be equivalent to more natural one-way functions: regular one-way functions [GKL89], functions one-way in a nonuniform model [ILL89] and functions one-way in a uniform model [Haa90].

Allender [All87] also discovered a connection between Kolmogorov complexity and pseudo-random number generators, which in turn implies a connection with one-way functions through the above equivalences. He showed that if there is a set of small deterministic complexity which contains only Kolmogorov complex strings, then there are bounds on how secure a pseudo-random generator can be. This paper offers a more direct connection between Kolmogorov complexity and one-way functions.

(Some of the results in this paper appeared in a different form in the first author's Ph.D. thesis [Lon86] and are still unpublished.)

2 Space Bounded Symmetry of Information

Symmetry of information holds in a time bounded environment.

Theorem 5 *[Lon86] Let $S(n) \geq n$ and $l(n)$ be a non-decreasing function computable in space $S(n)$. Then to within an additive term of $O(\log KS(xy \mid |xy|, S(n)))$, $\forall x, y$ such that $|y| = l(|x|)$,*

$$KS(xy \mid |xy|, O(S(n))) \asymp KS(x \mid |x|, O(S(n))) + KS(y \mid x, O(S(n)))$$

Proof. The proof technique for the unbounded symmetry can be applied to the space bounded case without much change. The proof will be included in the final paper.

3 Time bounded symmetry and One-Way functions

The proof for space bounds involved computing the rank of an element in a set. Since ranking elements does not seem to be possible in a time bounded environments, it would seem that we loose symmetry in polynomial time bounded environments.

In this section, we show that if Symmetry of Information holds for polynomial time bounds, then one-way functions cannot exist. Since many believe that there are one-way functions, this gives evidence to the effect that symmetry of information does not hold for polynomial time bounds.

But first we need to define the notion of one-way function that will be used in this section.

Definition 6 *A function f is defined to be* honest *if there is a polynomial p such that for every y in range(f) there is an x in domain(f) so that if $f(x) = y$ then $|x| \leq p(|y|)$.*

Definition 7 *A function f is* one-way *if it is computable in polynomial time, it is total, one-one, honest, and for any polynomial time computable function g, for some polynomial q, for all sufficiently large n, g fails to compute $f^{-1}(x)$ (more formally, $g(f(x)) \neq x$) on at least a fraction $1/q(n)$ of the strings x of size n.*

Many other definitions of one-way functions have been studied in the literature. In all definitions, the function f must be easy to compute and hard to invert. A more accepted definition allows probabilistic computation for the inversion and requires the inversion to fail everywhere except for a polynomial fraction of the strings. Since we take one of the weakest definition of one-way function, our result still holds when using the more accepted definitions of one-way function. In other words, we show that Symmetry of Information for polynomial time implies that every function can be inverted in deterministic time on almost all inputs, this implies that every function can be inverted in probabilistic time on a super polynomial fraction of the inputs.

Definition 8 *A function f is* non-decreasing on length *if $|x| \leq |y| \Rightarrow |f(x)| \leq |f(y)|$*

Theorem 9 *If the Polynomial time bounded symmetry hypothesis holds, then one-way functions do not exist.*

Lemma 9.1 *If f is a polynomial time computable function which is non-decreasing on length, then up to an additive constant, for all x, y such that $f(x) = y$,*

$$KT(yx \mid |yx|, \text{Poly}) \preceq KT(x \mid |x|, \text{Poly})$$

Proof. We show a method for printing yx given $|yx|$. For $1 \leq i \leq |yx|$, assume $|x| = i$. Compute x' from i, using the program that generates x from $|x|$. Compute y' from x' using f. Check to see if $|y'x'| = |yx|$. If so then print $y'x'$. Since f is non-decreasing on length, this process is guaranteed to find a unique i that works and will generate the correct yx. $\qquad\square$

Lemma 9.2 *If f is a 1-1, total function computable in polynomial time and f is non-decreasing on length, then for all but a polynomial fraction of strings x, if $y = f(x)$,*

$$KT(x \mid |x|, \text{Poly}) - KT(y \mid |y|, \text{Poly}) \preceq O(\log KT(yx \mid |yx|, \text{Poly})).$$

Proof. We need only consider $f(x) = y$ where

$$KT(y \mid |y|, \text{Poly}) \preceq KT(x \mid |x|, \text{Poly}).$$

(In any case, the K-complexity of a function cannot increase by more than a constant by applying a polynomial time function.) Consider any string x such that $K(x \mid |x|) > n - \log n$, where $n = |x|$. All but a polynomial fraction of the strings have this property. Now, considering the image of those strings through f, since only a polynomial fraction of the strings have Kolmogorov complexity less than $n - \log n$, and since f is 1-1, all but a polynomial fraction of the strings x map to strings y of high Kolmogorov complexity. For those strings,

$$KT(x \mid |x|, \text{Poly}) - KT(y \mid |y|, \text{Poly}) \preceq O(\log KT(yx \mid |yx|, \text{Poly})).$$

$\square$

Proof of theorem. One-way functions exist if and only if one-way functions that are non-decreasing on length exist. This can be seen by using a simple padding argument.

Let f be a 1-1, honest polynomial time computable function and assume that Symmetry of Information is true for polynomial time. Without loss of generality, assume f is non-decreasing on length. Let $f(x) = y$. The Symmetry of Information Hypothesis tells us that:

$$KT(y \mid |y|, \text{Poly}) + KT(x \mid y, \text{Poly})$$

$$\preceq KT(yx \mid |yx|, \text{Poly}) + O(\log KT(yx \mid |yx|, \text{Poly}))$$

Rewriting the right inequality

$$KT(x \mid y, \text{Poly}) \preceq KT(yx \mid |yx|, \text{Poly}) - KT(y \mid |y|, \text{Poly})$$

$$+ O(\log KT(yx \mid |yx|, \text{Poly}))$$

By Lemma 9.1 we note that the complexity of printing x given y is no more difficult than the difference between printing x given the length of x and printing y given the length of y plus $O(\log KT(yx \mid |yx|, \text{Poly}))$.

$$KT(x \mid y, \text{Poly}) \preceq KT(x \mid |x|, \text{Poly}) - KT(y \mid |y|, \text{Poly})$$

$$+ O(\log KT(yx \mid |yx|, \text{Poly}))$$

Using Lemma 9.2, we arrive at the conclusion that for all but a polynomial fractions of strings x,

$$KT(x \mid y, \text{Poly}) \preceq O(\log KT(yx \mid |yx|, \text{Poly})).$$

Now, to invert f, given any element y in the range of f we need only simulate up to polynomially many programs for some fixed polynomial time each to find an x such that $f^{-1}(y) = x$. Such an x can be verified by computing $f(x)$. This will work for all but a polynomial fraction of the strings x. $\square$

4 Open Problems

Obviously it would be interesting to know if the converse of Theorem 8 is true, i.e., under the assumption that one-way functions do not exist can it be shown that Symmetry of Information for polynomial time holds. This would make a very strong connection between one-way functions and symmetry of information.

Since this seems to be hard to prove, another possible direction would be to examine symmetry of information under the assumption that P = NP. This line of inquiry arises from the well known relationship between one-way functions and the assumption that P = NP. Under a very general definition of one-way functions, P = NP if and only if there are no one-way functions. A question for investigation is, does assuming P = NP imply symmetry of information holds for polynomial time?

If this still seems hard, then what assumption is needed for symmetry of information to hold for polynomial time? Surely, P = P$^{\#P}$ is sufficient...

References

[All87]   E. Allender. Some consequences of the existence of pseudorandom generators. In *Proc. 19th Annual ACM Symposium on Theory of Computing*, pages 151–159, 1987.

[Cha66]   G. Chaitin. On the length of programs for computing finite binary sequences. *J. Assoc. Comput. Mach.*, 13:547–569, 1966.

[Gac74]   P. Gács. On the symmetry of algorithmic information. *Soviet Math. Dokl.*, 15:1477, 1974.

[GKL89]   O. Goldreich, H. Krawczyk, and M. Luby. On the existence of pseudo-random generators. In *Proc. 21st Annual ACM Symposium on Theory of Computing*, pages 25–32, 1989.

[Haa90]   J. Håstad. Pseudo-random generators under uniform assumptions. In *Proc. 22nd Annual ACM Symposium on Theory of Computing*, pages 395–404, 1990.

[ILL89]   R. Impagliazzo, L. Levin, and M. Luby. Pseudo-random generation from one-way functions. In *Proc. 21st Annual ACM Symposium on Theory of Computing*, pages 12–24, 1989.

[Kol65]   A. Kolmogorov. Three approaches for defining the concept of information quantity. *Prob. Inform. Trans.*, 1:1–7, 1965.

[Kol68]   A. Kolmogorov. Logical basis for information theory and probability theory. *IEEE Trans. Information theory*, IT-14.5:662–664, 1968.

[Lev74]   L. Levin. Laws of information conservation (non-growth) and aspects of the foundation of probability theory. *Problems in Information Transmission*, 10:206–210, 1974.

[Lev87]   L. Levin. One-way functions and pseudorandom generators. *Combinatorica*, 7(4):357–363, 1987.

[Lon86]   L. Longpré. *Resource bounded Kolmogorov complexity, a link between computational complexity and information theory*. PhD thesis, Cornell University, 1986. Technical Report TR86-776.

[LV88] M. Li and P.M.B. Vitányi. Two decades of applied kolmogorov complexity. In *Proc. Structure in Complexity Theory third annual conference*, pages 80–101, 1988.

[Sha48] C.E. Shannon. A mathematical theory of communication. *Bell System Technical Journal*, 27:479–523 (Part I) and 623–656 (Part II), 1948.

[Sha49] C.E. Shannon. Communication theory of secrecy systems. *Bell System Technical Journal*, 28:656–715, 1949.

[Sol64] R. Solomonoff. A formal theory of inductive inference, part 1 and part 2. *Information and Control*, 7:1–22, 224–254, 1964.

[Yao82] A. Yao. Theory and applications of trapdoor functions. In *Proc. 23rd IEEE Symposium on Foundations of Computer Science*, pages 80–91, 1982.

[ZL70] A.K. Zvonkin and L.A. Levin. The complexity of finite objects and the development of the concepts of information and randomness by means of the theory of algorithms. *Russ. Math. Surv.*, 25:83–124, 1970.

A Linear Time Algorithm to Recognize the Double Euler Trail for Series–Parallel Networks

Lih–Hsing Hsu, J. Y. Hwang, T. Y. Ho and C. H. Tsai

Department of Information and Computer Science, National Chiao Tung University,
Hsinchu 30050, Taiwan, R.O.C.

1. Introduction

A *series–parallel network* (*network*, in short) N defined on a set Y of *type* t ∈ {L, S, P} (which respectively means *leaf*, *series* and *parallel*) is recursively defined as follows:

(1) N is a network of type L if $|Y| = 1$.

(2) N is a network either of type P or of type S if $|Y| > 1$, and there exist $k \geq 2$ *child subnetworks* $N_1, N_2, ..., N_k$ such that each N_i is defined on set Y_i of type t_i with $t_i \neq t$ and $\{Y_i\}$ forms a partition of Y.

Usually, every network is represented by a tree structure like Figure 1. Every node together with all of its descendants forms a *subnetwork* of N. The subnetwork of N formed by a child of the root is called a child subnetwork of N. The *leaf node* is labelled by y if it is a subnetwork of type L defined on {y}. Every *internal node* is labelled by either S or P according to the type of the subnetwork that it and all of its descendants form. Every node together with some of its descendants forms a *partial subnetwork* of N. The definition of the network is very natural because all boolean functions can be expressed in this form, with the series connection implementing logical–and and the parallel connection implementing logical–or. For example, the boolean function e∧(a∨b)∧(c∨d) corresponds to the network in Figure 1. Note that the order of every subtree of a given tree is immaterial because different orders of the subtrees will lead to the same boolean function.

Historically, every network is represented by a series–parallel graph (s.p. graph, in short), which is an edge–labelled graph with two distinct vertices. We can recursively represent a network by an s.p. graph as follows:

(1) Every network N defined on $Y = \{y\}$ of type L is represented by an edge–labelled graph G[N] with only one edge labelled by y and two vertices as distinct vertices.

(2) Let N be a network of type S with $N_1, N_2, ..., N_k$ as the child subnetworks of N and G[N_i] be an s.p. graph that represents N_i with distinct vertices $\{a_i, b_i\}$ for every i. We identify b_i with a_{i+1} for every $1 \leq i \leq k-1$. The resultant graph G[N] with distinct vertices $\{a_1, b_k\}$ represents the network N.

(3) Let N be a network of type P with $N_1, N_2, ..., N_k$ as the child subnetworks of N and G[N_i] be an s.p. graph that represents N_i with distinct vertices $\{a_i, b_i\}$ for every i. We identify all a_i's to get a new vertex a and identify all b_i's to get a new vertex b. The resultant graph G[N] with distinct vertices $\{a, b\}$ represents the network N.

However, such graph representation for a network is not unique because the order of its subnetworks and the order of the two distinct vertices are fixed in each graph representation. For example, there are two non–isomorphic s.p. graphs in Figure 2, both representing the network in

Figure 1. All black nodes in these two graphs indicate distinct vertices.

Let G[N] be a graph representation for network N and M be a partial subnetwork of N. Then $G_M[N]$ denotes the induced s.p. subgraph of G[N] induced by M. A *walk* in a graph $G = (V,E)$ is a finite non–null sequence $W = v_0, e_1, v_1, e_2, ..., e_k, v_k$ whose terms are alternately vertices and edges, such that the ends of e_i are v_{i-1} and v_i for $1 \leq i \leq k$. A *section* of the walk W is a walk that is a subsequence $v_i, e_{i+1}, v_{i+1}, e_{i+2}, ..., e_j, v_j$ of consecutive terms in W. If the edges $e_1, e_2, ..., e_k$ of walk W are distinct, W is called a *trail*. A trail that traverses every edge of G is called an *Euler trail* of G. A connected graph has an Euler trail if and only if either there does not exist any vertex of odd degree or there are exactly two vertices of odd degree. A network has an *Euler trail* if it has an Euler trail in some G[N].

We define the *dual network* N' for network N by interchanging the types of S and P for every internal node of the tree structure for N. For example, the network in Figure 3(a) is the dual of the network in Figure 1, and the corresponding s.p. graph is in Figure 3(b). Note that the corresponding boolean functions for N and N' are dual to each other.

Obviously, (N')' = N. Let G[N] and G[N'] be graph representations for network N and its dual N'. A sequence of edges $e_1, e_2, ..., e_m$ is a *common trail* for (G[N],G[N']) if there exist $x_0, x_1, ..., x_m \in V(G[N])$ and $x'_0, x'_1, ..., x'_m \in V(G[N'])$ such that $L = x_0, e_1, x_1, e_2, ..., e_m, x_m$ forms a trail in G[N] and $L' = x'_0, e_1, x'_1, e_2, ..., e_m, x'_m$ also forms a trail in G[N']. A network N has a *double Euler trail* (*DET*) if there is a common Euler trail for both some G[N] and some G[N']. In this case, we say that (G[N],G[N']) *realizes a DET pair* (L,L') for N where L and L' are the corresponding Euler trails in G[N] and G[N'], respectively. Figure 2(a) and Figure 3(b), for example, do not have a common Euler trail but Figure 2(b) and Figure 3(b) have a common Euler trail, say a, b, e, c, d.

Let G[N] and G[N'] be graph representations for network N and its dual N', and *DCT(G[N],G[N'])* be the minimum number of edge–disjoint common trails that cover all edges of G[N] and G[N']. Then *DCT(N)* is defined to be the minimum of DCT(G[N],G[N']) among all possible G[N] and G[N']. Hence DCT(N) = 1 if and only if some (G[N],G[N']) realizes a DET pair. The problem of finding DCT(N) is very important in VLSI layout [1,2,3,4,5].

Uehara and VanCleemput[5] show that optimizing the layout area of a CMOS functional cell for a network N is equivalent to solving DCT(N). They propose a very primitive method to find edge–disjoint common trails that cover (G[N],G[N']). Bruss et al. [1] solve DCT(G[N],G[N']) for any given pair (G[N],G[N']) and develop a linear time algorithm to recognize DET networks. But their approach is based on the concepts of representative graphs and finite state machines so that it is not easy to understand. Maziasz et al. [2] also present a linear time algorithm to compute DCT(G[N],G[N']) for any given pair (G[N],G[N']). In this paper, we use a different approach, survey some important properties of DET networks and then present a linear time algorithm to recognize DET networks.

2. Trail cover types

A set of edge–disjoint common trials is a *trail cover* of $(G[N],G[N'])$ if it covers all the edges of $(G[N],G[N'])$. If $K = \{(T_1,T_1'),\ (T_2,T_2'),...,(T_k,T_k')\}$ is a trail cover, $K' = \{(T_1',T_1),\ (T_2',T_2),...,(T_k',T_k)\}$ is defined as the *dual trail cover* for K. Let N be a DET network which realizes a DET (L,L') in $(G[N],G[N'])$ and M be a partial subnetwork of N such that both $G_M[N]$ and $G_{M'}[N']$ are connected s.p. subgraphs. Edges of $(G_M[N],\ G_{M'}[N'])$ form a set of maximal sections of (L,L'). Hence, a trail cover for $(G_M[N],\ G_{M'}[N'])$ is obtained. We use $K(M,L)$ to denote such trail cover for $(G_M[N],\ G_{M'}[N'])$ induced by (L,L'). Let z_1 and z_2 be the distinct vertices of $G_M[N]$, and z_1' and z_2' be the distinct vertices of $G_{M'}[N']$. If M is a network of leaf type, both $G_M[N]$ and $G_{M'}[N']$ contain only one edge. Then $K(M,L)$ may be a trail cover of the following types:

Type *1*: A trail cover contains exactly one common trail (T,T'), which satisfies one of the following conditions. (1) T begins at z_1 and terminates at z_2 ; T' begins at z_1' and terminates at z_2'. (2) T begins at z_2 and terminates at z_1 ; T' begins at z_2' and terminates at z_1'. In other words, both T and T' are in the same direction.

Type *2*: A trail cover contains exactly one common trail (T,T'), which satisfies one of the following conditions. (1) T begins at z_1 and terminates at z_2 ; T' begins at z_2' and terminates at z_1'. (2) T begins at z_2 and terminates at z_1 ; T' begins at z_1' and terminates at z_2'. In other words, T and T' are in opposite directions.

Obviously, the dual trail cover for a trail cover of type 1 (type 2) is one of type 1 (type 2). Any network is recursively constructed from its subnetworks of leaf type. Thus, if N is not a network of leaf type, $K(M,L)$ is just a trail cover induced by (L,L') and there are others adjacent to $K(M,L)$ in (L,L'). The other trail covers may be those of type 1 or 2 (or others that are unknown for the time being, but they are also generated from those of type 1 or 2).

For any $(G[N_1],G[N_1'])$ and $(G[N_2],G[N_2'])$, let x_1,x_2 be distinct vertices of $G[N_1]$, x_1',x_2' be distinct vertices of $G[N_1']$, y_1,y_2 be distinct vertices of $G[N_2]$, y_1',y_2' be distinct vertices of $G[N_2']$, $G[N_1 \, \sigma \, N_2]$ denote the series connection of $G[N_1]$ and $G[N_2]$ by identifying x_2 with y_1, and $G[N_1' \, \pi \, N_2']$ denote the parallel connection of $G[N_1']$ and $G[N_2']$ by identifying x_1' with y_1' and identifying x_2' with y_2'. In addition, let N be a DET network, which possesses a DET (L,L') in $(G[N],G[N'])$ such that G[N] contains $G[N_1 \, \sigma \, N_2]$ as a connected s.p. subgraph and G[N'] contains $G[N_1' \, \pi \, N_2']$ as a connected s.p. subgraph. If N_1 and N_2 are networks of leaf type, $K(N_i,L)$ is one of type 1 or 2 for $i = 1,\ 2$. The possible trail covers of $(G[N_1 \, \sigma \, N_2],\ G[N_1' \, \pi \, N_2'])$ in (L,L') can be calculated and some new types will be introduced. For example, assume that $K(N_i,L)$ is one of type 1 and (T_i,T_i') be the only common trail in $K(N_i,L)$ for $i = 1,\ 2$. Suppose that T_1 and T_2 are in opposite directions; say that T_1 begins at x_1 and terminates at x_2, and T_2 begins at y_2 and terminates at y_1 ($= x_2$). Then both T_1 and T_2 must be the ending section of L. But this is impossible. Hence, T_1 and T_2 are in the same direction. W.l.o.g., let us say that T_1 begins at x_1 and terminates at x_2, and T_2 begins at x_2 and terminates at y_2. In this case, T_1' begins at x_1' and terminates at x_2', and T_2' begins at x_1' and terminates at x_2'. Hence, (T_1, T_1') and (T_2, T_2') are the ending and the beginning section of (L, L'), respectively. Therefore, the trail cover $\{(T_1, T_1'), (T_2, T_2')\}$ in $(G[N_1 \, \sigma \, N_2], G[N_1' \, \pi \, N_2'])$ introduces a new type, which results from the connection of two trail covers of type 1, and the dual trail cover for a trail cover of the new type is one of this type.

Repeating the above procedure (calculate, introduce new types and take their dual) until no new type can be introduced, we will conclude that all trail covers induced by (L, L') must be those of 37 types, whose topological representations are showed in Table 1, and obtain a σ operation, which is defined on the trail covers of the 37 types and showed as Table 2, where a blank entry indicates void; i.e., it is impossible for any two trail covers of the types in the row and the column to be adjacent in (L, L'). (For this operation, the one in the row is placed above the one in the column.) For the topological representation, same common trails in a trail cover have the same color whereas different common trails have different colors. Besides, the symbols "6", "9", "ρ" and "∂" are used to represent the beginning or ending section of L or L' and for a trail cover, the arrows in the same common trails just indicate the relative direction between these common trails so that it is allowable to reverse the direction of them simultaneously.

3. Trail cover classes

By interchanging z_1 and z_2, or z_1' and z_2', all trail covers can be partitioned into 16 classes.

The 16 classes are:

a. $[2,2]$: includes trail covers of types 1 and 2.

b. $[2,0]$: includes trail covers of types 3 and 4.

c. $[0,2]$: includes trail covers of types 5 and 6.

d. $[\theta, \phi]$: includes trail covers of types 7.

e. $[\phi, \theta]$: includes trail cover of types 8.

f. $[x,x]$: includes trail covers of types 9, 10, 11 and 12.

g. $[(x,x)+(2,2)]$: includes trail covers of types 13, 14, 15 and 16.

h. $[(x,x)+(0,2)]$: includes trail covers of types 17, 18, 19 and 20.

i. $[(x,x)+(2,0)]$: includes trail covers of types 21, 22, 23 and 24.

j. $[\phi x, \theta x]$: includes trail covers of types 25 and 26.

k. $[\theta x, \phi x]$: includes trail covers of types 27 and 28.

l. $[\theta x, \theta x]$: includes trail covers of types 29 and 30.

m. $[(\theta x, \theta x)+(2,2)]$: includes trail covers of types 31 and 32.

n. $[(\phi x, \theta x)+(0,2)]$: includes trail covers of types 33 and 34.

o. $[(\theta x, \phi x)+(2,0)]$: includes trail covers of types 35 and 36.

p. $[y,y]$: includes trail covers of type 37.

The "2" means that a trail begins at a distinct vertex and terminates at the other distinct vertex.

The "0" means that a trail begins and terminates at the same distinct vertex. The "θ" means that two trails respectively behave like "0" and separate the graph into two parts, up and down. The "ϕ" means that two trails respectively behave like "2" and separate the graph into two parts, left and right. The "x" and "y" respectively mean that a trail is the beginning or ending section of L or L', and a trail is the beginning and ending section of L or L' so that any further connection with it will be flunked. Thus, "θx" ("ϕx") means that two trails are the beginning section and the ending section of L or L', and separate the graph into two parts, up and down (left and right). Moreover, the dual trail covers for the trail covers in class $a, b, c, d, e, f, g, h, i, j, k, l, m, n, o$ and p are in class $a, c, b, e, d, f, g, i, h, k, j, l, m, o, n$ and p respectively.

The interchange for distinct vertices means that the graph representation of networks can be properly chosen so that it is unnecessary to distinguish a trail cover from the others in same classes. Therefore, Table 2 can be easily transformed to Table 3. Having the further observation for Table 3, we can further partition the 16 classes into five groups, A, B, C, D and E , where A = $\{a,b,c\}$, B = $\{d,e\}$, C = $\{f,g,h,i\}$, D = $\{j,k,l,m,n,o\}$ and E = $\{p\}$. Any trail cover in classes of E reflects a DET trail but any further connection with it will be flunked out. Any trail cover in classes of D reflects that it may get a DET trail only if it is properly connected with the trail cover in classes of A or B. However, it will be in classes of D or E for any further connection. Others have similar properties.

Assigning a *score* for each trail cover K as follows: score(K) = 0 if K is in classes of A or B; score(K) = 1 if K is in classes of C; score(K) = 2 if K is in classes of D and score(K) = 3 if K is in classes of E, we have the following theorem.

Theorem 1. *Assume that $(G[N], G[N'])$ realizes a DET trail (L,L'). Let M be a subnetwork of N. Then the trail cover $K(M,L)$ can be hierarchically classified into 37 types, 16 classes and 5 groups; and the score of $K(M,L)$ is equal to the score of its dual. Moreover, if $M_1, M_2, ..., M_k$ are child subnetworks of M, then*

$$score(K(M,L)) \geq \sum_{i=1}^{k} score(K(M_i,L)) \quad and \quad \sum_{i=1}^{k} score(K(M_i,L)) \leq 2.$$ ∎

4. Trail decomposition

It is observed from Table 3 that there are two entries which indicate some series connections to get class a; say $a \in a \ \sigma \ b$, and $a \in b \ \sigma \ a$. In other words, $[2,2]_S = [2,2] \ \sigma \ [2,0]$ is the only way to derive $[2,2]_S$. Note that the subscript indicates a trail cover for some network of type S. Besides, the subscript P is used to indicate a trail cover for some network of type P. Similarly, $[2,0]_S = [2,2] \ \sigma$ $[2,2]$ and $[2,0]_S = [2,0] \ \sigma \ [2,0]$ are the only two methods to get $[2,0]_S$. Hence, for a trail cover K in class $[2,2]_S$, it must be a series connection of two trail covers, say K_1 in class $[2,2]$ and K_2 in class $[2,0]$. If K_1 is a trail cover for some network of type S, then K_1 is a series connection of two trail covers again; one is in class $[2,2]$ and the other is in class $[2,0]$. If K_2 is a trail cover for some network of type S, then K_2 is a series connection of two trail covers again; one is in class $[2,2]$ and the other is in class $[2,2]$, or one is in class $[2,0]$ and the other is in class $[2,0]$. The above procedure will repeat until K is a series connection of trail covers which are in class $[2,2]_P$ or $[2,0]_P$. Finally, any series connection of an odd number of trail covers in class $[2,2]_P$ together with any number of trail covers in class $[2,0]_P$ is the only method to get $[2,2]_S$. In term of equation, we get

$$[2,2]_S \leftarrow \{ \text{ odd number of } [2,2]_P \} \ \sigma \ \{\text{some}[2,0]_P\}.$$

Similarly, we may have

$$[2,0]_S \leftarrow \{ \text{ even number of } [2,2]_P \} \ \sigma \ \{\text{some}[2,0]_P\}.$$

In the above expressions, the number of $[2,0]_P$ is indicated by "some" which means that it can be zero or any positive integer. Repeating the above procedure for all the other trail cover classes, we have a result like Table 4. Actually, all rules besides those in Table 4 can be calculated not only by hand but also by a computer program to confirm their completeness.

A network N is *possible* if there exists a trail cover of type x where $x \in \{1, 2, ..., 37\}$, for some (G[N],G[N']). We use $\mathcal{L}$ to denote the set of all possible networks. For any network N in $\mathcal{L}$, Possible(N) = { (G[N],G[N'])| there exists a trail cover for (G[N],G[N']), which is one of type x where $x \in \{1, 2, ..., 37\}$ }, and Class(N) = { $y \mid y$ is a trail cover class for some (G[N],G[N']) in Possible(N) }.

Let $\mathcal{AB}$ = { $N \in \mathcal{L}$ | |Class(N) $\cap$ $\{a,b,c,d,e\}$| ≥ 1 }. (Note that the trail cover groups containing $\{a,b,c,d,e\}$ are group A and group B.) Similarly, let $\mathcal{C}$ = { $N \in \mathcal{L}$ | |Class(N) $\cap$ $\{f ,g,h,i\}$| ≥ 1 }. We have the following lemma.

Lemma 2. |*Class(N)* $\cap$ $\{a,b,c,d,e\}$| ≤ 1 *for any network N in* $\mathcal{L}$.

Proof. It suffices to prove that |Class(N) $\cap$ $\{a,b,c,d,e\}$| = 1 for any network N in $\mathcal{AB}$. The proof is by induction on the height of the tree structure for N and observing those equations that generate a,b,c,d and e in Table 4 for the fact that classes a,b,c,d and e are mutually exclusive, i.e. the derivation of one class is distinct from all the other classes. ∎

Again by induction, we have the following corollary.

Corollary 3. *Let (G[N],G[N']) be any graph representation for some network N in* $\mathcal{AB}$ *and x be the unique element in Class(N)* $\cap$ *$\{a,b,c,d,e\}$. We have the following properties for x.*
(1) If x = a, then there are exactly two vertices of odd degree in G[N] and G[N'], respectively, namely their two distinct vertices.
(2) If x = b, then there are exactly two vertices of odd degree in G[N], namely its two distinct vertices; and there is no vertex of odd degree in G[N'].
(3) If x = c, then there is no vertex of odd degree in G[N]; and there are exactly two vertices of odd degree in G[N'], namely its two distinct vertices.
(4) If x = d or e, then there is no vertex of odd degree in both G[N] and G[N']. ∎

Let us scan those equations that generate f, g, h and i. It is observed that for any network in $\mathcal{C}$, there is a subnetwork (not necessarily a child subnetwork) of N or N' which is a series connection of at least one network which contains a trail cover class in {[2,2],[2,0]} and at least one network which contains a trail cover class in {[0,2],[θ,ϕ],[ϕ,θ]}. As a result, if (G[N],G[N']) is any graph representation for some network N in $\mathscr{C}$, then there exists a non–distinct vertex which is of odd degree in G[N] or G[N']. Hence, by Corollary 3, we have the following corollary.

Corollary 4. There is no network in $\mathscr{AB} \cap \mathscr{C}$. ∎

5. Algorithm

Now, with tools in hand, we describe the algorithm that recognizes DET networks.

Algorithm RECOG_DET

Input: A network N in the tree representation.

Output: "Yes" if N possesses a DET; "No", otherwise.

Method: For each non–leaf node x, use N(x) to denote the subnetwork that it and all of its descendants form, and N(x)' to denote the dual of N(x). Define M(x) to be the network among N(x) and N(x)', which is one of type S. Hence, M(x)' denotes the dual of M(x). From leaves to the root and level by level, calculate Class(M(x)) by the following steps.

1. If x is a leaf node, Class(M(x)') = {[2,2]$_p$ }.

2. If x is a non–leaf node, calculate Class(M(x)) using a table more than Table 4.

3. If x is a non–leaf node and all equations in the table can not be applied, then output "No" and STOP.

4. If x is the root node and one of [2,2], [2,0], [0,2], [x,x] or [y,y] is in Class(M(x)), then output "Yes" and STOP. If x is the root node but Class(M(x)) does not include one of [2,2], [2,0], [0,2], [x,x] or [y,y], then output "No" and STOP.

5. If x is a non–leaf node, calculate Class(M(x)') by setting Class(M(x)') = { y | y is a trail cover class, which the dual trail cover for the trail cover in Class(M(x)) may be in }.

The correctness of our algorithm follows from Theorem 1 and the discussion in section 4. Moreover, by Theorem 1, Lemma 2 and corollary 4, the following theorem can be obtained.

Theorem 5. Algorithm RECOG_DET recognizes DET networks in linear time. ∎

References

[1] T. Lengauer and R. Muller, "Linear algorithm for optimizing the layout of dynamic CMOS cells", *IEEE Trans. on Circuits and System*, vol. 35, (1988), 279–285.

[2] R. Maziasz and J. Hayes, "Layout optimization of static CMOS functional cells", *IEEE Trans. on Computer-Aided Design.*, Vol.9, No.7, (1990), 708–719.

[3] R. Nair, A. Bruss and J. Reif, "Linear time algorithms for optimal CMOS layout", in *VLSI: Algorithms and Architectures*, P. Bertolazzi and F. Luccio (Editors), Elservier Science Publishers B.V. (North–Holland), (1985), 327–338.

[4] L. Thomas and M. Rolf, "Linear algorithms for optimizing the layout of dynamic CMOS cells", *IEEE Trans on circuits and systems*, vol. 35 no. 3 (1988) 279–285.

[5] T. Uehara and C. VanCleemput, "Optimal layout of CMOS functional arrays", *IEEE Trans. Comput.*, vol. C–30, (1981), 305–312.

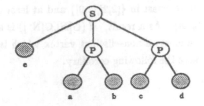

Figure 1

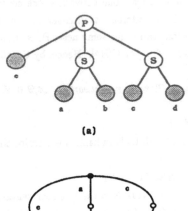

(a)

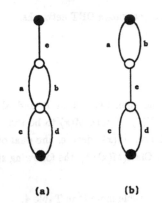

(a)   (b)

Figure 2

(b)

Figure 3

1.	$(\downarrow\,,\downarrow)$	20.	$(\xi,\mathcal{G})$
2.	$(\downarrow\,,\uparrow)$	21.	$(\mathcal{L}),\check{\mathcal{L}})$
3.	$(\downarrow\,,\cup)$	22.	$(\mathcal{G},\check{\mathcal{L}})$
4.	$(\downarrow\,,\cap)$	23.	$(\mathcal{L},\check{\mathcal{L}})$
5.	$(\cup\,,\downarrow)$	24.	$(\mathcal{G},\xi)$
6.	$(\cap\,,\downarrow)$	25.	$(\mathcal{LJ},\xi)$
7.	$(\times,\langle\rangle)$	26.	$(\mathcal{PI},\xi)$
8.	$(\langle\rangle,\times)$	27.	$(\xi,\mathcal{LJ})$
9.	$(\mathcal{G}\,,\mathcal{G})$	28.	$(\xi,\mathcal{PI})$
10.	$(\mathcal{G}\,,\mathcal{P})$	29.	(ξ,ξ)
11.	$(\mathcal{P}\,,\mathcal{G})$	30.	(ξ,ξ)
12.	$(\mathcal{P}\,,\mathcal{P})$	31.	(ξ,ξ)
13.	$(\mathcal{L},\mathcal{L})$	32.	(ξ,ξ)
14.	$(\mathcal{L},\mathcal{G})$	33.	(ξ,ξ)
15.	$(\mathcal{G},\mathcal{L})$	34.	(ξ,ξ)
16.	$(\mathcal{G},\mathcal{G})$	35.	(ξ,ξ)
17.	$(\times,\mathcal{L})$	36.	(ξ,ξ)
18.	$(\xi,\mathcal{G})$	37.	$(\beta\,,\beta)$
19.	$(\xi,\mathcal{L})$		

Table 1

set	A				B		C				D					E
σ	a	b	c	d	e	f	g	h	i	j	k	l	m	n	o	p
A a	b f	a f	f h	h	g	f f	i f	h f k	g		l	k				
b	a f	b f	f h	h	i	f f	g f	h f l	i		k	l				
c	f h	f h	d p		j n	h p	f h p		f n p				j		j	
d	h h	h h			d n		h h n		h h n				n		n	
B e	g	i	c j n	d e n	o	j	g h m	h n	i o	j			m n o			
f	f k l	f k l	h p		j	j	f k p	j	j							
C g	i l	g k	f k n	h g n m	j	l o	i k l n	k m								
h	h k l	h k l			h n		k l n		k l n							
i	g k l	i l	f l n	h i n o	j	k m	k l n	k l n o								
D j			n p		j											
k	l k	k l														
l	k l	k l														
m			j n m		n m											
n					n											
o			j n o		n o											
E p																

Table 3

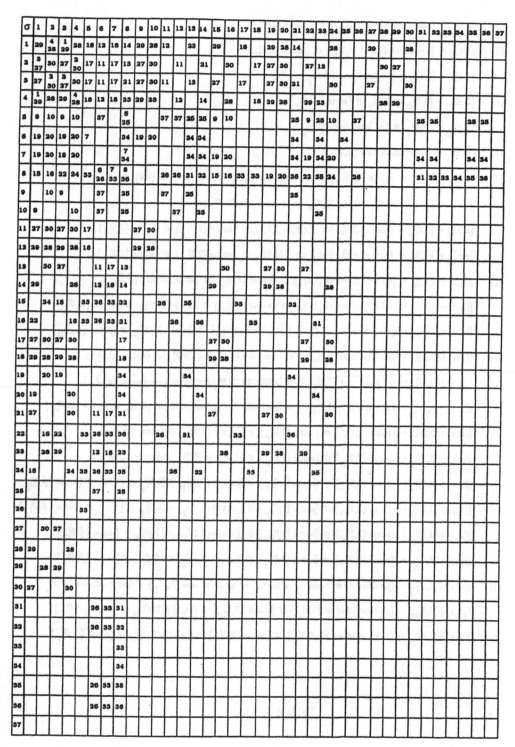

Table 2

(a) $[2,2]_S \leftarrow \{$odd number of $[2,2]_p\}$ σ $\{$ some $[2,0]_p\}$.

(b) $[2,0]_S \leftarrow \{$even number of $[2,2]_p\}$ σ $\{$ some $[2,0]_p\}$.

(c) $[0,2]_S \leftarrow [0,2]_p$ σ $\{$at least one $[\phi,\mathbb{0}]_p\}$.

(d) $[0,\phi]_S \leftarrow [0,2]_p$ σ $[0,2]_p$ σ $\{$some $[\phi,\mathbb{0}]_p\}$.

     $[0,\phi]_S \leftarrow [0,\phi]_p$ σ $\{$at least one $[\phi,\mathbb{0}]_p\}$.

(e) $[\phi,\mathbb{0}]_S \leftarrow \{$more than two $[\phi,\mathbb{0}]_p\}$.

(f) $[x,x]_S \leftarrow [0,2]_p$ σ $\{$some $[\phi,\mathbb{0}]_p\}$ σ $\{$at least one $[2,2]_p$ or at least one $[2,0]_p\}$.

     $[x,x]_S \leftarrow [x,x]_p$ σ $\{$at least one $[2,2]_p$ or at least one $[2,0]_p\}$.

     $[x,x]_S \leftarrow [0,2]_p$ σ $\{$some $[\phi,\mathbb{0}]_p\}$ σ $[(x,x)+(2,2)]_p$ σ $\{$some $[2,2]_p\}$ σ $\{$some $[2,0]_p\}$.

     $[x,x]_S \leftarrow [0,2]_p$ σ $\{$some $[\phi,\mathbb{0}]_p\}$ σ $[(x,x)+(2,0)]_p$ σ $\{$some $[2,2]_p\}$ σ $\{$some $[2,0]_p\}$.

(g) $[(x,x)+(2,2)]_S \leftarrow \{$odd number of $[2,2]_p\}$ σ $\{$some $[2,0]_p\}$ σ $\{$at least one $[\phi,\mathbb{0}]_p\}$.

     $[(x,x)+(2,2)]_S \leftarrow \{$odd number of $[2,2]_p\}$ σ $\{$some $[2,0]_p\}$ σ $[(x,x)+(2,0)]_p$ σ $\{$some $[\phi,\mathbb{0}]_p\}$.

     $[(x,x)+(2,2)]_S \leftarrow \{$even number of $[2,2]_p\}$ σ $\{$some $[2,0]_p\}$ σ $[(x,x)+(2,2)]_p$ σ $\{$some $[\phi,\mathbb{0}]_p\}$.

(h) $[(x,x)+(0,2)]_S \leftarrow \{$at least one $[2,2]_p$ or at least one $[2,0]_p\}$ σ $[0,2]_p$ σ $\{$some $[\phi,\mathbb{0}]_p\}$.

     $[(x,x)+(0,2)]_S \leftarrow \{$at least one $[2,2]_p$ or at least one $[2,0]_p\}$ σ $[0,\phi]_p$ σ $\{$some $[\phi,\mathbb{0}]_p\}$.

     $[(x,x)+(0,2)]_S \leftarrow \{$at least one $[2,2]_p$ or at least one $[2,0]_p\}$ σ $[0,2]_p$ σ $[0,2]_p$ σ $\{$some $[\phi,\mathbb{0}]_p\}$.

     $[(x,x)+(0,2)]_S \leftarrow \{$some $[2,2]_p\}$ σ $\{$some $[2,0]_p\}$ σ $[(x,x)+(0,2)]_p$ σ $\{$some $[\phi,\mathbb{0}]_p\}$.

     $[(x,x)+(0,2)]_S \leftarrow \{$some $[2,2]_p\}$ σ $\{$some $[2,0]_p\}$ σ $[x,x]_p$ σ $[0,2]_p$ σ $\{$some $[\phi,\mathbb{0}]_p\}$.

     $[(x,x)+(0,2)]_S \leftarrow \{$some $[2,2]_p\}$ σ $\{$some $[2,0]_p\}$ σ $[(x,x)+(2,2)]_p$ σ $[0,2]_p$ σ $[0,2]_p$ σ $\{$some $[\phi,\mathbb{0}]_p\}$.

     $[(x,x)+(0,2)]_S \leftarrow \{$some $[2,2]_p\}$ σ $\{$some $[2,0]_p\}$ σ $[(x,x)+(2,0)]_p$ σ $[0,2]_p$ σ $[0,2]_p$ σ $\{$some $[\phi,\mathbb{0}]_p\}$.

     $[(x,x)+(0,2)]_S \leftarrow \{$some $[2,2]_p\}$ σ $\{$some $[2,0]_p\}$ σ $[(x,x)+(2,2)]_p$ σ $[0,\phi]_p$ σ $\{$some $[\phi,\mathbb{0}]_p\}$.

     $[(x,x)+(0,2)]_S \leftarrow \{$some $[2,2]_p\}$ σ $\{$some $[2,0]_p\}$ σ $[(x,x)+(2,0)]_p$ σ $[0,\phi]_p$ σ $\{$some $[\phi,\mathbb{0}]_p\}$.

(i) $[(x,x)+(2,0)]_S \leftarrow \{$positive even number of $[2,2]_p$ or at least one $[2,0]_p\}$ σ $\{$at least one $[\phi,\mathbb{0}]_p\}$.

     $[(x,x)+(2,0)]_S \leftarrow \{$odd number of $[2,2]_p\}$ σ $\{$some $[2,0]_p\}$ σ $[(x,x)+(2,2)]_p$ σ $\{$some $[\phi,\mathbb{0}]_p\}$.

     $[(x,x)+(2,0)]_S \leftarrow \{$even number of $[2,2]_p\}$ σ $\{$some $[2,0]_p\}$ σ $[(x,x)+(2,0)]_p$ σ $\{$some $[\phi,\mathbb{0}]_p\}$.

Table 4

On Finding a Smallest Augmentation to Biconnect a Graph[1] (Extended Abstract)

Tsan-sheng Hsu   Vijaya Ramachandran

Department of Computer Sciences, University of Texas at Austin

Austin, TX 78712, USA

Abstract

The problem of finding a smallest set of edges whose addition biconnects an undirected graph is considered. An error in the earlier linear time sequential algorithm of Rosenthal & Goldner [19] is corrected and a new $O(\log^2 n)$ time parallel algorithm using a linear number of processors on an EREW PRAM is given, where n is the number of vertices in the input graph. The time bound is the best that is known for graph problems on an EREW PRAM, and the processor bound is close to the best possible.

1 Introduction

The problem of augmenting a graph to reach a certain connectivity requirement by adding edges has important applications in network reliability [7, 22] and fault-tolerant computing. One version of the augmentation problem is to augment the input graph to reach a given connectivity requirement by adding a smallest set of edges. We refer to this problem as the *smallest augmentation* problem.

The following results are known for solving the smallest augmentation problem on an undirected graph to satisfy a vertex connectivity requirement. Eswaran

[1] This work was supported in part by NSF Grant CCR-89-10707.

& Tarjan [5] gave a lower bound on the smallest number of edges for biconnectivity augmentation and proved that the lower bound can be achieved. Rosenthal & Goldner [19] developed a linear time sequential algorithm for finding a smallest augmentation to biconnect a graph. Watanabe & Nakamura [28, 30] gave an $O(n(n + m)^2)$ time sequential algorithm for finding a smallest augmentation to triconnect a graph where n and m are the number of vertices and edges of the given graph, respectively. Hsu & Ramachandran [13] recently gave a linear time algorithm for this problem. There is no efficient parallel algorithm known to find a smallest augmentation to k-vertex-connect a graph for $k \geq 2$.

For the problem of finding a smallest augmentation for a graph to reach other connectivity requirements, results can be found in [2, 5, 6, 8, 9, 14, 16, 21, 25, 26, 27, 28, 29, 31].

In this paper, we present an efficient parallel algorithm for finding a smallest augmentation to biconnect an undirected graph. In addition, we have discovered an error in the sequential algorithm of Rosenthal & Goldner [19]. In [12], we give a corrected linear time sequential algorithm for the problem. Our efficient parallel algorithm is based on this corrected sequential algorithm. However we have to utilize several insights into the problem to derive the parallel algorithm. The algorithm runs in $O(\log^2 n)$ time using a linear number of processors on an EREW PRAM where n is the number of vertices in the input graph. (For more on PRAM models and PRAM algorithms, see Karp & Ramachandran [15].)

Owing to space limitation, many proofs are omitted in this abstract. They can be found in the full paper [12].

2 Definitions

Let $G = (V, E)$ be an undirected graph with vertex set V and edge set E. Let $\{E_i | 1 \leq i \leq k\}$ be a partition of E into a set of k disjoint subsets such that two edges e_1 and e_2 are in the same partition if and only if there is a simple cycle in G containing e_1 and e_2 or e_1 is equal to e_2. Let q be the number of isolated vertices in G. Let $\{V_i | 1 \leq i \leq k+q\}$ be a collection of sets of vertices, where V_i is the set of vertices in E_i for each $1 \leq i \leq k$ and V_{i+k} contains only the ith isolated vertex for each $1 \leq i \leq q$. A vertex v is a *cutpoint* of a graph G if v appears in more than one vertex set V_i. G is *biconnected* if it has at least 3 vertices and contains no cutpoint or isolated vertex. The subgraph $G_i = (V_i, E_i)$, $1 \leq i \leq k$, is a *biconnected component* of G if V_i contains more than two vertices. Note that $E_i = \emptyset, \forall k < i \leq k + q$, since V_i contains an isolated vertex. The subgraph $G_i = (V_i, E_i), 1 \leq i \leq k + q$, is called a *block* of G. Given an undirected graph G, we can define its *block graph* $blk(G)$ as follows. Each block and each cutpoint of G is represented by a vertex of $blk(G)$. The vertices of $blk(G)$ which represent blocks are called *b-vertices* and those representing cutpoints are called *c-vertices*. Two vertices u and v of $blk(G)$ are adjacent if and only if u is a c-vertex, v is a b-vertex and the corresponding cutpoint of u is contained in the corresponding block of v or vice versa. It is well known that $blk(G)$ is

a forest and if G is connected, $blk(G)$ is a tree. If $blk(G)$ is a tree, it is also called a *block tree*.

Let n_c be the number of c-vertices in $blk(G)$. A vertex v_i represents a c-vertex of $blk(G)$ and d_i is the degree of v_i. We assume $d_i \geq d_{i+1}, \forall 1 \leq i < n_c$ throughout the discussion. For convenience, we define $a_i = d_i - 1$. If $blk(G)$ is a tree, let T be the rooted tree obtained from $blk(G)$ by rooting $blk(G)$ at the b-vertex connecting to v_1 and is on the path from v_1 to v_2. We use T_i to represent the subtree of T rooted at v_i for each i and we use T' to represent the subtree of T after deleting T_1. Let l_i be the number of leaves of T_i, for all $1 \leq i \leq n_c$. We also use T_v to represent the subtree rooted at a vertex v of $blk(G)$. The subgraph of T induced by deleting the vertex v is denoted by $T - v$.

In a forest, a vertex with degree 1 is called a *leaf*. Let l be the number of leaves in $blk(G)$. For a graph G', we use l' to denote the number of leaves in $blk(G')$. Let $d(v)$ be the degree of the vertex v in $blk(G)$ and let d be the largest degree of all c-vertices in $blk(G)$.

We also need the following definitions. A vertex v of $blk(G)$ is called *massive* if and only if v is a c-vertex with $d(v) - 1 > \lceil \frac{l}{2} \rceil$. A vertex v of $blk(G)$ is *critical* if and only if v is a c-vertex with $d(v) - 1 = \lceil \frac{l}{2} \rceil$. The graph $blk(G)$ is *critical* if and only if there exists a critical c-vertex in $blk(G)$. A block graph $blk(G)$ is *balanced* if and only if G is connected and contains no massive c-vertex. (Note that $blk(G)$ could have a critical c-vertex.) A graph G is balanced if and only if $blk(G)$ is balanced. Let v be a c-vertex of $blk(G)$. We call those components of $blk(G) - v$ containing only one vertex of degree 1 in $blk(G)$ *v-chains* [19]. A degree-1 vertex of $blk(G)$ in a v-chain is called a *v-chain leaf*. The following definition will be used later.

Definition 1
[The leaf-connecting condition]

Two leaves u_1 and u_2 of $blk(G)$ satisfy the leaf-connecting condition if and only if u_1 and u_2 are in the same tree of $blk(G)$ and the path between u_1 and u_2 in $blk(G)$ contains either (1) two vertices of degree more than 2, or (2) one b-vertex of degree more than 3.

In figures, we use a rectangle to represent a b-vertex and a circle to represent a c-vertex. A line denotes an edge.

3 Main Lemmas

In this section, we present results that will be crucial in the development of our efficient parallel algorithm. For detailed proofs of these results, please see [12].

Lemma 1 If $blk(G)$ has more than two c-vertices, then $a_1 + a_2 + a_3 - 1 \le l$. □

Corollary 1 (1) If $blk(G)$ has more than two c-vertices, then $a_3 \le \frac{l+1}{3}$. (2) There can be at most one massive vertex in $blk(G)$. (3) If there is a massive vertex in $blk(G)$, then there is no critical vertex in $blk(G)$. (4) There can be at most two critical vertices in $blk(G)$, if $l > 2$. □

Before introducing the next lemma, we have to study properties for updating the block tree. The following fact for obtaining $blk(G')$ from $blk(G)$ is given in Rosenthal & Goldner [19].

Fact 1 Given a graph G and its block tree $blk(G)$, adding an edge between two leaves u and v of $blk(G)$ creates a cycle C. Let G' be the graph obtained by adding an edge between u' and v' in G where u' and v' are non-cutpoint vertices in the blocks represented by u and v respectively. The following relations hold between $blk(G)$ and $blk(G')$. (1) Vertices and edges of $blk(G)$ that are not in the cycle C remain the same in $blk(G')$. (2) All b-vertices in $blk(G)$ that are in the cycle C contract to a single b-vertex b' in $blk(G')$. (3) Any c-vertex in

C with degree equal to 2 is eliminated. (4) A c-vertex x in C with degree greater than 2 remains in $blk(G')$ with edges incident on vertices not in the cycle. The vertex x also attaches to the b-vertex b' in $blk(G')$.

An example of getting $blk(G')$ from $blk(G)$ is illustrated in Figure 1.

Lemma 2 Let u_1 and u_2 be two leaves of $blk(G)$ satisfying the leaf-connecting condition (Definition 1). Let α and β be non-cutpoint vertices in blocks of G represented by u_1 and u_2 respectively. Let G' be the graph obtained from G by adding an edge between α and β and let P represent the path between u_1 and u_2 in $blk(G)$. The following three conditions are true. (1) $l' = l - 2$. (2) If v is a cutpoint in P with degree greater than 2 in $blk(G)$, then the degree of v decreases by 1 in $blk(G')$. (3) If v is a cutpoint in P with degree equal to 2, then v is eliminated in $blk(G')$.

Proof sketch: Parts (2) and (3) of the lemma follow from parts (3) and (4) of Fact 1. We now prove part (1) of the lemma. From part (2) in Fact 1, we know that every vertex of G that is in a component represented by a b-vertex in P is in a biconnected component Q of G'. Let Q be represented by a b-vertex b in $blk(G')$.

Case 1: Suppose that part (1) of the leaf-connecting condition (Definition 1) holds. Let w and y be two vertices of $blk(G)$ having degree more than 2 in $blk(G)$ and let $blk(G')$ be rooted at b. In $blk(G)$, let w' be a vertex adjacent to w and y' be a vertex adjacent to y, with neither w' nor y' in P. The vertex b has at least two children, w' and y', in $blk(G')$ and hence cannot be a leaf. Since leaves u_1 and u_2 are eliminated in $blk(G')$ and no new leaf is created, $l' = l - 2$.

Case 2: Suppose that part (2) of the leaf-connecting condition holds. Let w be a b-vertex of degree more than 3. We can find at least two c-vertices, y' and z', connected to w, but not in P. The same reasoning used in case 1 can be followed to prove this case. □

4 The Sequential Algorithm

The original linear time sequential algorithm in Rosenthal & Goldner [19] consists of three stages. We, however, have discovered an error in stage 3 of the algorithm in [19]. We describe this error in Section 4.3 below. We have a corrected version of stage 3 which is presented in [12]. Our parallel algorithm follows the structure of the corrected sequential algorithm. The first two stages are easy to parallelize and we describe them in Section 4.1 and Section 4.2. Stage 3, however, is highly sequential. Most of our discussion is on the parallel algorithm for stage 3.

We first state a lower bound on the number of edges needed to augment a graph to reach biconnectivity.

Theorem 1 *Eswaran & Tarjan [5]*
Let G be an undirected graph with h connected components and let q be the number of isolated vertices in $blk(G)$. Recall that l is the number of leaves in $blk(G)$ and $d+1$ is the largest degree of all c-vertices in $blk(G)$. Then at least $max\{d+h-2, \lceil \frac{l}{2} \rceil + q\}$ edges are needed to biconnect G, if $q+l > 1$. □

The following sections show that the lower bound proved in Theorem 1 can be achieved by an efficient parallel algorithm.

4.1 Stage 1

In this stage, we add edges to connect the input graph and make sure that the lower bound in Theorem 1 is reduced by the number of edges added. Let G be an undirected graph with h connected components. We can connect G by adding $h-1$ edges, which we may choose to be incident on non-cutpoint vertices in blocks corresponding to leaves or isolated vertices

in $blk(G)$. (For details, see [19].) Given $blk(G)$, stage 1 is easy to parallelize in time $O(\log n)$ optimally on an EREW PRAM by using the Euler tour technique on trees [24].

4.2 Stage 2

In this stage, we start with a connected unbalanced graph and we add edges to the graph to make the resulting graph balanced. The algorithm makes sure the lower bound in Theorem 1 is reduced by the number of edges added. Let G be connected and let v^* be a massive vertex in G. Let $\delta = d - 1 - \lceil \frac{l}{2} \rceil$. Then we can find at least $2\delta + 2$ v^*-chains [19]. Let Q be the set of v-chain leaves. By adding $2k, k \leq \delta$ edges to connect $2k+1$ vertices of Q, we can reduce both the degree of the massive vertex and the number of leaves in the block tree by k. By adding 2δ edges to connect $2\delta + 1$ vertices of Q, we can obtain a balanced block tree. (For details, see [19].) Given $blk(G)$, the finding of v^*-chain leaves and the whole stage 2 are easy to parallelize in time $O(\log n)$ optimally on an EREW PRAM by using the Euler tour technique on trees [24].

4.3 Stage 3

In this stage, we have to deal with a graph G where $blk(G)$ is balanced. The idea is to add an edge between two leaves y and z under the conditions that the path P between y and z passes through all critical vertices and the new block tree has two less leaves if $blk(G)$ has more than 3 leaves. Thus the degree of any critical vertex decreases by 1 and the tree remains balanced.

In Rosenthal & Goldner [19], $blk(G)$ is rooted at a b-vertex $\flat^*$. A path P is found that contains two leaves y and z such that if $blk(G)$ contains two critical vertices v and w, P contains both of them. If $blk(G)$ contains less than 2 critical vertices, P

contains b^* and a c-vertex with degree d (recall that d is the maximum degree of any c-vertex). It is possible that in the case that $blk(G)$ is balanced with more than three leaves and less than two critical vertices, P contains only one vertex of degree more than 2. If we add an edge between the two end points of P, it is possible that the new block tree has only one less leaf. An example of this is shown in Figure 2. Thus the lower bound cannot be achieved by this method. We present a corrected linear time sequential algorithm in [12]. The rest of the paper will deal with designing a parallel version of stage 3.

5 The Parallel Algorithm for Stage 3

In our parallel algorithm, we will find several pairs of leaves such that the path between any such pair of leaves passes through all critical c-vertices, if any. Thus the degrees of critical vertices in the new block tree decrease by the number of edges added to the original block tree. These pairs also satisfy the leaf-connecting condition (Definition 1), which guarantees that the number of leaves in the new block tree decreases by twice the number of edges added.

From Theorem 1 and the corrected linear time sequential algorithm given in [12] we know that exactly $\lceil \frac{l}{2} \rceil$ edges must be added to biconnect G if $blk(G)$ is balanced. That is, we have to eliminate l leaves during the computation. Our parallel algorithm runs in stages with at least $\frac{1}{4}$ of the current leaves eliminated in parallel time $O(\log n)$ using a linear number of processors during each stage, where n is the number of vertices in G. We call this subroutine $O(\log n)$ times to complete the augmentation.

Recall that $a_i + 1$ is equal to the degree of the ith c-vertex v_i and $a_i \geq a_{i+1}$. T'

is the subtree obtained from T by deleting the subtree rooted at v_1. Let $U_i = \{u|u$ is the leftmost leaf of T_y, where y is a child of $v_i\}$.

Depending on the degree distribution of vertices in the block tree, the parallel algorithm for stage 3 is divided into two cases. In case one, $a_1 > \frac{l}{4}$. We have one c-vertex with a very high degree. The first $min\{a_1 - 1, \lceil \frac{l}{2} \rceil - a_3\}$ leaves in U_1 are first matched with the first $a_2 - 1$ leaves in U_2, then matched with all remaining leaves but one in T' and finally properly matched within themselves, if necessary. In case two, $a_1 \leq \frac{l}{4}$. There is no c-vertex with a large degree. We show that we can find a vertex u^* with approximately the same number of leaves in each subtree rooted at a child of u^*. If u^* is a b-vertex, a suitable number of leaves between subtrees rooted at children of u^* are matched. Otherwise, u^* is a c-vertex and a suitable number of subtrees rooted at children of u^* are first merged into a single subtree rooted at u^*. Then leaves in the merged subtree are matched with leaves outside.

The algorithm first finds the matched pairs of leaves in each case. Then we add edges between matched pairs of leaves and update the block tree at the end of each case. The block tree and the sequence of cutpoints $v_1, \cdots, v_{n_c}$ will not be changed during the execution of each case.

We now describe the two cases in detail.

5.1 Case 1: $a_1 > \frac{l}{4}$

We root the block tree at the b-vertex b^* that is adjacent to v_1 and is on the path from v_1 to v_2. Let v_1 be the leftmost child of b^*. We permute the children of v_1 in non-increasing order (from left to right) of the number of leaves in subtrees rooted at them. We will call this procedure *tree-normalization* and the resulting tree T.

Recall that U_1 is the set of leftmost leaves in subtrees rooted at children of v_1. We select the first (from left to right)

$min\{a_1 - 1, \lceil \frac{l}{2} \rceil - a_3\}$ leaves from U_1 and call the set W_1. The order of the leaves as specified in the original tree is preserved. There are four phases for this case. In phases 1 and 2, leaves in W_1 are matched with leaves not in T_1. In phase 3, leaves in W_1 are matched with leaves in T_1 excluding those in W_1. In phase 4, the remaining leaves in W_1 are matched between themselves. The algorithm executes each phase in turn once until there is no leaf in W_1 left to be matched.

We now describe the four phases.

Phase 1: All leaves but the rightmost one in U_2 are matched with the rightmost $a_2 - 1$ leaves of W_1. The matched leaves are removed from W_1.

Phase 2: We match all remaining leaves but one in T' with the rightmost leaves of W_1 and remove matched leaves from W_1.

Phase 3: Recall that T is the original block tree before phase 1, l is the number of leaves in T, v_1 is a c-vertex with the largest degree in T, T_1 is the subtree of T rooted at v_1, l_1 is the number of leaves in T_1, T' is the tree obtained from T by removing T_1 and $U_1 = \{u | u$ is the leftmost leaf of T_y, where y is a child of $v_1\}$. Note that there are $min\{a_1 - 1, \lceil \frac{l}{2} \rceil - a_3\}$ $-(l - l_1 - 1)$ leaves remaining in W_1. Let the set of v_1-chain leaves in W_1 be Q_1. We denote by Q_2 the set of leaves other than the rightmost one of each subtree rooted at a child of v_1. In this phase, we match all leaves in Q_1 with an equal number of leaves in Q_2. Leaves in W_1 that are matched in phase 3 are removed from W_1.

Claim 1 shows that we can always find enough leaves in Q_2 to match all leaves in Q_1.

Claim 1 $|Q_2| \geq |Q_1|$, if $l > 3$. □

Phase 4: The remaining leaves of W_1 that are not matched during phase 3 are matched within themselves. If the number of remaining leaves in W_1 is odd, we match one of them with the rightmost leaf in the subtree rooted at v_1.

Claim 2 *The number of matched pairs k in case 1 satisfies $\lceil \frac{l}{2} \rceil - a_3 \geq k \geq \frac{l}{8}$, if $l > 3$.* □

Claim 3 *(1) Each pair of matched vertices found in case 1 satisfies the leaf-connecting condition (Definition 1). (2) Let us place an edge between each matched pair found in case 1 sequentially and update the block graph each time we add an edge. Critical vertices, if any, of the block graph are on the path between the endpoints of each edge placed.* □

Corollary 2 *By adding k new edges between k matched pairs of leaves found in case 1, the lower bound given in Theorem 1 for the resulting graph G' is decreased by k and G' remains balanced. The number of leaves in the new block tree is at most $\frac{3l}{4}$, if $l > 3$.* □

5.2 Case 2: $a_1 \leq \frac{l}{4}$

In this case, we take advantage of having no c-vertex with a large degree. Because there is no critical c-vertex, the algorithm can add at least $\lceil \frac{l}{2} \rceil - a_1$ edges between leaves that satisfy the leaf-connecting condition (Definition 1) without worrying about whether the path between them passes through a critical c-vertex. This gives a certain degree of freedom for us to choose the matched pairs. We first root the block tree such that no subtree other than the one rooted at the root has more than half of the total number of leaves.

Given any rooted tree T, we use l_v to denote the number of leaves in the subtree rooted at a vertex v. The following lemma shows that we can reroot T at a vertex u^* such that no subtree rooted at a child of u^* has more than half of the total number of leaves.

Lemma 3 *Given a rooted tree T, there exists a vertex u^* in T such that $l_{u^*} > \frac{l}{2}$, but none of the subtrees rooted at children of u^* has more than $\frac{l}{2}$ leaves.* □

We root the block tree at u^* and permute children of u^* from left to right in non-increasing order of the number of leaves in subtrees rooted at them. Let the rooted tree be T. Let $u_i, \forall 1 \leq i \leq r$, be the children (from left to right) of u^* and x_i be the number of leaves in the subtree rooted at u_i. Note that $x_i \leq \frac{l}{2}$, for each i. There are two subcases depending on whether u^* is a b-vertex or a c-vertex. We describe the two subcases in detail in the following paragraphs.

5.2.1 Subcase 2.1: u^* is a b-Vertex

We show that we can partition subtrees rooted at children of the root into two sets "evenly" such that we can match leaves between the two partitions. We first give a claim to show how to perform the partition.

Claim 4 *There exists p, such that $1 \leq p < r$ and $\frac{l}{2} \geq \sum_{i=1}^{p} x_i > \frac{l}{4}$.* □

We match $min\{(\sum_{i=1}^{p} x_i) - 1, \lceil \frac{l}{2} \rceil - a_1\}$ leaves in subtrees $T_{u_i}, \forall 1 \leq i \leq p$, with leaves outside them. From Claim 4, we know that the matching can be done and the number of pairs matched is at least $\frac{l}{4}$.

Claim 5 *Every matched pair found in case case 2.1 satisfies the leaf-connecting condition, if $l > 3$.* □

Corollary 3 *By adding k new edges between k matched pairs of leaves found in case 2.1, the lower bound given in Theorem 1 for the resulting graph G' decreases by k and G' remains balanced. The number of leaves in the new block tree is at most $\frac{l}{2}$, if $l > 3$.* □

5.2.2 Subcase 2.2: u^* is a c-Vertex

Recall that the $u_i, 1 \leq i \leq r$, are the children (from left to right) of u^* (the root). Let x_i be the number of leaves in the subtree rooted at u_i. We know that $\frac{l}{2} \geq x_i, 1 \leq i \leq r$, and $x_i \geq x_{i+1}, 1 \leq i < r$.

We partition the set of subtrees rooted at children of the root into two sets such that we can match leaves between two sets. We first give a claim to show how to partition the set of subtrees. The proof of the claim is similar to that of Claim 4.

Claim 6 *Let q be the largest integer with $x_q \geq 2$. There exists an integer p such that $1 \leq p \leq q$ and $\frac{l}{2} \geq \sum_{i=1}^{p} x_i > \frac{l}{8} + (p - 1)$.* □

Let T_{u_i} be the subtree rooted at u_i. We define the *merge* operation for the collection of subtrees $T_{u_i}, \forall 1 \leq i \leq p$, as follows. We first connect the rightmost leaf of T_{u_i} and the leftmost leaf of $T_{u_{i+1}}$ for all $1 \leq i < p$. This can be done by the fact that each T_{u_i}, $1 \leq i \leq p$, has at least 2 leaves.

Claim 7 *Let T^* be the block tree obtained from T by collapsing b-vertices that are in the same fundamental cycle created by the addition of new edges introduced by the merge operation. (1) The merge operation creates only one b-vertex b^*. (2) Vertex b^* is a child of the root and b^* is the root of the subtree that contains the updated portion of the block tree.* □

Note that if we root the updated block tree T^* given in Claim 7 at the b-vertex b^*, the situation is similar to that in case 2.1. Thus we can match an additional $min\{(\sum_{i=1}^{p} x_i) - 1, \lceil \frac{l}{2} \rceil - a_1\} - (p - 1)$ pairs of vertices by pairing up unmatched leaves in subtrees T_{u_i}, $1 \leq i \leq p$, and leaves in subtrees in subtrees T_{u_i}, $p < i \leq r$.

Claim 8 shows the correctness of procedures defined in case 2.2. The proof of the claim is similar to that of Claim 5.

Claim 8 *Each pair of vertices matched in case 2.2 satisfies the leaf-connecting condition, if $l > 3$.* □

Corollary 4 *By adding k new edges between k matched pair of leaves found in case 2.2, the lower bound given in Theorem 1 for the resulting graph G' decreases by k and G' remains balanced. The number of leaves in the new block tree is at most $\frac{3l}{4}$, if $l > 3$.* □

6 The Complete Parallel Algorithm and Its Implementation

The parallel algorithm for finding a smallest augmentation to biconnect an undirected graph G calls the procedure given in Section 4.1 if G is not connected. Then it calls the procedure given in Section 4.2 if G is not balanced. If G is balanced, the algorithm calls the procedures given in Section 5.1 (case 1) or Section 5.2 (case 2) $O(\log n)$ times. After each execution of the procedures given in case 1 or case 2, the algorithm updates the block tree. The correctness of the full parallel algorithm follows from the correctness we established earlier of the various cases (Corollary 2, Corollary 3 and Corollary 4).

In the previous sections, we have sketched each step in the parallel algorithm except the step for updating the block tree given the original block tree T and the set of edges S added to it. Note that the updated block tree can be obtained by using an algorithm for finding biconnected components. We, however, do not know how to implement algorithms to find biconnected components in a graph with n vertices on an EREW PRAM in time $O(\log n)$ using a linear number of processors. To describe an $O(\log n)$ time EREW parallel algorithm using a linear number of processors for updating the block graph T after adding a set of edges S, we first define the following equivalence relation $\mathcal{R}$ on the set of b-vertices B, where $B=\{v|v$ is a b-vertex in T and v is in a cycle created by adding the edges in $S\}$. A pair (x,y) is in $\mathcal{R}$ if and only if $x \in B$, $y \in B$ and vertices in blocks represented by x and y are in the same block after adding the edges in S. It is obvious that $\mathcal{R}$ is reflexive, symmetric and transitive. Since $\mathcal{R}$ is an equivalence relation, we can partition B into k disjoint subsets B_i, $1 \leq i \leq k$, such that for each

i, $x, y \in B_i$ implies $(x,y) \in \mathcal{R}$ and for any $(x,y) \in \mathcal{R}$, x and y both belong to the same B_i.

Claim 9 *Two b-vertices b_1 and b_2 are in the same equivalence class if and only if there exists a set of fundamental cycles $\{C_0, \cdots, C_q\}$ such that $b_1 \in C_0$, $b_2 \in C_q$ and C_i and C_{i+1} share a common b-vertex, for all $0 \leq i < q$.* □

From Claim 9 and the discussion of each case, we know that b-vertices in fundamental cycles formed by adding edges due to phase 1 and phase 2 of case 1 shrink into a single b-vertex in the new block tree. The b-vertices in fundamental cycles formed by adding edges due to phase 3 of case 1 which share a common child of v_1 shrink into a single b-vertex. The b-vertices in a fundamental cycle formed by adding edges due to phase 4 of case 1 shrink into a single b-vertices. The b-vertices in all fundamental cycles formed by adding edges due to subcase 2.1 or subcase 2.2 shrink into a single b-vertex. Thus we know how to compute the equivalence classes of $\mathcal{R}$ and implement the algorithm for updating the block tree given the original block tree and the set of edges S added to it. To perform the parallel algorithms needed for updating the block tree for case 1 and case 2, we need procedures for computing various functions on a tree. They can be done in $O(\log n)$ time using a linear number of processors on an EREW PRAM by using the Euler technique on trees [24] and computing the least common ancestors [20].

Given an undirected graph G with n vertices, we can find its block graph in time $O(\log^2 n)$ using a linear number of processors on an EREW PRAM by the parallel algorithm in Tarjan & Vishkin [24] for finding biconnected components and using some procedures in Nath & Maheshwari [17].

In stage 3, the children-permutation procedure can be done in time $O(\log n)$ using a linear number of processors on an EREW

PRAM by calling the parallel merge sort routine in Cole [3] and using the Euler tour technique on trees [24] to restructure and normalize the tree. This gives the following claim.

Claim 10 *The biconnectivity augmentation problem on an undirected graph G with n vertices can be solved in time $O(\log^2 n)$ using a linear number of processors on an EREW PRAM.* □

7 Conclusion

In this paper we have presented an efficient parallel algorithm to find a smallest augmentation to biconnect a graph. The algorithm runs in $O(\log^2 n)$ time using a linear number of processors on an EREW PRAM, where n is the number of vertices in the input graph. Our parallel algorithm is efficient. By using a more sophisticated algorithm in Cole & Vishkin [4] for finding connected components and the bucket sorting routine in Hagerup [10], the parallel algorithm can be implemented with the same time complexity $(O(\log^2 n))$ using $O(max\{\frac{n \log \log n}{\log n}, \frac{(n+m)\alpha(n,m)}{\log n}\})$ processors on an EREW PRAM, where $\alpha(n,m)$ is the inverse Ackermann's function.

References

[1] R. J. Anderson & G. L. Miller, "Deterministic parallel list ranking," *Proc. 3rd Aegean Workshop on Computing*, Springer-Verlag LNCS 319, 1988, pp. 81-90.

[2] Guo-Ray Cai & Yu-Geng Sun, "The minimum augmentation of any graph to a k-edge-connected graph," *Networks*, Vol. 19, 1989, pp. 151-172.

[3] R. Cole, "Parallel merge sort," *SIAM J. Comput.*, Vol. 17, 1988, pp. 770-785.

[4] R. Cole & U. Vishkin, "Approximate and exact parallel scheduling with applications to list, tree and graph problems," *Proc. 27th Annual IEEE Symp. on Foundations of Comp. Sci.*, 1986, pp. 478-491.

[5] Kapali P. Eswaran & R. Endre Tarjan, "Augmentation problems," *SIAM J. Comput.* Vol 5, No. 4, December 1976, pp. 653-665.

[6] András Frank, "Augmenting graphs to meet edge-connectivity requirements," *Proc. 31th Annual IEEE Symp. on Foundations of Comp. Sci.*, 1990, pp. 708-718.

[7] H. Frank & W. Chou, "Connectivity considerations in the design of survivable networks," *IEEE Trans. on Circuit Theory*, Vol. CT-17, No. 4, November 1970, pp. 486-490.

[8] Greg N. Frederickson & Joseph Ja'Ja', "Approximation algorithms for several graph augmentation problems," *SIAM J. Comput.*, Vol. 10, No. 2, May 1981, pp. 270-283.

[9] D. Gusfield, "Optimal mixed graph augmentation," *SIAM J. Comput.*, Vol. 16, No. 4, August 1987, pp. 599-612.

[10] Torben Hagerup, "Towards optimal parallel bucket sorting," *Information and Computation*, Vol. 75, 1987, pp. 39-51.

[11] D. Harel & R. E. Tarjan, "Fast algorithms for finding nearest common ancestors," *SIAM J. Comput.*, Vol. 13, 1984, pp. 338-355.

[12] Tsan-sheng Hsu & Vijaya Ramachandran, "On finding a smallest augmentation to biconnect a graph," Tech. Rep. TR-91-12, U.T. Austin, 1991.

[13] Tsan-sheng Hsu & Vijaya Ramachandran, "A linear time algorithm for triconnectivity augmentation," *Proc. 32th Annual IEEE Symp. on Foundations of Comp. Sci.*, 1991.

[14] Yoji Kajitani & Shuichi Ueno, "The minimum augmentation of a directed tree to a k-edge-connected directed graph," *Networks*, Vol. 16, 1986, pp. 181-197.

[15] Richard M. Karp & Vijaya Ramachandran, "Parallel algorithms for shared-memory machines," *Handbook of Theoretical Computer Science*, J. van Leeuwen, ed., North Holland, 1990, pp. 869-941.

[16] D. Naor, D. Gusfield & C. Martel, "A fast algorithm for optimally increasing the edge-connectivity," *Proc. 31th Annual IEEE Symp. on Foundations of Comp. Sci.*, 1990, pp. 698-707.

[17] Dhruva Nath & S. N. Maheshwari, "Parallel algorithms for the connected components and minimal spanning tree problems," *Information Processing Letters*, Vol. 14, No. 1, 27 March 1982, pp. 7-11.

[18] Vijaya Ramachandran, "Parallel open ear decomposition with applications to graph biconnectivity and triconnectivity," invited chapter for *Synthesis of Parallel Algorithms*, J. H. Reif, editor, Morgan-Kaufmann.

[19] Arnie Rosenthal & Anita Goldner, "Smallest augmentations to biconnect a graph," *SIAM J. Comput.*, Vol. 6, No. 1, March 1977, pp. 55-66.

[20] B. Schieber & U. Vishkin, "On finding lowest common ancestors: simplification and parallelization," *Proc. 3rd Aegean Workshop on Computing*, Springer-Verlag LNCS 319, 1988, pp. 111-123.

[21] Danny Soroker, "Fast parallel strong orientation of mixed graphs and related augmentation problems," *Journal of Algorithms*, 9, 1988, pp. 205-223.

[22] Kenneth Steiglitz, Peter Weiner & D. J. Kleitman, "The design of minimum-cost survivable networks," *IEEE Trans. on Circuit Theory*, Vol. CT-16, No. 4, November 1969, pp. 455-460.

[23] R. E. Tarjan, *Data Structures and Network Algorithms*, SIAM Press, Philadelphia, PA, 1983.

[24] R. E. Tarjan & U. Vishkin, "An efficient parallel biconnectivity algorithm," *SIAM J. Comput.*, Vol. 14, 1985, pp. 862-874.

[25] Shuichi Ueno, Yoji Kajitani & Hajime Wada, "Minimum augmentation of a tree to a k-edge-connected graph," *Networks*, Vol. 18, 1988, pp 19-25.

[26] T. Watanabe, "An efficient way for edge-connectivity augmentation," Tech. Rep. ACT-76-UILU-ENG-87-2221, Coordinated Science lab., University of Illinois, Urbana, IL, 1987.

[27] T. Watanabe, Y. Higashi & A. Nakamura, "Graph augmentation problems for a specified set of vertices," *Proceedings of the first Annual International Symposium on Algorithms*, 1990, pp. 378-387.

[28] T. Watanabe & A. Nakamura, "On a smallest augmentation to triconnect a graph," Tech. Rep. C-18, Department of Applied Mathematics, faculty of Engineering, Hiroshima University, Higashi-Hiroshima, 724, Japan (1983, revised 1987).

[29] T. Watanabe & A. Nakamura, "Edge-connectivity augmentation problems," *J. Comput. System Sci.*, 35(1987), pp. 96-144.

[30] T. Watanabe & A. Nakamura, "3-Connectivity augmentation problems," *Proceedings of 1988 IEEE International Symposium on Circuits and Systems*, 1988, pp. 1847-1850.

[31] T. Watanabe, T. Narita & A. Nakamura, "3-Edge-Connectivity augmentation problems," *Proceedings of 1989 IEEE International Symposium on Circuits and Systems*, 1989, pp.335-338.

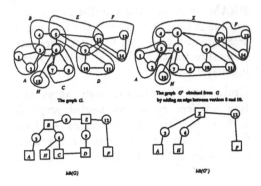

Figure 1: An example of obtaining $blk(G')$ from $blk(G)$. Vertices of G and G' circled with a dotted line are in the same block. Vertices B, C, D and E in $blk(G)$ are in a cycle if we add an edge between C and D. The cycle contracts into a new b-vertex X in $blk(G')$. The degree of a c-vertex in the cycle decreases by 1 in $blk(G')$, if the original degree is more than two. A degree-2 c-vertex in the cycle is eliminated in $blk(G')$.

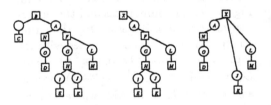

Figure 2: A counter example for the linear time sequential algorithm given by Rosenthal & Goldner. Vertex A is the c-vertex with the largest degree in the left tree. The middle tree is the new block tree after connecting two non-cutpoint vertices of G in the corresponding blocks represented by C and D. The number of leaves decreases by 1. The right tree is the new block tree after connecting two non-cutpoint vertices of G in the corresponding blocks represented by C and E. The number of leaves decreases by 2. The pair C and D could be chosen by the algorithm given by Rosenthal & Goldner while the pair C and E can be chosen to reduce the number of leaves by 2.

A Faster Algorithm for Edge-Disjoint Paths *
in Planar Graphs

Michael Kaufmann and Gerhard Klär

Max-Planck-Institut für Informatik

6600 Saarbrücken

Germany

Abstract: An efficient algorithm for the edge-disjoint paths problem in planar graphs is presented. Using Frederickson's [F84] decomposition method for planar graphs we improve the best bound for the running time of $O(n^2)$ [BM86, MNS86] for the edge-disjoint paths problem to $O(n^{5/3}(\log\log n)^{1/3})$.

Introduction and Basic Algorithm

The planar edge-disjoint paths problem (PEP) is given by

a) a planar undirected graph $G = (V, E), |V| = n$, embedded into the plane,

b) a set $N = \{N_1, .., N_k\}$ of nets, specified each by a pair of vertices on the boundary of the infinite face of G.

A solution to a PEP is given by a set of pairwise edge-disjoint paths P_i, such that path P_i in G connects the two vertices of net $N_i, 1 \leq i \leq k$.

The edge-disjoint paths problem is significant from theoretical and practical points of view. The practical side comes from VLSI-design. It can be regarded as a routing problem. Routing is a very important step during the design of integrated circuits, where the interconnections between different prefabricated modules or cells have to be produced. In this context, the two boundary vertices of each net are called terminals or pins, and the path of each net corresponds to the electrical connection of the terminals by wires. In several routing models, edge-disjointness of the paths is required to avoid long overlaps of the wires (crosstalking!).

In direct applications of this theory to layout algorithms for VLSI, mostly restricted versions of the problem have been considered. Since the wires in VLSI-layouts usually run horizontally and vertically, the corresponding planar graph is a finite subgraph of

---

* This work was supported by the Deutsche Forschungsgemeinschaft, SFB 124, VLSI-Entwurfsmethoden und Parallelität.

the integer grid. Under these restrictions, efficient algorithms for problems like channel routing [MPS86], rectangle routing [F82], routing in convex polygons [NSS85], [K90] and routing in general switchboxes [KM86] could be found. They all have a running time linear or almost linear to the size of the graph.

The theory that forms the basis of all these algorithms comes from combinatorial optimization. Namely the problem can be seen as a multicommodity flow problem. The theory on multicommodity flow is a generalization of the well-known theory on network flow [FF56] allowing multiple sources, sinks and commodities. If we consider PEPs as multicommodity flow problems, each net represents the demand to send one unit of flow of a certain kind of commodity from one terminal of the net to the other. We have the additional constraint that a commodity has to be sent along a single path and cannot be split into pieces. If integer flows are required the problem is NP-hard in nonplanar graphs, even if there are only two commodities or if all edge capacities are 1 [EIS76]. Okamura/Seymour [OS81] give necessary and sufficient conditions for the solvability of multicommodity flow problems in planar graphs, and in particular for PEPs. To complete the picture, we also mention the papers of Okamura [O83] and Schrijver [S87]. They extend the theorem of Okamura/Seymour, allowing that the terminals lie on two or even more faces. In these cases, several restrictions on the net configurations have been made.

At this stage, we need some definitions: A subset $X \subset V$ is called a cut. The capacity $cap(X)$ of X is defined as the number of edges having exactly one endpoint in X and the density $dens(X)$ is the number of nets having exactly one terminal in X. The free capacity $fcap(X)$ of cut X is given by $fcap(X) = cap(X) - dens(X)$, and finally we call a cut (over-) saturated, if the free capacity is zero (negative). A cut X crosses an edge e if exactly one vertex incident to this edge lies in X. X is called simple if it crosses exactly two boundary edges. A problem is even if the free capacity of every cut is even. Okamura/Seymour's Theorem deals with even problems.

Theorem 1: *Let $P = (V, E, N)$ be an even PEP. Then P is solvable iff there is no oversaturated cut.*

This necessary and sufficient condition for the solvability is called cut condition.

Theorem 1 is proved by induction on the size of the graph and provides the basic algorithm [OS81].

In the induction step, the problem P is transformed into a 'smaller' problem P' by eliminating one edge, such that P' is even and solvable iff P is even and solvable, and a solution of P' easily supplies a solution of P. The algorithm stops either if an oversaturated cut is found or if there is no edge left. In the second case, the solution is found.

The algorithm works as follows:

while $E \neq \emptyset$

    (* Invariant: P is even and the cut condition holds *)

    do let $e_0 = (a, b)$ be one of the boundary edges;

        let $e_0, e_1, e_2, ... e_k$ be the boundary edges in clockwise order;

        let X be the simple cut crossing e_0 with minimal free capacity

        and among these cuts having minimal cardinality;

        case $fcap(X)$ is

        positive: delete e_0 and add net $\{a, b\}$;

        zero: let $N = \{s, t\}$ be a net that contributes to the density of X, such that

            s is as close as possible to a in a counterclockwise traversal of the

            boundary of the infinite face;

            reserve edge e_0 for net N;

            delete e_0 and replace net N by nets $N_1 = \{s, a\}$ and $N_2 = \{b, t\}$;

        negative: stop;   (* The problem has no solution.*)

    od;

basic algorithm

The proof of correctness mainly consists of a proof of the invariant. It has to be shown that the free capacity of every cut remains even and nonnegative during each traversal of the while-loop.

An efficient implementation of this algorithm has been given by Becker/Mehlhorn [BM86]. $O(n)$ edge elimination steps have to be executed, one for each edge. In their implementation, each step takes $O(n)$ time, namely $O(n)$ for a shortest-path computation from e_0 to all other boundary edges to determine the capacity of simple cuts extending from e_0. Note that the shortest-path computation can be done by an easy application of BFS. Then the density of every simple cut crossing e_0 is found by running around the boundary, and the same method also works to find the appropriate net to be routed across e_0, if necessary. The total running time of this method is $O(n \cdot n) = O(n^2)$.

An alternative method with approximately the same running time has been given by Matsumoto/Nishizeki/Saito [MNS86], but it is also essentially based on Okamura/Seymour's Theorem. Their method even works for general planar multicommodity flow problems in time $O(n^2 \sqrt{\log n} + kn)$, where k is the number of source-sink pairs. The main result of both papers relevant for PEPs can be formulated in the following theorem:

Theorem 2: *Let* $P = (V, E, N)$ *be an even PEP. In time* $O(n^2)$ *a solution for* P *can be constructed if there is any or unsolvability is indicated.*

In this paper we refine the basic algorithm and give a faster implementation. Theorem 3 summarizes the result of this paper.

Theorem 3: *Let $P = (V, E, N)$ be an even PEP. In time $O(n^{5/3}(\log\log n)^{1/3})$ a solution for P can be constructed if there is one or unsolvability is indicated.*

The New Algorithm

Our algorithm makes use of a division of planar graphs into regions, which has been introduced by G. Frederickson [F84]. A region will contain two types of vertices: interior vertices, which are contained in exactly one region, and separator vertices, which are shared among at least two regions.

Let $G = (V, E), |V| = n$ be a graph and let $r \leq n$ be a parameter. An r-division of G is a division of G into $O(\frac{n}{r})$ regions of $O(r)$ vertices each and $O(\sqrt{r})$ separator vertices each.

The regions of the r-division are not necessarily connected, but as we will see later we may assume that they are.

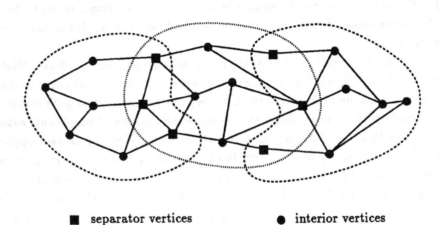

■ separator vertices   ● interior vertices

Figure 1 :   An example of an r-division $(r = \frac{n}{3})$.
The regions are indicated by dashed closed curves.

G. Frederickson proved the following lemma [F84]:

Lemma 1: *Let $G = (V, E), |V| = n$ be a graph and let $r \leq n$ be a parameter. An r-division of G can be found in time $O(n \log n)$.*

Using this r-division we change the algorithm as follows:

Construct an r-division of graph $G = (V, E)$;
while $E \neq \emptyset$ **do**
    choose a region $R = (V_R, E_R)$ that borders on the infinite face of G;
    while $E_R \neq \emptyset$ **do** choose an edge e on the boundary of the infinite face of R;
                delete e according to the basic algorithm;
                $E_R := E_R - e$;
    od;
    $E := E - E_R$;
od.

new algorithm

Dependent on the choice of R we have to distinguish between two cases:

Case 1: No separator vertex of R lies on the boundary of the infinite face of G, i.e. R encloses $G - R$.

We choose an arbitrary edge e on the boundary of R and have to determine the simple cut X of minimal free capacity and minimal cardinality that crosses e. The density of every simple cut crossing e can be computed by a clockwise traversal of the boundary of G. Since R encloses $G - R$, the boundary of G consists of at most $O(r)$ vertices. Thus the densities of these cuts can be determined in time $O(r)$.

The capacities of these cuts can be computed by means of the multiple source dual graph. In the dual graph there is a dual edge for every edge of the original graph. The dual edge connects vertices which are located in the faces separated by the edge. In every finite face we position only one dual vertex, but in the infinite face the vertices incident to the dual edges are kept distinct (cf. Figure 2).

Let X be a simple cut crossing the edges e, e' and let v, v' be the dual vertices in the infinite face incident to the edges dual to e, e'. The capacity of X is equal to the length of the shortest path from v to v' in the dual graph. Hence the capacities can be computed in time $O(n)$ by BFS on the multiple source dual graph starting from v.

We achieve a time bound of $O(r \log \log n)$ by compressing the dual graph in the following way (cf. Figure 3):

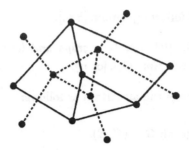

Figure 2 : A planar graph and its multiple-source dual graph.
Dual edges are shown as dashed lines.

Let $G_d = (V_d, E_d)$ be the dual graph of G.

  i) Mark the vertices of G_d that correspond to interior faces of G which border on R. Let M be the set of these vertices.

 ii) Let $v \in M$ and let $E_v = E_1 \cup E_2$ be the edges incident to v, where E_1 (E_2) is the set of edges that correspond to edges of R $(G - R)$. Substitute v by two vertices v_1, v_2, such that v_1 is incident to E_1 and v_2 is incident to E_2. Add v_1 to the set M_1 and v_2 to the set M_2. (The sets M_1, M_2 are initialized by $\emptyset$.)

iii) By performing step ii) G_d is split into two parts $R_d = (V_{R_d}, E_{R_d})$ and $(G - R)_d$. Construct the complete graph $C_{|M_1|} = (M_1, M_1 \times M_1)$ and compute the function $dist : M_1 \times M_1 \to \mathbf{Z} \cup \{\infty\}$, $dist(u,v) =$ the length of the shortest path from u to v in graph $(G - R)_d$, if the path exists, and $dist(u,v) = \infty$ otherwise.

iv) Combine R_d and $C_{|M_1|}$ to a new graph $G_d{}' = (V', E')$ by uniting the corresponding nodes of M_1 and M_2. Multiple edges are deleted. Provide the distance function $dist' : E' \to \mathbf{Z} \cup \{\infty\}$, $dist'(u,v) = dist(u,v)$ if $u, v \in M_1$ and $dist'(u,v) = 1$ otherwise.

Let u, v be vertices on the boundary of the infinite face of G_d and let u', v' be the corresponding vertices of $G_d{}'$. Obviously the length of the shortest path from u to v in G_d is equal to the length of the shortest path from u' to v' in $G_d{}'$. Since region R has $O(\sqrt{r})$ separator vertices, it borders on $O(\sqrt{r})$ interior faces of G. Thus the cardinality of M is $O(\sqrt{r})$ and $|M_1 \times M_1| = O(r)$. To get $G_d{}'$ from G_d, we essentially have to do a single shortest path computation for every vertex in M_1. Note that this computation is done only once for region R in a preprocessing step, which takes time $O(n\sqrt{r})$. Hence we can compute the capacities of all cuts crossing edge e in time $O(r \log r)$ by a standard shortest path computation in $G_d{}'$, or in time $O(r \log \log n)$ using the priority queue of van Emde-Boas [EKZ77]. This is possible since the length of each edge is a positive integer $\leq n$. (Edges of length ∞ can be removed.)

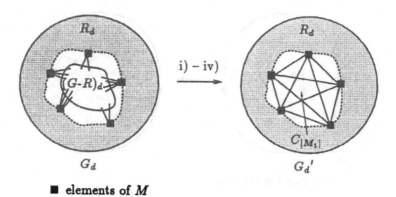

■ elements of M

Figure 3 : The compression of graph G_d.

After cut X is found we have to detect a net $N = \{s, t\}$ to be routed across edge $e = (a, b)$. According to the basic algorithm net N is determined by two conditions :

i) N must contribute to the density of X and

ii) terminal s has to be situated as close as possible to a in a counterclockwise traversal of the boundary of the infinite face.

Thus N can be found by a counterclockwise traversal of the boundary of G in time $O(r)$.

Since region R consists of $O(r)$ vertices, the deletion of R takes time $O(n\sqrt{r})$ for the preprocessing, $O(r^2 \log\log n)$ for the shortest path computations and $O(r^2)$ to find the nets to be routed across the edges.

This result is summarized by Lemma 2:

Lemma 2: *Let $P = (V, E, N)$ be an even PEP with an r-division, such that there is only one region R on the boundary of the infinite face of G.*
Then region R can be removed in time $O(r^2 \log\log n + n\sqrt{r})$.

Case 2: At least two separator vertices of R lie on the boundary of the infinite face of G.

Let Y be the cut crossing exactly the edges of R which are incident to the separator vertices of R. We separate $R' := R - \{\text{separator vertices of } R\}$ from G by processing the edges crossed by Y according to the basic algorithm. Since R' has $O(r)$ vertices, a solution of problem R' can be found in time $O(r^2)$ using the algorithm of Becker/Mehlhorn. The remaining problem is to separate R' from G in a target time $O(n\sqrt{r} + r^2 \log\log n)$.

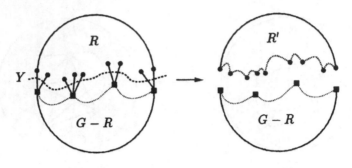

■ separator vertices

Figure 4 : Separating R' from G.

Let e be the rightmost edge in clockwise order that lies on the boundary of the infinite face of R and is crossed by Y. Again we have to determine a simple cut X of minimal cardinality and minimal free capacity that crosses e and a net N_i to be routed across e.

Let us turn to the first task. If the second boundary edge crossed by X is an element of E_R, then the free capacity of X can be computed in the same way as in Case 1. Construct the graph G'_d and the distance function $dist'$ (time $O(n\sqrt{r})$). The capacity of every simple cut crossing e is determined by a shortest path computation in G'_d. Inspecting the boundary of R the densities of these cuts can be ascertained in time $O(r)$. Thus, having constructed G'_d and $dist'$ once, it takes time $O(r \log \log n)$ to find the required cut X.

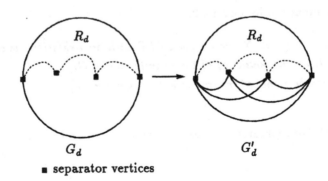

■ separator vertices

Figure 5 : Compression of graph G_d.

If the second boundary edge e' crossed by X lies in $G - R$, then the situation is more difficult. Let v, v' be the vertices in the infinite face of G_d which are dual to e, e' and let M, M_1, M_2 be defined as in Case 1. X corresponds to a shortest path p from v to v' in G_d. There is at least one element of M on this path. Let v_m be the last vertex on this path that belongs to M. v_m divides p into two pieces p_1, p_2, where p_1 is the path from v to v_m. p_1 is the shortest path from v to v_m in G'_d and p_2 is the shortest path from v_m to v' in $(G - R)_d$. Since $(G - R)_d$ never changes as long as the basic algorithm deals with edges of R, the length of p_2 can be computed in a preprocessing step. Because of the choice of edge e, the density of X is determined by an inspection of the boundary of $(G - R)_d$.

To compute the free capacity of cut X efficiently we make use of the above observations. Let B be the set of vertices on the boundary of the infinite face of $(G - R)_d$ which are dual to boundary edges of G. In a preprocessing step we compute the functions $lsp : M_2 \times B \rightarrow \mathbf{Z} \cup \{\infty\}$, $lsp(u, w) =$ length of the shortest path from u to w in $(G - R)_d$, and $dens : B \rightarrow \mathbf{Z}$, $dens(u) =$ density of a simple cut that crosses e and the boundary edge of $G - R$, which is dual to u. The functions lsp and $dens$ can be determined by $|M|$ BFS computations each.

For every vertex $u \in M$ we hold the set $S_u := \{(w, lsp(u, w) - dens(w))/w \in B\}$ in a priority queue, such that the tupel with minimal second component will be found in $O(1)$. (The data structure is described in the next section.) The preprocessing takes time $O(n\sqrt{r})$, since $|M| = O(\sqrt{r})$, and has to be executed only once for region R.

To find cut X we determine the length of the shortest paths in G'_d from v to every vertex $s \in M$, denoted by lsp' (time $O(r \log \log n)$). Let u be an element of M. If u lies on the path in G_d that corresponds to X, then the free capacity of X is given by the equation: $fcap(X) = lsp'(v, u) + lsp(u, w) - dens(w)$, where $(w, lsp(u, w) - dens(w))$ is the element of S_u with minimal second component. Hence X can be found by visiting the elements of M and computing the above term.

We have shown how to find the appropriate cut X as required in the basic algorithm. To find the net N_i which has to be routed across e is easier and will be described in the next section. The result is stated in

Lemma 3: *There is a data structure which supports the search for the appropriate net N_i to be routed across edge e in time $O(r + \log n)$.*

After net N_i is found and routed across e the densities of many cuts will change, such that the priority queues have to be updated. In the next section we will show that this can be done in time $O(\log n)$ per priority queue.

We summarize Case 2 by adding up the running times for the single steps. It takes time $O(n\sqrt{r})$ to compute graph G'_d, $O(r^2 \log \log n)$ to compute the functions lsp, lsp', $dens$,

$O(n\sqrt{r})$ to build up the priority queues, $O(r\sqrt{r}\log n)$ to update the priority queues, $O(r\cdot(r+\log n))$ to find the appropriate nets N_i for at most $O(r)$ edges and finally $O(r^2)$ to remove R'. Thus, the running time is $O(r^2\log\log n + n\sqrt{r})$ as in Case 1.

Lemma 4: *Let $P = (V, E, N)$ be an even PEP with an r-division, such that there is a region R bordering on the infinite face which does not enclose $G - R$. Then R can be removed in time $O(r^2\log\log n + n\sqrt{r})$.*

Now we can compute the total running time of our algorithm.

It takes time $O(n\log n)$ to build up the r-division and $O(r^2\log\log n + n\sqrt{r})$ to process Case 1 or Case 2. Since there are $O(\frac{n}{r})$ regions in the r-division, the total running time of our algorithm is $O(n\cdot r\cdot\log\log n + \frac{n^2}{\sqrt{r}} + n\log n)$. This term is minimized if $n\cdot r\cdot\log\log n = \frac{n^2}{\sqrt{r}}$. Hence we choose $r = (\frac{n}{\log\log n})^{2/3}$ and achieve a time bound of $O(n^{5/3}(\log\log n)^{1/3})$. This proves Theorem 3.

During the description of the algorithm we assumed that region R is connected. If R is disconnected, the algorithm can be applied to the connected components of R. Let $R_1, .., R_k$ be the connected components of R and let $sep(R_i)$, $in(R_i)$ denote the number of separator vertices and inner vertices of R_i. In accordance with Lemma 2, Lemma 4 respectively, it takes time $T = \sum_{i=1}^{k} in(R_i)^2\log\log n + n\cdot sep(R_i)$ to process the connected components of R. Since $\sum_{i=1}^{k} sep(R_i) = O(\sqrt{r})$ and $\sum_{i=1}^{k} in(R_i) = O(r)$ it follows immediately that $T \le O(r^2\log\log n + n\sqrt{r})$. Hence the time bound of our algorithm holds even if the regions are not connected.

This concludes the description of the algorithm. The missing data structure is given in the next section.

Data Structures

In this section we introduce an extended priority queue that supports the operations postulated in the last section in time $O(\log n)$, namely how to find the appropriate net for each edge and how to implement the sets S_u for every vertex $u \in M$ (Case 2).

Let $U = \{1, 2, .., n\}$ and $S \subseteq U \times U$, such that for every $x \in U$ there is at most one ordered pair $(x, y) \in S$. The extended priority queue is a leaf oriented height-balanced searchtree for set U. Each inner node v contains three pieces of information: $split(v), prio(v)$ and $change(v)$. $split(v)$ directs the search, i.e. $split(v)$ is smaller than every node in the right subtree of v and bigger than every node in the left subtree of v. $prio(v)$ is used to store a copy of an element of S. $change(v)$ indicates how much the second components of the elements of S which are stored in the subtree rooted at v have been changed by update operations. The extended priority queue is constructed as follows: Initialize the *change*-fields to 0. Store $(x,y) \in S$ in $prio(l)$, where l is the leaf that corresponds to x. If there is a leaf w whose *prio*-field is still empty, then initialize $prio(w)$ to $(split(w), \infty)$. Next consider a node v such that the *prio*-fields of its children are already defined. Among these *prio*-fields choose the one with the smallest second component (if they are equal, take the one of the right child) and move a copy to $prio(v)$. Continuing in this way we finally fill every *prio*-field. Obviously the construction takes time $O(n)$ and the copies of an ordered pair $(x,y) \in S$ are stored at the searchpath from the root to leaf x.

The most important property of the extended priority queue is that the *prio*-fields together with the *change*-fields implement a priority queue for the second components of the elements of S. I.e. let v be a node of the extended priority queue, such that $prio(v) = (x,y)$ and $prio(elter(v)) = (x',y')$. Then $y' \leq y + change(v)$.

Lemma 5: *The extended priority queue supports the operation $MinYinXrange(x_1, x_2)$ $= min\{y; \exists x : x_1 \leq x \leq x_2$ and $(x,y) \in S\}$ in time $O(\log n)$.*

Proof: Let p_{x_1}, p_{x_2} be the searchpaths to the leaves x_1, x_2 and let $C_{max}(x_1, x_2) = \{v; split(v) \in [x_1, x_2]$ and $split(elter(v)) \notin [x_1, x_2]\}$. The elements of $C_{max}(x_1, x_2)$ are the roots of subtrees of maximum size whose leaves are in the interval $[x_1, x_2]$. A copy of $MinYinXrange(x_1, x_2)$ is stored either at a node on p_{x_1}, p_{x_2} respectively or at an element of $C_{max}(x_1, x_2)$. Hence $MinYinXrange(x_1, x_2)$ can be determined by inspecting p_{x_1} and p_{x_2} and the children of the nodes of p_{x_1}, p_{x_2}. This takes time $O(\log n)$ because the extended priority queue is height-balanced. ∎

This operation can be used in Case 2 to find net N_i to be routed across edge e. Let k be the number of boundary vertices of $G - R$. Number the boundary vertices of $G - R$ in counterclockwise order from 1 to k. Let N be a net whose terminals are situated at the boundary vertices a and $b, 1 \leq a \leq b \leq k$. Store (b,a) in an extended priority queue. Let $\{x_1, .., k\}$ be the boundary vertices that lie in cut X. One terminal of N_i is situated in $\{x_1, .., k\}$. If the second terminal lies in $G - R$, then N_i is determined by $MinYinXrange(x_1, k)$. Otherwise N_i can be found by inspecting the boundary of R. Hence it takes time $O(r + \log n)$ to detect net N_i. This proves Lemma 3.

It remains to show how to implement the sets S_u. If the boundary of $(G - R)_d$ is numbered as described above, each set S_u can be implemented as an extended priority queue, which clearly fulfills the demands of the last section.

If net $N_i = \{s, t\}$ is routed across edge $e = (a, k)$, it is split into two nets $\{s, a\}, \{k, t\}$, such that the density of every simple cut that crosses exactly one boundary edge between k and t increases by one. To update the extended priority queue S_u, we need the operation $increase(t, k, 1)$ which increases the second component of every tupel $(x, y) \in S$, $t \leq x \leq k$ by one.

Lemma 6: *The extended priority queue supports the operation $increase(x_1, x_2, c)$ in time $O(\log n)$.*

Proof: Add c to the *change*-fields of the nodes in $C_{max}(x_1, x_2)$. Let $v \in C_{max}(x_1, x_2)$. Since every element of S which is stored in the subtree rooted at v is affected by the operation, the subtree rooted at v still implements a priority queue. Hence we only have to update the elements of S which are situated at nodes $w \notin C_{max}(x_1, x_2)$ on the searchpaths from the root to x_1, x_2, respectively. Empty the *prio*-fields of these nodes and refill them in the following way. Let v be a node whose *prio*-field is empty, but the *prio*-fields of its children are defined. Let v_l, v_r be the left child and the right child of v and let $prio(v_l) = (x_l, y_l)$, $prio(v_r) = (x_r, y_r)$. $prio(v)$ is determined by :

$$\text{if } y_r + change(v_r) \leq y_l + change(v_l)$$
$$\text{then } prio(v) := (x_r, y_r + change(v_r))$$
$$\text{else } prio(v) := (x_l, y_l + change(v_l))$$

Applying the last step repeatedly finishes the operation $increase(x_1, x_2, c)$.

Since the extended priority queue is height-balanced, $increase(x_1, x_2, c)$ can be performed in time $O(\log n)$. ∎

This finishes the description of the data structures.

Conclusion

We have presented a new technique to get a faster algorithm for the planar edge-disjoint paths problem. Some problems remain:

1) The new algorithm can also be applied to non-even problems and provides an improvement on $O(n^{5/3}(\log \log n)^{1/3} + b \cdot n)$, where b denotes the size of the boundary. This bound might still be $O(n^2)$. It is not clear how to remove the b-factor.
2) Apply the new technique to the general multicommodity flow problem. Note that a direct application to the algorithm of Matsumoto/Nishizeki/Saito is not possible, since the edges are considered in prescribed orders in both techniques and the orders are incompatible.
3) The running time of this algorithm might be open to further improvements. A running time of $O(n^{3/2})$ should be achievable using the separation techniques.

References

[BM86]   M. Becker, K. Mehlhorn: "Algorithms for Routing in Planar Graphs", Acta Informatica 23 (1986), pp. 163–176.

[EIS76]   S.Even, A.Itai, A.Shamir: "On the complexity of timetable and multicommodity flow problems", SIAM J. Comp. 5(4) (1976), pp. 691–703.

[EKZ77]   P. Van Emde Boas, R. Kaas, and E. Zijlstra: "Design and implementation of an efficient priority queue", Math. Sys. Theory 10 (1977), pp. 99–127.

[F82]   A. Frank: "Disjoint Paths in Rectilinear Grids", Combinatorica 2, 4 (1982), pp. 361–371.

[F84]   G. N. Frederickson: "Fast Algorithms For Shortest Paths In Planar Graphs, With Applications ", Technical report (1984), Purdue University, Indiana.

[FF56]   L.R. Ford, D.R. Fulkerson: "Maximal flow through a network", Canadian Journal of Mathematics, Vol. 8, pp. 399-404 (1956).

[K90]   M.Kaufmann: "A linear-time algorithm for routing in a convex grid", IEEE Trans. on Computer-Aided Design 2 (1990), pp. 180-184.

[KM86]   M. Kaufmann, K. Mehlhorn: "Routing through a Generalized Switchbox", J. Algorithms 7, pp. 510-531 (1986).

[MNS86]   K. Matsumoto, T. Nishizeki, N. Saito: "An Efficient Algorithm for Finding Multicommodity Flows in Planar Networks", SIAM Journal of Computing, Vol 15:2 (1986), pp. 495–510.

[MPS86]   K. Mehlhorn, F.P. Preparata, M. Sarrafzadeh: "Channel routing in knock-knee mode: Simplified algorithms and proofs", Algorithmica 1 (1986), pp. 213-221.

[NSS85]   T. Nishizeki, N. Saito, K. Suzuki: "A linear-time routing algorithm for convex grids", IEEE Trans. on Computer Aided Design, CAD-4 (1985), pp. 68–76.

[O83]   H. Okamura: "Multicommodity Flows in Graphs", Discrete Applied Mathematics 6 (1983), pp. 55–62.

[OS81]   H. Okamura, P.D. Seymour: "Multicommodity Flows in Planar Graphs", Journal of Combinatorial Theory 31, Series B (1981), pp. 75–81.

[S87]   A. Schrijver: "Decomposition of graphs on surfaces and a homotopic circulation theorem", tech. report (1987), to appear in J. Comb. Theory, Ser. B.

An Optimal Construction Method for Generalized Convex Layers

Hans-Peter Lenhof   Michiel Smid*

Max-Planck-Institut für Informatik
W-6600 Saarbrücken, Germany

1 Introduction

We consider the following problem: Let P be a set of n points in the Euclidean plane and let C be a convex figure. Preprocess P such that for any query point q, the points of P in the translate $C + q$ can be retrieved efficiently. In 1985, Chazelle and Edelsbrunner [5] provided a space and query time optimal solution for so called computable figures C (for the definition of computable figures see section 2). Their solution uses $O(n)$ space. A query with output size k takes $O(\log n + k)$ time. The preprocessing step, however, has time complexity $O(n^2)$. In a few special cases alternative solutions have been developed:

- Fixed radius neighbor problem: In this case C is a disk. The previously best known solutions to this problem are:

	preprocessing	query	space
[4]	$O(n(\log n)^5(\log\log n)^2)$	$O(\log n + k)$	$O(n(\log n \log\log n)^2)$
	$O(n\log n)$	$O(k(\log n)^2)$	$O(n\log n)$
prob.	?	$O(\log n + k)$	$O(n(\log n)^2)$
[5]	$O(n^2)$	$O(\log n + k)$	$O(n)$
[1]	polyn. time	$O(\log n + k)$	$O(n\log n)$
prob.	$O(n(\log n)^2)$	$O(\log n + k)$	$O(n\log n)$

  The algorithms in [1] and [4] also handle queries with non-fixed radius.

- In 1986, Klein et al. [6] presented an optimal dynamic solution for polygonal figures with a constant number of boundary edges. Their dynamic data structure has size $O(n)$. Insertions and deletions can be carried out in $O(\log n)$ time. A query with output size k can be done in $O(\log n + k)$ time.

As mentioned above, Chazelle and Edelsbrunner presented the first space and query time optimal solution, but the preprocessing step of their algorithm has time complexity

*This author was supported by the ESPRIT II Basic Research Actions Program, under contract No. 3075 (project ALCOM).

$O(n^2)$. During this preprocessing step the Euclidean plane is decomposed into cells. Then, for every non-empty cell, i.e., a cell that contains points of P, a family of so-called layers is constructed. In the worst case, the layers construction method of Chazelle and Edelsbrunner requires $\Omega(n^2)$ operations.

In this paper, we introduce a kind of "dual or mirror layers" with respect to the layers, which Chazelle and Edelsbrunner use to build their query data structures. A family of these "dual layers" can be constructed in $O(n \log n)$ time. By walking across the "dual layers" of such a family, we can determine a family of layers in $O(n)$ time. Hence the whole layer computation takes $O(n \log n)$ time. Therefore, the main result of this paper is the following:

Theorem 1 *Let P be a set of n points in the Euclidean plane and C a convex computable figure. There exists a data structure that stores the point set P, such that $O(\log n + k)$ time suffices to retrieve all k points lying inside a query translate C_q. The data structure has size $O(n)$ and can be constructed in $O(n \log n)$ time.*

We thus provide the first space, query time and preprocessing time optimal solution for this class of point retrieval problems. We want to emphasize that all these problems can now be solved optimally by one general technique. Besides, this layer construction method gives new dynamic data structures for the above class of point retrieval problems:

Theorem 2 *Let P be a set of n points in the Euclidean plane and C a convex computable figure. There exists a dynamic data structure which stores the point set P and which has size $O(n)$, such that $O(\log n + k(\log n)^2)$ time suffices to retrieve all k points of P lying inside a query translate C_q. The data structure can be dynamically maintained at a worst-case cost of $O((\log n)^2)$ per insertion and deletion.*

In Section 2, we describe the algorithm of Chazelle and Edelsbrunner given in [5]. The new construction method for layers will be explained in Section 3.

2 The algorithm of Chazelle and Edelsbrunner

First of all, we introduce some geometric notions: The Euclidean plane is denoted by E^2. Let $v = (v_x, v_y), w = (w_x, w_y) \in E^2$, and $A, B \subseteq E^2$. We use the following notations:

$$int(A) := \text{interior of A} \qquad cl(A) := \text{closure of A} \qquad bd(A) := \text{boundary of A}$$
$$v + w := (v_x + w_x, v_y + w_y) \qquad A_v := \{a + v | a \in A\} \qquad A + B := \{a + b | a \in A, b \in B\}$$

Definition 1 *A convex closed figure C is called computable, if (1) constant time suffices to test for any point $p \in E^2$ whether or not p is contained in C, and (2) constant time suffices to compute the intersection $bd((-C)_v) \cap bd((-C)_w)$ for any two points v and w in E^2.*

Throughout this paper, C is a bounded computable figure in E^2.

Definition 2 *Let v be a point, $r_u(v)$ the vertical ray with v as lower endpoint and $r_d(v)$ the vertical ray with v as upper endpoint. We call the set $S(v) := -C + r_u(v)$ the silo of v and the set $RS(v) := C + r_d(v)$ the rotated silo of v (see Figure 1).*

We now describe the algorithm: In a first step the computable figure C is decomposed in two computable figures C_a, the "upper" part of C, and C_b, the "lower" part of C. This step can be carried out in constant time. We assume that $C = C_a$ and explain the query algorithm for $C = C_a$.

After a basis transformation the Euclidean plane E^2 is decomposed into cells, such that any translate C_q intersects at most 9 cells lying in three consecutive rows and columns. Then, the non-empty cells are determined, sorted in lexicographical order and stored in a balanced binary search tree. Hence the non-empty cells, which are intersected by a translate C_q, can be found in $O(\log n)$ time. Determining and storing the non-empty cells takes $O(n \log n)$ time.

Definition 3 *Let CE be a cell and let $N = north$, $E = east$, $S = south$ and $W = west$ denote the four edges of $bd(CE)$. We say that C_q is D-grounded, if $C_q \cap D$ equals the orthogonal projection of $C_q \cap cl(CE)$ onto D, for $D \in \{N, S, W, E\}$. C_q is said to be grounded if it is D-grounded for at least one assignment of D to N, S, W or E.*

Chazelle and Edelsbrunner prove the following statements: (1) Any query translate C_q, which intersects a cell CE, is grounded with respect to CE. (2) We can determine in constant time, whether C_q is N, S, W or E grounded with respect to cell CE.

For every non-empty cell CE we build four data structures, one for every assignment of D to N, S, W, E. Since all four data structures are built in the same way, we only show how the query data structure of cell CE for S-grounded queries is built. In such a query, we want to compute all points $p \in P_{CE} := P \cap CE$ which lie in C_q.

Lemma 1 ([5]) *Let p be a point of P_{CE}. Then, p is in C_q if and only if $q \in S(p)$.*

Lemma 1 tells us, that we can compute all points $p \in P_{CE} \cap C_q$ in the following way: Compute all silos $S(p)$ for $p \in P_{CE}$ which contain q. Note that these are exactly those silos, whose boundaries are intersected by the vertical ray from q towards $y = -\infty$.

How can we find all these silos $S(p)$? We assume that the points $P_{CE} = \{p_1, \cdots, p_m\}$ are sorted in order of increasing x-coordinates. Since any pair $S(p_i), S(p_j)$ of silos with $p_i, p_j \in P_{CE}$ intersect each other, the set $U := \bigcup_{1 \le i \le m} S(p_i)$ is connected. The boundary $L(P_{CE}) = bd(U)$, which we call the *S-layer* of P_{CE}, is an unbounded, connected, x-monotone curve. Any vertical line intersects this S-layer in at most one point or ray. We call $e_i := [L(P_{CE}) \cap bd(S(p_i))] \setminus \bigcup_{1 \le j < i} bd(S(p_j))$ the *edge* of p_i. Since e_i can be empty, not every point $p_i \in P_{CE}$ contributes to a part of $L(P_{CE})$. If e_i is empty, then p_i is called *redundant* and we define $ext(P_{CE}) := \{p \in P_{CE} | p \text{ non-redundant}\}$.

Lemma 2 ([5]) *Let $e_{k_1}, \cdots, e_{k_t}$ be the sequence of non-empty edges of $L(P_{CE})$ ordered from left to right, C_q a translate of C that is S-grounded with respect to CE and $r := r_d(q)$ the vertical ray with upper point q.*
(i) $k_i < k_{i+1}$ for $1 \le i \le t - 1$, i.e., the ordering of the edges coincides with that of P_{CE}.
(ii) The ray r intersects $L(P_{CE})$, if and only if $C_q \cap P_{CE}$ is not empty.
(iii) If $e_{k_l} \cap r \ne \emptyset$, then p_{k_l} lies in C_q and there are indices i and j, with $i \le l \le j$, such that $C_q \cap ext(P_{CE}) = \{p_{k_\alpha} | i \le \alpha \le j\}$.

The algorithm for finding all points of $C_q \cap P_{CE}$ works as follows: First, we search for the points of $C_q \cap ext(P_{CE})$. In order to find these points, we search for the edge

e_{k_l} that intersects the ray $r := r_d(q)$. (If there is no such edge then $C_q \cap P_{CE} = \emptyset$ and the algorithm stops.) Note that e_{k_l} can be found in $O(\log t)$ time. Then we start at e_{k_l} and walk along $L(P_{CE})$ to the left, until we find an edge $e_{k_{i-1}}$ such that $q \notin S(p_{k_{i-1}})$. Analogously, we walk to the right, again starting at e_{k_l}, until we find an edge $e_{k_{j+1}}$ such that $q \notin S(p_{k_{j+1}})$. During these walks, we report all points $p_{k_i}, p_{k_{i+1}}, \cdots, p_{k_j}$. In this way, we have determined the points of $C_q \cap ext(P_{CE})$ in $O(\log t + |C_q \cap ext(P_{CE})|)$ time.

Let $P_{CE}^1 := P_{CE}$ and $P_{CE}^i := P_{CE}^{i-1} \setminus ext(P_{CE}^{i-1})$ for $i > 1$. Suppose we have constructed the S-layers $L(P_{CE}^i)$, for $i \geq 1$. If $C_q \cap ext(P_{CE}^1) \neq \emptyset$, we search for all points $p \in C_q \cap ext(P_{CE}^2)$ by testing, if the ray r intersects $L(P_{CE}^2)$. If r intersects the second S-layer, we walk across $L(P_{CE}^2)$ in the same way as described above. We continue to test the S-layers, until we find an S-layer, which is not intersected by r, or until we have checked all non-empty S-layers. Since the family $L_S(P_{CE}) = (L(P_{CE}^1), \cdots, L(P_{CE}^z))$ of non-empty S-layers is nested, an S-layer which lies above a non-intersected S-layer, cannot be intersected by the ray r. Hence, no point represented by such an S-layer, can lie in C_q.

All k_{CE} points $p \in C_q \cap P_{CE}$ are reported in $O(|\text{visited S-layers}| \log m + k_{CE})$ time. Since we have found at least one point in every S-layer, the query time is $O(k_{CE} \log m)$. By applying Chazelle's hive graph [2] to $L_S(P_{CE})$, the query time can be improved to $O(\log m + k_{CE})$. The hive graph connects $L(P_{CE}^i)$ with $L(P_{CE}^{i+1})$ in such a way that the knowledge of the edge in $L(P_{CE}^i)$ that intersects r, allows us to find the intersecting edge in $L(P_{CE}^{i+1})$ in constant time. $O(m)$ space suffices to store the hive graph of $L_S(P_{CE})$. The hive graph can be constructed in $O(m)$ time (see [2]). Hence the cost of the preprocessing step is dominated by the $O(m^2)$ operations required to construct the family of S-layers. (Note that there are at most m S-layers, each of which is constructed in $O(m)$ time.) We can now cite the main result of [5]:

Theorem 3 *Let P be a set of n points in the Euclidian plane E^2 and C a convex closed computable figure. There exists a data structure, such that $O(k + \log n)$ time suffices to retrieve all k points of P lying in a query translate C_q. The data structure has size $O(n)$ and can be constructed in $O(n^2)$ time.*

3 Fast construction method for S-layers

Let $P_{CE} = P \cap CE = \{p_1, \cdots, p_m\}$ be the sequence of points in cell CE, sorted in order of increasing x-coordinates. We assume that there are no two points in P_{CE} with the same x-coordinate. We consider again only S-grounded queries. In this section we describe a method to construct the family of S-layers for the point set P_{CE}, which takes only $O(m \log m)$ time. In Subsection 3.1 we present a new geometric concept for convex curves, the so-called dual or mirror S-layers. The family of S-layers can be constructed from the family of dual S-layers, in $O(m)$ time. In Subsection 3.2 we describe an $O(m(\log m)^2)$ time dual S-layer construction algorithm, which is a modification of Overmars and van Leeuwen's convex layers construction algorithm (see [7]). Besides, we show that this dual S-layer construction method gives new dynamic query data structures. In Subsection 3.3 we discuss a modified version of Chazelle's convex layers construction algorithm [3] and we show that this algorithm enables the construction of the dual S-layers with only $O(m \log m)$ operations.

3.1 Dual S-layers

We start with the definition of minimal representations for the S-layer $L(P_{CE})$. Recall that the points of $P_{CE} = \{p_1, \cdots, p_m\}$ are sorted by their x-coordinates.

Definition 4 *a) A sequence $R(P_{CE}) = <p_{k_1}, \cdots, p_{k_t}> \subseteq P_{CE}$, which satisfies the following properties, is called a representation system or r-system of the S-layer $L(P_{CE})$:*

- *$k_i < k_{i+1}$ for all $i = 1, \cdots, t-1$*

- *$bd(S(p_{k_i})) \cap L(P_{CE}) \neq \emptyset$ for all $i = 1, \cdots, t$*

- *$L(P_{CE}) \subseteq \bigcup_{j=1}^{t} bd(S(p_{k_j}))$.*

b) An r-system $R(P_{CE}) = <p_{k_1}, \cdots, p_{k_t}>$ of the S-layer $L(P_{CE})$ with minimal length t is called a minimal r-system of the S-layer $L(P_{CE})$.

In order to prevent the number of variables becoming too large, we redefine the edges e_{k_i} and sets P_{CE}^i. If $R(P_{CE}) = <p_{k_1}, \cdots, p_{k_t}>$ is a minimal r-system for $L(P_{CE})$, then the S-layer $L(P_{CE})$ can be represented by the union

$$L(P_{CE}) = \bigcup_{i=1}^{t} e_{k_i} \quad \text{of the edges} \quad e_{k_i} := [L(P_{CE}) \cap bd(S(p_{k_i}))] \setminus \bigcup_{1 \leq j < i} bd(S(p_{k_j})).$$

We say that edge e_{k_i} represents point p_{k_i}. Let $P_{CE}^1 := P_{CE}$ and $P_{CE}^i := P_{CE}^{i-1} \setminus R_{i-1}$, where $R_{i-1} := R(P_{CE}^{i-1})$ is a minimal r-system of the S-layer $L(P_{CE}^{i-1})$. Furthermore let $R_S(P_{CE}) = (R_1, \cdots, R_z)$ be a family of minimal r-systems defined recursively in the above way, such that every point p of P_{CE} is contained in one P_{CE}^i for some $1 \leq i \leq z$. The assertions of Lemma 2 are also valid for each S-layer $L(P_{CE}^i)$ defined as above. Furthermore each family of S-layers that is constructed in the above way, consists of nested S-layers. Hence, if we know such a family of minimal r-systems for $L(P_{CE})$, we can construct the data structures used in [5] in $O(m)$ time.

We need some more notions: Let a and b be two points in cell CE with different x-coordinates (we assume $a_x < b_x$). We call the point in the intersection $bd(S(a)) \cap bd(S(b))$ having the smallest x-coordinate the *si-point* of a and b and the point having the largest x-coordinate the *SI-point* of a and b. We use the notations $si(a,b)$ and $SI(a,b)$ for these points. If the intersection is a unique point, then $si(a,b) = SI(a,b)$. We define $int_{si}(a,b) := int(RS(si(a,b)))$, $int^{si}(a,b) := int(RS(SI(a,b)))$ and $int_{si}^{si}(a,b) := int(RS(si(a,b))) \cup int(RS(SI(a,b)))$. The intersection

$$e(a,b) := bd(RS(SI(a,b))) \cap \{p = (p_x, p_y) | a_x \leq p_x \leq b_x\}$$

will be called the *dual-edge* or *d-edge* of a and b. The d-edge of a and b is x-monotone and connects a and b. We call the set

$$reg(a,b) := \{p = (p_x, p_y) | a_x < p_x < b_x \wedge p \text{ lies (strictly) above } e(a,b)\}$$

the *region* of a and b (see Figure 2). The following lemma is proved in the full paper:

Lemma 3 *Let a, b, c be three points in cell CE with different x-coordinates. Then $bd(S(c)) \cap bd(S(a) \cup S(b)) = \emptyset$ if and only if $c \in reg(a,b)$.*

We now define a kind of "dual layer" to the S-layer $L(P_{CE})$:

Definition 5

a) Let $DL(P_{CE})$ be the lower envelope of the set of d-edges $\{e(p_i, p_j)|p_i, p_j \in P_{CE}\}$. We call $DL(P_{CE})$ the *dual S-layer* or *dS-layer* of P_{CE}.

b) A sequence $DR(P_{CE}) =< p_{k_1}, \cdots, p_{k_t} >\subseteq P_{CE}$, which satisfies the following properties, is called a *dual representation system* or *dr-system* of $DL(P_{CE})$:

- $k_i < k_{i+1}$ for all $i = 1, \cdots, t-1$

- $e(p_{k_i}, p_{k_{i+1}}) \subseteq DL(P_{CE})$ for all $i = 1, \cdots, t-1$

- $\bigcup_{i=1}^{t-1} e(p_{k_i}, p_{k_{i+1}}) = DL(P_{CE})$.

c) A dr-system $DR(P_{CE}) =< p_{k_1}, \cdots, p_{k_t} >$ of $DL(P_{CE})$ with minimal length t is called a *minimal dr-system of the dS-layer* $DL(P_{CE})$.

See Figure 3 for an example of a dS-layer. The dS-layer is an x-monotone curve. Using Lemma 3 we can easily show, that $< p_{k_1}, \cdots, p_{k_t} >$ is a minimal dr-system for $DL(P_{CE})$ if and only if $< p_{k_1}, \cdots, p_{k_t} >$ is a minimal r-system for the S-layer $L(P_{CE})$.

Let $P_{CE}^1 := P_{CE}$ and $P_{CE}^i := P_{CE}^{i-1} \setminus DR_{i-1}$, where $DR_{i-1} = DR(P_{CE}^{i-1})$ is a minimal dr-system for $DL(P_{CE}^{i-1})$. By computing a family $DL_S(P_{CE}) = (DL(P_{CE}^1), \cdots, DL(P_{CE}^z))$ of dS-layers resp. a corresponding minimal dr-system $DR_S(P_{CE}) = (DR_1, \cdots, DR_z)$, we get a family $R_S(P_{CE}) = DR_S(P_{CE})$ of minimal r-systems, which enables us to build the query data structure for S-grounded queries in $O(m)$ time.

3.2 Convex layers and dS-layers

In this subsection we show that there are important similarities between lower convex hulls and dS-layers. These similarities enable us to construct dS-layers with the same methods that are used to construct lower convex hulls. The dS-layer consists of curves, which connect points of the corresponding point set. All points of this point set lie on or above the dS-layer. Given two dS-layers, where all points of the first dS-layer are to the left of all points of the second dS-layer, there is exactly one new d-edge on the dS-layer of all points, which connects the two dS-layers and shares only start- and endpoint with them.

We transfer the terms *concave*, *reflex* and *supporting* from the theory of convex hulls to the theory of dS-layers: Let $P_{CE}^{[l]} = \{p_1, \cdots, p_h\}$ and $P_{CE}^{[r]} = \{p_{h+1}, \cdots, p_m\}$ with

$$(p_1)_x < (p_2)_x < \cdots < (p_h)_x < (p_{h+1})_x < \cdots < (p_m)_x$$

be a partition of P_{CE}. Let $DR(P_{CE}^{[l]}) =< p_1^l, \cdots, p_s^l >$ be a minimal dr-system of $DL(P_{CE}^{[l]})$ and $DR(P_{CE}^{[r]}) =< p_1^r, \cdots, p_t^r >$ a minimal dr-system of $DL(P_{CE}^{[r]})$. Let u and v be such that $p_i^l = p_u$ and $p_j^r = p_v$. The d-edge $e(p_i^l, p_j^r)$ is called

- *si-supporting* in p_i^l, if for all $f \in \{1, \cdots, h\}$, $p_f \notin int_{si}(p_i^l, p_j^r)$,

- *si-supporting* in p_j^r, if for all $g \in \{h+1, \cdots, m\}$, $p_g \notin int_{si}(p_i^l, p_j^r)$,

- *si-concave* in p_i^l, if there is a point $p_f \in int_{si}(p_i^l, p_j^r)$, where $f \in \{u+1, \cdots, h\}$,

- *si-concave* in p_j^r, if there is a point $p_g \in int_{si}(p_i^l, p_j^r)$, where $g \in \{h+1, \cdots, v-1\}$,

- *si-reflex* in p_i^l, if there is a point $p_f \in int_{si}(p_i^l, p_j^r)$, where $f \in \{1, \cdots, u-1\}$,

- *si-reflex* in p_j^r, if there is a point $p_g \in int_{si}(p_i^l, p_j^r)$, where $g \in \{v+1, \cdots, m\}$.

Analogously we define *SI-supporting*, *SI-concave* and *SI-reflex*, by replacing $int_{si}(p_i^l, p_j^r)$ by $int^{si}(p_i^l, p_j^r)$ in the above definition. The three cases supporting, concave and reflex are mutually exclusive. The d-edge $e(p_i^l, p_j^r)$ is called a *supporting d-edge* for $DL(P_{CE}^{[l]})$ and $DL(P_{CE}^{[r]})$, if for all $f \in \{1, \cdots, m\}$, $p_f \notin int^{si}_{si}(p_i^l, p_j^r)$. Hence the d-edge $e(p_i^l, p_j^r)$ is a supporting d-edge if and only if $e(p_i^l, p_j^r)$ is si-supporting and *SI-supporting* in both endpoints p_i^l and p_j^r.

Note that every d-edge $e(p_{k_i}, p_{k_{i+1}})$, which belongs to two neighboring points $p_{k_i}, p_{k_{i+1}}$ of a minimal dr-system, is a supporting d-edge with respect to the corresponding point set.

In order to compute a minimal dr-system of $DL(P_{CE})$, we have to determine a "longest" supporting d-edge for $DL(P_{CE}^{[l]})$ and $DL(P_{CE}^{[r]})$. If we know a supporting d-edge $e(p_i^l, p_j^r)$, we can determine a "longest" supporting d-edge resp. a minimal dr-system of $DL(P_{CE})$ in constant time (see the full paper). Before we show how a supporting d-edge can be computed efficiently, we prove that we can determine in constant time, if a d-edge $e(p_i^l, p_j^r)$ is si-supporting, si-concave or si-reflex (resp. *SI-supporting*, *SI-concave* or *SI-reflex*).

Lemma 4 a) A d-edge $e(p_i^l, p_j^r)$ is

- *si-supp.* in p_i^l, if $p_{i-1}^l, p_{i+1}^l \notin int_{si}(p_i^l, p_j^r)$
- *si-concave* in p_i^l, if $p_{i+1}^l \in int_{si}(p_i^l, p_j^r)$
- *si-reflex* in p_i^l, if $p_{i-1}^l \in int_{si}(p_i^l, p_j^r)$
- *si-supp.* in p_j^r, if $p_{j-1}^r, p_{j+1}^r \notin int_{si}(p_i^l, p_j^r)$
- *si-concave* in p_j^r, if $p_{j-1}^r \in int_{si}(p_i^l, p_j^r)$
- *si-reflex* in p_j^r, if $p_{j+1}^r \in int_{si}(p_i^l, p_j^r)$.

b) If we replace $int_{si}(p_i^l, p_j^r)$ by $int^{si}(p_i^l, p_j^r)$ in a), we get analogous conditions for the properties SI-supporting, SI-concave and SI-reflex.

Proof: We only show the first statement of part a) for property si-supporting, because all other statements can be shown analogously or follow by definition. Let $p_i^l = p_u$. So assume that $p_{i-1}^l, p_{i+1}^l \notin int_{si}(p_i^l, p_j^r)$. We assume that a point $p_f \in int_{si}(p_i^l, p_j^r)$ with $f < u$ exists. Since $int^{si}_{si}(p_{i-1}^l, p_i^l)$ contains the part of $int_{si}(p_i^l, p_j^r)$, which lies to the left of the vertical line through p_i^l, the point p_f lies in $int^{si}_{si}(p_{i-1}^l, p_i^l)$. But this is a contradiction to the fact that $e(p_{i-1}^l, p_i^l)$ is a supporting d-edge in $DL(P_{CE}^{[l]})$. Therefore a point $p_f \in int_{si}(p_i^l, p_j^r)$ with $f < u$ does not exist. A similar argument implies, that there is no point $p_f \in int_{si}(p_i^l, p_j^r)$ with $f > u$. Hence the d-edge $e(p_i^l, p_j^r)$ is si-supporting in p_i^l. ∎

In the following lemma we show how we can determine a supporting d-edge for the dS-layers $DL(P_{CE}^{[l]})$ and $DL(P_{CE}^{[r]})$ efficiently. Using this lemma, the dS-layer $DL(P_{CE})$ can be computed by a divide-and-conquer algorithm.

Lemma 5
Given two minimal dr-systems $DR(P_{CE}^{[l]}) = <p_1^l, \cdots, p_s^l>$ and $DR(P_{CE}^{[r]}) = <p_1^r, \cdots, p_t^r>$ we can compute a supporting d-edge of the corresponding dS-layers in $O(\log s + \log t)$ time.

Proof: We assume that the two dr-systems are stored in two arrays. Let $1 \leq i \leq s$ and $1 \leq j \leq t$. We consider the d-edge $e(p_i^l, p_j^r)$. Each of the two points p_i^l and p_j^r can be classified (1) as either si-reflex or si-supporting or si-concave and (2) as either SI-reflex or SI-supporting or SI-concave with respect to the d-edge $e(p_i^l, p_j^r)$. As in the case of ordinary convex hull construction we classify nine possible cases, which are schematically illustrated in Figure 4.

In all cases, in which we have not found a supporting d-edge, we can reduce the number of candidates for the left or right endpoints of the supporting d-edge. The dashed parts of the dS-layers are those, which can be eliminated from further consideration for containing a supporting point. If $si(p_i^l, p_j^r) \neq SI(p_i^l, p_j^r)$, we classify the d-edge $e(p_i^l, p_j^r)$ with respect to both intersection endpoints and eliminate the parts of the S-layers given by the two classifications. For both intersection endpoints we have nine possible cases, which are illustrated in Figure 4 for si-classification. Since the table for SI-classification is the same as for si-classification and since the dashed parts of the dS-layers, which can be eliminated from further considerations, are also the same, we only consider the si-classification. We prove here only two of the nine cases:

$(p_i^l, p_j^r) = $ (si-concave, si-supporting): In this case the set $\{e(p_f^l, p_g^r) | f < i \wedge 1 \leq g \leq t\}$ of d-edges cannot contain a supporting d-edge, because $e(p_i^l, p_j^r)$ lies below or on these d-edges in the range spanned by the x-coordinates of the dS-layer $DL(P_{CE}^{[t]})$ and, hence, all the d-edges in the above set are also si-concave in the left endpoint. Therefore the point set $\{p_f^l | f \leq i\}$ can be removed from the candidate list of left supporting points.

$(p_i^l, p_j^r) = $ (si-concave, si-concave): This is the difficult case. Let l_1, l_2 be the vertical lines with $(l_1)_x = (p_s^l)_x$ and $(l_2)_x = (p_1^r)_x$. Let further

$$\Delta = bd(RS(SI(p_i^l, p_{i+1}^l))) \cap bd(RS(si(p_{j-1}^r, p_j^r))).$$

<u>Case 1:</u> $\Delta = \emptyset$. Consider Figure 5. If $bd(RS(SI(p_i^l, p_{i+1}^l)))$ does not intersect line l_2, any rotated silo, whose boundary contains a point of the set $\{p_f^l | f \leq i\}$ and whose interior does not contain a point of the set $P_{CE}^{[t]}$, lies on the left side of l_2. Hence the boundary of such a rotated silo does not touch the shaded right region, which contains all points of the set $P_{CE}^{[r]}$. Therefore the set $\{p_f^l | f \leq i\}$ cannot contain a left endpoint of a supporting d-edge. If $bd(RS(si(p_{j-1}^r, p_j^r)))$ does not intersect l_1, we can eliminate the set $\{p_g^r | g \geq j\}$ from the candidate list of right "supporting points".

<u>Case 2:</u> $\Delta = (\Delta_x, \Delta_y)$ is a point or a vertical line segment with x-coordinate Δ_x.
Case 2a: If $\Delta_x \leq (l_1)_x$, an argument similar to Case 1 implies, that the set $\{p_f^l | f \leq i\}$ cannot contain a left endpoint of a supporting d-edge.
Case 2b: If $(l_1)_x < \Delta_x < (l_2)_x$, the set $\{p_f^l | f \leq i\}$ does not contain a left supporting point and the set $\{p_g^r | g \geq j\}$ does not contain a right supporting point.
Case 2c: If $\Delta_x \geq (l_2)_x$, then the set $\{p_g^r | g \geq j\}$ does not contain a right supporting point.

<u>Case 3:</u> If Cases 1 and 2 do not apply, then Δ is a line segment with startpoint $a = (a_x, a_y)$ and endpoint $b = (b_x, b_y)$, where $a_x < b_x$. We distinguish again three subcases:
Case 3a: If $b_x < (l_2)_x$, then the set $\{p_g^r | g \geq j\}$ does not contain a right supporting point.
Case 3b: If $a_x > (l_1)_x$, then the set $\{p_f^l | f \leq i\}$ does not contain a left supporting point.
Case 3c: $b_x \geq (l_2)_x$ and $a_x \leq (l_1)_x$. Assume $e(p_f^l, p_g^r)$ with $g > j$ is a supporting d-edge. Since $(p_i^l, p_j^r) = $ (si-concave, si-concave), p_i^l must lie outside the rotated silo $RS(si(p_{j-1}^r, p_j^r))$. Therefore the part of $bd(RS(si(p_{j-1}^r, p_j^r)))$ lying to the left of the vertical line through a is contained in $int(RS(SI(p_i^l, p_{i+1}^l)))$. Hence the index f must be greater

or equal to i and p'_f must lie on that part of the line segment from a to b, that is on the left side of l_1 (see Figure 6). The fact that all rotated silos are translates of a convex figure, implies the following statement: Consider any rotated silo, whose interior does not contain a point of the set $P_{CE}^{[r]}$ and whose boundary contains the point p_g^r. The part of the boundary of such a rotated silo, lying to the left of the vertical line through p_{j-1}^r, is contained in $int(RS(si(p_{j-1}^r, p_j^r)))$. Hence the boundary of such a silo does not touch the shaded left region, in which all points of the left dr-system lie. Thus p_g^r cannot be the right endpoint of a supporting d-edge and we get a contradiction to our assumption. Therefore we can eliminate the set $\{p_g^r | g > j\}$ from the candidate list of right supporting points. The same argument implies, that the set $\{p'_f | f < i\}$ does not contain a left supporting point.

In all cases, in which we do not find a supporting d-edge, a portion of one or both dr-systems can be eliminated. By testing always a pair of points lying in the "middle" of the remaining chains, we can find a supporting d-edge in $O(\log s + \log t)$ time. ∎

We construct a family of minimal dr-systems by using a slightly modified version of the (lower) convex layer construction algorithm of Overmars and van Leeuwen (see [7]). We only have to exchange the elementary test for the three properties supporting, concave, reflex and the elimination rules for the layer construction. The elimination rules for the dS-layer construction are given in the proof of Lemma 5. We use the same augmented balanced binary search tree T to construct the dS-layers. This dynamic data structure has size $O(m)$. Since all elementary operations can be carried out in constant time, the dS-layer construction costs as much as the convex layer construction. Hence the whole dS-layer construction for the cell CE with $|P_{CE}| = m$ can be done in $O(m(\log m)^2)$ time. See [7] for details.

Theorem 4 *Let P be a set of n points in the Euclidian plane E^2. There exists a data structure of size $O(n)$, such that $O(\log n + k)$ time suffices to retrieve all k points of P lying inside a query translate C_q of a convex computable figure C. The data structure can be constructed in $O(n(\log n)^2)$ time.*

Proof: Decomposing the Euclidean plane, distributing the points of P in their corresponding cells and storing the non-empty cells in sorted order in a binary search tree takes $O(n \log n)$ time. Then we have to construct the four query data structures for every non-empty cell CE. If $|P_{CE}| = m$, the construction of a family of d*-layers ($* \in \{N, S, O, W\}$) costs $O(m(\log m)^2)$ time. Building the four query data structures for cell CE with the help of these dual layer families can be done in $O(m)$ time. Hence the preprocessing for all non-empty cells can be carried out in $O(n(\log n)^2)$ time.

Since the new data structures, which we use to construct the dual layers, have size $O(n)$, all data structures together use $O(n)$ space. The query time is the same as in [5], because we use the same query data structure. ∎

Note that we also get dynamic data structures, because we can search for all points of $CE \cap C_q$, where C_q is S-grounded with respect to CE, in the dynamic data structure T, that was used to construct the dS-layers. Insertions and deletions in the dynamic data structure T can be done in $O((\log |P_T|)^2)$ time, where P_T is the point set stored in T (see [7]). We describe now a simple way to find all points stored in T, which lie in the query translate C_q: Search the edge of the S-layer $L(P_T)$, which lies below or above the query point q. This search can be carried out in the concatenable queue associated to the root of T with $O(\log |P_T|)$ operations. If the point p, which belongs to the above edge, does not lie in C_q, we are ready in this cell. If the point lies in C_q, we store p in a queue called REMEMBER, delete p from T and search again. We continue to do this until we have found all points $p \in C_q$ stored in the original tree T. Afterwards we restore T by inserting all points $p \in$ REMEMBER in T again. This retrieval operation can be done in $O(\log |P_T| + k(\log |P_T|)^2)$ time, where $k = |P_T \cap C_q|$. This proves Theorem 2.

3.3 An improved dS-layers construction algorithm

Let $P_{CE} = P \cap CE = \{p_1, \cdots, p_m\}$ be the sequence of points in cell CE, sorted in order of increasing x-coordinates. We consider again only S-grounded queries.

During the construction of the dS-layers all points stored in the tree T mentioned above are deleted one after another. All deletions together cost $O(m(\log m)^2)$ time. Chazelle showed in [3] that the deletions involved in the computation of the convex layers can be batched together, such that all deletions can be done in $O(m \log m)$ time. We prove now, that we can use a slightly modified version of Chazelle's convex layer construction algorithm to reduce the preprocessing time of the data structure to $O(m \log m)$.

We store $p_1, \cdots, p_m$ in the leaves of the balanced binary search tree T. The set of points stored at the leaves of a subtree $T(u)$ with root u, is denoted by $P(u)$. Let $DL(P(u))$ be the dS-layer of $P(u)$ and $DR(P(u))$ a minimal dr-system of $DL(P(u))$. Connecting P_{CE} by the set $\mathcal{E}_d = \bigcup_{u \in T} \{e(a,b) | a, b \text{ neighboring points in } DR(P(u))\}$ of d-edges, we get a planar embedding of the graph $G = (V, \mathcal{E})$ with nodes $V = P_{CE}$ and edges $\mathcal{E} = \{\{a,b\} | e(a,b) \in \mathcal{E}_d\}$. The connected acyclic planar graph G is called the dS-graph of P_{CE}. We use the notation G also for the two-dimensional embedding of the graph. The d-edges of G are in one-to-one correspondence with the nodes of the tree T (see Figure 8). Each node $u \in T$ corresponds to the "longest" supporting d-edge of G, which connects the dS-layers of node u's children. Assume v and w are the children of node $u \in T$, then u corresponds to a "longest" supporting d-edge of $DL(P(v))$ and $DL(P(w))$.

The dS-graph G is represented by an adjacency list structure. We endow each vertex $p \in P_{CE}$ with a list $V(p)$, which contains the names of the adjacent vertices. Each list $V(p)$ consists of two sublists $VL(p)$ and $VR(p)$, defined as follows: $VL(p)$ (resp. $VR(p)$) contains the vertices adjacent to p, which have smaller (resp. larger) x-coordinates than p. The points in $VL(p)$ and $VR(p)$ are sorted with respect to the corresponding d-edges from bottom d-edge to top d-edge. Each vertex p has a pointer to the bottom d-edge of $VL(p)$ and a pointer to the bottom d-edge of $VR(p)$ (see Figure 9). The sorting of the d-edges in $VL(p)$ and $VR(p)$ can be carried out by considering the intersection points of the d-edges with vertical lines near vertex p. If we have d-edges $e(a,p)$ and $e(b,p)$, which deliver the same intersection point with the selected vertical line, we sort these d-edges by considering the x-coordinates of the corresponding points a and b.

By using the simple algorithm given in [3], we can compute the dS-graph G in $O(m \log m)$ time. Starting in the leftmost vertex of the dS-graph G and following the extra pointers to the bottom d-edges, we can find a minimal dr-system $DR(P_{CE})$ for $DL(P_{CE})$. We store this dr-system and then we remove all points $p \in DR(P_{CE})$ from the dS-graph G.

Before we describe in detail how points will be deleted from G, we briefly introduce the geometrical concept, on which Chazelle's deletion method is based. In order to remove the vertex p from G and in order to reshape the dS-graph G, we move the vertex $p = (p_x, p_y)$ on the vertical ray $l_p := \{(p_x, y) | y \geq p_y\}$ towards $y = \infty$. By moving the point towards $y = \infty$, the d-edges adjacent to p will be pulled up and will be removed one by one. The d-edges adjacent to p have to be considered in the order in which they appear as supporting d-edges in the path from leaf p to the root of T (see Figure 10).

The deletion of a point p lying on the current dS-layer will now be described in detail. Let $v_1, \cdots, v_l$ be the nodes of T, which lie on the path from leaf p to the root of T. Every node v_i corresponds to a "longest" supporting d-edge of two dS-layers. Since p lies in the current dS-layer of all points present in G, p lies on one of these two dS-layers. Let $w_1, \cdots, w_k$ be the subsequence of $v_1, \cdots, v_l$, that corresponds to supporting d-edges with p as an endpoint (see Figure 10). We distinguish between *p-left supporting* d-edges $e(w_i, p)$, where $w_i \in VL(p)$, and *p-right supporting* d-edges $e(p, w_j)$, where $w_j \in VR(p)$. Let $e_{z_1}, \cdots, e_{z_k}$ be the sequence of supporting d-edges corresponding to the sequence $w_1, \cdots, w_k$ of nodes. Since the d-edges with endpoint p have to be removed in this "leaf-to-root" order, we have to merge the lists $VL(p)$ and $VR(p)$. This merging can be done in $O(l)$ steps.

For any node w_i we let $G(w_i)$ denote the dS-subgraph of G, that belongs to the subtree of T rooted at w_i. In order to remove p from G, we have to update the sequence of dS-graphs $G(w_1), \cdots, G(w_k)$ in this order. We assume that we have already removed p from the dS-graphs $G(w_1), \cdots, G(w_{i-1})$ and we want to update $G(w_i)$. Furthermore we assume wlog, that the supporting d-edge $e(p, c)$ corresponding to w_i is a p-right supporting d-edge. Let $e(a, p)$ and $e(p, b)$ be the last p-left supporting d-edge and the last p-right supporting d-edge, which we have pulled up. We assume wlog, that $e(a, p)$ and $e(p, b)$ exist. Then the new part of $G(w_{i-1})$ lies between a and b above the composed curve $(e(a, p), e(p, b))$ (see Figure 11).

Let $e(a, a'), \cdots, e(b', b)$ be the sequence of d-edges between a and b, which lie on the dS-layer of all points represented in $G(w_{i-1})$. Hence a' is the vertex of the current dS-layer $DL(P(w_{i-1}))$ following a in counterclockwise order and b' is the vertex following b in clockwise order. Let c' denote the vertex of the right dS-subgraph following c in clockwise order (see Figure 11). Updating $G(w_i)$ means pulling up the vertex p until it disappears from the dS-layer $DL(P(w_i))$. During this process we stop at every point on the vertical line l_p, where a supporting d-edges has to be exchanged by another "longest" supporting d-edge, and change the dS-graph.

Let IS_a, IS_b, IS_c be the intersection points of the vertical line from p towards $y = \infty$ with the three boundaries $bd(RS(SI(a, a')))$, $bd(RS(si(b', b)))$ and $bd(RS(si(c', c)))$. (We consider $SI(\,,\,)$, if both points lie to the left of l_p, and $si(\,,\,)$, if both points lie to the right of l_p. If one point lies to the left and the other to the right of l_p, we can choose $si(\,,\,)$ or $SI(\,,\,)$. If the intersection is a line segment, we choose the point with maximal y-coordinate.) The first placement for p, where a supporting d-edge has to be exchanged, is that point of these three points, which has minimal y-coordinate. Therefore we compute the three points and sort them in order of increasing y-coordinate. If for example IS_c is the point with smallest y-coordinate, the point p reaches first IS_c on his way towards $y = \infty$. At this point we have to replace the d-edge $e(p, c)$ by the d-edge $e(p, c')$ in the dS-graph. Hence we replace c by c', compute the intersection point $IS_{c'}$ of the vertical line through p with the boundary $bd(RS(si(c', c'')))$ and insert the new point in the sorted sequence of intersection points (see Figure 12). The first element in this sorted sequence always determines the next placement for p. At every placement of p with corresponding endpoints a, b, c we have to test, if $e(a, c)$ or $e(b, c)$ is a supporting d-edge for the two dS-layers. If we have found a "longest" supporting d-edge, we change the dS-graph $G(w_i)$ and start to remove p from $G(w_{i+1})$. In this way we handle all supporting d-edges with endpoint p. If $v_i \notin \{w_1, \cdots, w_k\}$, then p is not an endpoint of the supporting d-edge associated to v_i. Therefore this supporting d-edge does not change, and no additional work is required.

In this way we remove all points $p \in DR(P_{CE})$ from G. Then we determine a minimal dr-system $DR(P_{CE}^2)$ of the dS-layer $DL(P_{CE}^2)$, where $P_{CE}^2 := P_{CE} \setminus DR(P_{CE})$. This can be done by running across the path of the current dS-graph G, which starts in the leftmost or rightmost vertex, and following the extra pointers to the bottom d-edges. After we have determined and stored $DR(P_{CE}^2)$, we remove all points in $DR(P_{CE}^2)$ from the dS-graph G. Then we determine a minimal dr-system $DR(P_{CE}^3)$ of the dS-layer $DL(P_{CE}^3)$, where $P_{CE}^3 := P_{CE}^2 \setminus DR(P_{CE}^2)$. We continue in this way, until we get $P_{CE}^{z+1} = \emptyset$.

It can easily be seen that the above deletion procedure and the whole dS-layer construction algorithm work correctly. We leave out here the complexity analysis, because it is almost identical to that given in [3]. Since the computation of a family of minimal dr-systems for the set P_{CE} with $|P_{CE}| = m$ takes only $O(m \log m)$ operations and $O(m)$ space, the whole preprocessing costs $O(n \log n)$ operations and $O(n)$ space. Hence, we get Theorem 1, which is the main result of this paper.

Remarks: (1) Examples of computable shapes, which C may assume, are circles, triangles, rectangles, ellipses or hybrid convex figures bounded by a constant number of analytic curves.
(2) The algorithm of Chazelle and Edelsbrunner also works for non-bounded convex computable figures. In such cases we only have to modify the decomposition of the Euclidean plane.
(3) If C is a convex m-gon, the primitive operations like intersection computing and point-inside-or-outside-test can be done in $O(\log m)$ time.

Corollary 1 *Let P be a set of n points in the Euclidean plane and let C be a convex m-gon. In $O((n \log n + m) \log m)$ time we can preprocess P so that for any query translate C_q, all k points of P, which lie in C_q, can be retrieved in $O(\log n + (k + 1) \log m)$ time. The query data structure has size $O(n + m)$.*

A similar result was given by Klein et al.[6]. Their query algorithm, however, takes $O(m \log n + k)$ time.
(4) By using persistent data structures, it might be possible to extend the retrieval possibilities of the above algorithm, such that queries for homothets $\lambda C + q$, where $\lambda_{min} \leq \lambda \leq \lambda_{max}$, can be carried out.

References

[1] A. Aggarwal, M. Hansen and T. Leighton. *Solving Query-Retrieval Problems by Compacting Voronoi Diagrams.* PROC. OF THE 16TH STOC (1990), pp. 331-340.

[2] B. Chazelle. *Filtering Search: A New Approach to Query-Answering.* SIAM J. OF COMP. vol. 15 (1986), pp. 703-724.

[3] B. Chazelle. *On the Convex Layers of a Planar Set.* IEEE TRANSACTIONS ON INFORMATION THEORY, vol. 31, no. 4 (1985), pp. 509-517.

[4] B. Chazelle, R. Cole, F.P. Preparata and C. Yap. *New Upper Bounds for Neighbor Searching.* INFORMATION AND CONTROL, vol. 68 (1986), pp. 105-124.

[5] B. Chazelle and H. Edelsbrunner. *Optimal Solutions for a Class of Point Retrieval Problems.* J. SYMBOLIC COMPUTATION, vol. 1 (1985), pp. 47-56.

[6] R. Klein, O. Nurmi, T. Ottmann and D. Wood. *A Dynamic Fixed Windowing Problem.* ALGORITHMICA, vol. 4 (1989), pp. 535-550.

[7] M.H. Overmars and J. van Leeuwen. *Maintenance of Configurations in the Plane.* J. COMPUT. SYST. SCI. vol. 23 (1981), pp. 166-204.

[8] F.P. Preparata and M.I. Shamos. *Computational Geometry: an Introduction.* Springer-Verlag New York-Berlin-Heidelberg-Tokyo (1985).

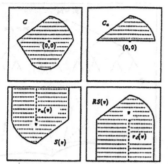

Figure 1: Example of a convex computable figure C, the upper part C_u of C, the silo $S(v)$ and rotated silo $RS(v)$ with respect to C_u.

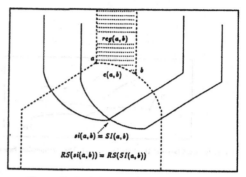

Figure 2: SI-point, d-edge and region of two points a and b.

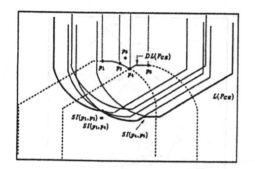

Figure 3: S-layer and dS-layer of a point set $P_{GE} = \{p_1, \cdots, p_t\}$. In the above situation $\{p_1, p_4, p_5\}$ is the minimal dr-systems for $DL(P_{GE})$.

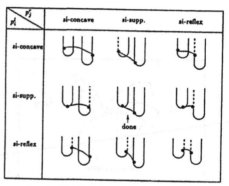

Figure 4: The nine possible cases.

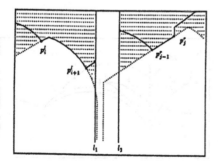

Figure 5: A possible situation in Case 1.

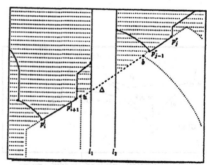

Figure 6: Example for Case 3c.

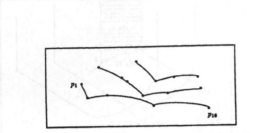

Figure 7: Dual S-layers of 16 points $\{p_1, \cdots, p_{16}\}$. The corresponding figure C is a disk.

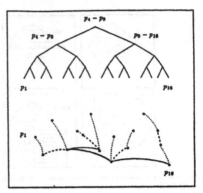

Figure 8: Tree T and the dS-graph for the point set of Figure 7.

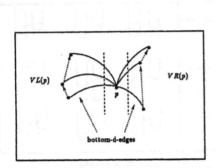

Figure 9: Adjacency list structure for vertex p and the bottom d-edges.

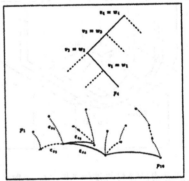

Figure 10: Leaf-to-root path for point p_4 of Figure 7 and the corresponding d-edges in the dS-graph.

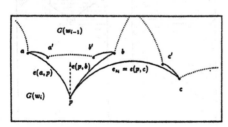

Figure 11: Schematical illustration of the deletion process.

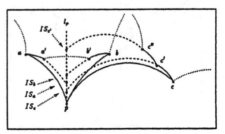

Figure 12: Determining the next placement for point p.

Rectangular point location and the dynamic closest pair problem

Michiel Smid[*]

1 Introduction

The dynamic closest pair problem has received much attention recently. In this problem, we want to maintain a closest pair—or its distance—in a set of n points in k-dimensional space, if points are inserted and/or deleted. Distance are measured in an L_t-metric, where $1 \leq t \leq \infty$. In this metric, the distance $d_t(p, q)$ between two points $p = (p_1, \ldots, p_k)$ and $q = (q_1, \ldots, q_k)$ is defined by $d_t(p, q) := (\sum_{i=1}^{k} |p_i - q_i|^t)^{1/t}$, if $1 \leq t < \infty$, and $d_t(p, q) := \min_{1 \leq i \leq k} |p_i - q_i|$ if $t = \infty$. Throughout this paper, we fix t and measure all distances in the L_t-metric. We write $d(p, q)$ for $d_t(p, q)$.

Supowit [11] considers this problem for the case where there are only deletions. He obtains a data structure of size $O(n(\log n)^{k-1})$ having an amortized deletion time of $O((\log n)^k)$.

If both insertions and deletions have to be supported, the best linear size data structure is obtained by combining results of Smid [9], Dickerson and Drysdale [1] and Salowe [6]. The resulting structure has a worst-case update time of $O(\sqrt{n}\log n)$. In Smid [10], an amortized update time of $O((\log n)^k \log\log n)$ is obtained using a structure of size $O(n(\log n)^k)$.

In this paper, we first consider the case where there are only insertions. We only consider the planar case. In order to solve this problem, we need a data structure for the rectangular point location problem. In this problem, we have to store a set of n non-overlapping axes-parallel rectangles—called *boxes*—such that for a given query point p, we can find the box that contains p. Edelsbrunner, Haring and Hilbert [3] considered the static version of this problem and introduced the *skewer tree* for solving it. In Section 2, we show how this skewer tree can be adapted such that boxes can be inserted and split. The balance condition for this data structure is non-standard, although it resembles that of BB[α]-trees. The method of rebalancing the skewer tree is a new variation of the partial rebuilding technique. We also equip the skewer tree with dynamic fractional cascading (Mehlhorn and Näher [4]) to speed up the query algorithm. The result is as follows:

Theorem 1 *For the problem of point location in a set of n non-overlapping planar axes-parallel rectangles, there exists a data structure that has a query time of $O(\log n)$, in which*

[*]Max-Planck-Institut für Informatik, D-6600 Saarbrücken, Germany. This work was supported by the ESPRIT II Basic Research Actions Program, under contract No. 3075 (project ALCOM).

insert and split operations take $O((\log n)^2)$ amortized time, that can be built in $O(n \log n)$ time, and that has size $O(n)$.

In this theorem, the main difficulty is in proving that the amortized time for insert and split operations is $O((\log n)^2)$, meanwhile guaranteeing a query time of $O(\log n)$.

As mentioned, we need the skewer tree to obtain an efficient data structure for the closest pair problem with insertions. In Section 3, we give a data structure that maintains a collection of boxes having sides of length at least the current minimal distance. Each box contains $O(\log n)$ points. If a point is inserted, we only have to compare the new point with the points that are contained in a constant number of surrounding boxes. If a box contains too many points, it is split into a constant number of boxes, each of which has sides of length at least the current minimal distance. The result is as follows:

Theorem 2 *There exists a data structure of size $O(n)$, that maintains the closest pair in a set of n points in the plane in $O(\log n)$ amortized time per insertion. As a result, the closest pair in a set of n planar points can be be computed on-line in $O(n \log n)$ time. This is optimal.*

Theorems 1 and 2 can be generalized to higher dimensions. The k-dimensional version of Theorem 2 gives a data structure of size $O(n)$ that maintains the closest pair in $O((\log n)^{k-1})$ amortized time per insertion. In Section 4, we obtain a better result. In fact, this better result is obtained even for *semi-online* updates, as introduced by Dobkin and Suri [2]. A sequence of updates is called semi-online, if the insertions are on-line, but with each inserted point p, we get an integer d which says that p will be deleted d updates from the moment of insertion. Dobkin and Suri show that in the planar case, the minimal distance of a point set can be maintained in $O((\log n)^2)$ amortized time when semi-online updates are performed. This method was made worst-case in Smid [8]. Since the data structure of [2, 8] is based on Voronoi diagrams, it does not generalize efficiently to higher dimensions.

In Section 4, we adapt the method of [2, 8]. As a result, we do not need Voronoi diagrams. The result is the following:

Theorem 3 *There exists a data structure that maintains the minimal distance of a set of n points in k-dimensional space in $O((\log n)^2)$ worst-case time per semi-online update. The size of the data structure is $O(n)$. If there are only insertions, the update time can be reduced to $O((\log n)^2 / \log\log n)$. As a result, the closest pair in a set of n points in k-dimensional space can be be computed on-line in $O(n(\log n)^2 / \log\log n)$ time.*

2   The rectangular point location problem

We are given a set V of n non-overlapping boxes of the form $[a_1 : b_1] \times [a_2 : b_2]$. Boxes may be infinite, i.e., $a_1, a_2 \in \mathcal{R} \cup \{-\infty\}$ and $b_1, b_2 \in \mathcal{R} \cup \{+\infty\}$. The boxes do not necessarily partition the entire plane. The following four operations have to be supported.

Point location: Given a query point p, find the boxes—if any—that contain p. If p lies on the boundary of a box, there may be several boxes that contain p. Since the boxes do not overlap, a query point is contained in at most four boxes.

Insertion: Insert a box into V. The boxes in the new set must still be non-overlapping.

Vertical split: The operation 1-*split*(*s*) replaces box $[a_1 : b_1] \times [a_2 : b_2]$ by the boxes $[a_1 : s] \times [a_2 : b_2]$ and $[s : b_1] \times [a_2 : b_2]$.

Horizontal split: The operation 2-*split*(*t*) replaces box $[a_1 : b_1] \times [a_2 : b_2]$ by the boxes $[a_1 : b_1] \times [a_2 : t]$ and $[a_1 : b_1] \times [t : b_2]$.

The skewer tree [3]: The skewer tree is recursively defined as follows. If V is empty, the skewer tree is also empty.

Assume that V is non-empty. Let l be a vertical line that intersects at least one box of V. The skewer tree for the set V consists of a binary tree—called the *skeleton tree*—in which each node contains additional information:

1. In the root r, we store the size of V, the line l and a balanced binary search tree T_r that is defined as follows. Let V_r be the boxes in V that are intersected by l in their interiors or that touch l with their right boundaries. Let W_r be the y-coordinates of the top and bottom sides of the boxes in V_r. (Each y-coordinate is represented exactly once.) For convenience, we add $-\infty$ and $+\infty$ to W_r. Then T_r stores the values of W_r in increasing order. With each value s, we store the box $below(s)$ resp. $above(s)$, which is the box in V_r that has its top resp. bottom side at height s. If $below(s)$ or $above(s)$ does not exist, the value of this variable is *nil*.

2. The root r has two subtrees. The left subtree is a skewer tree for all boxes in V that lie completely to the left of l. Similarly, the right subtree is a skewer tree storing all boxes in V that lie completely to the right of l or that touch l with their left boundaries.

Balance condition: Let α be a real number such that $1/2 \leq \alpha < 1$. For each node v of the skeleton tree, let n_v be the total number of boxes that are stored in the subtree of v (including node v itself), and let $d(v)$ be the depth of v in the skeleton tree. (The root has depth 0.) Then we require that $n_v \leq \alpha^{d(v)} n$, where n is the current number of boxes that are stored in the entire data structure. Such a skewer tree is called α-*balanced*. If $\alpha = 1/2$, the skewer tree is called *perfectly balanced*. The *height* of a skewer tree is defined as the height of its skeleton tree.

Point location in an α-balanced skewer tree: Given a query point $p = (p_1, p_2)$, we start in the root r of the skewer tree. We perform a binary search with p_2 in the search tree T_r that is stored with r. This gives two values s and t, such that $s \leq p_2 < t$. If $p_2 > s$ and $p \in above(s)$, then we report this box $above(s)$. Otherwise, if $p_2 = s$, we report those boxes of $below(s)$ and $above(s)$ that contain p.

If p is contained in the interior of a reported box, then the search procedure is finished. Otherwise, we proceed recursively: If p lies to the left of the line l that is stored in the root, we proceed our search in the left subtree. Otherwise, p lies on or to the right of l, in which case we proceed in the right subtree.

This algorithm needs $O((\log n)^2)$ time. We improve this query time to $O(\log n)$, using dynamic fractional cascading. (See Mehlhorn and Näher [4].) We use the terminology of [4].

Dynamic fractional cascading: The *catalogue graph* is the skeleton tree and the range $R(e)$ of each edge e in this catalogue graph is the set of real numbers. The *catalogue* $C(v)$ of a node v in the catalogue graph is the list of y-coordinates that are stored in the binary search tree that is stored with v, extended with $-\infty$ and $+\infty$. Instead of this binary search tree, we store in each node v an *augmented catalogue* $A(v)$ as described in [4]. Note that $C(v) \subseteq A(v)$. These augmented catalogues are implemented as balanced binary search trees. Elements in $C(v)$ are called *proper*, those in $A(v) \setminus C(v)$ are called *non-proper*. Each proper element s contains the values $below(s)$ and $above(s)$. There are *bridges* between the augmented catalogues of adjacent nodes of the catalogue graph, as described in [4]. Finally, we store in each node v a data structure that solves the SPLIT-FIND problem. This structure is used for finding proper elements that are next to non-proper elements. See [4]. The following lemma is proved in the full paper.

Lemma 1 *An α-balanced skewer tree, equipped with fractional cascading, has size $O(n)$ and can be built in $O(n \log n)$ time. In this data structure, a point location query can be solved in $O(\log n)$ time.*

Insertion: To insert a box R into a skewer tree, we do the following: We start in the root of the skeleton tree, and follow a path until we reach the first node v such that the vertical line that is stored in this node intersects R in its interior or touches the right boundary of R. In each node that is visited during this walk, we increase the number of boxes that are stored in its subtree by one. Then we do a binary search in the augmented catalogue $A(v)$ to locate the position where R has to be inserted. Then, we insert the y-coordinates of the top and bottom sides of R—together with the appropriate values for *below* and *above*—into $A(v)$, as described in [4]. If these y-coordinates are present already, we only update the appropriate *below*- and *above*-values.

If we do not find node v, we end our search in a node w of the skeleton tree, one of whose sons is missing—namely the one to which the search wants to proceed. In this case, we give w a new son, i.e., we insert a node u with an empty catalogue $C(u)$ and an edge (w, u) into the catalogue graph, as described in [4]. This will give node u a non-empty augmented catalogue $A(u)$. Then, we store in u a vertical line that intersects the interior of R, and the number of boxes that are stored in u—which is equal to one. Finally, we insert into $A(u)$, the y-coordinates of the top and bottom sides of the box R and the values $-\infty$ and $+\infty$, as described in [4], together with the appropriate *above*- and *below*-values.

Vertical split: To perform the operation $1\text{-}split(s)$ on the box $R = [a_1 : b_1] \times [a_2 : b_2]$, we search for the node v in the skeleton tree, whose augmented catalogue "contains" R. Let l be the vertical line that is stored in v. Assume w.l.o.g. that l intersects the left part of the box R in its interior or touches its right boundary. We search in the augmented catalogue $A(v)$ for the values a_2 and b_2, and replace the boxes $above(a_2)$ and $below(b_2)$ by the left part of R. Then, we use the above insertion algorithm to insert the right part of R into the skewer tree.

Horizontal split: To perform the operation $2\text{-}split(t)$ on the box $R = [a_1 : b_1] \times [a_2 : b_2]$, we search for the node v in the skeleton tree, whose augmented catalogue "contains" R. In each node that is visited during this walk, we increase the number of boxes that are stored in its subtree by one. Then, we do a binary search in the augmented catalogue $A(v)$ to locate the positions of a_2 and b_2. We replace the box $above(a_2)$ (resp. $below(b_2)$) by the lower (resp. upper) part of the box R. Finally, we insert the y-coordinate t into $A(v)$, as described in [4], together with the appropriate values $below(t)$ and $above(t)$.

Rebalancing the skewer tree: After an insert or split operation, the skewer tree might not satisfy the balance condition anymore. To keep the skewer tree balanced, we use a variation of the partial rebuilding technique (see e.g.[5]):

During the update operation, we have inserted y-coordinates in the augmented catalogue of exactly one node. Starting in that node, we walk back to the root and find the highest node w that does not satisfy the balance condition of the α-balanced skewer tree anymore. Then we rebuild the complete subtree rooted at the *father* of w as a perfectly balanced skewer tree. More precisely, we do the following:

If the father of w is the root, then we rebuild the complete data structure as a perfectly balanced skewer tree. Otherwise, let v be the father of w and let u be the father of v. We delete the edge between u and v from the catalogue graph. Then, we rebuild the SPLIT-FIND structure corresponding to $A(u)$.

Next, we construct the skeleton tree of a perfectly balanced skewer tree for the boxes that are stored in the subtree of v. At each node of this skeleton tree, we store a sorted list of the y-coordinates that have to be stored there.

We add this skeleton tree S_v to the old skewer tree, as follows: We insert the root of S_v and an edge between u and this root into the catalogue graph. Then we insert the nodes and edges of S_v into the catalogue graph, as described in [4]. (At this moment, all catalogues of the nodes in S_v are empty. The augmented catalogues, however, are non-empty.) For each node of S_v, we insert the y-coordinates of the top and bottom sides of the boxes that are stored there, and the values $-\infty$ and $+\infty$, into its augmented catalogue, one after another, in increasing order. We also insert the appropriate *above*- and *below*-values.

In the rest of this section, we give the main lemmas that are used to prove the update time in Theorem 1. We mention here that the above update algorithms correctly maintain the α-balanced skewer tree. In particular, after a rebalancing operation, the resulting skewer tree is again α-balanced. The proofs of these lemmas are technical and can be found in the full paper.

Lemma 2 *Let v be a node in the skeleton tree of an α-balanced skewer tree and let $d(v)$ be the depth of this node. Then the entire subtree of v, together with the augmented catalogues that are stored in the nodes of this subtree, has size $O(\alpha^{d(v)}n)$. Here, n is the number of boxes that are stored in the entire skewer tree.*

Next, we bound the time for a rebalancing operation. This time bound is *not* a function of the number of boxes that are stored in the rebuilt subtree. This number can be much smaller.

Lemma 3 *Suppose that during an insert or split operation, we rebuild a subtree with root v. This rebuilding takes $O(\alpha^{d(v)}n\log n)$ time.*

In a similar way as for BB[α]-trees, it can be shown that expensive rebuilding operations seldom occur. That is, in the amortized sense, the value of $d(v)$ in Lemma 3 is small. Using this, the amortized update time in Theorem 1 can be proved.

In the full paper, it is shown how Theorem 1 can be generalized to higher dimensions. It is also shown there how boxes can be deleted and how boxes can be merged (this is the inverse of a split). Of course, in these generalized structures, the time bounds are different.

3 Maintaining the closest pair in a point set

Lemma 4 *Let V be a set of points in k-dimensional space, and let δ be the distance of a closest pair in V. Then any k-dimensional cube C having sides of length δ contains at most $(k+1)^k$ points of V.*

Proof: Partition C into $(k+1)^k$ subcubes having sides of length $\delta/(k+1)$. Assume that C contains at least $(k+1)^k + 1$ points of V. Then one of the subcubes contains at least two points of V. These two points have a distance that is at most equal to the L_t-diameter of this subcube. This diameter, however, is at most $k \times \delta/(k+1) < \delta$. This contradicts the fact that the minimal distance of V is δ. ∎

The closest pair structure: Let V be a set of n points in the plane. We assume that n is sufficiently large. We maintain the following information:

1. A pair (P, Q) that maintains the closest pair and a variable $\delta = d(P, Q)$.

2. At each moment, the plane is partitioned into non-overlapping boxes. Each box in this partition has sides of length at least δ. Each box of the partition contains at least one and at most $16 \log n$ points of V.

3. The boxes of the partition are stored in an α-balanced skewer tree. With each box, we store a list of those points in V that are contained in this box. (These points are stored in an arbitrary order. If a point is on the boundaries of several boxes, then it is stored in only one of these boxes.)

First, we show how this data structure can be built. In [12], it is shown how the closest pair can be computed in $O(n \log n)$ time, using $O(n)$ space. In Theorem 1, we saw that a skewer tree can be built in $O(n \log n)$ time. So it remains to be shown how the partition of the plane into boxes can be computed. We give a recursive algorithm that computes this partition. We define a 1-box as an interval on the real line and a 2-box as a box in the plane.

The partitioning algorithm: Let V be a set of n points in k-space, where $k = 1, 2$. Let δ be the distance of a closest pair in V. This variable δ is a global variable, i.e., in recursive calls it does not get a new value.

If $|V| = 1$, then the partition consists of one k-box, namely the entire space. So assume that $|V| > 1$.

Order the points of V with respect to their last coordinates. Let p be the smallest point in this ordered set. Let a_1 be the last coordinate of point p. Let $i \geq 1$, and assume that $a_1, \ldots, a_i$ are computed already.

If there is a point in V with last coordinate in $(a_i : a_i + \delta]$, then we set $a_{i+1} := a_i + \delta$. Otherwise, we set a_{i+1} to the last coordinate of the first point in the ordered set V that lies to the right of the "line" $x_k = a_i$. If there are no points to the right of this line, then a_{i+1} is not defined, and the construction of the a_j's stops.

This gives a sequence of intervals $(-\infty : a_1], (a_1 : a_2], \ldots, (a_l : \infty)$ for some l. Let $a_0 := -\infty$ and $a_{l+1} := \infty$. Partition V into subsets $V_0, \ldots, V_l$, where V_i contains those points of V that have their last coordinates in the interval $(a_i : a_{i+1}]$.

If $k = 1$, this is the desired partition of 1-space into 1-boxes, together with the corresponding partition of V.

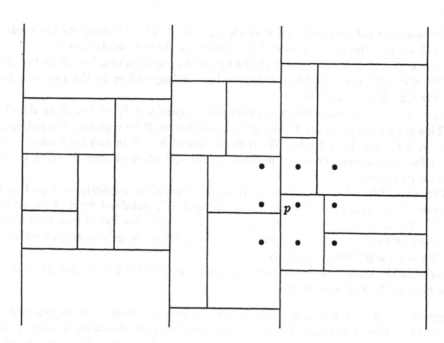

Figure1: The 9 point location queries.

Assume that $k = 2$. For $i = 0, 1, \ldots, l$, do the following. Use the same algorithm recursively, with $k - 1 = 1$, for the first coordinates of the points in V_i. (Note that in this recursive call, the value of δ remains equal to the minimal distance in the 2-dimensional set V.) This gives a partition of the real numbers into 1-boxes of the form $(b_1 : c_1]$, together with a corresponding partition of V_i. Replace each such 1-box by the 2-box $(b_1 : c_1] \times (a_i : a_{i+1}]$. The resulting boxes—for all i together—form the desired partition of the plane, together with the partition of V.

Lemma 5 *For $k = 2$, the 2-boxes, that are computed by the above algorithm, are non-overlapping and form a partition of the plane. Each box has sides of length at least δ. Each box contains at least one and at most $16 \log n$ points of V.*

Lemma 6 *The closest pair structure has size $O(n)$ and can be built in $O(n \log n)$ time.*

The insert algorithm: Let $p = (p_1, p_2)$ be the point to be inserted. Then we perform 9 point location queries in the skewer tree, with query points $(p_1 + \epsilon_1, p_2 + \epsilon_2)$, for $\epsilon_1, \epsilon_2 \in \{-\delta, 0, \delta\}$. Each query gives at most 4 answers. So all queries together give at most 36 different boxes. For each of these boxes, we walk through its list of points. For each point q in these lists, if $d(p, q) < \delta$, we set $(P, Q) := (p, q)$ and $\delta := d(p, q)$. (See Figure 1.)

Next, we insert p into the list of a box that contains p. If afterwards this list contains more than $16 \log n$ points, we perform a *split operation* on its box:

Split operation: Suppose we want to split a box $B = [a_1 : b_1] \times [a_2 : b_2]$. Let V' be the set of points that are stored in the list of B. For $i = 1, 2$, we compute the values m_i and M_i, which are the minimal (resp. maximal) i-th coordinate of any point of V'. If $M_i - m_i \leq 2\delta$, for all $i = 1, 2$, the algorithm stops.

Otherwise, we take an index i for which $M_i - m_i > 2\delta$. We compute the median c_i of the i-th coordinates of the points of V'. There are three possible cases:

If $a_i + \delta \leq c_i \leq b_i - \delta$, we perform the operation $i\text{-}split(c_i)$ on box B in the skewer tree. We also split the list of box B in two lists corresponding to the two new boxes. Then, the algorithm is finished.

If $a_i \leq c_i < a_i + \delta$, we perform the operation $i\text{-}split(a_i + \delta)$ on box B in the skewer tree. This gives two new boxes B' and B'', obtained from B by replacing the i-th interval by $[a_i : a_i + \delta]$ resp. $[a_i + \delta : b_i]$. We split the list of box B in two lists corresponding to these two new boxes. Then, we perform a split operation on box B' using the same algorithm recursively.

Otherwise, if $b_i - \delta < c_i \leq b_i$, we perform the operation $i\text{-}split(b_i - \delta)$ on box B in the skewer tree. This gives two new boxes B' and B'', obtained from B by replacing the i-th interval by $[a_i : b_i - \delta]$ resp. $[b_i - \delta : b_i]$. We split the list of box B in two lists corresponding to these two new boxes. Then, we perform a split operation on box B'' using the same algorithm recursively.

Remark: In the full paper, it is shown that there is an index i such that $M_i - m_i > 2\delta$ at the start of the split operation.

Lemma 7 *Let B be a box whose list contains m points, where m is sufficiently large. Let δ be the minimal distance of V at the moment the split algorithm is carried out on B. After this algorithm, the sides of all boxes that have been created have length at least δ, and each such box contains at least one and at most $\lceil m/2 \rceil$ points of V. Moreover, the split algorithm takes $O(m + (\log n)^2)$ amortized time.*

Proof: The proof uses Theorem 1 and can be found in the full paper. ∎

Lemma 8 *The insert algorithm correctly maintains the closest pair structure in $O(\log n)$ amortized time per insertion.*

Proof: Let δ be the minimal distance just before the insertion of point p. If this minimal distance changes, there must be a point inside the L_t-ball of radius δ centered at p. This ball is contained in the box $[p_1 - \delta : p_1 + \delta] \times [p_2 - \delta : p_2 + \delta]$. Therefore, it suffices to compare p with all points of the current set V that are in this box. Let $W := V \cap ([p_1 - \delta : p_1 + \delta] \times [p_2 - \delta : p_2 + \delta])$ be the set of these points, and let W' be the set of points that are contained in the lists corresponding to the at most 36 boxes that result from the 9 point location queries. The algorithm compares p with all points in W'. Since all boxes in the partition of the plane have sides of length at least δ, it can be shown that $W \subseteq W'$. It follows that the algorithm correctly maintains the closest pair.

Since δ can only decrease, the side lengths of a box that is not split remain at least equal to δ. Clearly, if such a box contains at least one point before the insertion, so it does afterwards. Also, the box still contains at most $16 \log n$ points, because the value of n only increases. If a box is split, then Lemma 7 guarantees that the new boxes have sides of length at least δ, that they contain at least one and at most $\lceil (1/2) 16 \log n \rceil \leq 16 \log n$ points. This proves the correctness of the insert algorithm.

By Theorem 2, it takes $O(\log n)$ time to perform the 9 point location queries in the skewer tree. For each of the at most 36 found boxes, we walk through its list of points and compare these with the new point. Since each such list contains $O(\log n)$ points, this step of the insert algorithm takes $O(\log n)$ time.

If a box is split, it contains $16 \log n$ points. By Lemma 7, this operation takes $O((\log n)^2)$ amortized time. It follows from Lemma 7 that each of the boxes that are created during a split operation contains at most $8 \log n$ points. Therefore, at least $8 \log n$ points must be inserted into such a box before it is split again. Charging the $O((\log n)^2)$ time for the split operation to these $8 \log n$ insertions, it follows that a split operation adds an amount of time to the overall amortized insertion time that is bounded by $O(\log n)$. ∎

4   The higher-dimensional case

As mentioned already in Section 1, all results we have obtained so far can be generalized to higher dimensions. In this section, we show that better results can be obtained—even for semi-online updates—by adapting the algorithm of [2, 8].

In [2, 8], the following is shown. Let $g : T \times T \to \mathcal{R}$ be a symmetric function and let V be a subset of T of size n. Suppose we have a static data structure that stores V, such that for a given $p \in T$, we can compute $\min\{g(p,q) : q \in V, p \neq q\}$ in $Q(n)$ time. Suppose this data structure has size $S(n)$ and can be built in $P(n)$ time.

Then there is a structure of size $O(S(n))$, that maintains a partition $V_0, V_1, \ldots, V_m$ of V, values $\min\{g(p,q) : q \in V_j, p \neq q\}$ for $0 \leq i \leq m$, $p \in V_i$, $i \leq j \leq m$, and the value

$$\min_{0 \leq i \leq m} \min_{p \in V_i} \min_{i \leq j \leq m} \min_{q \in V_j} g(p,q), \tag{1}$$

in $O((P(n)/n) \log n + Q(n) \log n + (\log n)^2)$ worst-case time per semi-online update.

In our case, let $g(p,q) = d(p,q)$. We need a data structure that computes $\min\{d(p,q) : q \in V, p \neq q\}$. This is the post-office problem, for which no efficient methods are available in higher dimensions. Since we are interested in the overall minimal distance, however, we do not have to know the value $\min\{d(p,q) : q \in V, p \neq q\}$ if this minimum is "large". That is, it suffices to consider a restricted post-office problem:

A restricted post-office problem: If A is a set of points in k-space, p a point in k-space and σ a real number, then we define $\delta(A) := \min\{d(p,q) : p, q \in A, p \neq q\}$, $d(p,A) := \min\{d(p,q) : q \in A, q \neq p\}$ and $f(p,A,\sigma) := \min\{d(p,A), \sigma\}$.

Let V be a set of n points in k-space and σ a positive real number such that $\sigma \leq \delta(V)$. We want to store V in a static data structure such that for a given query point p, the value $f(p,V,\sigma)$ can be computed.

This structure is constructed as follows. Partition k-space into hypercubes with sides of length σ: For integers $i_1, i_2, \ldots, i_k$, let $\mathbf{i} := (i_1, i_2, \ldots, i_k)$ and $S_{\mathbf{i}} := [i_1\sigma : (i_1+1)\sigma] \times \ldots \times [i_k\sigma : (i_k+1)\sigma]$.

Partition V into subsets $V_{\mathbf{i}} := V \cap S_{\mathbf{i}}$. Points that are on the boundary of a hypercube are put in only one (arbitrary) subset. Let L be the set of indices $\mathbf{i}$ for which $V_{\mathbf{i}}$ is nonempty. Then we store the elements of L in a balanced binary search tree, sorted in lexicographical order. With each index $\mathbf{i}$, we store a list containing—in arbitrary order—the points of subset $V_{\mathbf{i}}$.

To answer a query, we get a point $p = (p_1, \ldots, p_k)$, and we have to compute the value of $f(p,V,\sigma)$. For $j = 1, \ldots, k$, let $i_j := \lfloor p_j/\sigma \rfloor$. Then we search in the tree for the 3^k indices $(i_1 + \epsilon_1, i_2 + \epsilon_2, \ldots, i_k + \epsilon_k)$, where $\epsilon_1, \epsilon_2, \ldots, \epsilon_k \in \{-1, 0, 1\}$.

If none of these indices is stored in the tree, we set $f(p,V,\sigma) := \sigma$. Otherwise, for each index $\mathbf{i}'$ that we find, we walk through the list of points of $V_{\mathbf{i}'}$, and compare them with the query point p. Of all the points we encounter in this way, we select a point $q \neq p$ that is closest to p and set $f(p, V, \sigma) := \min\{d(p,q), \sigma\}$.

The result is given in the following lemma. The proof is similar to that of Lemma 8 and uses Lemma 4.

Lemma 9 *Let V be a set of n points in k-space and let σ be a positive real number such that $\sigma \leq \delta(V)$. There exists a data structure, such that for each point p in k-space, we can find the value of $f(p, V, \sigma)$ in $O(\log n)$ time. The data structure has size $O(n)$ and can be built in $O(n \log n)$ time.*

We apply the algorithm of [2, 8] with the function f and the static structure of Lemma 9. If this structure stores a subset V_i, then we take $\sigma = \delta(V_i) \leq \delta(V)$. Note that hence the function f depends on the subset V_i. It can be shown, however, that the algorithm still works. It maintains, among other things, the value (see (1))

$$\min_{0 \leq i \leq m} \; \min_{p \in V_i} \; \min_{i \leq j \leq m} \; f(p, V_j, \delta(V_j)).$$

But this is just the value we want, namely $\delta(V)$:

Lemma 10 *Let V be a set of points in k-space and let $V_0, V_1, \ldots, V_m$ be a partition of V. Then*

$$\delta(V) = \min_{0 \leq i \leq m} \; \min_{p \in V_i} \; \min_{i \leq j \leq m} \; f(p, V_j, \delta(V_j)).$$

Now we can sketch the proof of Theorem 3. The minimal distance of a set V of size n can be computed in $O(n \log n)$ time, see Vaidya [12]. Hence, the data structure of Lemma 9 with $\sigma = \delta(V)$ can be built in $P(n) = O(n \log n)$ time. It has size $S(n) = O(n)$. A query "given point p, compute $f(p, V, \delta(V))$" can be solved in $Q(n) = O(\log n)$ time. Then the method of [2, 8] implies the existence of a data structure of size $O(S(n)) = O(n)$ that maintains the minimal value of the function f in $O((P(n)/n) \log n + Q(n) \log n + (\log n)^2) = O((\log n)^2)$ worst-case time per semi-online update. By Lemma 10, the minimal value of f is equal to the minimal distance of the entire set V.

If only insertions take place, we can use a structure that is also based on Lemma 9. The method is based on a dynamization method for decomposable searching problems. Now, rebuilding of structures can be done in $O(n)$ instead of $O(n \log n)$ time. Therefore, we write the current number of points in the number system with base $\log n$, instead of the binary system. The details are given in the full paper. See also [5, pages 109-111].

Note: After this paper was written, Schwarz and Smid designed an algorithm for maintaining the minimal distance in a set of n points in k-space—under insertions— in $O(\log n \log \log n)$ amortized time. See [7].

References

[1] M.T. Dickerson and R.S. Drysdale. *Enumerating k distances for n points in the plane.* Proc. 7-th ACM Symp. on Comp. Geom., 1991, pp. 234-238.

[2] D. Dobkin and S. Suri. *Dynamically computing the maxima of decomposable functions, with applications.* Proc. 30th Annual FOCS, 1989, pp. 488-493.

[3] H. Edelsbrunner, G. Haring and D. Hilbert. *Rectangular point location in d dimensions with applications*. The Computer Journal 29 (1986), pp. 76-82.

[4] K. Mehlhorn and S. Näher. *Dynamic fractional cascading*. Algorithmica 5 (1990), pp. 215-241.

[5] M.H. Overmars. *The Design of Dynamic Data Structures*. Lecture Notes in Computer Science, Vol. 156, Springer-Verlag, Berlin, 1983.

[6] J.S. Salowe. *Shallow interdistance selection and interdistance enumeration*. To appear in Proceedings WADS, 1991.

[7] C. Schwarz and M. Smid. *An $O(n \log n \log \log n)$ algorithm for the on-line closest pair problem*. In preparation.

[8] M. Smid. *Algorithms for semi-online updates on decomposable problems*. Proc. 2nd Canadian Conf. on Computational Geometry, 1990, pp. 347-350.

[9] M. Smid. *Maintaining the minimal distance of a point set in less than linear time*. Algorithms Review 2 (1991), pp. 33-44.

[10] M. Smid. *Maintaining the minimal distance of a point set in polylogarithmic time (revised version)*. Report MPI-I-91-103, Max-Planck-Institut für Informatik, Saarbrücken, 1991. See also: Proc. 2nd SODA, 1991, pp. 1-6.

[11] K.J. Supowit. *New techniques for some dynamic closest-point and farthest-point problems*. Proc. 1st SODA, 1990, pp. 84-90.

[12] P.M. Vaidya. *An $O(n \log n)$ algorithm for the all-nearest-neighbors problem*. Discrete Comput. Geom. 4 (1989), pp. 101-115.

Parallel algorithms for some dominance problems based on a CREW PRAM

Ip-Wang CHAN † and Donald K. FRIESEN ‡

† Department of Information Systems and Computer Science, National University of
Singapore, Lower Kent Ridge Road, Singapore 0511

‡ Department of Computer Science, Texas A&M University, College Station, TX 77843, USA

Abstract

Two parallel geometric algorithms based on the idea of point domination are presented. The first algorithm solves the d-dimensional isothetic rectangles intersection counting problem of input size $N/2^d$, where $d > 1$ and N is a multiple of 2^d, in $O(\log^{d-1} N)$ time and $O(N)$ space. The second algorithm solves the direct dominance reporting problem for a set of N points in the plane in $O(\log N + J)$ time and $O(N \log N)$ space, where J denotes the maximum of the number of direct dominances reported by any single point in the set. Both algorithms make use of the CREW PRAM (Concurrent Read Exclusive Write Parallel Random Access Machine) consisting of $O(N)$ processors as the computational model.

1. Introduction

Geometric problems involving such objects as points, segments, polygons, etc. frequently arise in many applications (see [8], [9], [10] for more discussions). In this paper, parallel algorithms for solving two geometric problems involving the idea of point domination are presented. All results in the present paper are based on a CREW PRAM (Concurrent Read Exclusive Write Parallel RAM). A CREW PRAM is a shared memory parallel random access machine which allows concurrent reads but no two processors can simultaneously write into the same memory location.

Given a point set S in a d-dimensional space, for any two points p and q in S, we say p dominates q (denoted by $q \prec p$) if $p \neq q$ and for each $j = 1, \ldots, d$, $x_j(q) \leq x_j(p)$. Furthermore, if $q \prec p$ and there exists no $r \in S$ such that p dominates r and r dominates q, then we say that p directly dominates q (denoted by $q \lhd p$). In this paper, the following problems are addressed:

1. **Isothetic rectangles intersection counting (RIC) problem** - An isothetic rectangle in d dimensions is the Cartesian product of d intervals, one on each of the d coordinate-axes and such that all edges of the rectangle are axis-parallel. Given a set S of $N/2^d$ d-dimensional isothetic rectangles, where N is a multiple of 2^d, determine for each rectangle the number of rectangles in S that intersect it.

2. **All-points direct dominance reporting (DDR) problem** - Given a set S of N points in the plane, for each point p in S, report all the other points in S that directly dominate p.

The isothetic rectangles intersection counting problem is closely related to the two-set dominance counting problem. Given a set $A = \{p_1, \ldots, p_n\}$ and a set $B = \{q_1, \ldots, q_m\}$ of points in Euclidean d-space, where $n + m = N$, the two-set dominance counting problem asks for the determination for each point r from A (similarly, B) the number of points from $B(A)$ that are dominated by r. Recently efficient parallel solution for the two-set dominance counting problem in the plane has been obtained [1]. The algorithm in [1] improves upon the result in [2] and solves the planar two-set dominance counting problem in $O(\log N)$ time and $O(N)$ space using $O(N)$ processors on a CREW PRAM. In [1], [2], the solution for the two-set dominance counting problem in two dimensions was applied to yield solution for the multiple range-counting problem in the plane. The same authors did not consider either the two-set dominance counting problem or the multiple range-counting problem in higher dimensions. In this paper, we show that for $d > 1$, the two-set dominance counting problem can be solved under the general framework of multidimensional divide-and-conquer described in [3] in $O(\log^{d-1} N)$ time and $O(N)$ space using $O(N)$ processors on a CREW PRAM. Furthermore, we are able to apply our parallel results for the two-set dominance counting problem to obtain a solution for the isothetic rectangles intersection counting problem in d-space, for $d > 1$. Under the sequential searching or querying mode of operation, the paper in [5] shows that the general d-dimensional isothetic rectangles intersection counting problem is equivalent to the d-dimensional one-set dominance counting problem. The present paper shows that under the all-points computation mode and in a parallel computing environment, the d-dimensional isothetic rectangles intersection counting problem is also equivalent to the d-dimensional two-set dominance counting problem.

Sequentially, the work in [6] shows that the direct dominance reporting problem can be solved in $O(N \log N + k)$ time and $O(N)$ space, where N denotes the size of the input set and k represents the number of direct dominances in the set. The direct dominance searching and reporting problem addressed in [6] differs from some of the geometric searching or querying problems discussed in the literature (e.g. [3]) in that in the latter problems the computational procedures often involve the processing of external query elements whereas in the former problem the existence of elements other than the original input set is not assumed. In this paper, we propose a parallel solution to solve the same planar problem studied in [6]. Given a set S of N points in the plane, the goal is to list out or report, for each point p in S, all the other points in S that directly dominate p. Our result shows that the problem can be solved in $O(\log N + J)$ time, where J denotes the maximum of the number of direct dominances reported by any single point in the original input set of size N, and $O(N \log N)$ space using $O(N)$ processors on a CREW PRAM.

Throughout this paper, whenever sorting is called forth, it is assumed that the parallel merge sort procedure by [4] is to be used. The sorting scheme by [4] sorts N given numbers in $O(\log N)$ time and $O(N)$ space using $O(N)$ processors on a CREW PRAM. The remainder of this paper is organized as follows. Section 2 describes our parallel algorithm for solving the RIC problem. Section 3 presents our parallel solution for the DDR problem. Section 4 summarizes the main results in this paper.

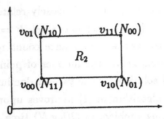

Figure 1: Labelling of vertices and pairing of the v's and N's for a given rectangle R_2 in the plane (Note: N_j represents the number of points of type v_j that are within the dominance region of vertex v_i. The indices i and j are complement to each other).

2. Algorithm for the RIC problem

In the isothetic rectangles intersection counting (RIC) problem, we represent each rectangle R_d in d-space, $d > 1$, by its 2^d distinct vertices and write $R_d = \{v_0, \ldots, v_{2^d-1}\}$. Among the 2^d distinct vertices in R_d, v_0 is closest to and v_{2^d-1} is farthest away from the origin. For $d = 2$, Figure 1 illustrates the labelling of the four vertices in binary notation for a given rectangle R_2. In d dimensions, the index associated with each vertex will be comprised of d binary digits. The labelling for each vertex can be done starting at the lowest level and extending all the way to dimension d. In any isothetic rectangle R_d, where $d > 2$, vertices with a 1 as the leading digit will all have a larger d dimensional value than the remaining vertices in R_d. For convenience, we also use the label of a vertex to denote the vertex type. Thus given a set of d-dimensional rectangles, 2^d different types of vertices can be distinguished. Using the idea of dominance, a distinct dominance region $\delta_d(v)$ can be defined for each vertex v within a rectangle R_d in d-space as follows:

$$\delta_d(v) = \{p \mid p \prec v \text{ for all possible } p \text{ in the } d\text{-space}\}$$

Physically $\delta_d(v)$ represents that portion, say Λ, of the d-space defined for v such that for any point p in Λ, p is dominated by v.

Let a set S of $N/2^d$ isothetic rectangles be given in the d-space, where $d > 1$ and N is a multiple of 2^d. It is observed that if we let each individual vertex v within a rectangle R_d in S in d-space compute its dominance count involving only vertices of a specified type that are inside the dominance region of v, there will be a direct simple relationship between the 2^d dominance counts computed for R_d and the number of other rectangles in S that intersect R_d. Throughout this paper, for any rectangle R_d in an input set S, we denote by N_j the number of vertices of type v_j that are within the dominance space of v_i in R_d. The relationship between the subscripts i and j is that in binary notation they are merely complement to each other (see Figure 1 for an illustration when $d = 2$). In $d > 1$ dimensions, there will be $N/2^d$ isothetic rectangles in the input set S, where N is a multiple of 2^d. For any rectangle R_d in S, let I_d denote the total number of rectangles in S that intersect R_d. The following relationship can be established.

Lemma 1 *In d dimensions, where $d > 1$, for any rectangle R_d in S, an expression for I_d in terms* *of the well defined N_j's associated with the vertices of R_d can be given by:*

$$I_d = \sum_{k=0}^{2^d-1} (-1)^{|k|} N_k - 1 \tag{1}$$

where $|\,l\,| =$ number of 1's in the binary representation of l.

Proof: The result follows from the direct application of the Inclusion-Exclusion Principle in combinatorics. □

Based on the above observations, an algorithm known as Solve-RIC can be constructed for solving the RIC problem. Algorithm Solve-RIC can be described below.

Algorithm Solve-RIC

Initialization: On a CREW PRAM, $O(1)$ processors will be assigned to each input vertex. Initially each vertex v_i within a rectangle R_d will set its own dominance counter N_j to zero, where the subscripts i and j in binary form are complement to each other.

Step 1. Partition the set of vertices of rectangles in S into 2^{d-1} separate subsets. Each subset will be of size $N/2^{d-1}$ and consists of two types of vertices only. If vertex v_i has been assigned to compute N_j (where in binary form i and j are complement to each other), then vertices of types v_i and v_j will be placed in the same subset.

Step 2. Within each of the 2^{d-1} subsets, vertex v_i determines its dominance count involving vertices of type v_j only using the solution for the two-set dominance counting problem.

Step 3. After obtaining the dominance counts by individual vertices within each rectangle R_d, the problem solution for R_d is to be computed by evaluating I_d associated with R_d using all the processors assigned to R_d. The expression for I_d is given in lemma 1.

End of Algorithm Solve-RIC

Theorem 1 *Given a set S of $N/2^d$ isothetic rectangles in d-space, where $d > 1$ and N is a multiple of 2^d, algorithm Solve-RIC correctly computes, for each rectangle R_d in S, the number of rectangles in S that intersect R_d in time $O(\log^{d-1} N)$ and $O(N)$ space using $O(N)$ processors on a CREW PRAM.*

Proof: To verify the correctness of algorithm Solve-RIC, we need only to show that for any rectangle R_d in S, the expression I_d for R_d in terms of the dominance counts computed for R_d correctly gives the number of rectangles in S that intersect R_d. This follows directly from lemma 1. □

It can be seen that in algorithm Solve-RIC, the initialization requires only $O(1)$ parallel time. The partitioning of S in Step 1 can be done by the repeated use of a parallel type of computation as discussed in [7]. For a given set of N vertices, Step 1 takes $O(2^{d-1} \log N)$ time and $O(N)$ space using $O(N)$ processors on a CREW PRAM. The computing of the dominance count by each individual point in each subset in Step 2 represents merely another instance of the two-set dominance counting problem. Since each subset is of size $N/2^{d-1}$, we know that Step 2 can be done in $O(\log^{d-1} N)$ time and $O(N)$ space using $O(N)$ processors in the CREW PRAM model. After obtaining the necessary dominance counts, $O(\log N)$ parallel time will suffice for the evaluation of I_d in Step 3. Thus algorithm Solve-RIC solves the isothetic rectangles intersection counting problem as claimed in Theorem 1.

3. Algorithm for the direct dominance reporting problem

Let S be a set of N points in the plane. For any $a \in S$, let $y(a)$ denote the y-coordinate value of a. Also for each element a in S, the following can be defined.

(a) One-dominance set $\quad S'(a) \quad = \quad \{p \mid p \in S,\ y(p) \geq y(a)\}$,
(b) Minima set $\quad\quad\quad M_S(a) \quad = \quad \{p \mid p, q \in S'(a),\ q \not\prec p\}$,
(c) Direct dominance set $DD_S(a) = \{p \mid p \in S'(a),\ a \prec p\}$,
(d) Window size $\quad\quad\quad W_S(a) \quad = \quad$ minimum $\{y(q) \mid q \in DD_S(a)\}$

From the above definitions, the following properties are immediate:

1. The sets $S'(a)$ and $M_S(a)$ are non-empty since they each contain at least one element which is a. Thus for a set of N input points there will be N minima sets.

2. Elements within $M_S(a)$ for any $a \in S$ are always ordered in non-increasing y-coordinates with increasing x-coordinates.

3. For any $a \in S$, if $DD_S(a)$ is nonnull, then $DD_S(a)$ will be the union of the subsets from one or more minima sets in S.

It can be observed that to be able to solve the all-points direct dominance reporting (DDR) problem, one needs only to identify and list out, for each point a in a given input set S, all members in the direct dominance set of a (that is, $DD_S(a)$). By property 3, this is equivalent to locating and reporting, for each $a \in S$, the corresponding elements in some appropriate minima set in S.

The DDR problem can be solved in a parallel environment using the method of divide-and-conquer. Let S be a given set of N points in the plane. Depending on the distribution of S, it is possible for an individual point in S to be within the minima sets or the direct dominance sets of a large number of points in S. Thus it becomes impractical for each input element to keep all members of its own minima set in separate storage. We avoid the need of creating and storing multiple copies of any input point in S by associating with each point two sets of pointers called parent and target respectively. Both sets of pointers are each of size $\log N$. The parent pointers are used to indicate membership within each minima set. The target pointers are headers to be used for retrieving direct dominance pairs. Specifically for any $a \in S$, its arrays of parent and target pointers will be referred to as parent$_a$ and target$_a$ respectively. In a parallel cascading divide-and-conquer environment, for any input set of N points, there will be $\log N$ subproblems combination levels which correspond to the $\log N$ levels of the binary tree. At the leaf level, parent$_a$[0] and target$_a$[0] for each input element a will be initialized to null. Whenever an internal tree node A at level j, where $j = 1, \ldots, \log N$, of the tree becomes full, each individual element in A will locate its target point in A and update the information relevant to its own minima set as well as direct dominance set. Let B and C be the two sons of node A such that all elements in B are x-dominated by elements in C. Also for any $a \in A$, let r be the target of a. Recall that by convention, if a is from B (similarly, C), then r must be from $C(B)$. Depending on whether a is from B or C, two cases may arise.

Case 1. Element a is from B. If there exists at least one point from B that are in between a and r along the y-coordinate and at the same time dominate a, then no points in the minima set of r can directly dominate a. When this happens, target$_a[j]$ can be set to null to signal that no reporting of direct dominance pair for a is to be carried out in the present combination level j. If r is found to be directly dominating a, it is possible that other members in the minima set of r may also directly dominate a (see Figure 2 for an illustration). To be able to decide

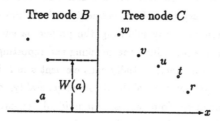

Figure 2: Checking for direct dominances in C by element a in node $B \bigcup C$ (Note: element r is the target point of a. Even though elements $r, t, u, v,$ and w are in $M_C(r)$, only $r, t,$ and u are directly dominating a in $B \bigcup C$).

exactly which elements in the minima set of r are directly dominating a, it is necessary for each individual element a in A to keep a working variable called window size $W(a)$. If a is from B (similarly, C), then $W(a)$ will contain the y-coordinate of the point in $B(C)$ that has the minimum y-coordinate among all elements in $DD_B(a)$ (similarly, $DD_C(a)$). By checking $W(a)$ against the y-coordinate of individual elements in the minima set of r, elements in the minima set of r that are directly dominating a can be effectively identified. For example, in Figure 2 after checking elements $r, t,$ and u in the minima set of r will be found to be contained in the direct dominance set of a with respect to $A = B \bigcup C$. Since there could be a large discrepancy in the number of direct dominances among individual points in a given node or subproblem, points with only a few direct dominance pairs to report may be kept idling while waiting for others to finish. Thus instead of having each a in A at the jth tree level to report its direct dominances found in A right away, each a simply enters the position of its target r in target$_a[j]$. Members other than r in the minima set of r in C that are also directly dominating a can be retrieved later on simply by following the $(j-1)th$ component of the parent pointer of these points in sequence, starting at element r. Since all elements from node B are all x-dominated by elements from C, the minima set of a found in B is correct with respect to the combined set A. Thus the parent$_a[j]$ will be set equal to parent$_a[j-1]$. The window size of a, $W(a)$, in A will need to be updated to $y(a)$ if $y(r) < W(a)$. Otherwise $W(a)$ is correct with respect to A.

Case 2. Element a is from C. Since the target r of a cannot possibly dominate a, it follows that the direct dominance set of a found in C is correct with respect to A. Thus $target_a[j]$ can be set to null, signalling that no reporting of direct dominance pair for a will be needed at the current jth combination stage. The window size of a, $W(a)$, established at node C is also correct with respect to A. The determination of $parent_a[j]$ for a can be done according to the following: (i) If r does not exist, then $parent_a[j]$ will be set to null, (ii) If r exists and $parent_a[j-1]$ is null (that is, the minima set of a in C is a singleton), then $parent_a[j]$ will be set pointing to r, and (iii) If r exists and a has a parent t pointed to by $parent_a[j-1]$, then the following is checked. If $y(t) > y(r)$, then $parent_a[j]$ will be set pointing at r. Otherwise, $parent_a[j]$ will be directing at t.

At the end of the cascading merge procedure, all elements from the original input problem S will be at the root node of the binary tree. Each element in S will have its arrays of parent and target pointers properly determined. By following the proper parent and target pointers established during the cascading merge procedure, the checking and reporting of direct dominance pairs can be carried out in an orderly manner. Initially each element a in S initializes its working window, $W(a)$, to be arbitrarily large (say, ∞). Without loss of generality, let r denote the target of a. The whole process of reporting direct dominances by individual elements will consist of $\log N$ iterations. At the jth iteration, where $j = 1, \ldots, \log N$, two situations may occur:

Situation 1. If r does not exist or $x(r) < x(a)$, no reporting is needed at that iteration. In either case, $target_a[j]$ should have been set to null.

Situation 2. If $x(a) < x(r)$ and $y(r) < W(a)$, then (a, r) will be output as a direct dominance pair by a. After this, the parent of r which is pointed to by $parent_r[j-1]$ is sought. Suppose r has a parent denoted by t. Then $y(t)$ will be checked against $W(a)$ for possible reporting. This checking and reporting process goes on until there is an ancester, say u, in the minima set of r such that $y(u) > W(a)$ or that all elements in the minima set of r have been examined. When this happens, the reporting of direct dominances for a is complete at the current jth iteration and $W(a)$ will be set equal to $y(r)$.

At the end of the $(\log N)th$ iteration, all direct dominances in S will have been reported.

From the above observations, it can be seen that an algorithm known as Solve-DDR can be constructed for solving the DDR problem. Algorithm Solve-DDR will be comprised of two phases: (1) linkage construction, and (2) direct dominance reporting. During the linkage construction phase, individual elements at each internal tree node establish their own parent and target pointers to keep track of the key elements found in specific minima sets while performing the cascading merge procedure. In the direct dominance reporting phase, all input elements retrieve and report their respective direct dominances via the pointer information collected during the linkage construction phase. To implement the two phases, algorithm Solve-DDR will invoke two subprocedures known respectively as BUILDUP and REPORT. Routine BUILDUP will carry out the task of linkage construction while routine REPORT will be solely responsible for the direct dominance reporting. A presentation for each of the two subprocedures, BUILDUP and REPORT, is given below.

Subprocedure BUILDUP

Preprocessing: All input points are presorted on the x-coordinate so that $x(p_i) \leq x(p_{i+1})$ for $i = 1, \ldots N - 1$. These points are then assigned to the leaf nodes of the binary tree in the order of increasing x-coordinate.

Initialization Each element a at the leaf level of the tree initializes both $\text{parent}_a[0]$ and $\text{target}_a[0]$ to null and sets $W(a)$ to be arbitrarily large (say, ∞).

Description The procedure described in [1] for performing cascading merge for an input set is brought into action. Initially $O(1)$ processors are assigned to each leaf node. Each time when p elements are passed from a node V to the parent of V for performing the cascading merge at the parent, $O(p)$ processors will also be passed to the parent node. When the parent node of V becomes full, any processors formerly assigned to V will be released. Without loss of generality, let B and C denote the two sons of an internal node A in the tree. For any a in A, let the target of a be r. Whenever a node A at the jth tree level, where $j = 1, \ldots, \log N$, becomes full, the following is to be performed:

1. If a does not have a target in A, then $\text{target}_a[j]$ will be set to null. The pointer $\text{parent}_a[j]$ will have the same contents as in $\text{parent}_a[j - 1]$. The window size $W(a)$ will remain unchanged.

2. If a has a target r and that a is from B, then $\text{parent}_a[j]$ will be set equal to $\text{parent}_a[j-1]$. If $y(r) < W(a)$, then $\text{target}_a[j]$ will be directing at r and $W(a)$ will be updated to $y(r)$. If $y(r) \geq W(a)$, then $\text{target}_a[j]$ will be set to null and $W(a)$ will remain unchanged.

3. If a has a target r and a is from C, then $\text{target}_a[j]$ will be set to null and no change will be needed for $W(a)$. If $\text{parent}_a[j - 1]$ is null, then $\text{parent}_a[j]$ will be set pointing at r. If $\text{parent}_a[j - 1]$ is nonnull, let t be the element pointed at by $\text{parent}_a[j - 1]$. If $y(t) > y(r)$, then $\text{parent}_a[j]$ will be directing at r. Otherwise both $\text{parent}_a[j - 1]$ and $\text{parent}_a[j]$ will be pointing at t.

Termination Subprocedure BUILDUP terminates when the root node of the binary tree becomes full and the parent and target pointers associated with individual elements in the input set have all been properly determined.

Lemma 2 *Based on the method of cascading divide-and-conquer, subprocedure BUILDUP successfully performs the task of linkage construction for a set S of N points in the plane in $O(\log N)$ time and $O(N \log N)$ space using $O(N)$ processors on a CREW PRAM.*

Proof: The correctness of the cascading merge technique in a parallel environment has been established in [1]. For a given set of N items, the technique works in $O(\log N)$ time and $O(N)$ space on a CREW PRAM using $O(N)$ processors. Since subprocedure BUILDUP requires each input element to carry two sets of pointers of size $\log N$ each, the overall storage for BUILDUP is $O(N \log N)$. During the implementation of the cascading merge procedure, whenever an internal node in the binary tree is full, all elements inside the node are required to locate their respective target and to establish their parent and target linkages. Using the the cross ranking information computed at the two sons of a full node, the location of target by each individual element inside the node takes only $O(1)$ time on a CREW PRAM. The determination of the parent and target pointers can also be done in $O(1)$ parallel time. Thus the claim in the lemma follows. □

Subprocedure REPORT

Description: The reporting of direct dominances by individual points in parallel essentially recapitulates, from level to level, the same information established during the execution of subprocedure BUILDUP. Initially each element a in the input set S sets its own working window size $W(a)$ to be arbitrarily big. The process of retrieving and reporting the direct dominance pairs by each a in S is to be repeated $\log N$ times. During the jth round, where $j = 1, \ldots, \log N$, two cases may arise: (without loss of generality, let r denote the target of a for the current jth round)

1. If r does not exist (that is, $target_a[j]$ is null) or $x(a) > x(r)$, then no reporting is needed for the current jth round.

2. If a has a target r and $x(a) < x(r)$, two cases will be considered. If $W(a) < y(r)$, then no reporting is needed. If $W(a) \geq y(r)$, then (a, r) will be output as a direct dominance pair by a. After this, the parent of r (which is pointed to by $parent_r[j-1]$) is sought. If $parent_r[j-1]$ is nonnull, let t be the element pointed to by $parent_r[j-1]$. Then $y(t)$ will be checked against $W(a)$ for possible reporting. This checking and reporting process is repeated until there is an ancestor, u, in the minima set of r such that $y(u) > W(a)$ or all elements in the minima set of r have been examined. When this happens, reporting for a is complete for the current jth round and $W(a)$ will be set equal to $y(r)$.

Termination: Subprocedure REPORT terminates when all elements have completed their direct dominance reporting at the end of the $(\log N)$th round.

End of REPORT

Lemma 3 *Given a set S of N points in the plane, let J denote the maximum of the number of direct dominances associated with any single point in S. Subprocedure REPORT correctly reports all the direct dominances for individual elements in S in $O(\max(J, \log N))$ time and $O(N \log N)$ space using $O(N)$ processors on a CREW PRAM.*

Proof: Even though each element in S has to retrieve and report, level by level, its associated direct dominance pairs, there is no time dependence among individual elements in S when subprocedure REPORT is executed. Within each of the $\log N$ levels, the process of checking and reporting can be done in $O(1)$ time by following the appropriate pointers correctly set up during the execution of subprocedure BUILDUP. Since the arrays of parent and target pointers for individual points are each of size $\log N$, searching through these arrays may take up $O(\log N)$ time. Thus the total time required for the execution of subprocedure REPORT is $O(\max(J, \log N))$ as claimed. □

Theorem 2 *Given a set S of N points in the plane, algorithm Solve-DDR correctly solves the direct dominance reporting problem in time $O(\log N + J)$ and $O(N \log N)$ space on a CREW PRAM using $O(N)$ processors, where J denotes the maximum of the number of direct dominances associated with any single point in S.*

Proof: The correctness of algorithm Solve-DDR follows directly from the correctness of subprocedures BUILDUP and REPORT. Since it has been shown in Lemmas 2 and 3 that subprocedures BUILDUP and REPORT require $O(\log N)$ and $O(\max(J, \log N))$ time respectively, running the two subprocedures in tandem will result in an overall time that is $O(\log N + J)$ for algorithm Solve-DDR. Thus algorithm Solve-DDR solves the DDR problem as stated in Theorem 2. $\Box$

4. Conclusion

In this paper, we found that the RIC problem of input size $N/2^d$, where $d > 1$ and N is a multiple of 2^d, can be solved in $O(\log^{d-1} N)$ time and $O(N)$ space using $O(N)$ processors on a CREW PRAM. This supports the notion that under the all-points computational mode and in a parallel computing environment, the d-dimensional RIC problem is equivalent to the two-set dominance counting problem. Whether the RIC problem can be solved in $o(\log^{d-1} N)$ time on a CREW PRAM is still an open question. In the present paper, we showed that the DDR problem consisting of N points in the plane can be solved in $O(\log N + J)$ time and $O(N \log N)$ space, where J denotes the maximun of the number of direct dominances reported by any single point in the input set. It would be of interest to find out if linear space is sufficient for solving the DDR problem on a CREW PRAM.

References

[1] M. J. Atallah, R. Cole, and M. T. Goodrich, Cascading divide-and-conquer: A technique for designing parallel algorithms, *SIAM J. Comput.* **18** (1989) 499–532.

[2] M. J. Atallah and M. T. Goodrich, Efficient plane sweeping in parallel, in: *Proc. of the Second ACM Symposium on Computational Geometry* (1986) 216–225.

[3] J. L. Bentley, Multidimensional divide-and-conquer, *Commun. ACM* **22** (1980) 214–229.

[4] R. Cole, Parallel merge sort, *SIAM J. Comput.* **17** (1988) 770–785.

[5] H. Edelsbrunner and M. H. Overmars, On the equivalence of some rectangle problems, *Inform. Process. Lett.* **14** (1982) 124–127.

[6] R. H. Güting, O. Nurmi, and T. Ottmann, The direct dominance problem, in: *Proc. of the First ACM Symposium on Computational Geometry* (1985) 81-88.

[7] R. E. Ladner and M. J. Fischer, Parallel prefix computation, *JACM* **27** (1980) 831–838.

[8] D. T. Lee and F. P. Preparata, Computational geometry – A survey, *IEEE Trans. Comput.* **C-33** (1984) 1072–1101.

[9] K. Mehlhorn, *Data Structures and Algorithms 3: Multi-dimensional Searching and Computational Geometry.* (Springer Verlag, New York, 1984).

[10] F. P. Preparata and M. I. Shamos, *Computational Geometry: An Introduction.* (Springer Verlag, New York, 1985).

Parallel algorithms for finding maximal k-dependent sets and maximal f-matchings

Krzysztof Diks [*]   Oscar Garrido[†]   Andrzej Lingas [†]

Abstract

Let k be a positive integer, a subset Q of the set of vertices of a graph G is k-dependent in G if each vertex of Q has no more than k neighbours in Q. We present a parallel algorithm which computes a maximal k-dependent set in a graph on n nodes in time $\mathcal{O}(\log^4 n)$ on an EREW PRAM with $\mathcal{O}(n^2)$ processors. In this way, we establish the membership of the problem of constructing a maximal k-dependent set in the class NC. Our algorithm can be easily adapted to compute a maximal k-dependent set in a graph of bounded valence in time $\mathcal{O}(\log^* n)$ using only $\mathcal{O}(n)$ EREW PRAM processors.

Let f be a positive integer function defined on the set V of vertices of a graph G. A subset F of the set of edges of G is said to be an f-matching if every vertex $v \in V$ is adjacent to at most $f(v)$ edges in F. We present the first NC algorithm for constructing a maximal f-matching. For a graph on n nodes and m edges the algorithm runs in time $\mathcal{O}(\log^4 n)$ and uses $\mathcal{O}(n + m)$ EREW PRAM processors. For graphs of constantly bounded valence, we can construct a maximal f-matching in $\mathcal{O}(\log^* n)$ time on an EREW PRAM with $O(n)$ processors.

1 Introduction

A problem belongs to the NC class introduced in [12] if it can be solved in poly-logarithmic time on a PRAM with a polynomial number of processors. If we succeed in showing a problem to be in NC, the next step is to design an algorithm for the problem running in poly-log time such that the product of the number of processors used by the algorithm and its time complexity is asymptotically as close to the time complexity of the fastest known sequential algorithm as possible.

The two problems whose parallel complexity is the subject of this paper are natural generalizations of two well known graph problems.

A subset of the vertices of a graph is *independent* if there are no adjacent vertices in the subset. An independent set is *maximal* if it is not a proper subset of any other independent set. While the problem of constructing a maximum cardinality independent set is NP-hard [4], the problem of constructing a maximal independent set (MIS for short) can be trivially solved in linear time. However, the problem of constructing an efficient NC algorithm for MIS is non-trivial. Karp and Widgerson were the first who proved MIS to be in NC [9]. Presently, the most efficient deterministic NC algorithm is due to Goldberg and Spencer [5]. It runs in time $\mathcal{O}(\log^4 n)$ and uses an EREW PRAM with a linear number of processors. Luby has constructed a randomized parallel algorithm for MIS that runs in $\mathcal{O}(\log n)$ time and uses a CRCW PRAM with a linear number of processors [11].

A subset of the set of edges of a graph G is a *matching* if no two edges in the subset are incident. In other words, a matching is a subset of the set of edges of G in one-to-one correspondance with an independent set in the edge graph induced by G. In the edge graph, the vertices correspond

---

[*]Institute of Informatics, Warsaw University, PKiN VIII p., 00-901 Warsaw, Poland.

[†]Department of Computer Science, Lund University, Box 118, S-221 00 Lund, Sweden.

to the edges of G and two such vertices are adjacent if and only if the corresponding edges of G are incident in G. A matching is maximal if it is not a proper subset of any other matching. While there are no known subquadratic-time sequential algorithms for constructing a maximum cardinality matching, the problem of finding a maximal matching (MM for short) admits trivial linear-time sequential solutions. Also, the known NC algorithmic solutions to MIS can be specialized to MM. However, as the edge graph may have a quadratic number of edges with respect to the size of the input graph, more efficient NC algorithms for MM can be derived directly. Presently, the most efficient deterministic algorithm for MM is due to Israeli and Shiloach [7]. It can be implemented in time $\mathcal{O}(\log^3 n)$ on a CRCW PRAM with a linear number of processors and consequently in time $\mathcal{O}(\log^4 n)$ on an EREW PRAM with a linear number of processors [8]. Both notions of independent set and matching can be naturally generalized by weakening the requirement on vertex non-adjacency or edge no-incidency.

For example, Djidjev et al considered the so called k-dependent sets in [3]. They call a subset of the set of vertices of a graph a k-dependent if no vertex in the subset is adjacent to more than k vertices in the subset. Note that a 0-dependent set in G is an independent set of vertices of G. A 1-dependent set is in general a set of independent vertices and edges, while a 2-dependent set is a set of independent paths (possibly degenerated) and cycles such that no two nonconsecutive vertices of these paths or cycles are adjacent. Such paths are called permissible in [1]. In [3], Djidjev et al showed that the problem of constructing a maximum cardinality k-dependent set is NP-hard for each fixed k and they presented practical linear-time algorithms for solving this problem for trees.

Also, there exists a well known notion of f-matching in the literature [10]. Let f be an integer function defined on the set of vertices of a graph G. A subset of the set of edges of G forms an f-matching if for any vertex v of G there are at most $f(v)$ edges in the subset that are incident to the vertex. Note that if $f(v) = 1$ for all vertices v in G then f-matching is just a matching. The reader may observe that the generalization of matching to f-matching is more general than that of independent set to k-dependent set. Indeed, one could also define an f-dependent set as a subset of the set of vertices such that no vertex v in the subset is adjacent to more than $f(v)$ vertices in the subset [1]. We can naturally generalize the notion of maximal independent sets and maximal matchings to include maximal k-dependent sets and maximal f-matchings. Again, the problems of finding a maximal k-dependent set and a maximal f-matching can be solved by trivial greedy algorithms in linear time. At this point it is natural to ask whether these two generalized problems admit NC algorithms.

In this paper we give an affirmative answer to the above question in both cases. We present the first NC algorithm for constructing a maximal k-dependent set in a graph G on n nodes. The algorithm runs in time $\mathcal{O}(\log^4 n)$ using an EREW PRAM with $\mathcal{O}(n^2)$ processors. We also observe that if G has bounded valence then it can be modified to run in time $\mathcal{O}(\log^* n)$ on an EREW PRAM with a linear number of processors. The algorithm can be also easily generalized to find a maximal f-dependent set in G in time $\mathcal{O}((\log^4 n) \cdot (\max_{v \in V} f(v)^2))$ using an EREW PRAM with $\mathcal{O}(n^2)$ processors. We also present the first NC algorithm for constructing a maximal f-matching. It is a non-trivial generalization of the advanced algorithm due to Israeli and Shiloach [7] for maximal matching and as their algorithm it runs in time $\mathcal{O}(\log^3 n)$ on a CRCW PRAM with a linear number of processors. Moreover, we show that a maximal f-matching in a graph of bounded valence can be constructed in time $\mathcal{O}(\log^* n)$ on an EREW PRAM with a linear number of processors.

---

[1]In order to generalize the correspondance between matchings and independent sets in the edge graphs we would have to split the set of neighbours of each vertex in the edge graph into two subsets and assign to v two upper bounds $f_1(v)$ and $f_2(v)$ on the number of neighbours from the first and the second subset respectively.

2 Terminology

Let $G = (V, E)$ be a loop-free graph. We denote the number of vertices of the graph by n ($|V| = n$), and the number of edges by m, ($|E| = m$). For any set $S \subseteq V$, we define the *neighbourhood* of S in the graph G as : $N_G(S) = \{w \in V \mid \text{there exists } u \in S \text{ such that } (u, w) \in E\}$. In the same way, the set of *neighbours* of a node v in the graph G is the set $N_G(v) = N_G(\{v\})$.

When it is clear in the context which graph we are refering to, we use the notation $N(S)$ and $N(v)$ instead of $N_G(S)$ and $N_G(v)$. The number of neighbours of a vertex v in a graph G is called the *degree* of v in G and denoted as $d(G, v)$. In a similar manner, for a given subgraph H of G, and for any vertex v of G we define the *degree* of v in H as the number of neighbours of v which are vertices of the subgraph H, and denote it as $d(H, v)$. Note that in the last definition the vertex v is not necesarily in H.

If $S \subseteq V$ then $\gamma(S)$ will denote the *subgraph induced by* S. This subgraph has vertex set S, and its edge set $E_{\gamma(S)}$ consists of these edges in E that are incident only to vertices in S.

3 The NC algorithm for maximal k-dependent set

Our parallel algorithm for maximal k-dependent ($k > 0$) set can be seen as an NC Turing-like reduction to the problem of constructing a maximal 0-dependent (independent) set (MIS). Recall that the best known parallel algorithm for MIS runs in $\mathcal{O}(\log^4(n))$ time using $\mathcal{O}(n + m)$ EREW processors [5]. We shall analyse our algorithm also in terms of the EREW shared memory model, where simultaneous reads and writes into the same memory locations are not permitted.

```
Algorithm  Maximal-k-dependent-set(G)
input :    A graph G = (V, E).
output :   A maximal k-dependent set Q for G.
method :

           Q ← Maximal-Independent-set(G);
           R ← V \ Q;
           B ← {v ∈ R | d(γ(Q), v) ≤ k};
           while B ≠ ∅ do
                     H ← the graph whose set of vertices is B
                         such that (v, w) is an edge of H iff
                         v and w have a common neighbour in Q
                         or (v, w) is also an edge in G;
                     M ← Maximal-Independent-Set(H);
                     Q ← Q ∪ M;
                     S ← {v ∈ Q | d(γ(Q), v) = k};
                     R ← R \ (M ∪ N_{γ(R)}(S));
                     B ← {v ∈ R | d(γ(Q), v) ≤ k};
           endwhile
           output Q;
end  Maximal-k-dependent-set
```

Lemma 1 ALGORITHM MDS *is partially correct, i.e. if it stops then the set Q to output is a maximal k-dependent set in G.*

Proof: It is sufficient to observe that the augmentation of Q by M is correct since M is in particular independent in G, no two vertices in M share a common neighbour in Q and no vertex in M is a neighbour of a vertex in Q that has already k neighbours in Q. □

Lemma 2 *The block under the while statement is iterated $\mathcal{O}(k^2)$ times.*

Proof: For a vertex $v \in G$, let $cap(v) = \min(d(G, v), k) - |N_{\gamma(Q)}(v)|$ at a given stage of performance of ALGORITHM MDS. Consider a vertex $v \in B$ at the beginnig of the ith iteration of the block. Next, let $g(v) = \sum_{w \in N_{\gamma(Q)}(v)} \text{cap}(w)$ Note that $g(v)$ is always bounded by k^2 from above. Also, if v disappears from B in some of the next iterations then it never can reappear in B. On the other hand, after each iteration, if v remains in B then there exists a vertex w newly inserted into Q that either shares a neighbour in Q with v or it is itself a neighbour of v in G. In the first case, $g(v)$ decreases at least by 1. In the second case $g(v)$ increases by $\text{cap}(w)$, i.e. at most by k. However, the second case can occur at most k times since $\text{cap}(v)$ cannot be negative if v is to stay in B. Since $\text{cap}(w)$ for each neighbour of v in Q has to be positive in order to keep v in B, we conclude that after the $k^2 + k$ iterations v has to disappear from B. Suppose that v is not a member of the original set B. It means that v has no neighbour in the original set Q. Therefore, it could be added to the original set Q preserving its independence property which would contradict the maximality of the set. We obtain a contradiction. Thus, we can conclude that all vertices that appear in the sets B are members of the original set B. Combining this conclusion with the shown fact that no member of B can survive more than the $k^2 + k$ iterations we obtain the thesis of the lemma. $\qquad \square$

Combining the two above lemmas, we obtain the correctness of ALGORITHM MDS.

Theorem 1 ALGORITHM MDS *is correct.*

Lemma 3 *Suppose that a maximal independent set in a graph on n nodes can be found in time $T(n)$ using an EREW PRAM with $P(n)$ processors. ALGORITHM MDS can be implemented in time $\mathcal{O}(\log n + T(n))$ using a PRAM with $\mathcal{O}(n^2 + P(n))$ processors.*

Proof: A maximal independent set in G as well as a maximal independent set in the auxiliary graph H can be found in time $T(n)$ using $P(n)$ processors. By Lemma 2, we can replace the while stament by a "for" loop with the number of iterations $\mathcal{O}(k^2)$, in this way avoiding the test for emptiness of B. All other instructions except for the construction of the auxiliary graph H can be easily implemented in time $\mathcal{O}(\log n)$ using an EREW PRAM with $\mathcal{O}(n^2)$ processors. For instance, filtering B out of R can be implemented using the sorting algorithm due to Cole [2] running in a logarithmic time on an EREW PRAM with a linear number processors ($\mathcal{O}(n^2)$ processors in our application). We can implement the set operations in constant time by representing all the involved sets B, Q, R, S with n element vectors, each of them with 1 on the ith position if and only if the ith vertex in G is currently in the set. The construction of the auxiliary graph H, in particular finding all pairs (v, w) such that v and w have a common neighbour in Q, seems to be more costly. It immediately reduces to the following problem:
Given a bipartite graph $F = (V_1, V_2, E)$, where the degree of each vertex in V_2 is bounded by k, construct the graph $F' = (V_2, E')$ such that (v, w) is in E' if and only if v and w have a common neighbour in V_1. Suppose that $V_2 = \{1, 2, \ldots, s\}$. Construct a matrix of size $s \times s$ such that $W(i, j)$, $1 \le i \le j \le s$, is set to the list of neighbours of i in F (i.e. in V_1). Note that such a list can have at most k elements. For this reason, the matrix can be constructed in time $\mathcal{O}(\log s)$ using an EREW PRAM with $\mathcal{O}(s^2)$ processors. Now the construction of the graph F' becomes easy. For each pair i, j where $1 \le i \le j \le s$, we check whether the lists $W(i, j)$ and $W(j, i)$ have at least one element in common. If so we augment E' by (i, j), i.e. we set to one the corresponding entry of the adjancy matrix of F'. Note that comparing two such lists takes time $\mathcal{O}(k)$. We conclude that F' can be constructed in time $\mathcal{O}(\log s)$ using an EREW PRAM with $\mathcal{O}(s^2)$ processors. Consequently, the auxiliary graph H can be constructed in time $\mathcal{O}(\log n)$ using an EREW PRAM with $\mathcal{O}(n^2)$ processors. As by Lemma 2, all instructions within the loop are executed only $\mathcal{O}(k^2)$ times, we conclude that ALGORITHM MDS can be implemented in time $\mathcal{O}(\log n + T(n))$ using $\mathcal{O}(n^2 + P(n))$ processors. $\qquad \square$

Theorem 2 *Let k be a nonnegative integer. A maximal k-dependent set in a graph on n nodes can be computed in time $\mathcal{O}(\log^4 n)$ using an* EREW PRAM *with $\mathcal{O}(n^2)$ procesors.*

Proof: As a maximal independent set in a graph on n nodes can be computed in time $\mathcal{O}(\log^4 n)$ using an EREW PRAM with $\mathcal{O}(n^2)$ processors we obtain the thesis by Lemma 3. □

Corollary 4 *The problem of constructing a maximal k-dependent set is in* NC.

A maximal independent set in a graph of bounded valence on n nodes can be computed in time $\mathcal{O}(\log^* n)$ using an EREW PRAM with $\mathcal{O}(n)$ procesors [6]. We can use this fact to speed-up Algorithm MDS in the case of graphs of bounded valence.

Theorem 3 *Let k be a nonnegative integer. A maximal k-dependent set in a graph of bounded valence on n nodes can be computed in time $\mathcal{O}(\log^* n)$ using an* EREW PRAM *with $\mathcal{O}(n)$ procesors.*

Proof: We specialize Algorithm MDS to the bounded degree case. Let Δ denote the maximum vertex degree in the input graph G. It is easy to see that the maximum vertex degree in the auxiliary graph H is bounded by Δ^2 from above. For this reason, we can find a maximum independent set not only in G but also in H in time $\mathcal{O}(\log^* n)$ using an EREW PRAM with $\mathcal{O}(n)$ procesors. As in the proof of Lemma 3, we replace the while statement by a "for" loop with the number of iterations $\mathcal{O}(k^2)$ to avoid the test for emptiness of B. From the proof of Lemma 3 we know that the set instructions can be implemented in constant time. It remains to show that all the other instructions in Algorithm MDS can be implemented in constant time on an EREW PRAM with $\mathcal{O}(n)$ processors. To achieve this, we form an n element vector W such that $W(i)$ contains the list of the at most Δ neighbours of the ith vertex in G. Now, to implement each of the considered instructions, we assign a single processor to each vertex of G. Suppose for a moment that we can use concurrent read. Then, it is easy to see that each of the instructions takes constant time. For instance, to create H, the ith processor scans $W(i)$ and each $W(j)$ where j is a member of the neighbour list $W(i)$. It takes $\mathcal{O}(\Delta^2)$ time. Another example: to determine B, the ith processor scans $W(i)$ accesing the vector corresponding to Q in order to count the number of neighbours in Q, etc. Observe that each of the entries of the vectors is accesed only $\mathcal{O}(\Delta)$ times during the implementation of any of these instructions. For this reason, each of the concurrent read steps can be simulated by $\mathcal{O}(\Delta)$ exclusive read steps. □

Algorithm MDS can be also easily generalized to construct a maximal f-dependent set in G (see Introduction) provided an integer function f defined on the set of vertices of G is given. It is enough to redefine B as the set of all vertices in V or R respectively such that $f(v)$ minus the number of neighbours of v in Q is nonnegative. By reasoning analogously as in the proof of Lemma 2 we conclude that the while block is iterated $\mathcal{O}(\max_{v \in V}(f(v))^2)$ times. Hence, we obtain the following generalization of Theorem 2 leaving the proof details to the reader.

Theorem 4 *Let G be a graph on n nodes and m edges, and let f be a positive integer function defined on the set of vertices of G. A maximal f-dependent set in G can be computed in time $\mathcal{O}((\log^4 n) \cdot (\max_{v \in V} f(v)^2))$ using an* EREW PRAM *with $\mathcal{O}(n^2)$ processors.*

4 NC algorithms for maximal f-matching

In this section we present two parallel algorithms for maximal f-matching. The first one applies to graphs of bounded valence and it is a simple reduction to the problem of constructing a maximal matching in graphs of bounded valence. The second algorithm applies in the general case and is a generalization of the algorithm due to Israeli and Shiloach for maximal matching

[7]. The analysis of our algorithm is different from that of the Israeli-Shiloach algorithm in crucial points (e.g. in the proof of the key lemma).

4.1 f-matching for bounded degree graphs

A maximal matching in a graph of bounded valence can be computed by reduction to the problem of finding a maximal independent set in the corresponding edge graph (which is also of bounded valence) in time $\mathcal{O}(\log^* n)$ time on an EREW PRAM with $\mathcal{O}(n)$ processors [6]. In the following algorithm we use this fact to achieve the same asymptotic resource-bounds for maximal f-matching in the bounded valence case.

```
Algorithm  Maximal-f-Matching(G, f)
input :     A bounded valence graph G = (V, E),
            its matching function f.
output :    A maximal f-matching M of G.
method :

    M ← ∅;  G₀ = (V₀, E₀) ← G = (V, E);
    f₀ ← f;  i ← 0;
    while Eᵢ ≠ ∅ do
        U ← {v ∈ Vᵢ | fᵢ(v) > 0};
        F ← {(u, v) | u, v ∈ U} ∪ Eᵢ;
        Max ← Maximal-1-Matching((U, F));
        M ← M ∪ Max;
        i ← i + 1;
        (Vᵢ, Eᵢ) ← (U, F \ Max);
        for each u ∈ U in parallel do
            if u is incident to an edge in Max then
                fᵢ(u) ← fᵢ₋₁(u) - 1
            else
                fᵢ(u) ← fᵢ₋₁(u);
        endfor
    endwhile
    output M;
end  Maximal-f-Matching
```

Theorem 5 *Let G be a graph of bounded valence on n nodes. A maximal f-matching in G can be found in time $\mathcal{O}(\log^* n)$ time on an EREW PRAM with $\mathcal{O}(n)$ processors.*

Proof: All steps of the while loop except the computation of the maximal matching Max take constant time on an EREW PRAM with n processors. The computation of Max can be done in time $\mathcal{O}(\log^* n)$ using an EREW PRAM with $\mathcal{O}(n)$ processors [6]. Observe that if (u, v) is an edge in the graph $G_i = (V_i, E_i)$ before the ith iteration of the while loop then either at least one of vertices u, v does not appear in the graph $G_{i+1} = (V_{i+1}, E_{i+1})$ or the sum od degrees of u and v is at least 1 smaller than the sum of their degrees in G_i. We conclude that the number of the iterations of the while statement is at most 2 times max degree of the input graph G. □

4.2 f-matching for general graphs

The algorithm due to Israeli and Shiloach constructs a maximal matching in a graph of on n nodes and m edges in time $\mathcal{O}(\log^3 m)$ time on an CRCW PRAM with $\mathcal{O}(n+m)$ processors [7]. In this subsection we present an advanced generalization of the above algorithm to include maximal f-matching achieving the same asymptotic resource-bounds. The generalized algorithm consists of several procedures. In order to present them we need the following notation. For each vertex $v \in V$ we define its weight $w(G, v, f)$ with respect to f as follows:

$$w(G, v, f) = \min_{s \in Z_+} \{s \mid 2^s f(v) \geq d(G, v)\}.$$

Let $W(G, f) = \max_{v \in V} w(G, v, f)$. The number $W(G, f)$ will be called the weight of the graph G with respect to f.

Observe that if $W(G, f) = 0$ then all edges of the graph belong to a maximal f-matching.

A vertex v is said to be an active one in G with respect to a matching function f if and only if $2^{W(G,f)-1} f(v) \leq d(G, v) \leq 2^{W(G,f)} f(v)$. A procedure REDUCE, defined below, reduces G (removing some of its edges) to a graph G' such that $W(G', f) \leq W(G, f) - 1$.

```
Procedure  REDUCE(G, f, G')
input :        a simple graph G = (V, E).
               a matching function f : V → Z₊ such that
               W(G, f) ≥ 1.
               (* We assume that d(G, v), w(G, v, f),  ∀v ∈ V, *)
               (* and W(G, f) are computed. *)
output :       a subgraph G' = (V, E') of the graph G such that
               W(G', f) ≤ W(G, f) − 1.
method :
               Construct an auxiliary graph H induced by all those edges of G for which
                       at least one end-point is an active vertex (called later real edges)
                       and then for every vertex of odd degree in this graph add an edge
                       to an introduced vertex u;
               Find an Eulerian circuit in each connected component of H;
               Trace the Eulerian circuit in each component of H and
                       label the edges 0 and 1 alternately;
               if the number of real edges labelled 0 < the number of
                                                     edges labelled 1 then
                       G' ← the graph obtained by the removal of
                            all real edges labelled with 0 from G ;
               else
                       G' ← the graph obtained by the removal of
                            all real edges labelled with 1 from G ;
               endif
               output G';
end  REDUCE
```

Lemma 5 If $W(G, f) \geq 1$ then $W(G', f) \leq W(G, f) - 1$.

Proof: First of all let us observe that if $w(G, v, f) \leq W(G, f) - 1$, for some $v \in V$, then also $w(G', v, f) \leq w(G, f)$.

Consider now a vertex v for which $w(G, v, f) = W(G, f)$. Because v is an active vertex then $d(G', v) \leq \lceil \frac{1}{2} d(G, v) \rceil$ and $d(G, v) \leq 2^{W(G,f)} f(v)$. Hence $d(G', v) \leq \lceil \frac{1}{2} d(g, f) \rceil \leq 2^{W(G,f)-1} f(v)$. It implies $W(G', f) \leq W(G, f) - 1$. $\qquad \Box$

Lemma 6 Let $W(G, f) \geq 2$. If v is an active vertex in G then v remains active in G'.

Proof: It suffices to show that if v is an active vertex in G then $2^{W(G,f)-2} f(v) \leq d(G', v) \leq 2^{W(G,f)-1} f(v)$. The procedure REDUCE reduces the input graph in such a way that $\lfloor \frac{1}{2} d(G, v) \rfloor \leq d(G', v) \leq \lceil \frac{1}{2} d(G, v) \rceil$. If v is an active vertex then the following inequality holds:

$$f(v) 2^{W(G,f)-1} \leq d(G, v) \leq f(v) 2^{W(G,f)}.$$

Hence

$$f(v) 2^{W(G,f)-2} \leq d(G', v) \leq f(v) 2^{W(G,f)-1}.$$

$\qquad \Box$

We define a procedure f-MATCHING which computes some "large" f-mathing M for a given input graph G.

```
Procedure  f-MATCHING(G, f, M)
input :      a graph G = (V, E).
             a matching function f : V → N
output :     a large f-matching for G.
method :
             i ← 0;
             G₀ ← G;
             repeat
                      for  all vertices v ∈ V in parallel  do
                           compute d(Gᵢ, v);
                           compute w(Gᵢ, v, f);
                      endfor
                      compute W(Gᵢ, f);
                      j ← i;
                      if  W(Gᵢ, f) ≥ 1  then
                           REDUCE(Gᵢ, f, Gᵢ₊₁);
                           i ← i + 1;
                      endif
             until j = i;
             M ← the set of edges of the graph Gᵢ;
end  f-MATCHING
```

Lemma 7 *The set of edges M computed by the procedure f-MATCHING is a an f-matching in the input graph G.*

Proof: Let i_{max} be the maximal value of the variable i. The weight of the graph $G_{i_{max}}$ with respect to the function f is 0. Because for each vertex $v \in V, d(G_{i_{max}}, v) \le f(v)$ then all edges in the graph $G_{i_{max}}$ belong to M. $\qquad\Box$

Let M be a f-matching in a graph G. The procedure MODIFY, defined below, deletes the edges of M from the graph producing a graph K and next computes a matching function h such that any maximal h-matching of K extended with the edges of the set M is a maximal f-matching in G:

```
Procedure  MODIFY(G, f, M, K, h)
input :      a graph G = (V, E),
             its matching function f,
             a f-matching in M in G.
output :     a subgraph K = (V, E') of G,
             a matching function h for K such that any
             maximal h-matching H extended with the edges
             of M is a maximal f-matching of the graph G.
method :
             M' ← {(u, v) | (u, v) ∈ M or u is incident to f(u) edges in M
                   or v is incident to f(v) edges in M};
             E' ← E \ M';
             for v ∈ V in parallel  do
                  if d(K, v) = 0  then
                       h(v) ← 1;
                  else
                       h(v) ← f(v) − |{(v, u) | (v, u) ∈ M}|;
                  endif
             endfor
             K ← (V, E');
             output K, h;
end  MODIFY
```

Lemma 8 *Let H be a maximal h-matching in K then $H \cup M$ is a maximal f-matching in G.*

Proof: It is sufficient to observe that:

- E' consists of all possible edges which can extend the f-matching M, and

- for each non-isolated vertex v in H its new matching value $h(v)$ is equal to the old matching value $f(v)$ decreased by the number of edges belonging to the f-matching M. □

Now we are ready to write the entire algorithm for finding maximal f-matchings in graphs:

```
Algorithm  Maximal-f-Matching(G, f)
input :     a graph G = (V, E),
            a matching function f : V → Z₊.
output :    a maximal f-matching MaxM in the graph G.
method :

            i ← 0;
            G₀ = (V₀, E₀) ← G = (V, E);
            f₀ ← f;
            MaxM ← ∅;
            while |Eᵢ| > 0 do
                    f-MATCHING(Gᵢ, fᵢ, Mᵢ);
                    MaxM ← MaxM ∪ Mᵢ;
                    MODIFY(Gᵢ, fᵢ, Mᵢ, Gᵢ₊₁, fᵢ₊₁);
                    i ← i + 1;
            endwhile
end  Maximal-f-Matching
```

Consider a graph G and its matching function f. Let A be a subset of the set of vertices of the graph. By $\text{cost}(G, A, f)$ we will denote a cost of the set A with respect to the function f defined as follows:

$$\text{cost}(G, A, f) = \sum_{\substack{u \text{ is non-isolated} \\ \text{in } A}} f(u).$$

Lemma 9 (KEY LEMMA) : *Let G be a graph, f its matching function and C a vertex cover of the graph G. Next, let M be the f-matching which is the result of the call f-MATCHING(f, G, M). Finally, let H, h be the graph and its matching function, respectively, obtained as the result of the call* MODIFY(G, f, M, H, h). *Then there exists a vertex cover A in the graph H such that $\text{cost}(H, A, h) \leq \frac{3}{4}\text{cost}(G, C, f)$.*

proof: Let us consider the application of the procedure f-MATCHING to the graph G. Let $i_{\max}$ be the maximum value of the variable i. If $i_{\max} = 0$ then all edges of the graph belong to M. In this case it suffices to take as the set A simply the empty set. Let us assume now that $i_{\max} > 0$. Let B denote a set of all active vertices in the graph $G_{i_{\max}-1}$. Each edge $e \in E$ has either both endpoints in $V \setminus B$ or at least one of its endpoints belongs to B. If both endpoints belong to $V \setminus B$ then naturally e is an edge in the graph $G_{i_{\max}}$ and hence it is in M. This and the fact that each active vertex in G_j, for each $j \leq i_{\max-1}$, remains active in $G_{i_{\max}-1}$ (Lemma 6) imply that B is a vertex cover of the graph H. Let us observe that

$$\text{cost}(G, B, f) = \sum_{u \in B} f(u) \leq \sum_{u \in B} d(G_{i_{\max}-1}, u).$$

Let l denote the number of edges in $G_{i_{\max}-1}$ with exactly one end-point in B and k the number of edges with both end-points in B. Then $\text{cost}(G, B, f) \leq l + 2k$. The f-matching M contains at least $\frac{1}{2}(l + k)$ edges incident with vertices in B. Because $\frac{1}{2}(l + k) \geq \frac{1}{4}(l + 2k)$ then we have the following:

$$\begin{aligned}
\text{cost}(H, B, h) &\leq \text{cost}(G, B, f) - \tfrac{1}{2}(l + k) \\
&\leq \text{cost}(G, B, f) - \tfrac{1}{4}(l + 2k)
\end{aligned}$$

$$\leq \quad \text{cost}(G, B, f) - \tfrac{1}{4}\text{cost}(G, B, f)$$
$$\leq \quad \tfrac{3}{4}\,\text{cost}(G, B, f).$$

Let us consider now two cases:

CASE 1 : $\text{cost}(G, B, f) \leq \text{cost}(G, C, f)$. If we take as the set A the set B then
$\text{cost}(H, A, h) \leq \tfrac{3}{4}\text{cost}(G, C, f)$.

CASE 2 : $\text{cost}(G, B, f) > \text{cost}(G, C, f)$. Let us observe that M contains at least $\tfrac{1}{4}\text{cost}(G, B, f)$
edges. If we take C as the set A the the following holds:

$$
\begin{aligned}
\text{cost } (H, A, h) &\leq \quad \text{cost } (G, C, f) - \tfrac{1}{4}\text{cost}(G, B, f) \\
&\leq \quad \text{cost } (G, C, f) - \tfrac{1}{4}\,\text{cost } (G, C, f) \\
&\leq \quad \tfrac{3}{4}\,\text{cost } (G, C, f). \qquad \square
\end{aligned}
$$

Theorem 6 *Let* $G = (V, E)$ *be an n-vertex graph with m edges and let f be its matching function. A maximal f-matching in G can be computed in $\mathcal{O}(\log^3 n)$ time using a CRCW PRAM with $\mathcal{O}(n+m)$ processors or in time $\mathcal{O}(\log^4 n)$ using an EREW PRAM with $\mathcal{O}(n+m)$ processors.*

Proof: We can compute a maximal f-matching using the algorithm MAXIMAL-f-MATCHING. It follows directly from the lemmas 7 and 8 that if the algorithm stops then $MaxM$ is a maximal f-matching in G. It suffices to show how to implement efficiently the algorithm MAXIMAL-f-MATCHING. We assume that the input graph G is represented by adjacency lists. Each of the steps of the procedure REDUCE consists of some computations on the adjacency lists which can be performed using the doubling technique. In the second step the Israeli and Shiloach algorithm [7] can be applied. Hence the procedure REDUCE takes $\mathcal{O}(\log n)$ time on $\mathcal{O}(n+m)$ processors of a CRCW PRAM or $\mathcal{O}(\log^2 n)$ time on $\mathcal{O}(n+m)$ processors of a EREW PRAM. Each iteration of the repeat loop in the procedure f-MATCHING consists of some computations on the adjacency lists and of the call of the procedure REDUCE. Hence it takes at most $\log n$ time. It follows from Lemma 5 that the number of iterations of the repeat loop can not be larger than $\lceil \log(n-1) \rceil$. Hence the procedure f-MATCHING runs in time $\mathcal{O}(\log^2 n)$ using only $\mathcal{O}(n+m)$ processors. The procedure MODIFY consists only of simple computations on the adjacency lists. It takes $\mathcal{O}(\log n)$ times. Let us observe that the cost of each vertex cover in the input graph G is bounded by n^2 from above. Taking into account the key lemma 9 we infer that the number of iterations of the while loop of the algorithm is $\mathcal{O}(\log n)$. Hence the algorithm stops and runs in time $\mathcal{O}(\log^3 n)$ using only the processors associated with the vertices and the edges of the graph. $\qquad \square$

The above theorem yields immediately the following corollary.

Corollary 10 *The problem of computing a maximal f-matching is in NC.*

5 Final remarks

For non-dense graphs our NC algorithm for maximal k-dependent set is far from being optimal in the sense of the time-processor product. It seems that more processor-efficient NC algorithms for maximal k-independent set can be derived in the special cases of $k = 1, 2$ and for planar graphs. However, the ultimate goal here would be to derive an NC algorithm for maximal k-dependent set in the general case such that the time-processor product would be within a logarithmic factor from the size of the input graph. The generalization of our algorithm for maximal k-dependent set to include maximal f-dependent set (see Theorem 4) runs in poly-log time only if the maximum value of f is poly-logarithmic in the input size. Thus, the problem of whether one can construct a maximal f-dependent set in the general case of f using an NC algorithm is also open.

References

[1] S. Carlsson, Y. Igarashi, K. Kanai, A. Lingas, K. Miura and Ola Petersson, *Information Disseminating Schemes for Fault Tolerance in Hypercubes*. Technical report, Gunma University.

[2] R. Cole, *Parallel merge sort*. SIAM J. Comput., vol 17, No. 4, 1988, pp 770-785.

[3] H. Djidjev, O. Garrido, C. Levcopoulos and A. Lingas, *On the maximum q-dependendent set problem*. in the Proc. of the International Conf. for Young Computer Scientists ICYCS91. 271-274.

[4] M. R. Garey and D.S. Johnson, *Computers and Intractability. A Guide to the Theory of NP-Completeness*. W.H. Freeman and Company, San Francisco, 1979.

[5] M. Goldberg and T. Spencer, *A New Parallel Algorithm for the Maximal Independent Set Problem*. In Proc. 28th Symp. on Foundations of Computer Science, 1987.

[6] A. V. Goldberg and S. A. Plotkin, *Parallel ($\Delta+1$)-Coloring of Constant-degree Graphs*. Information Processing Letters 25 (1987) 241-245.

[7] A. Israeli and Y. Shiloach, *An improved parallel algorithm for maximal matching*. Information Processing Letters 22 (1986) 57-60.

[8] R. M. Karp and V. Ramachandran, *A Survay of Parallel Algorithms for Shared-Memory Machines*. Report No. UCB/CSD 88/403 Computer Science Division (EECS), University of California, Berkeley, California 94720.

[9] R. M. Karp and A. Wigderson, *A Fast Parallel Algorithm for the Maximal Independent Set Problem*. In Proceedings of the 16th Annual ACM Symposium on Theory of Computing, 1984.

[10] L. Lovász and M.D. Plummer, *Matching Theory, Annals of Discrete Mathematics* (29). North-Holland Mathematics Studies 121. Elsevier Science Publishers B.V. ISBN 0444 879161.

[11] M. Luby, *A simple parallel algorithm for the maximal independent set problem*. In SIAM J.Comput. 15, 3 (1986) pp. 1036-1053.

[12] N. J. Pippenger, *On simultaneous resource bounds*. In the Proc. 20th. Annual Symp. on Foundation of Computer Science, 1979, pp 307-311.

Author Index

Lecture Notes in Computer Science

For information about Vols. 1–466
please contact your bookseller or Springer-Verlag